BLOCK COPOLYMERS

958/14

BLOCK COPOLYMERS

Synthetic Strategies, Physical Properties, and Applications

NIKOS HADJICHRISTIDIS
STERGIOS PISPAS
GEORGE FLOUDAS

A John Wiley & Sons, Inc., Publication

Published by John Wiley & Sons, Inc., Hoboken, New Jersey.
Published simultaneously in Canada.

Library of Congress Cataloging-in-Publication Data:

Hadjichristidis, Nikos, 1943-
Block copolymers : synthetic strategies, physical properties, and applications / Nikos Hadjichristidis, Stergios Pispas, George Floudas.
p. cm.
Includes index.
ISBN 0-471-39436-X (cloth : acid-free paper)
1. Block copolymers. I. Pispas, Stergios, 1967- II. Floudas, George, 1961- III. Title.

QD382.B5 H33 2003
547′.84–dc21 2002014989

Printed in the United States of America.

10 9 8 7 6 5 4 3 2 1

To Our Wives
Dina, Hara, and Maria

CONTENTS

PREFACE

Block copolymers are a fascinating class of polymeric materials belonging to a big family knwon as "soft materials." This class of polymers is made by the covalent bonding of two or more polymeric chains that, in most cases, are thermodynamically incompatible giving rise to a rich variety of microstructures in bulk and in solution. The length scale of these microstructures is comparable to the size of the block copolymers molecules (typically 5–50 nm); therefore, the microstructures are highly coupled to the physical and chemical characteristics of the molecules. The variety of microstructures gives rise to materials with applications ranging from thermoplastic elastomers and high-impact plastics to pressure-sensitive adhesives, additives, foams, etc. In addition, block copolymers are very strong candidates for potential applications in advanced technologies such as information storage, drug delivery, and photonic crystals.

There is a rapid development of research activity on the synthesis, physical properties, and applications of block copolymers with an enormous number of scientific papers published every year in leading journals in the fields of polymer and material science. Excellent reviews on the physical properties of block copolymers by F. S. Bates and G. H. Fredrickson and a book by Ian Hamley have served many in the scientific community.

This book deals with the synthesis, characterization methods, physical properties, and applications of block copolymers. It is meant to serve as an advanced introductory book for scientists who will be working with block copolymers. The book provides guidelines to strategies that can be employed in the synthesis of a variety of block copolymer architectures with well-defined molecular characteristics; an overview of some fundamental physical properties; and the description of how the information on chemical structure and physical properties can be utilized in designing block copolymers with particular functions.

In Chapter 1 the possible ways for synthesizing block copolymers by anionic polymerization are presented. This is one of the oldest and more versatile polymerization mechanisms to date to synthesize block copolymers with a plethora of complex architectures.

The possibilities for block copolymer preparation by cationic polymerization are given in Chapter 2. This polymerization methodology is considered as the second more advanced route for obtaining well-defined block copolymers from monomers that cannot be polymerized by anionic means.

The synthetic capabilities of controlled radical polymerization are outlined in Chapter 3. This newly invented technique has emerged as a very valuable tool for obtaining block copolymers using simple experimental protocols and a large variety of the available monomers.

The synthesis of block copolymers, derived mainly from (meth)acrylic monomers, by group transfer polymerization is described in Chapter 4. Some examples of the applicability of catalysts and coordination chemistry in the synthesis of block copolymers are given in Chapter 5, in the form of ring opening metathesis polymerization of multicyclic olefins.

The possibilities of obtaining block copolymers by combining more than one polymerization mechanisms are outlined in Chapter 6. These combination methods are invaluable in cases where block copolymers comprised of monomers that cannot be polymerized by the same polymerization mechanism, have to be synthesized.

Some examples of obtaining new block copolymers from preformed precursor block copolymers are presented in Chapter 7. By the use of post polymerization derivatization reactions, adapted to polymers from classical organic chemistry, new materials with substantially different properties can be obtained, sometimes even from commercially available copolymers.

Synthetic strategies for the preparation of nonlinear block copolymers with a variety of complex architectures are presented in Chapter 8. Macromolecular architecture has recently emerged as another factor of controlling macromolecular properties in the molecular level. The imagination of polymer chemists, together with new emerging synthetic methodologies, led to the synthesis of complex macromolecules that aided in the establishment of the fundamental structure-properties relationships in these systems.

The experimental protocols and available methods for the molecular characterization of all classes of block copolymers are given in Chapter 9. Primary chemical structure identification, molecular weight, composition, and size determination are essential elements in understanding the properties of complex materials like block copolymers.

A brief description of the behavior and conformation of simple and complex block copolymers in their isolated molecular state, i.e., in dilute solutions, is given in Chapter 10.

The interesting property, both from academic as well as from the application point of view, of block copolymers to form micelles in solvents selective for one of the blocks, like low molecular weight surfactants, is discussed in Chapter 11.

The equally important phenomenon of block copolymer adsorption onto solid surfaces from solutions is presented in Chapter 12, together with some of the available techniques for investigating this phenomenon, as well as results from theoretical and experimental studies.

The theoretical aspects related to the structure factor and the phase state of bulk block copolymers in the weak and strong segregation limits are discussed in Chapter 13.

In Chapter 14 the effects of block copolymer architecture on the compatibility of individual blocks in complex copolymers are presented. The phase diagrams of families of block copolymers are discussed in Chapter 15. Emphasis is given to the effects of conformational asymmetry of the blocks comprising the block copolymers.

The viscoelastic properties of block copolymers are discussed in Chapter 16. The viscoelastic behavior of ordered block copolymers, the possibility of shear-induced orientational order, and the determination of the order-to-disorder transition temperature are included in the discussion.

The equilibrium order-to-disorder transition temperature, together with the phase transformation kinetics (i.e., disorder-to-order kinetics, kinetics of the transition between different ordered phases, epitaxy, activation barriers, etc.) are investigated in Chapter 17. Chapter 18 is devoted to the phase state, viscoelastic properties, and dynamics of block copolymers containing strongly interacting groups in their chains. The discussion is focused on the effects and changes in the phase state of block copolymers due to the presence of these groups.

The experimentally determined morphologies of a variety of block copolymer architectures (linear diblock and triblock copolymers, star block and miktoarm star copolymers, etc.) are presented in Chapter 19. The effects of block sequence and overall block copolymer architecture are also described. Chapter 20 deals with the dynamics of block copolymer chains in concentrated solution and in the melt state.

Finally, Chapter 21 is devoted to the commercialized and potential applications of block copolymers as thermoplastic elastomers, structural materials: in encapsulation technologies, in nanotechnology, and in other areas.

Nikos Hadjichristidis
Stergios Pispas
George Floudas

ABBREVIATIONS AND SYMBOLS

A_2, A_3	Second and third virial coefficient
AcOVE	Acetoxyvinylether
AGTR	Aldol group transfer polymerization
AIBN	α,α'-azo-diisobutylnitrile
AN	Acetonitrile
A_nB_m	Star copolymer with n A arms and m B arms
ATRP	Atom transfer radical polymerization
9-BN	9-borabicyclo[3.3.1]nonane
BPO	Benzoylperoxide
Bpy	2,2′-bipyridyl
Cmc	Critical micelle concentration
CMC	Constant mean curvature interfaces
Cmt	Critical micelle temperature
D	Diffusion coefficient
d	Microdomain period
D(f,x)	Debye function
DLS	Dynamic light scattering
DNbpy	4,4′-di-(5-nonyl)-2,2′-bipyridyl
DPE	1,1-Diphenylethylene
DPMK	Diphenylmethylpotassium
F	Free energy
f	Volume fraction
F(x,f)	Combination of Debye functions
f_H	Free energy density (Hartree approximation)
G*	Complex shear modulus

G′	Storage modulus
G″	Loss modulus
GTP	Group transfer polymerization
IB	Isobutylene
Is	Isoprene
ISL	Intermediate segregation limit
L, l	Grain size
LAOS	Large amplitude oscillatory shear
M	Molecular weight
M_{app}	Apparent molecular weight
M_e	Entanglement molecular weight
MeVE	Methylninylether
MFT	Mean-field theory
Miktoarm star	Star polymer with different arms
M_n	Number-average molecular weight
M_w	Weight-average molecular weight
n	Avrami exponent
$\bar{N}$	Fluctuation corrected N
N	Degree of polymerization
P	Total scattering power
PαMeS	Poly(α-methylstyrene)
P2VP	Poly(2-vinyl pyridine)
PBd	Polybutadiene
PCL	Poly(ε-caprolactone)
PCPPHMA	Poly{6[4-(4′-cyanophenyl)phenoxy]hexyl methacrylate}
PDI	Polydispersity index
PDMS	Poly(dimethylsiloxane)
PEB	1,3(or 1,4)bis(1-phenylethenyl)benzene
Pentablock copolymer	Five blocks, two of them different (f.e. ABABA)
Pentablock terpolymer	Five blocks, three of them different (f.e. ABCBA)
PEO	Poly(ethylene oxide)
PI	Poly(isoprene)
PIB	Polyisobutylene
PMMA	Poly(methyl methacrylate)
PpMeS	Poly(p-methylstyrene)
PS	Polystyrene
PS-PI or PS-b-PI	Diblock copolymer of styrene and Isoprene
PtBOS	Poly(p-tert-butoxystyrene)
PtBu(M)A	Poly(tert-butyl methacrylate)
PtBuA	Poly(tert-butyl acrylate)
PTHF	Poly(tetrahydrofurane)
q, Q	Scattering wavevector
R_θ	Rayleigh ratio
R, R_g	Radius of gyration
R_h	Hydrodynamic radius

RIE	Reactive ion etching
ROMP	Ring opening metathesis polymerization
RPA	Random phase approximation
S(q)	Structure factor
SANS	Small angle neutron scattering
SAXS	Small angle x-ray scattering
s-BuLi	sec-Butyllithium
SCFT	Self-consistent field theory
SEC	Size exclusion chromatography
SFRP	Stable free radical polymerization
SLS	Static light scattering
SSL	Strong segregation limit
St	Styrene
T	Temperature
$t_{1/2}$	Half-time
$\tan\delta$	Loss tangent
$TASHF_2$	Tris(dimethylamino)sulfonium difluoride
TBABB	Tetra-n-butylammonium bibenzoate
t-BOS	p-tert-butoxystyrene
TEM	Transmission electron microscopy
TEMPO	2,2,6,6-tetramethylpiperidinoxy
Tetrablock copolymer	Four blocks, two of them different (f.e. ABAB)
Tetrablock quaterpolymer	Four different blocks (ABCD)
Tetrablock terpolymer	Four different blocks, three of them different (f.e. ABCA)
T_g	Glass transition temperature
THF	Tetrahydrofurane
T_m	Melting temperature
TMEDA	N,N,N′,N′-Tetrametylethylenediamine
T_m^o	Equilibrium melting temperature
T_{ODT}	Order-to-disorder transition temperature
T_{ODT}^o	Equilibrium order-to-disorder transition temperature
TPA	Thermoplastic polyamide
TPE	Thermoplastic elastomer
TPES	Thermoplastic polyesters
TPU	Thermoplastic polyurethane
Triblock copolymer	Three blocks, two of them different (ABA)
Triblock terpolymer	Three different blocks (ABC)
WSL	Weak segregation limit
x	$q^2R_g^2$
z	Rate constant
α-MeSt	α-methylstyrene
α	Statistical segment length
Δ	Interfacial thickness
δ	Undercooling parameter

ΔH	Heat of fusion
$\Delta\varepsilon$	Dielectric strength
ε'	Dielectric permittivity
ε''	Dielectric loss
ζ	Conformational asymmetry parameter
ζ_{eff}	Effective friction coefficient
$[\eta]$	Intrinsic viscosity
κ_o	Composition polydispersity
σ	Interfacial tension
σ_e	Fold-surface free energy
$\varphi(r)$	Local volume fraction
χ	Interaction parameter
$\psi(r)$	Order parameter
ω	Frequency
ω_c, ω_d	Characteristic frequencies in the viscoelastic response

PART I

BLOCK COPOLYMER SYNTHESIS

CHAPTER 1

BLOCK COPOLYMERS BY ANIONIC POLYMERIZATION

Anionic living polymerization has been known for almost fifty years. Since its discovery in the 1950s, it has emerged as the most powerful synthetic tool for the preparation of well-defined polymers, i.e., narrow molecular weight distribution polymers with controlled molecular characteristics including molecular weight, composition, microstructure, and architecture. Its ability to form well-defined macromolecules is mainly due to the absence of termination and chain transfer reactions, under appropriate conditions (Young 1984, Hsieh 1996).

Anionic polymerization proceeds via organometallic sites, carbanions (or oxanions) with metallic counterions. Carbanions are nucleophiles; consequently, the monomers that can be polymerized by anionic polymerization are those bearing an electroattractive substituent on the polymerizable double bond. Initiation of polymerization is accomplished by analogous low molecular weight organometallic compounds (initiators). A wide variety of initiators has been used so far in order to produce living polymers. Among them, the most widely used are organolithiums (Hadjichristidis 2000). The main requirement for the employment of an organometallic compound as an anionic initiator is its rapid reaction with the monomer at the initiation step of the polymerization reaction and, specifically, with a reaction rate larger than that of the propagation step. This leads to the formation of polymers with narrow molecular weight distributions because all active sites start polymerizing the monomer almost at the same time. Propagation proceeds through nucleophilic attack of a carbanionic site onto a monomer molecule with reformation of the first anionic active center. The situation is similar in the case of the ring opening polymerization of cyclic monomers containing heteroatoms (oxiranes, lactones, thiiranes, siloxanes, etc). The role of the solvent and additives in the polymerization mechanism is important and has been studied extensively in several cases (Young 1984, Hsieh 1996).

Under appropriate experimental conditions (Hadjichristidis 2000), due to the absence of termination and chain transfer reactions, carbanions (or, in general, anionic sites) remain active after complete consumption of monomer, giving the possibility of block copolymer formation, in the simplest case, by introduction of a second monomer into the polymerization mixture. However, a variety of different synthetic strategies have been reported for the preparation of linear block copolymers by anionic polymerization.

1. SYNTHESIS OF AB DIBLOCK COPOLYMERS

Linear AB block copolymers are the simplest block copolymer structures where two blocks of different chemical structures are linked together through a common junction point.

The most general method for the preparation of AB block copolymers is sequential monomer addition. In this method one of the monomers is polymerized first. After its complete consumption, the second monomer is added, and the polymerization is again allowed to proceed to completion. At this point an appropriate terminating agent is added, and the diblock copolymer can then be isolated (usually by precipitation in a nonsolvent).

The most important conditions (mechanistic and experimental features) that must be fulfilled in order to synthesize well-defined block copolymers are:

i) The carbanion formed by the second monomer must be more, or at least equally, stable than the one derived from the first monomer. In other words the first monomer carbanion must be able to initiate the polymerization of the second monomer, i.e., the first monomer carbanion must be a stronger nucleophile than the second one.
ii) The rate of the crossover reaction, i.e., the initiation of polymerization of the second monomer by the anion of the first, must be higher then the rate of propagation of monomer B. This ensures narrow molecular weight distribution for block B and absence of A homopolymer in the final block copolymer that can arise from incomplete initiation.
iii) The purity of the second monomer must be high. Otherwise, partial termination of the living A anions can take place leading to the presence of A homopolymer in the final product. Additionally, loss of molecular weight and composition control of the second block and of the whole copolymer will occur because the concentration of the active centers will be decreased.

Taking into account these mechanistic and experimental features, a large number of AB diblock copolymers have been synthesized by sequential addition of monomers. Some representative examples are given in Table 1.1. The list is by

TABLE 1.1. AB Diblock Copolymers Formed by Sequential Addition of Monomers Using Anionic Polymerization

1st Monomer	2nd Monomer	Reference
Styrene	Isoprene	Corbin 1976
	Butadiene	Quirk 1992
	Cyclohexadiene	Hong 2001
	Methylmethacrylate	Varshney 1990
	tert-Butyl methacrylate	Varshney 1990
	tert-Butyl acrylate	Hautekeer 1990
	2,3-glycidyl methacrylate	Hild 1998
	Stearyl methacrylate	Pitsikalis 1999
	2-Vinylpyridine	Schindler 1969
	4-Vinylpyridine	Grosius 1970
	Ethylene Oxide	Finaz 1962
	ε-Caprolactone	Paul 1980
	Hexamethylcyclotrisiloxane	Zilliox 1975
	Ferrocenyldimethylsilane	Ni 1996
	Hexyl isocyanate	Chen 1995
α-Methyl styrene	Butadiene	Elgert 1973
Isoprene	2-Vinyl pyridine	Matsushita 1986
	Ethylene oxide	Forster 1999
	Hexamethylcyclotrisiloxane	Almdal 1996
Butadiene	ε-Caprolactone	Balsamo 1998
	Ethylene oxide	Forster 1999
Methyl methacrylate	*tert*-Butyl methacrylate	Allen 1986
2-Vinyl pyridine	*tert*-Butyl methacrylate	Yin 1994
	ε-Caprolactone	Diuvenroorde 2000
	Ethylene oxide	Martin 1996

no means exhaustive. More examples are given in the general literature on anionic polymerization (Hsieh 1996, Morton 1983).

A wide variety of diblock copolymers of styrene and isoprene or butadiene, having predictable molecular weight and composition as well as narrow molecular weight and compositional distribution, have been synthesized by sequential addition of monomers (Morton 1983, Hsieh 1996). Synthesis of these diblocks starts with styrene, and then the diene is added to the reaction mixture because it is well established that PSLi active centers can initiate efficiently the polymerization of dienes in hydrocarbon solvents (Scheme 1.1) and not vice versa. The use of

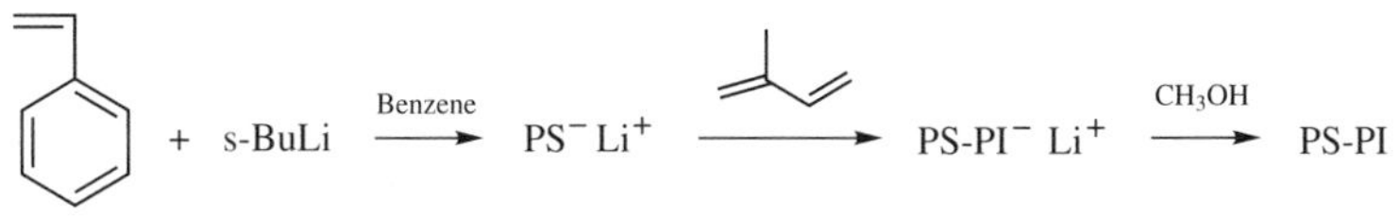

$$\text{Styrene} + \text{s-BuLi} \xrightarrow{\text{Benzene}} \text{PS}^-\,\text{Li}^+ \xrightarrow{\text{isoprene}} \text{PS-PI}^-\,\text{Li}^+ \xrightarrow{CH_3OH} \text{PS-PI}$$

Scheme 1.1

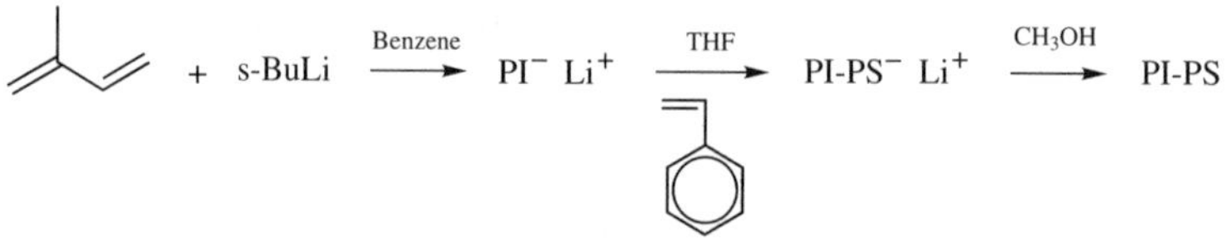

+ s-BuLi $\xrightarrow{\text{Benzene}}$ $PI^- Li^+$ $\xrightarrow{\text{THF}}$ $PI\text{-}PS^- Li^+$ $\xrightarrow{CH_3OH}$ PI-PS

Scheme 1.2

hydrocarbon solvents and Li as counterion in the initiator are essential for the production of polydienes having high 1,4 microstructure, leading to block sequences with low T_g and good elastomeric properties. Alternatively, the polymerization of the diene can be performed first in hydrocarbon solvents, followed by that of styrene in the presence of a small amount of a polar compound (usually THF) (Scheme 1.2). The presence of polar compounds alters the stereochemistry and activity of the polydiene active centers through complexation, enabling a fast crossover reaction, giving diblocks with low polydispersity in molecular weight and composition (Antokowiak 1972). Diblock copolymers containing styrene and isoprene having different types and numbers of polar groups at specific sites along the copolymer chain have been synthesized using appropriate functionalization techniques (functional initiators, living end capping, and postpolymerization reactions) proving the great versatility of these comonomer system (Pispas 1994, 2000a,b, Schadler 1997 and 1998, Schops 1999).

The synthesis of a number of block copolymers containing blocks of functionalized styrenic and other types of monomers have been reported. The polymerization of these monomers becomes possible after protection of the functional groups that interfere with the anionic polymerization process (Nakahama 1990, Hirao 1998, Ishizone 1993).

Diblock copolymers containing styrene (or styrenic monomers) or dienes (isoprene, butadiene), and (meth)acrylic monomers have been reported in the

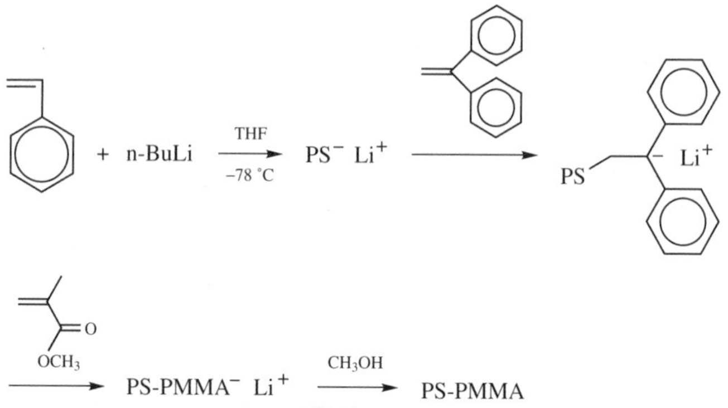

+ n-BuLi $\xrightarrow[-78\,^\circ C]{\text{THF}}$ $PS^- Li^+$ ⟶ PS Li^+

$\xrightarrow{}$ $PS\text{-}PMMA^- Li^+$ $\xrightarrow{CH_3OH}$ PS-PMMA

OCH$_3$ O

Scheme 1.3

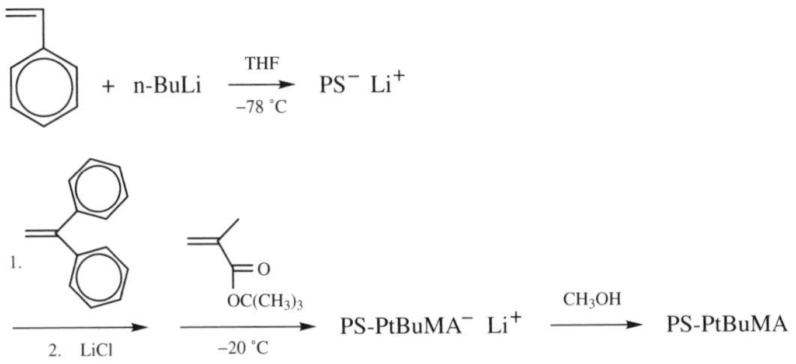

+ n-BuLi $\xrightarrow[-78\ ^\circ C]{THF}$ $PS^-\ Li^+$

1. (DPE) 2. LiCl; $OC(CH_3)_3$, −20 °C → $PS\text{-}PtBuMA^-\ Li^+$ $\xrightarrow{CH_3OH}$ PS-PtBuMA

Scheme 1.4

literature. These diblocks are synthesized by polymerizing first the most reactive monomer (styrenic or diene) and then the (meth)acrylic monomer. Polymerization of the (meth)acrylic monomers requires low polymerization temperatures (−78°C), polar solvents (usually THF) (Allen 1986, Hadjichristidis 2000), and relatively less active and sterically hindered initiators in order to avoid reaction of active anions with the carbonyl group of the (meth)acrylic monomers. For this reason the more active styrenic or dienic anion is usually transformed to a less active more sterically hindered one by reaction with diphenylethylene, DPE, a nonhomopolymerizable monomer (Scheme 1.3). In this way polymerization of the (meth)acrylic monomers proceeds in a controlled way, giving diblock copolymers with well-defined molecular characteristics. The use of additives like LiCl (Varshney 1990), in conjuction with DPELi transformation in polar solvents and low temperatures, has been proven to enhance control of the polymerization of (meth)acrylic monomers, and, in some cases, the polymerization can be contacted at higher temperatures (e.g., *tert*-butylmethacrylate polymerization can be performed at −20°C) (Scheme 1.4). It has been shown that complexation of LiCl to the active chain ends leads to the formation of complexes with different stereochemistry and reactivity able to perform a more controllable polymerization of the (meth)acrylic monomers. If high 1,4 content is desired in the dienic block, the polymerization of the diene should be performed in nonpolar hydrocarbon solvents (e.g., benzene). After the polymerization of the diene, the polarity of the solvent is changed by addition of THF (with or without prior removal of the hydrocarbon solvent), the temperature is lowered to −78°C, DPE and LiCl are added, followed by slow distillation of the (meth)acrylic monomer into the reaction flask. For the synthesis of poly(styrenic monomer-b-methacrylic monomer) copolymers, where the problem of microstructure is not existent, polymerization of both monomers can be carried out in THF at low temperatures from the beginning. In the case of dienic monomers, this procedure will result in diblocks with a low 1,4 microstructure in the dienic block.

A large number of diblock copolymers containing methacrylic monomers have been prepared by anionic polymerization (including pure methacrylic diblock copolymers, Scheme 1.5) due to the fact that a wide variety of methacrylic monomers can be synthesized having different side groups. Side groups can be chosen appropriately depending on the desired properties of the final methacrylic

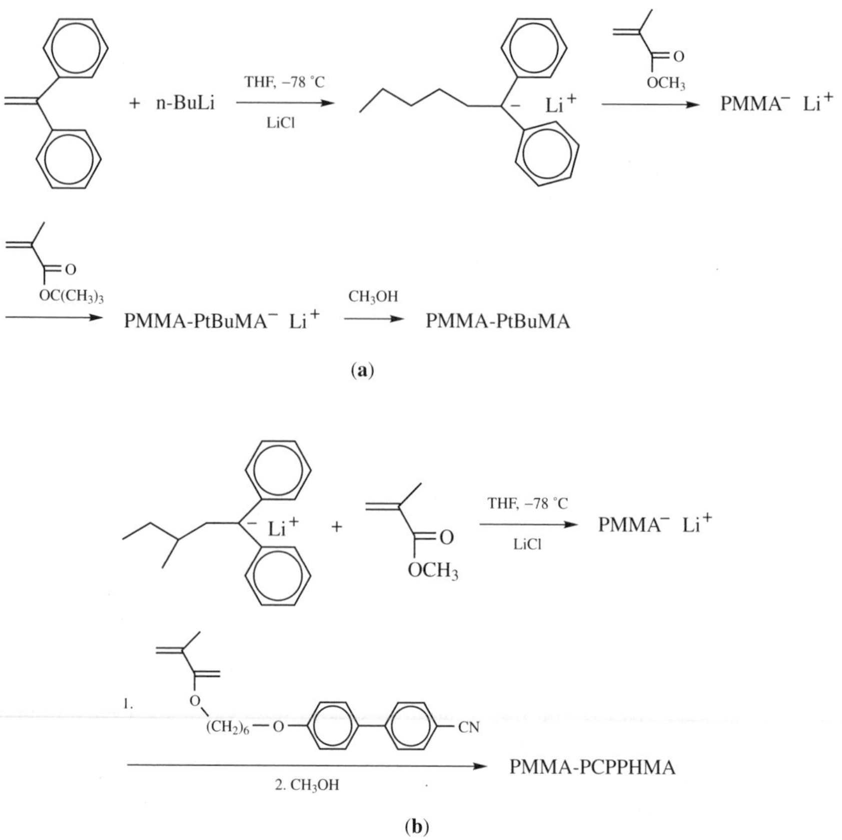

+ n-BuLi $\xrightarrow[\text{LiCl}]{\text{THF, -78 °C}}$ Li^+ $\xrightarrow{\text{OCH}_3}$ $PMMA^- Li^+$

$OC(CH_3)_3$ $\longrightarrow$ $PMMA\text{-}PtBuMA^- Li^+$ $\xrightarrow{CH_3OH}$ PMMA-PtBuMA

(a)

Li^+ + OCH_3 $\xrightarrow[\text{LiCl}]{\text{THF, -78 °C}}$ $PMMA^- Li^+$

1. $(CH_2)_6$—O— —CN

$\xrightarrow[\text{2. } CH_3OH]{}$ PMMA-PCPPHMA

(b)

Scheme 1.5

block in the block copolymer. Long hydrocarbon (Pitsikalis 2000) or fluorocarbon chains (Ishizone 1999, Imae 2000) or mesogenic groups (Lehmann 2000, Yamada 1999 and 1998) have been attached to methacrylic monomers, and these monomers were copolymerized anionically, giving block copolymers with amphiphilic, liquid crystalline properties and different solubility in various solvents.

Diblock copolymers containing 2- or 4-vinylpyridine and styrenic or dienic blocks are synthesized by sequential monomer addition in polar solvents and low temperatures (Matsushita 1986) in order to avoid the attack of the reactive anionic centers to the pyridine ring, which is activated due to the presence of the nitrogen atom. The nitrogen atom is also responsible for the lower reactivity of the vinylpyridine anion; therefore, vinylpyridines are added as the second monomer (Scheme 1.6). The crossover reaction is fast and well-defined; diblock copolymers can be produced in this way. The complexing ability of the pyridine ring to various metal cations and the possibility of formation of polyelectrolyte blocks with appropriate transformation reactions on the nitrogen atom of the pyridine ring give this kind of copolymer a number of interesting properties and potential applications.

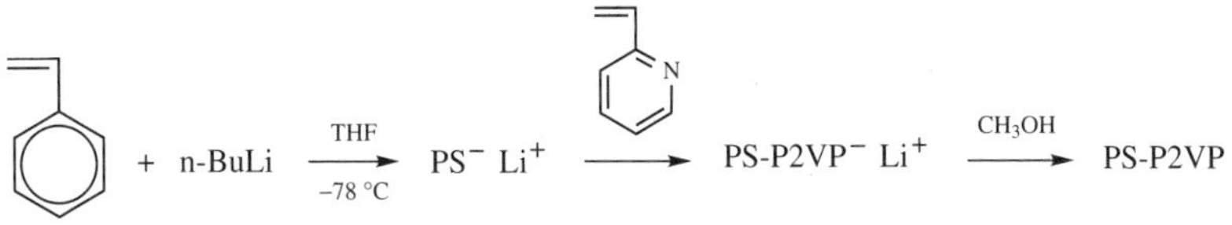

+ n-BuLi $\xrightarrow[-78\,^{\circ}C]{THF}$ $PS^-\ Li^+$ $\longrightarrow$ $PS\text{-}P2VP^-\ Li^+$ $\xrightarrow{CH_3OH}$ PS-P2VP

Scheme 1.6

Diblock copolymers containing ethylene oxide have been synthesized using a variety of initiators containing Na or K counterions (Barker 1984, Morton 1983) (Scheme 1.7). This type of copolymers presents interesting properties due to the

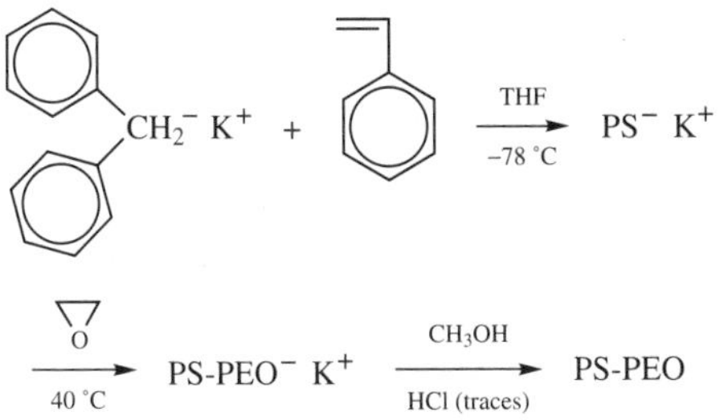

$CH_2^-\ K^+$ + $\xrightarrow[-78\,^{\circ}C]{THF}$ $PS^-\ K^+$

$\xrightarrow[40\,^{\circ}C]{}$ $PS\text{-}PEO^-\ K^+$ $\xrightarrow[HCl\ (traces)]{CH_3OH}$ PS-PEO

Scheme 1.7

solubility of PEO in water and its ability to crystallize. Since the C-O-M centers has lower reactivity than C-M centers, where M is Na or K, ethylene oxide is used as the second monomer. Under appropriate conditions (polar solvents and temperatures higher than room temperature), propagation of EO proceeds rather slowly compared with initiation and diblock copolymers with low polydispersities, and, essentially free of PEO homopolymer can be prepared. Initiators having Li as the counterion cannot be used because the $C\text{-}O^-\ Li^+$ pair is very tight, due to the localization of the negative charge on the oxygen, and insertion of monomers in the O-Li bond is impossible. Recently, a synthetic procedure involving initiating moieties having Li as counterion, in the presence of the strong phosphazene base $tBuP_4$, have been reported for the synthesis of diblock copolymers containing styrene or dienes and ethylene oxide (Förster 1999). The ability of phosphazine to complex with Li weakens the strength of O-Li bonding, and insertion of EO monomer is possible. By this way, well-defined block copolymers were prepared with narrow molecular weight distributions having a range of compositions and molecular weights and free of PEO homopolymer (Scheme 1.8).

The synthesis of diblock copolymers containing ethylene oxide and propylene or butylene oxide has also been accomplished by anionic polymerization using initiators with Na or K as counterions (Deng 1994, Booth 2000). The sequential addition method has also been used here, and, because all monomers belong to the same family, interconversion of the active chain ends is possible.

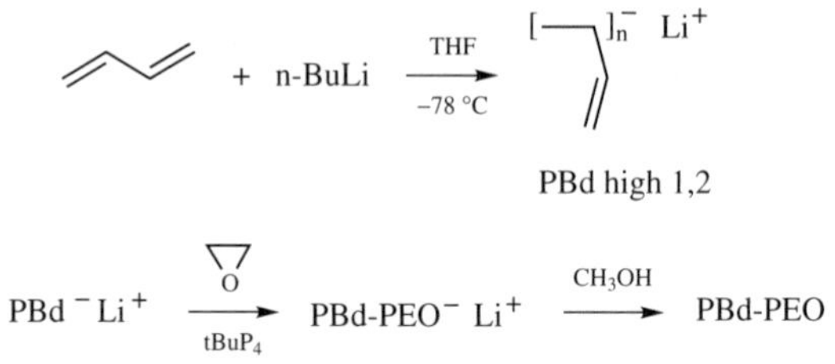

+ n-BuLi $\xrightarrow[-78\,^{\circ}C]{THF}$ $[\ ]_n^-$ Li^+

PBd high 1,2

PBd $^-$ Li^+ $\xrightarrow{tBuP_4}$ PBd-PEO$^-$ Li^+ $\xrightarrow{CH_3OH}$ PBd-PEO

Scheme 1.8

Another monomer that forms crystallizable polymers is ε-caprolactone. Diblock copolymers containing styrene and ε-caprolactone have been synthesized by anionic polymerization (Paul 1980). Sequential addition of monomers is also here the method of choice, and ε-caprolactone is used as the second monomer (Scheme 1.9). However, polymerization of ε-caprolactone is fast and must be kinetically controlled. Thus, termination must be carried out in a few minutes after the addition of the monomer because undesired back-biting reactions can occur, leading to degradation of the ε-caprolactone block.

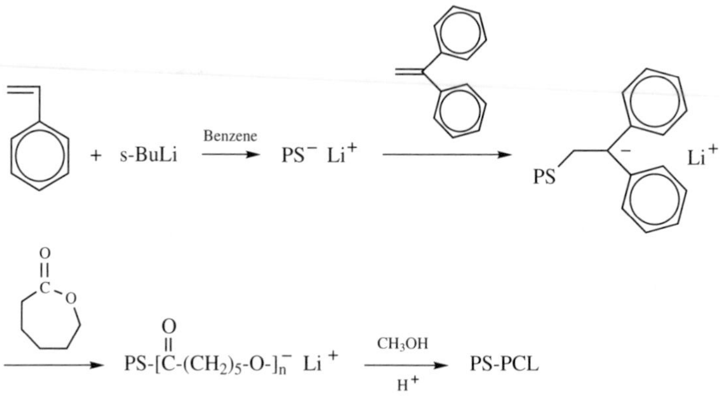

+ s-BuLi $\xrightarrow{Benzene}$ PS^- Li^+ ⟶ PS Li^+

⟶ PS-[C(=O)-$(CH_2)_5$-O-]$_n^-$ Li^+ $\xrightarrow[H^+]{CH_3OH}$ PS-PCL

Scheme 1.9

Another class of diblock copolymers are those containing siloxane monomers. The anionic polymerization of cyclosiloxanes, especially hexamethylcyclotrisiloxane (dimethylsiloxane is the monomeric unit in the polymer chain), has been investigated. Cyclosiloxanes have been used as second monomers in diblock copolymers' synthesis where styrene and dienes were the first monomers to be polymerized (Zilliox 1976, Almdal 1996). Initiators containing Li as the counterion can be used in this case. Well-defined block copolymers can be produced if sufficient care is exercised in the purification of the siloxane monomer and in minimizing the back biting reactions involving the Si-OLi active centers and

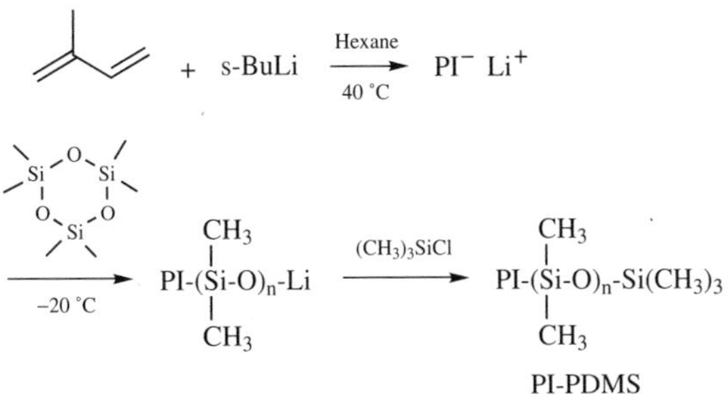

$$\text{isoprene} + \text{s-BuLi} \xrightarrow[40\,^{\circ}\text{C}]{\text{Hexane}} \text{PI}^- \text{Li}^+$$

$$\xrightarrow[-20\,^{\circ}\text{C}]{\text{D}_3} \text{PI-(Si(CH}_3)_2\text{-O)}_n\text{-Li} \xrightarrow{(\text{CH}_3)_3\text{SiCl}} \text{PI-(Si(CH}_3)_2\text{-O)}_n\text{-Si(CH}_3)_3$$

PI-PDMS

Scheme 1.10

the Si-O-Si bonds already formed (i.e., low polymerization temperatures, conversions lower than 80%) (Bellas 2000) (Scheme 1.10). Block copolymers containing two different siloxane monomers have also been reported (Nagase 1983).

Diblock copolymers containing polyferrocenylsilane blocks originating from the anionic ring opening polymerization of silicon bridged [1] ferrocenophanes have been synthesized (Ni 1996). These hybrid block copolymers containing a hydrocarbon (PS) or polysiloxane block and an iron-containing block was found to have very interesting bulk properties.

Diblock copolymers containing PS or PI and poly(hexylisocyanate) have been prepared by anionic polymerization (Chen 1995). Extreme care should be exercised in the purification of the isocyanate monomer in order to minimize deactivation of the first living block or depolymerization of polyisocyanate. Under the appropriate conditions, well-defined block copolymer having a flexible and a rod like block have been obtained, and their interesting bulk properties were investigated (Scheme 1.11).

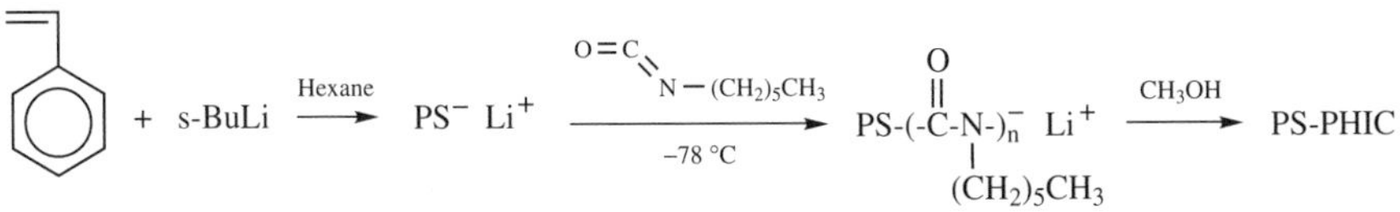

$$\text{styrene} + \text{s-BuLi} \xrightarrow{\text{Hexane}} \text{PS}^- \text{Li}^+ \xrightarrow[-78\,^{\circ}\text{C}]{\text{O=C=N}-(\text{CH}_2)_5\text{CH}_3} \text{PS-(-C(=O)-N((CH}_2)_5\text{CH}_3)\text{-)}_n^- \text{Li}^+ \xrightarrow{\text{CH}_3\text{OH}} \text{PS-PHIC}$$

Scheme 1.11

2. SYNTHESIS OF TRIBLOCK COPOLYMERS

A variety of triblock copolymer architectures, i.e., block copolymers containing three sequences of monomers, are possible because they can be comprised of two (ABA copolymers) or three (ABC terpolymers) different monomers. Each type of triblock can be synthesized according to an appropriate synthetic pathway depending on the monomers used and their sequence in the triblock chain.

2.1. Symmetric Triblock Copolymers Containing Two Different Monomers (ABA Triblocks)

Linear triblock copolymers consisting of two chemically different monomers of the symmetric type contain three blocks of A and B monomers arranged in a way that the first and the third block have the same chemical nature and molecular weight, whereas the middle block differs in chemical nature. There are three possible procedures to synthesize block copolymers of this type.

2.1.1. Sequential Monomer Addition. In this reaction scheme, the first monomer is polymerized followed by the polymerization of the second one. After complete consumption of the second monomer, an equal amount of the first monomer is added to the reaction mixture resulting in an ABA triblock copolymer (Avgeropoulos 1997).

This approach involves three monomer additions, and, therefore, the probability of partial termination of growing chains during the second or the third reaction step increases, due to impurities present in the monomers used. This can result in the presence of undesirable homopolymer A and/or diblock AB in the final product. Furthermore, small differences in the quantity of monomer A used in the first and third step may result in the synthesis of a triblock, which is not perfectly symmetric.

Another point that must be considered is the ability of monomer B to initiate polymerization of monomer A. If this criterion is not fulfilled, ill-defined products will be obtained. Sometimes even monomer A may not be able to initiate polymerization of monomer B appropriately. A typical example is the preparation of PI-PS-PI triblock copolymers. Initiation of styrene by PILi can be successful if a small amount of THF is used before the addition of styrene, but one should have in mind that the presence of THF will alter the microstructure of the third PI block. It is obvious that the outlined method cannot be used in the present example if high 1,4-microstructure of the PI blocks is desired. However, it can be applied in the case where the 1,4-microstructure is not essential. In this case THF can be added from the beginning of the polymerization and both end PI blocks will have the same microstructure.

2.1.2. Coupling of Living AB Chains. In this methodology a living diblock AB copolymer, having the same composition but half the molecular weight of the final triblock copolymer, is synthesized by sequential addition of monomers. Then an appropriate coupling agent, i.e., a compound having two functional groups able to react with the active anions forming covalent bonds, is used to connect two AB chains producing the desired symmetric triblock copolymer.

This technique has the advantage of an exactly symmetric triblock being formed. In addition, two, instead of three, steps are required. However, caution should be exercised in the stoichiometry of the coupling reaction. Usually, excess of living anions is used to ensure complete reaction of both groups of the coupling agent. This makes necessary an additional fractionation step in order to separate the ABA triblock from excess AB. Finally, the coupling reaction may be completed in days

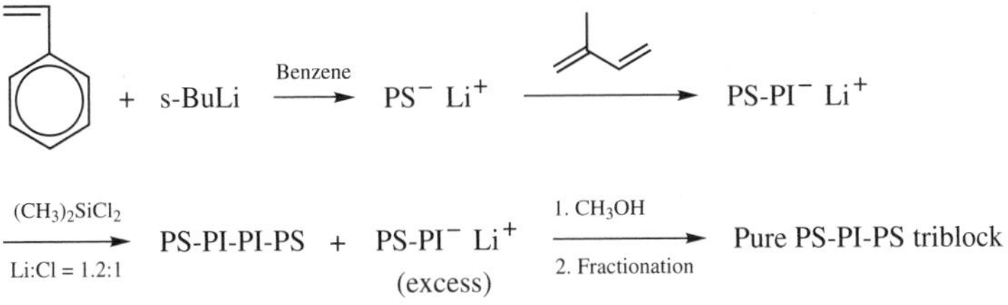

+ s-BuLi —Benzene→ $PS^- Li^+$ ——→ $PS\text{-}PI^- Li^+$

—$(CH_3)_2SiCl_2$, Li:Cl = 1.2:1→ PS-PI-PI-PS + $PS\text{-}PI^- Li^+$ (excess) —1. CH_3OH, 2. Fractionation→ Pure PS-PI-PS triblock

Scheme 1.12

(Pitsikalis 1999). Obviously, this synthetic route is more time consuming than the sequential addition method.

PS-PI (or PBd)-PS triblock copolymers have been synthesized by the coupling method by Morton and coworkers (Morton 1983) (Scheme 1.12). A PS-PD diblock is formed first where the length of the polydiene (PI or PB) block is half of that in the final triblock copolymer. Then the living diblocks are coupled using $(CH_3)_2SiCl_2$ as the coupling agent. A small excess of the living diblock is used in order to ensure complete coupling. Solvent/nonsolvent fractionation of the crude product is performed in order to isolate the pure triblock copolymers. These triblocks have found applications as thermoplastic elastomers in everyday life. PD-PS-PD symmetric triblock copolymers can also be made in the same way. $(CH_3)_2SiCl_2$ can be used as a coupling agent for PS (or PD)-polysiloxane living diblocks in order to produce triblocks with a siloxane central block (Bellas 2000, Hahn 1999). When the outer blocks are PBd with high 1,4 microstructure, PE-PDMS-PE triblocks with semicrystalline outer and elastomeric inner blocks are obtained by hydrogenation of the parent PBd-PDMS-PBd materials (Hahn 1999).

Bis(bromomethyl)benzene has been used as a coupling agent in cases where living (meth)acrylate and vinylpyridine anions were involved (Pitsikalis 1999). Due to the greater reactivity of the C-Br bond, this agent is efficient for coupling less reactive anions at lower temperatures (Scheme 1.13). Well-defined symmetric

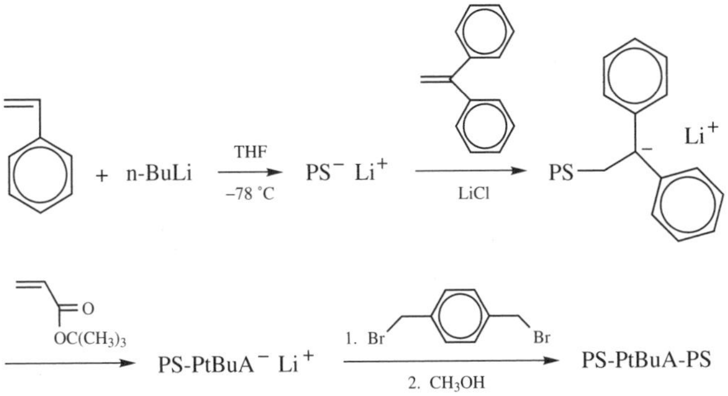

+ n-BuLi —THF, −78 °C→ $PS^- Li^+$ —LiCl→ PS Li^+

—O, $OC(CH_3)_3$→ $PS\text{-}PtBuA^- Li^+$ —1. Br Br, 2. CH_3OH→ PS-PtBuA-PS

Scheme 1.13

triblock copolymers with pyridine or (meth)acrylate central blocks can be prepared in both cases.

Varshney et al. (Varshney 1999) used terephthaloyl chloride as the coupling agent for the synthesis of ABA symmetric triblock copolymers with PS, P2VP, or polydiene of high 1,4 microstructure end blocks, and poly(*tert*-butyl acrylate) middle blocks. The PtBuA blocks could be converted to other types of acrylic blocks by transesterification reactions, leading to a larger variety of ABA triblock copolymers.

2.1.3. Use of a Difunctional Initiator. In this synthetic methodology, the symmetric triblock is formed by the use of a difunctional initiator, i.e., an organometallic compound having two anionic sites able to initiate polymerization, in a two-step addition of monomers.

The middle B block is formed first followed by the polymerization of A monomer. A number of difunctional initiators soluble in polar and/or nonpolar solvents have been reported in the literature (Hsieh 1996, Guyot 1982, Tung 1994a,b, Lo 1994, Bandermann 1985). The points that deserve attention are:

i) The quality of the difunctional initiator used, which is essentially its ability to form pure difunctional polymer. This depends on the precise chemical structure of the initiator and sometimes the solvent medium as well as the chemical nature of the monomer used. Not all known difunctional initiators behave in the desired manner in all solvents commonly used for anionic polymerization and towards any available monomer. If the initiator/solvent/monomer system cannot generate pure difunctional polymers, it is obvious that the final product will be a mixture of homopolymer, diblock, and triblock copolymer.
ii) The purity of the monomers must be high in order to avoid deactivation of one or both initiators' active sites or lead to premature termination of growing chains, which will result in a mixture containing the desired triblock and other undesired impurities that are difficult to be eliminated by fractionation or other separation methods.

Well-defined triblock copolymers with a dienic middle block and styrenic outer blocks have been synthesized in hydrocarbon solvents using a difunctional initiator derived from 1,3-bis(1-phenylethenyl)benzene (PEB). In this way high 1,4 diene microstructure is attained, and the mechanical properties of the materials are similar to those of triblock copolymers obtained with the coupling method (Lo 1994, Quirk 1991, Cunninham 1972) (Scheme 1.14).

The synthesis of triblock copolymers with diene inner block and methacrylic outer blocks using a difunctional initiator has been reported (Yu 1997a,b and 1996a,b). The use of methacrylic monomers gives the possibility of synthesizing thermoplastic elastomers with different than styrenic end blocks. The use of difunctional initiators in order to prepare all methacrylic thermoplastic elastomers, i.e., triblocks where all blocks belong to the methacrylic family of monomers, has

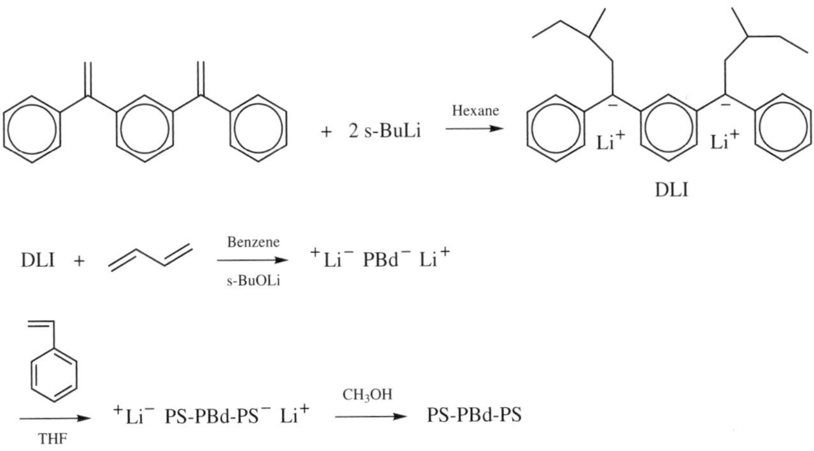

+ 2 s-BuLi —Hexane→ Li^+ Li^+

DLI

DLI + —Benzene / s-BuOLi→ $^+Li^-$ PBd$^-$ Li^+

—THF→ $^+Li^-$ PS-PBd-PS$^-$ Li^+ —CH_3OH→ PS-PBd-PS

Scheme 1.14

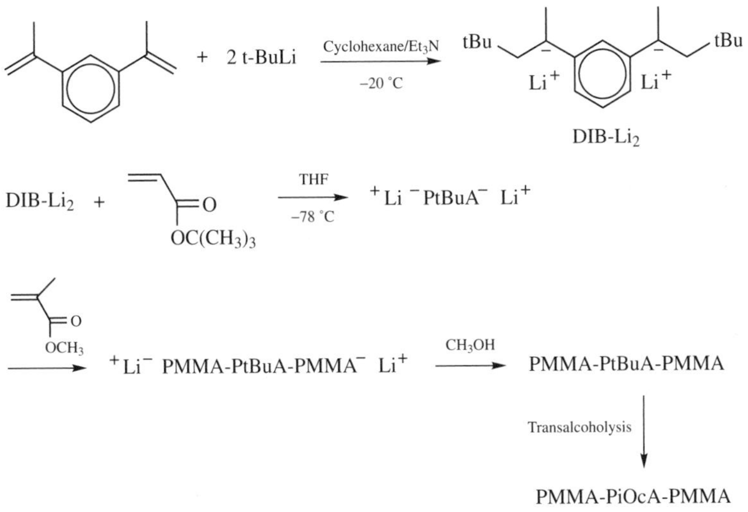

+ 2 t-BuLi —Cyclohexane/Et_3N, −20 °C→ tBu Li^+ Li^+ tBu

DIB-Li_2

DIB-Li_2 + O, $OC(CH_3)_3$ —THF, −78 °C→ ^+Li $^-$PtBuA$^-$ Li^+

O, OCH_3

→ $^+Li^-$ PMMA-PtBuA-PMMA$^-$ Li^+ —CH_3OH→ PMMA-PtBuA-PMMA

Transalcoholysis ↓

PMMA-PiOcA-PMMA

Scheme 1.15

also been reported (Tong 2000a,b,c) (Scheme 1.15). In this case the methacrylic monomers comprising the middle block are chosen to have a low T_g (lower than room temperature, i.e., iso-octyl acrylate), and those comprising the end blocks are chosen to have a high T_g (i.e., methylmethacrylate). This can be done due to the great variety of methacrylic monomers with different side chains and, therefore, different T_gs. The final products are expected to have better oxidative stability than the dienic middle blocks, but their mechanical properties may be inferior to

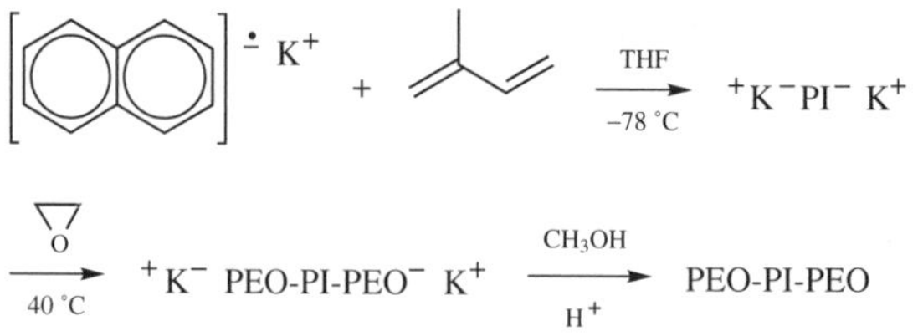

THF, −78 °C → $^{+}K^{-}$PI^{-} K^{+}

40 °C → $^{+}K^{-}$ PEO-PI-PEO^{-} K^{+} → (CH_3OH, H^{+}) PEO-PI-PEO

Scheme 1.16

classical PS/PD thermoplastic elastomers due to greater miscibility of the blocks comprising the triblock.

PEO-PI-PEO symmetric triblock copolymers were synthesized using sodium or potassium naphthalene as the difunctional initiator (Batra 1997). Isoprene was polymerized first followed by addition of EO (Scheme 1.16). The copolymers had a variety of block compositions and total molecular weights. Their molecular weight distributions were narrow and monomodal. Since Na or K naphthalenide is soluble only in polar solvents (e.g., THF), the microstructure of PI obtained was mostly 3,4.

Symmetric ABA block copolymers where A is poly(4-vinylpyridine) and B is PBd have been prepared by the use of the difunctional initiator derived from m-diisopropenylbenzene (Li 2000). The reaction took place in a mixture of THF and toluene at temperatures higher than room temperature for the polymerization of Bd, whereas, for the polymerization of 4-vinylpyridine, additional THF was added, and the temperature was lowered to −78°C. P4VP content was kept at 30% in order to minimize solubility problems coming from the poor solubility of P4VP in the reaction mixture (Scheme 1.17). In this case too, the microstructure of PBd was mostly 1,2.

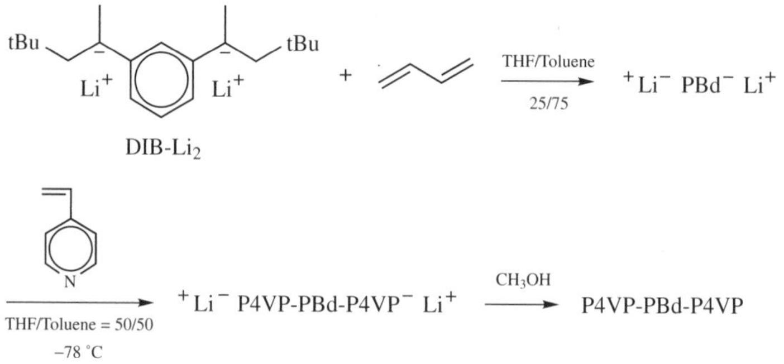

DIB-Li_2 + → (THF/Toluene 25/75) $^{+}Li^{-}$ PBd^{-} Li^{+}

→ (THF/Toluene = 50/50, −78 °C) $^{+}Li^{-}$ P4VP-PBd-P4VP^{-} Li^{+} → (CH_3OH) P4VP-PBd-P4VP

Scheme 1.17

ABA triblocks with a PI inner block and poly(tert-butyl methacrylate) end blocks were synthesized in a similar manner (Long 1989). End-capping of the

TABLE 1.2. ABA Triblock Copolymers Synthesized by Anionic Polymerization

Monomer A	Monomer B	Reference
Styrene	Butadiene	Morton 1983
Styrene	Isoprene	Morton 1983
p-Methylstyrene	Butadiene	Quirk 1986
α-Methyl styrene	Isoprene	Fetters 1969
tert-Butyl styrene	Butadiene	Fetters 1977
4-Vinyl pyridine	Butadiene	Li 2000
Ethylene oxide	Isoprene	Batra 1997
tert-Butyl acrylate	Styrene	Hautekeer 1990
Styrene	*tert*-Butyl acrylate	Varshney 1999 Pitsikalis 1999
Isoprene	*tert*-Butyl acrylate	Varshney 1999
2-Vinylpyridine	*tert*-Butyl acrylate	Varshney 1999
tert-Butyl methacrylate	Isoprene	Long 1989
Methyl methacrylate	Butadiene	Yu 1996a
	iso-Octyl acrylate	Tong 2000b
	n-Butyl acrylate	Tong 2000a,c
Hexamethylcyclotrisiloxane	Tetramethyl-p-silphenylene siloxane	Nagase 1983
Butadiene	Hexamethylcyclotrisiloxane	Hahn 1999
Glycidyl methacrylate	Butadiene	Yu 1997a
Isobornyl methacrylate	Butadiene	Yu 1996b
Hexamethylcyclotrisiloxane	Silicon Bridged [1] Ferrocenophanes	Ni 1996
Silicon Bridged [1] Ferrocenophanes	Styrene	Ni 1996
Ethylene oxide	Styrene	Hirata 1977

difunctional PI formed in the first stage of the reaction with DPE and addition of LiCl in the reaction system is essential in the controlled polymerization of the second monomer. The resulting triblocks had monomodal distributions and relatively low polydispersities.

2.2. Asymmetric Triblock Copolymers Containing Two Different Monomers (ABA′ Triblocks)

The synthesis of this type of triblocks can be accomplished only by sequential addition of monomers. However, if PI-PS-PI' triblock copolymers with high 1,4 microstructure is desired, laborious synthetic routes may be employed.

As an example, in the case of PI-PS-PI' triblocks, a PI-PSLi diblock formed in benzene and having a PI block with high 1,4 microstructure can be reacted with excess Me_2SiCl_2 in order to give a diblock having a SiCl terminal group. After elimination of excess Me_2SiCl_2 under vacuum, the macromolecular linking agent

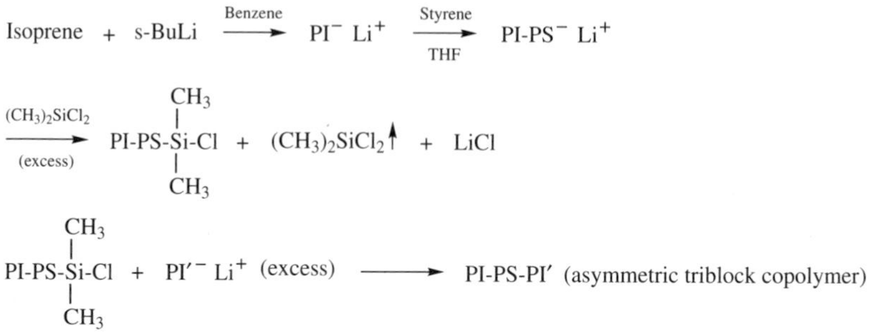

$$\text{Isoprene} + \text{s-BuLi} \xrightarrow{\text{Benzene}} \text{PI}^- \text{Li}^+ \xrightarrow[\text{THF}]{\text{Styrene}} \text{PI-PS}^- \text{Li}^+$$

$$\xrightarrow[\text{(excess)}]{(CH_3)_2SiCl_2} \text{PI-PS-}\overset{\overset{CH_3}{|}}{\underset{\underset{CH_3}{|}}{\text{Si}}}\text{-Cl} + (CH_3)_2SiCl_2\uparrow + \text{LiCl}$$

$$\text{PI-PS-}\overset{\overset{CH_3}{|}}{\underset{\underset{CH_3}{|}}{\text{Si}}}\text{-Cl} + \text{PI}'^- \text{Li}^+ \text{ (excess)} \longrightarrow \text{PI-PS-PI}' \text{ (asymmetric triblock copolymer)}$$

Scheme 1.18

can be reacted with excess of PILi, also formed in benzene and having the 1,4 microstructure, giving the desired triblock copolymer, which can be isolated in the pure form after elimination of excess PI block by fractionation (Scheme 1.18).

A similarly tedious synthetic strategy was followed for the preparation of asymmetric PS-P2VP-PS' triblock copolymers (Dhoot 1994). Living PS-P2VP diblock chains were synthesized first by sequential monomer addition starting from styrene. The living diblock was reacted with bis(bromomethyl)benzene (in excess) at low temperature, resulting in a PS-P2VP diblock with a terminal bromomethyl group. After purification, this group was used as binding site where living PSLi (with different molecular weight compared with the PS block of the diblock) was linked, giving the desired asymmetric triblock (Scheme 1.19).

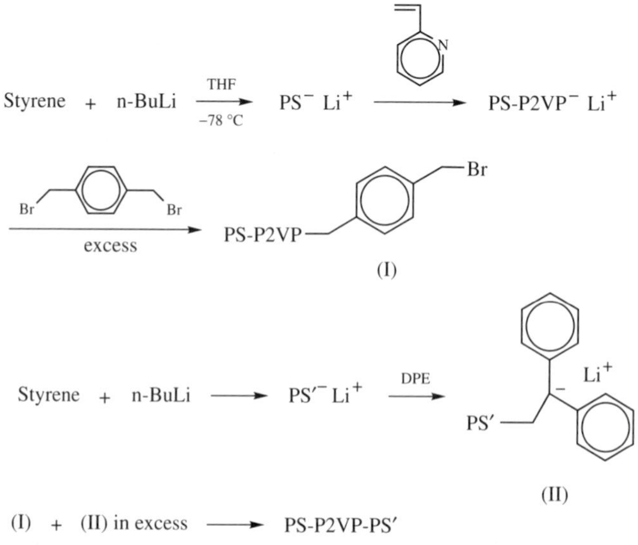

$$\text{Styrene} + \text{n-BuLi} \xrightarrow[-78\,^\circ C]{\text{THF}} \text{PS}^- \text{Li}^+ \xrightarrow{\text{2-vinylpyridine}} \text{PS-P2VP}^- \text{Li}^+$$

$$\xrightarrow[\text{excess}]{\text{Br}CH_2\text{-}C_6H_4\text{-}CH_2\text{Br}} \text{PS-P2VP-}CH_2\text{-}C_6H_4\text{-}CH_2\text{Br} \quad \text{(I)}$$

$$\text{Styrene} + \text{n-BuLi} \longrightarrow \text{PS}'^- \text{Li}^+ \xrightarrow{\text{DPE}} \text{PS}'\text{-}CH_2\text{-}C^-(C_6H_5)_2\ \text{Li}^+ \quad \text{(II)}$$

$$\text{(I)} + \text{(II) in excess} \longrightarrow \text{PS-P2VP-PS}'$$

Scheme 1.19

2.3. Triblock Terpolymers Containing Three Different Monomers (ABC Triblocks)

The usual method of choice for the synthesis of linear triblock copolymers consisting of three different monomers (terpolymers) by anionic polymerization is the sequential three-step addition of monomers.

As in the case of simple AB diblocks, the most important part of the design of the synthetic methodology to be followed is the order of monomer addition. Active chain-end stability as well as monomer purity should be highly considered. Extra care should be exercised in the amounts of monomers added in order to keep the desired stoichiometry and final composition of the triblock copolymer. Some examples of ABC triblock copolymers reported in the literature are given below.

ABC triblocks containing PS and polydienes (PI, PBd) of different microstructures have been synthesized by anionic polymerization using the sequential monomer addition method (Neumann 1998). The authors were aiming for the fine-tuning of polymer morphology by taking advantage of the changes in incompatibility between the three blocks resulting from the changes in the microstructure of the polydienes.

Triblock terpolymers containing PS, PI (or PBd), and P2VP have been synthesized for morphological studies (Mogi 1992, Matsushita 1994, Huckstadt 2000). Polymerizations were conducted in THF. The order of addition can be changed between styrene and isoprene, but 2-vinylpyridine was always the final monomer to be added to the polymerization mixture (Scheme 1.20). The molecular weights of

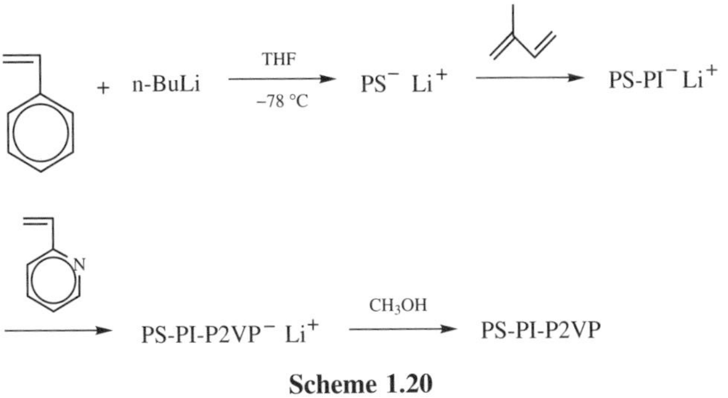

+ n-BuLi $\xrightarrow[-78\,^{\circ}C]{THF}$ $PS^{-}\ Li^{+}$ ⟶ $PS\text{-}PI^{-}\ Li^{+}$

⟶ $PS\text{-}PI\text{-}P2VP^{-}\ Li^{+}$ $\xrightarrow{CH_3OH}$ PS-PI-P2VP

Scheme 1.20

the final products were close to the expected ones, and their molecular weight distributions were narrow, indicating a well-controlled polymerization scheme. Triblocks having a block sequence of PS-P2VP-PBd were prepared by Watanabe et al. (Watanabe 1993). PS-P2VP living diblocks were reacted with excess p-xylene dichloride. The excess dichloride was removed from the PS-P2VP-Cl diblock by repetitive dissolution/isolation of the polymer under vacuum. The pure PS-P2VP-Cl functional diblocks were subsequently reacted with stoichiometric excess of living

PBd end-capped with DPE. Fractionation was required in order to isolate the PS-P2VP-PBd triblock terpolymers. After this synthetic route, well-defined triblocks were obtained.

Triblock terpolymers containing PS, PBd, and PMMA were synthesized by anionic polymerization using the sequential addition method (Krappe 1995, Stadler 1995). In this case styrene was polymerized, first in THF, followed by the addition of butadiene, and, finally, the methylmethacrylate monomer was added after transformation of the living chain-end to DPELi-end by reaction with DPE (Scheme 1.21). The products showed monomodal distributions although the molecular weights of the individual blocks employed were relatively high in some cases.

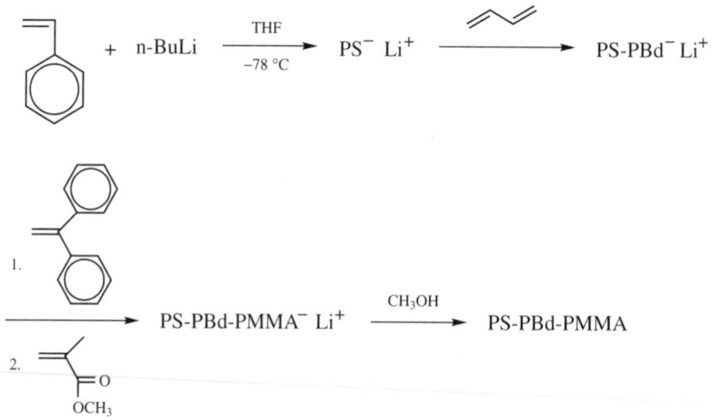

+ n-BuLi —THF, −78 °C→ $PS^- Li^+$ ——→ $PS\text{-}PBd^- Li^+$

1. (DPE) 2. (MMA, OCH$_3$) ——→ $PS\text{-}PBd\text{-}PMMA^- Li^+$ —CH_3OH→ PS-PBd-PMMA

Scheme 1.21

This indicates high purity of monomers used and low degree of deactivation of active center after each monomer addition. Interconversion of styrene and butadiene, in the order of monomer addition, is also possible here. Their hydrogenated analogues were prepared by postpolymerization hydrogenation of the diene block.

In another case PS-PBd-PCL triblock terpolymers were synthesized (Balsamo 1996). The order of addition was first styrene, then butadiene, and, lastly, caprolactone. Before the addition of the third monomer, the nucleophilicity of the living chain ends was reduced by addition of DPE in order to avoid side reactions. Also, inter- and intramolecular transesterification reactions were minimized by reducing the polymerization time of the caprolactone monomer. Benzene was used as the solvent at room temperature, resulting in high 1,4 addition in the PBd block (Scheme 1.22). Triblocks with monomodal distributions and the predicted molecular weights were obtained, having two amorphous and one crystallizable block in each molecule. Hydrogenation of the diene block resulted in triblock terpolymers, having two adjutant semicrystalline blocks and one amorphous block (Balsamo 1998).

PI-PS-poly(dimethylsiloxane) triblock terpolymers were obtained by sequential monomer addition using n-BuLi as the initiator and THF as the solvent (Malhotra

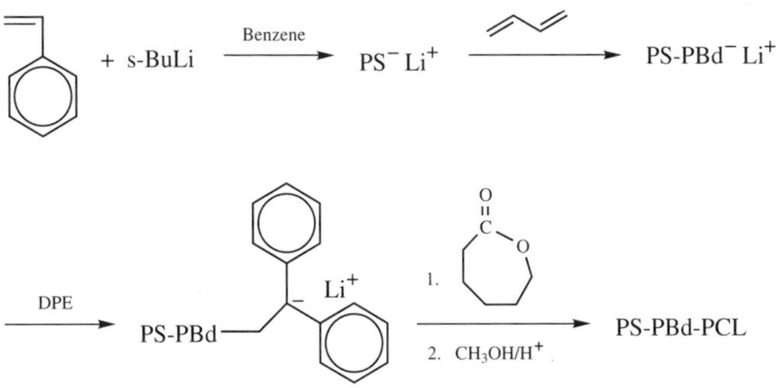

+ s-BuLi $\xrightarrow{\text{Benzene}}$ $PS^- Li^+$ $\longrightarrow$ $PS\text{-}PBd^- Li^+$

$\xrightarrow{\text{DPE}}$ PS-PBd— Li^+ $\xrightarrow[\text{2. } CH_3OH/H^+]{\text{1.}}$ PS-PBd-PCL

Scheme 1.22

1986). The use of n-BuLi and THF gives PS with narrow molecular weight distributions (Scheme 1.23). The final products had a monomodal molecular weight

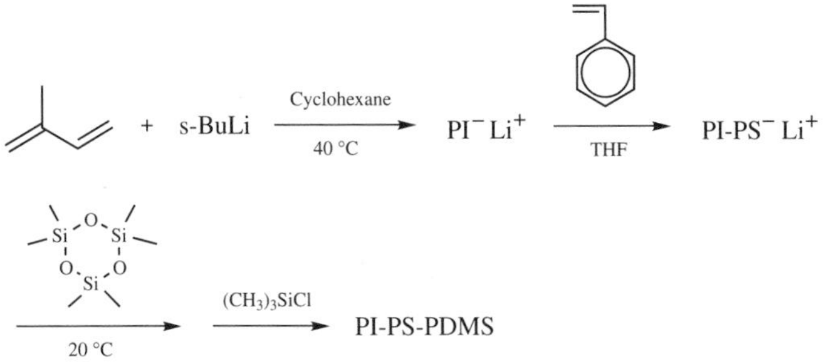

+ s-BuLi $\xrightarrow[\text{40 °C}]{\text{Cyclohexane}}$ $PI^- Li^+$ $\xrightarrow{\text{THF}}$ $PI\text{-}PS^- Li^+$

$\xrightarrow{\text{20 °C}}$ $\xrightarrow{(CH_3)_3SiCl}$ PI-PS-PDMS

Scheme 1.23

distribution. Their molecular characterization, determined using a variety of techniques, showed their low compositional heterogeneity. PI-PS-PDMS triblock terpolymer has also been synthesized because interconversion of the isoprene and the styrene monomers is possible, but siloxane monomer must be the last one. In the latter case, the PI possessed a high 1,4 microstructure due to the use of a hydrocarbon solvent for the polymerization of isoprene (Shefelbine 1999).

Another example is the synthesis of PI-P2VP-PEO triblock terpolymer in THF using benzyl potassium as the initiator (Ekizoglou 2001). Monomers were added sequentially starting with isoprene followed by 2-vinylpyridine. The polymerization of these monomers was carried out at −78°C. Finally, ethylene oxide was added and was left to polymerize at 50°C for a few days. Molecular characterization of the final product by a number of techniques proved its low compositional and molecular weight heterogeneity.

ABC triblock terpolymers containing three different methacrylic monomers were synthesized by living anionic polymerization. Using 2-(perfluorobutyl)ethyl methacrylate, tert-butylmethacrylate, and 2-(trimethylsililoxy)ethyl-methacrylate,

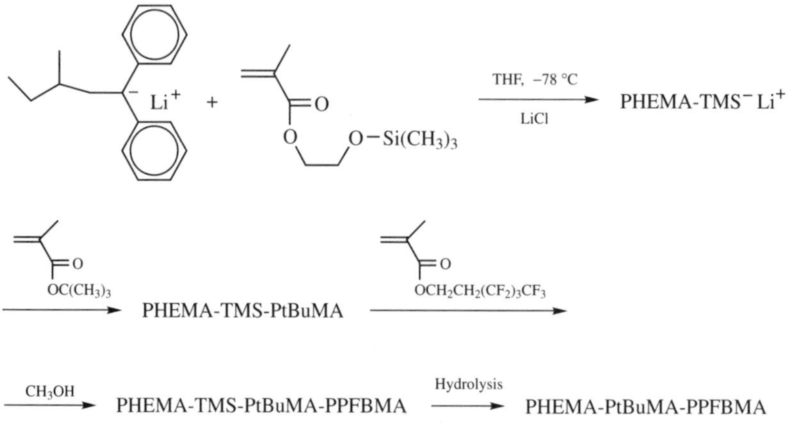

Li^+ + O O O−Si(CH_3)_3 —THF, −78 °C / LiCl→ PHEMA-TMS$^-$ Li^+

O OC(CH_3)_3 → PHEMA-TMS-PtBuMA → O OCH_2CH_2(CF_2)_3CF_3

—CH_3OH→ PHEMA-TMS-PtBuMA-PPFBMA —Hydrolysis→ PHEMA-PtBuMA-PPFBMA

Scheme 1.24

triblock copolymers having hydrophobic and perfluoalkyl groups were synthesized in THF in the presence of LiCl (Tanaka 1999). By changing the order of monomer addition, triblocks with different sequence of blocks were prepared. In all cases well-defined polymers were obtained (Scheme 1.24). Removal of the trimethylsi-

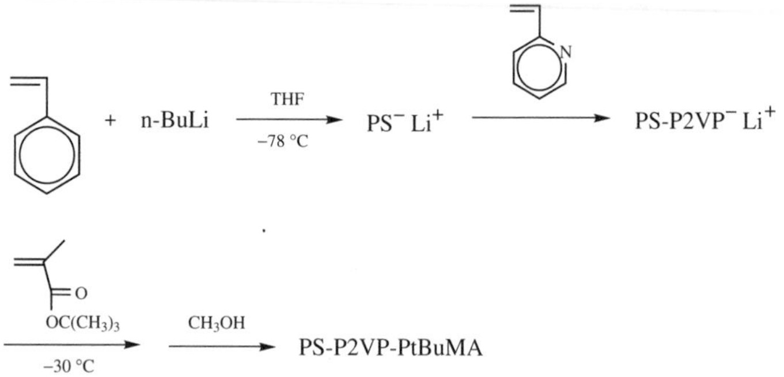

+ n-BuLi —THF / −78 °C→ PS$^-$ Li^+ —N→ PS-P2VP$^-$ Li^+

O OC(CH_3)_3 / −30 °C → —CH_3OH→ PS-P2VP-PtBuMA

Scheme 1.25

lane group by acidic hydrolysis gave triblocks with a hydrophilic block, a hydrophobic block, and a block containing perfluoroakyl groups.

Linear triblock terpolymers containing a hydrophobic, a polyacid, and a polybase block were synthesized by Stadler et al. (Giebeler 1997). The hydrophobic block was PS, whereas the polyacid was a hydrolyzed poly(*tert*-butyl methacrylate) block, and the polybase was poly(2-vinylpyridine) or poly(4-vinylpyridine). Sequential polymerization in THF gave PS-P2VP-PtBMA and PS-P4VP-PtBMA block copolymers with narrow molecular weight distribution. This synthetic approach is a good example of the way that the reactivity of each anion has to be taken into account in a sequential anionic polymerization procedure. The final products had monomodal and narrow molecular weight distributions (Scheme 1.25).

3. LINEAR BLOCK COPOLYMERS WITH MORE THAN THREE BLOCKS

The synthesis of block copolymers with a number of blocks equal to or larger than four is possible in principle. If the synthesis of a linear ABCD tetrablock quaterpolymer having four chemically different blocks is demanded, this can be accomplished by sequential monomer addition. Although this route involves a four-step addition process, it can be successful if special attention has been paid to the purification of the reagents used. An alternative method would be the preparation of the two diblocks AB and CD and their subsequent connection using the appropriate coupling agent. In the coupling step, one of the two functional groups of the linking agent must be reacted with one, e.g., AB, of the living diblocks. Excess of linking agent and careful choice of the reaction conditions with respect to reaction temperature, state of dilution, and mode of addition of the reagents can lead to selective substitution of one of the linking sites present in the molecule. This end functional macromolecule can then react with an excess of the CD living diblock to give the desired tetrablock.

Linear pentablock terpolymers of the ABCBA type were synthesized by anionic polymerization using a difunctional initiator (Yu 1996a, Yu 1997b). In this case A = alkylmethacrylate (PIBMA or PMMA), B = styrene, and C = butadiene. The preparation of these polymers started with the formation of the difunctional PBd inner block, followed by the formation of the PS block, and, finally, the synthesis of the polymethacrylate block (Scheme 1.26). Intermediate products were isolated, by

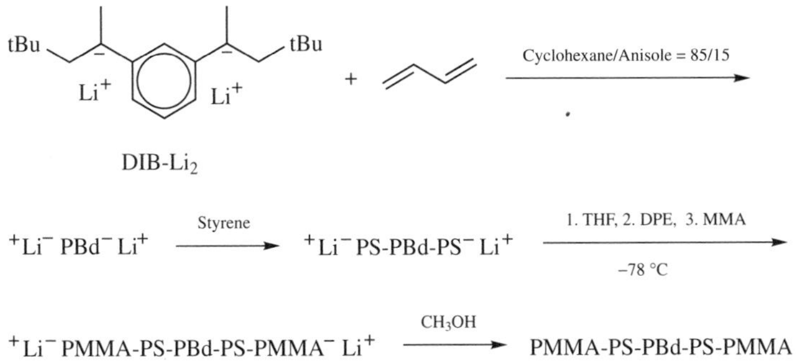

tBu Li$^+$ Li$^+$ tBu

DIB-Li_2 + butadiene $\xrightarrow{\text{Cyclohexane/Anisole = 85/15}}$

$^+Li^-$ PBd$^-$ Li^+ $\xrightarrow{\text{Styrene}}$ $^+Li^-$ PS-PBd-PS$^-$ Li^+ $\xrightarrow[-78\ °C]{\text{1. THF, 2. DPE, 3. MMA}}$

$^+Li^-$ PMMA-PS-PBd-PS-PMMA$^-$ Li^+ $\xrightarrow{CH_3OH}$ PMMA-PS-PBd-PS-PMMA

Scheme 1.26

sampling, before the addition of the next monomer and characterized by SEC. The final products showed monomodal distributions.

The synthesis of PEO-PS-PI-PS-PEO pentablock terpolymers was accomplished by the use of a difunctional initiator. Their dynamic and thermal properties were investigated (Benson 1985).

Pentablock terpolymers of the type PtBuS-PS-PB-PS-PtBuS were synthesized by sequential addition of monomers and by coupling preformed living PtBuS-PS-PtBuS using Me_2SiCl_2 as the coupling agent (Chen 1987). Copolymers prepared by both methods showed narrow distribution peaks in SEC analysis. They were evaluated as thermoplastic elastomers showing good phase separation and mechanical properties.

The synthesis of a pentablock terpolymer by sequential addition of monomers has been reported. A PI-PS-PI-poly[(4-vinylbenzyl)dimethylamine]-PI terpolymer was obtained using s-BuLi as the initiator and benzene as the solvent. The intermediate and final products were characterized by SEC, ultracentrifugation, and membrane osmometry. All samples showed one peak in the SEC chromatogram as well as in their sedimentation pattern, indicating the successful synthesis of the complex materials.

REFERENCES

Allen R. D., Long T. E., McGrath J. E. (1986) Polym. Bull. 15, 127.

Almdal K., Mortensen K., Ryan A. J., Bates F. S. (1996) Macromolecules 29, 5940.

Antkowiak T. A., Oberster A. E., Halasa A. F., Tate D. P. (1972) J. Polym. Sci. Part A-1 10, 1319.

Avgeropoulos A., Dair B. J., Hadjichristidis N., Thomas E. L. (1997) Macromolecules 30, 5634.

Balsamo V., Muller A. J., von Gyldenfeldt F., Stadler R. (1998) Macromol. Chem. Phys. 199, 1063.

Balsamo V., von Gyldenfeldt F., Stadler R. (1996) Macromol. Chem. Phys. 197, 1159.

Bandermann F., Speikamp H-D., Weigel L. (1985) Makromol. Chem. 186, 2017.

Batra U., Russel W. B., Pitsikalis M., Sioula S., Mays J. W., Huang J. S. (1997) Macromolecules 30, 6120.

Bellas V., Iatrou H., Hadjichristidis N. (2000) Macromolecules 33, 6993.

Benson R. S., Wu Q., Ray A. R., Lyman D. J. (1985) J. Polym. Sci. Polym. Chem. Ed. 23, 399.

Booth C., Attwood D. (2000) Macromol. Rapid Commun. 21, 501.

Burker M. C., Vincent B. (1984) Colloids & Surfaces 8, 289.

Chen J. C., Fetters L. J. (1987) J. Vinyl Technol. 9, 1300.

Chen J. T., Thomas E. L., Ober C. K., Hwang S.S. (1995) Macromolecules 28, 1688.

Corbin N., Prud'homme J. (1976) J. Polym. Sci. Polym. Chem. Ed. 14, 1645.

Cunningham R. E., Huerbach M., Floyd W. J. (1972) J. Appl. Polym. Sci. 16, 113.

Deng Y., Price C., Booth C. (1994) Eur. Polym. J. 30, 103.

Dhoot S., Watanabe H., Tirrell M. (1994) Colloids & Surfaces A: Physicochem. Eng. Aspects 86, 47.

Duivenvoorde F. L., Steven van Es J. J. G., van Nostrum C. F., van der Linde R. (2000) Macromol. Chem. Phys. 201, 656.

Ekizoglou N., Hadjichristidis N. (2001) J. Polym. Sci. Part A: Polym. Chem. 39, 1198.

Elgert K. F., Seiler E., Puschendort G., Kantow H. J. (1973) 165, 245.

Fetters L. J., Firer E. M., Dafauti E. M. (1977) Macromolecules 10, 1200.

Fetters L. J., Morton M. (1969) Macromolecules 2, 453.

Finaz G., Rempp P., Parrod J. (1962) Bull. Soc. Chim. Fr. 262.

Förster S., Kramer E. (1999) Macromolecules 32, 2783.

Funabashi H., Isono Y., Fujimoto T., Matsushita Y., Nagasawa M. (1983) Macromolecules 16, 1.

Giebeler E., Stadler R. (1997) Macromol. Chem. Phys. 198, 3815.

Grosius P., Gallot Y., Skoulios A. (1970) Eur. Polym. J. 6, 355.

Guyot P., Favier J. C., Fontanille M., Sigwalt P. (1982) Polymer 23, 73.

Guyot P., Favier J. C., Uytterhoeven H., Fontanille M., Sigwalt P. (1981) Polymer 22, 1724.

Hadjichristidis N., Iatrou H., Pispas S., Pitsikalis M. (2000) J. Polym. Sci. Part A: Polym. Chem. 38, 3211.

Hahn S. F., Vosejpka P. C. (1999) Polym. Prepr. 40(1), 970.

Hauteker J-P., Varshney S. K., Fayt R., Jacobs C., Jérôme R., Teyssié P. (1990) Macromolecules 23, 3893.

Hild G., Lamps J. P. (1998) Polymer 39, 2637.

Hirata E., Ijitsu T., Soen T., Hashimoto T., Kawai H. (1977) Copolymers, Polyblends and Composites, Advances in Chemical Series 142, ACS, New York, p. 288.

Hong K., Mays J. W. (2001) Macromolecules 34, 3540.

Hsieh H. L., Quirk R. P. (1996) Anionic Polymerization: Principles and Practical Applications, Marcel Dekker Inc., New York.

Huckstadt H., Gopfert A., Abetz V. (2000) Polymer 41, 9089.

Imae T., Tabuchi H., Funayama K., Sato A., Nakamura T., Amaya N. (2000) Colloids & Surfaces A: Physicochem. Eng. Aspects 167, 73.

Ishizone T., Hirao A., Nakahama S. (1993) Macromolecules 26, 6964.

Ishizone T., Sugiyama K., Sakano Y., Mori H., Hirao A., Nakahama S. (1999) Polymer J. 31, 983.

Krappe V., Stadler R., Voigt-Martin I. (1995) Macromolecules 28, 4558.

Lehmann O., Forster S., Springer J. (2000) Macromol. Rapid Commun. 21, 133.

Li H-J., Tsiang R. C-C. (2000) Polymer 41, 5601.

Lo G., Y-S., Otterbacher E. W., Gatzke A. L., Tung L. H. (1994) Macromolecules 27, 2233.

Long T. E., Broske A. P., Bradley D. J., McGrath J. E. (1989) J. Polym. Sci. Part A: Polym. Chem. 27, 4001.

Malhotra S. L., Bluhm T. L., Delsandes Y. (1986) Eur. Polym. J. 22, 391.

Martin T. J., Prochazka K., Munk P., Webber S. E. (1996) Macromolecules 29, 6071.

Matsushita Y., Shimizu N., Nakao Y., Choshi H., Nagasawa M. (1986) Polymer J. 18, 361.

Matsushita Y., Yamura M., Noda I. (1994) Macromolecules 27, 3680.

Miyaki Y., Iwata M., Fujita Y., Tanisugi H., Isono Y., Fujimoto T. (1984) Macromolecules 17, 1907.

Mogi Y., Kotsuji H., Kaneko Y., Mori Y., Matsushita Y., Noda I. (1992) Macromolecules 25, 5408.

Morton M. (1983) Anionic Polymerization: Principles and Practice, Academic Press, New York.

Nagase Y., Nakamura T., Misawa A., Ikeda K., Sekine Y. (1983) Polymer 24, 457.

Nakahama S., Hirao A. (1990) Prog. Polym. Sci. 15, 299.

Neumann C., Abetz V., Stadler S. (1998) Coll. Polym. Sci. 276, 19.

Ni Y., Rulkens R., Manners I. (1996) J. Am. Chem. Soc. 118, 4102.

Paul D. R., Barlow J. W. (1980) J. Macromol. Sci., Rev. Macromol. Chem. C18(1), 109.

Pispas S., Hadjichristidis N. (1994) Macromolecules 27, 1891.

Pispas S., Hadjichristidis N. (2000a) J. Polym. Sci. Part A: Polym. Chem. 39, 3219.

Pispas S., Hadjichristidis N. (2000b) Macromolecules 33, 6396.

Pitsikalis M., Siakali-Kioulafa E., Hadjichristidis N. (2000) Macromolecules 33, 5460.

Pitsikalis M., Sioula S., Pispas S., Hadjichristidis N., Cook D. C., Li J., Mays J. W. (1999) J. Polym. Sci. Part A: Polym. Chem. 37, 4337.

Quirk R. P., Lee R. (1992) Polym. Int. 27, 359.

Quirk R. P., Ma J-J. (1991) Polym. Int. 24, 197.

Quirk R. P., Sarkis M. T., Meier D. J. (1986) in Advances in Elastomers and Rubber Elastisity, Lal J., Mark J. E. (Eds), Plennum Press, New York, p. 143.

Schadler V., Kniese V., Thurn-Albrecht T., Wiesner U., Spiess H. W. (1998) Macromolecules 31, 4828.

Schadler V., Wiesner U. (1997) Macromolecules 30, 6698.

Schindler A., Williams J. L. (1969) Polym. Prepr. 10(2), 832.

Schops M., Leist H., DuChesne A., Wiesner U. (1999) Macromolecules 32, 2806.

Shefelbine T. A., Vigild M. E., Matsen M. W., Hajduk D. A., Hillmayer M. A., Cussler E. L., Bates F. S. (1999) J. Am. Chem. Soc. 121, 8457.

Stadler R., Auschra C., Beckmann J., Krappe U., Voigt-Martin I., Leibler L. (1995) Macromolecules 28, 3080.

Tanaka Y., Hasegawa H., Hashimoto T., Ribbe A., Sugiyama K., Hirao A., Nakahama S. (1999) Polymer J. 31, 989.

Tong J. D., Jérôme R. (2000a) Polymer 41, 2499.

Tong J. D., Leclerc P., Doneux C., Bredas L., Lazzaroni R., Jérôme R. (2000b) Polymer 41, 4617.

Tong J. D., Moineau G., Leclerc P., Bredas L., Lazzaroni R., Jérôme R. (2000c) Macromolecules 33, 470.

Tung L. H., Lo G. Y-S. (1994a) Macromolecules 27, 1680.

Tung L. H., Lo G. Y-S. (1994b) Macromolecules 27, 2219.

Varshney S. K., Hautekeer J. P., Fayt R., Jérôme P., Teyssié P. (1990) Macromolecules 23, 2618.

Varshney S. K., Kesani P., Agarwal N., Zhang J. X., Rafailovich M. (1999) Macromolecules 32, 235.

Watanabe H., Shimura T., Kotaka T., Tirrell M. (1993) Macromolecules 26, 6338.

Worsfold D. J., Bywater S. (1982) Makromol. Chem. Rapid Commin. 3, 239.

Yamada M., Iguchi T., Hirao A., Nakahama S., Watanabe J. (1998) Polymer J. 30, 23.

Yamada M., Itoh T., Nakagawa R., Hirao A., Nakahama S., Watanabe J. (1999) Macromolecules 32, 282.

Yin R., Hogen-Esch T. E. (1994) J. Polym. Sci. Part A: Polym. Chem. 32, 363.

Young R. N., Quirk R. P., Fetters L. J. (1984) Adv. Polym. Sci. 56, 1.

Yu J. M., Dubois P., Jérôme R. (1996b) Macromolecules 29, 7316.

Yu J. M., Dubois P., Jérôme R. (1997a) J. Polym. Sci. Part A: Polym. Chem. 35, 3507.

Yu J. M., Dubois P., Teyssié P., Jérôme R. (1996a) Macromolecules 29, 6090.

Yu J. M., Yu Y., Dubois P., Teyssié P., Jérôme R. (1997b) Polymer 38, 3091.

Yu Y. S., Dubois P., Jérôme R., Teyssié P. (1996) J. Polym. Sci. Part A: Polym. Chem. 34, 2221.

Zilliox J. G., Roovers J. E. L., Bywater S. (1975) Macromolecules 8, 573.

CHAPTER 2

BLOCK COPOLYMERS BY CATIONIC POLYMERIZATION

Advances in cationic polymerization methodology, starting in the middle 80s with the discovery of the true living cationic polymerization of vinyl ethers by Higashimura et al. (Miyamoto 1984), have shown their real potential for the synthesis of tailor-made macromolecules. In recent years many investigations have demonstrated that almost all classes of cationically polymerizable vinyl and alkene-type monomers can be polymerized in a controllable way (Faust 1997, Matyjaszewski 1996, Sawamoto 1991). The formation of polymers having predictable molecular weight and narrow molecular weight distributions gives unambiguous experimental evidence for elimination or suppression of termination and chain transfer reactions in these systems. These studies opened the way for block copolymer synthesis using cationically polymerizable monomers, extending the range of block copolymers available for basic research and for possible technological applications. Many important monomers like isobutylene and alkyl vinyl ether can be polymerized only by cationic polymerization.

Cationic polymerization can be described as an additional polymerization reaction where chain propagation is achieved through a carbocation, which can be generated by a cationic initiator and a vinyl monomer (Sawamoto 1991, Kennedy 1992).

Carbocations, in general, are very reactive and unstable; consequently, they can participate in a number of side reactions like termination, chain transfer, and carbocation rearrangement. The major side reaction is chain transfer to monomer. Since the positive charge is present on the α–carbon of the double bond, the hydrogen atom on the β-carbon is acidic. The monomers used in cationic polymerization are nucleophilic; therefore, this kind of side reaction is intrinsic to most systems and difficult to eliminate. However, several methods have been proposed in order to overcome this difficulty. The most successful strategy is the stabilization of the carbocationic intermediate using (Higashimura 1988):

i) An appropriate counterion.
ii) A carefully selected Lewis base.

Both methods aim to decrease the positive charge on α-C and, as a result, decrease β-H acidity, suppressing the chain transfer reaction. A typical example of case (i) is the combination HI/I_2 and of case (ii) are systems containing cationogenic compounds, like tertiary esters, ethers and alcohols, with organometallic sustances, like $EtAlCl_2$, BCl_3, $TiCl_4$, which produce weak nucleophilic counteranions. A variety of initiating systems have been reported in the literature. The two methods have to be considered as complementary. Because there are no systems that can be used with all available monomers, each initiating system must be chosen according to the monomer used, i.e., its reactivity and stability of the carbocation formed.

The most common cationically polymerizable monomers are isobutylene, vinyl ethers, styrene and its derivatives with electron donating groups, N-vinylcarbazole, furan, and some other heterocyclic monomers.

1. SYNTHESIS OF AB DIBLOCK COPOLYMERS

The synthesis of linear block copolymers containing two chemically different blocks by cationic polymerization can be accomplished by sequential monomer addition in a simple and convenient way. The successful synthesis of diblock copolymers is based on the appropriate selection of experimental polymerization conditions, like Lewis acid, additives, solvent, and temperature. The most important part of the synthetic design is the selection of the appropriate order of monomer addition. Efficient crossover reaction occurs when the two monomers have almost equal reactivities or when the more reactive monomer is polymerized first followed by the addition of the less reactive. Monomer reactivity can be estimated using the nucleophilicity parameter (N) (Matyjaszewski 1996).

Diblock copolymers of styrene and isobutylene can be prepared by polymerizing either monomer first because these two monomers have similar reactivities. The order of monomer reactivity is so close that identical reaction conditions can also be used (i.e., $TiCl_4$ as the Lewis acid, a temperature of $-80°C$) in order to accomplish the polymerization of both monomers. However, another parameter has to be taken into account in this system, i.e., the stability of living cations close to the end of polymerization (Faust 1999). Since PS cations are less stable than PIB cations, undergoing a slow decomposition at monomer conversions close to 100%, IB must be added, when used as the second monomer, at a styrene conversion of ~95%, in order to avoid partial termination of living PS. Another way is to start with the formation of the PIB block first, because PIB carbocations are stable even after complete monomer conversion, followed by the addition and polymerization of St (Schemes 2.1 and 2.2).

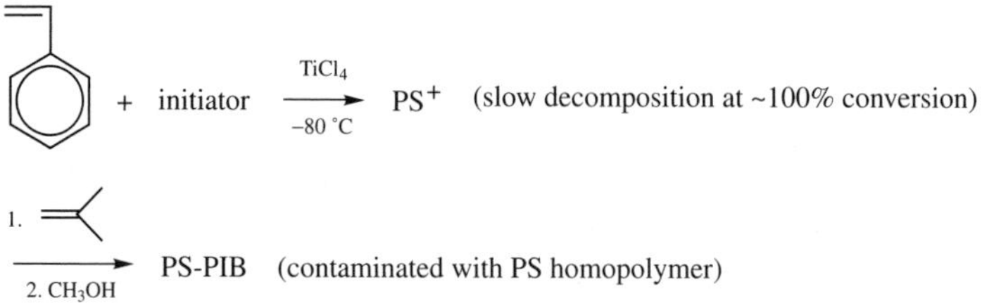

+ initiator $\xrightarrow[-80\ ^\circ C]{TiCl_4}$ PS^+ (slow decomposition at ~100% conversion)

$\xrightarrow[2.\ CH_3OH]{1.}$ PS-PIB (contaminated with PS homopolymer)

Scheme 2.1

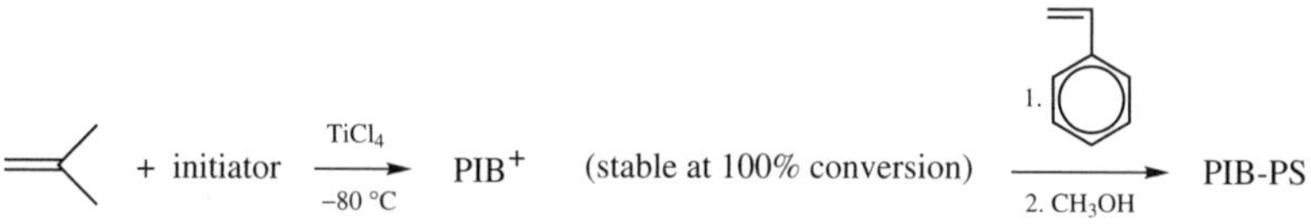

+ initiator $\xrightarrow[-80\ ^\circ C]{TiCl_4}$ PIB^+ (stable at 100% conversion) $\xrightarrow[2.\ CH_3OH]{1.}$ PIB-PS

Scheme 2.2

The pair α-methylstyrene (α-MeSt) and IB shows differences in reactivity. α-MeSt is more reactive than IB. So the synthesis of a PαMeS-PIB diblock must start by the formation of the α-MeSt block. As a consequence, a different Lewis acid must be used for the polymerization of each monomer (Li 1995). Thus, α-MeSt is polymerized in the presence of the weaker Lewis acid BCl_3, which is inefficient to promote the living polymerization of IB. The second monomer IB is polymerized in the presence of a stronger Lewis acid, i.e., $TiCl_4$, and well-defined diblock copolymers are obtained (Scheme 2.3).

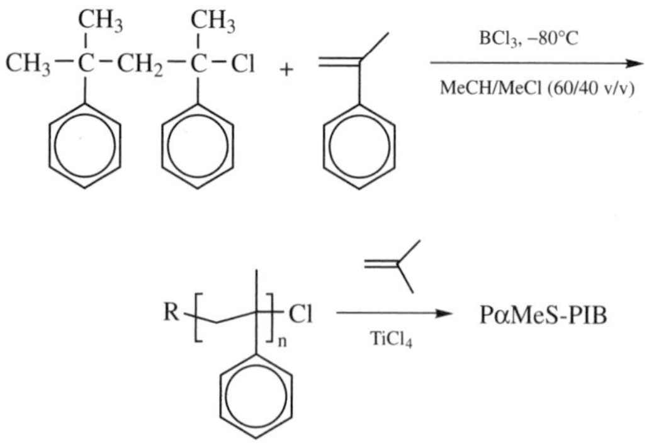

$CH_3-C(CH_3)(Ph)-CH_2-C(CH_3)(Ph)-Cl$ + $\xrightarrow[MeCH/MeCl\ (60/40\ v/v)]{BCl_3,\ -80^\circ C}$

$R-[CH_2-CH(Ph)]_n-Cl$ $\xrightarrow[TiCl_4]{}$ PαMeS-PIB

Scheme 2.3

Sequential addition of monomers in a reverse order compared with their reactivity may sometimes be desirable, especially if diblock copolymer formation is a part of the synthesis of a more complicated architecture (a triblock or a starblock copolymer). In this case the diblock copolymer and the final product is contaminated by the less reactive living homopolymer. For instance, block polymerization of IB with more reactive monomers, p-methylstyrene (pMeSt) isobutyl, or methyl vinyl ether (MeVE) using IB as the first monomer results in the formation of diblock copolymers contaminated with PIB homopolymer (Li 1995, Fodor 1994, Hadjikyriacou 1995 and 1996). This is a result of a slower crossover reaction compared to the rate of the propagation reaction of the second monomer.

Faust et al. (Faust 2000, Hadjikyriacou 2000) recently developed a synthetic scheme that allows the preparation of diblock copolymers in the case where the second monomer is more reactive than the first. The method involves end-capping of living PIB chains with 1,1-diphenylethylene (DPE) or 1,1-ditolylethylene (DTE) and subsequent modification of cation reactivity, using the appropriate Lewis acid, towards the second monomer. Tuning cation activity enables the generation of stronger nucleophilic counterions, ensuring in this way the increase of the ratio Rc/Rp, where Rc is the rate of the crossover reaction and Rp is the rate of the propagation reaction of the second monomer. By this procedure diblock copolymers with well-defined molecular characteristics can be prepared with no homopolymer contamination due to the controlled polymerization of the second monomer.

This synthetic procedure has been employed in the synthesis of diblocks where the first block was comprised of IB segments and the second one of α-MeSt (Scheme 2.4), p-MeSt, or MeVE segments (Fodor 1994, Hadjikyriacou 1996). In

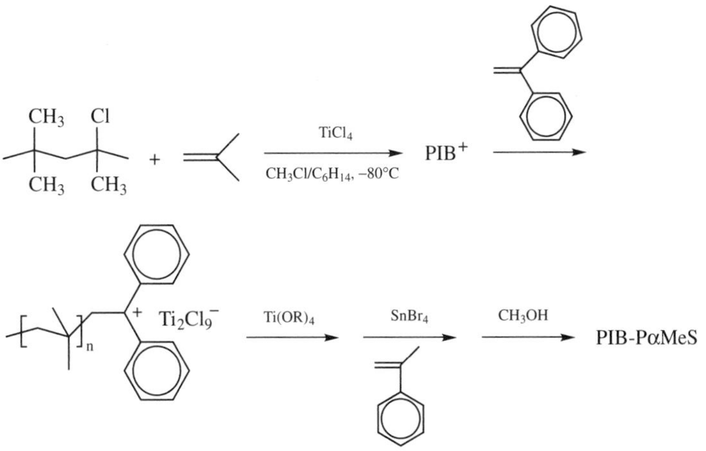

CH_3 Cl / CH_3 CH_3 + (isobutylene) $\xrightarrow[CH_3Cl/C_6H_{14}, -80°C]{TiCl_4}$ PIB^+ $\xrightarrow{DPE}$

PIB–DPE$^+$ $Ti_2Cl_9^-$ $\xrightarrow{Ti(OR)_4}$ $\xrightarrow[\alpha\text{-MeSt}]{SnBr_4}$ $\xrightarrow{CH_3OH}$ PIB-PαMeS

Scheme 2.4

these cases titanium (IV) alkoxides were used as Lewis acids after the capping reaction. Diblock copolymers of IB and MeVE could also be produced by capping of PIB chains with 2-alkylfurans instead of DPE or DTE (Hadjikyriacou 1999). Alkylfurans present a new class of nonhomopolymerizable monomers used in the modification of crossover reactions. In the case of α-MeSt, weaker Lewis acids like $SnBr_4$ or $SnCl_4$ could be used. Isobutylvinylether (IBVE) could also be used as the second monomer if $(n\text{-}Bu)_4NCl$ was employed as the Lewis acid (Hadjikyriacou 1995).

The general rules that correlate monomer reactivity with order of monomer addition in the polymerization mixture also hold for the case of diblock copolymers with styrenic blocks. Well-defined block copolymers of p-tert-butoxy styrene (t-BOS) and p-methyl styrene could be obtained only in the case where t-BOS was added as the first monomer (Kojima 1991) (Scheme 2.5). The HI/ZnX_2 initiating

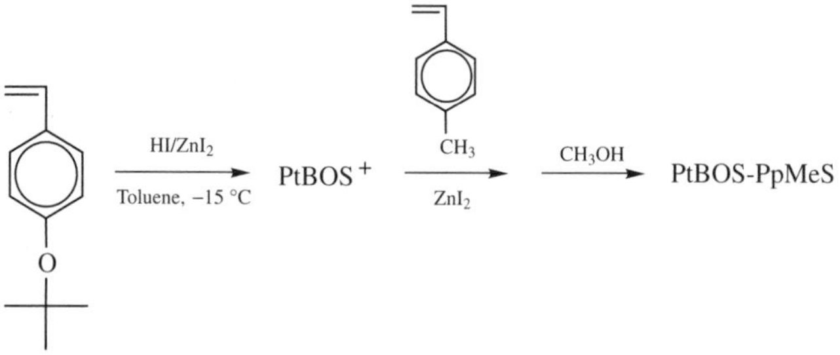

HI/ZnI_2, Toluene, −15 °C → $PtBOS^+$; CH_3, ZnI_2 → ; CH_3OH → PtBOS-PpMeS

Scheme 2.5

system was used in this synthetic procedure in toluene as the solvent and at low temperature (−15°C). Further addition of activator (ZnX_2) is needed in order to accelerate the polymerization of p-MeS. In the case of the pair t-BOS/p-methoxy-styrene (p-MOS), the order of monomer addition can be reversed due to the similar reactivity of the monomers without any deleterious effects in the molecular weight distribution of the final copolymers. The diblocks obtained by the two methods have narrow molecular weights and predictable molecular weights and compositions (Scheme 2.6).

Recently, the synthesis of amphiphilic block copolymers of p-hydroxystyrene (p-HS) and p-methoxystyrene was achieved by direct polymerization of the unprotected hydroxyl group containing monomer (Satoh 2000). These block copolymers were obtained by polymerizing p-HS first using the initiating system $CH_3CH(Ph OCH_3)OH/BF_3OEt_2/H_2O$ in acetonitrile at 0°C. After completion of the polymerization, a predetermined amount of CH_2Cl_2 was added as a cosolvent followed by addition of the second monomer (p-MOS). The polymerization of p-MOS was done in a controlled way, and the final materials were essentially free of terminated homopolymer (Scheme 2.7). The later observation leads to the conclusion that complete crossover from the active p-HS chain ends to the p-MOS living ends has occurred. NMR analysis of the final products showed that they had the desired composition (similar to that obtained by the stoichiometric ratio of the monomers used).

A large variety of AB block copolymers containing vinylethers have been synthesized by sequential addition of monomers (Sawamoto 1991). Alkyl substituted

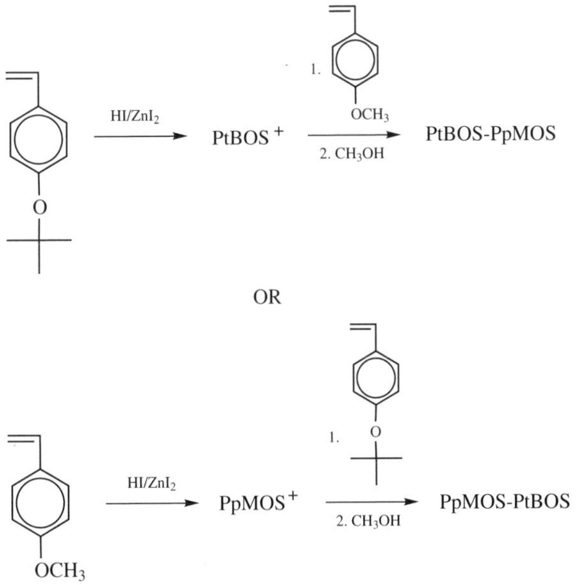

HI/ZnI$_2$ PtBOS$^+$ 1. OCH$_3$ 2. CH$_3$OH PtBOS-PpMOS

OR

HI/ZnI$_2$ PpMOS$^+$ 1. O 2. CH$_3$OH PpMOS-PtBOS

OCH$_3$

Scheme 2.6

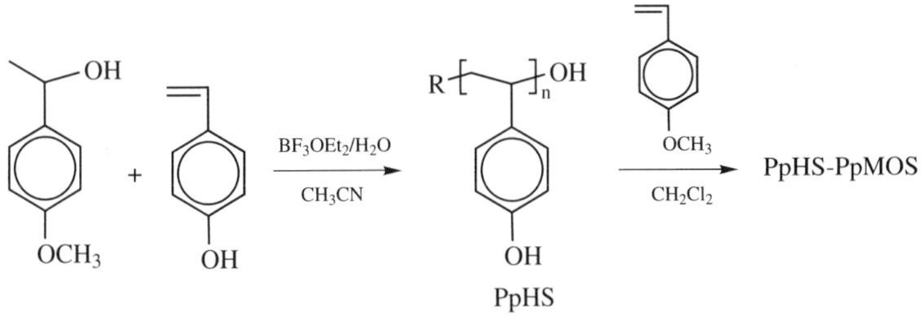

OH + BF$_3$OEt$_2$/H$_2$O CH$_3$CN R OH n OCH$_3$ CH$_2$Cl$_2$ PpHS-PpMOS

OCH$_3$ OH OH PpHS

Scheme 2.7

vinyl ethers have, in most cases, similar reactivities, so the order of monomer addition can be changed accordingly (Scheme 2.8a). The alkyl substituents cover a wide range of the number of carbon atoms (from one, methyl, to 16, hexadecyl) (Miyamoto 1984). Vinyl ethers with polar pendant groups have different reactivity from their alkyl-substituted counterparts, and interchanges in monomer addition cannot be made in most cases without adjusting the reaction conditions appropriately (Scheme 2.8b). The most widely used initiating systems are HI/I$_2$ and HX/ZnX$_2$ (where X is a halogen atom). Combinations of nonpolar and polar vinylethers lead to the formation of amphiphilic block copolymers with or without postpolymerization modification of one of the blocks. For instance, acetoxyvinylether (AcOVE) is less reactive than ethyl vinyl ether (EVE). Therefore, AcOVE is used as the second monomer if a EVE-AcOVE diblock copolymer must be synthesized

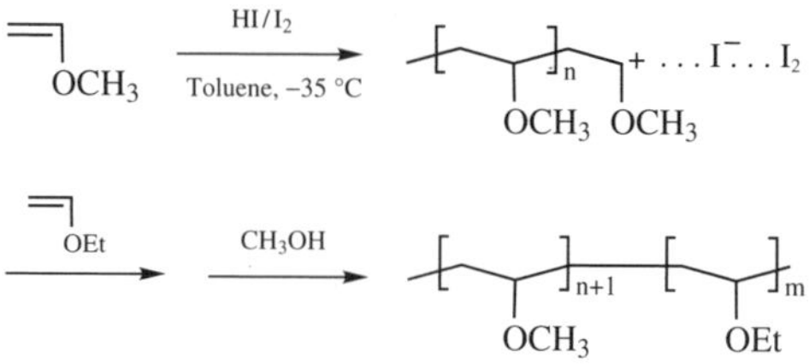

(a)

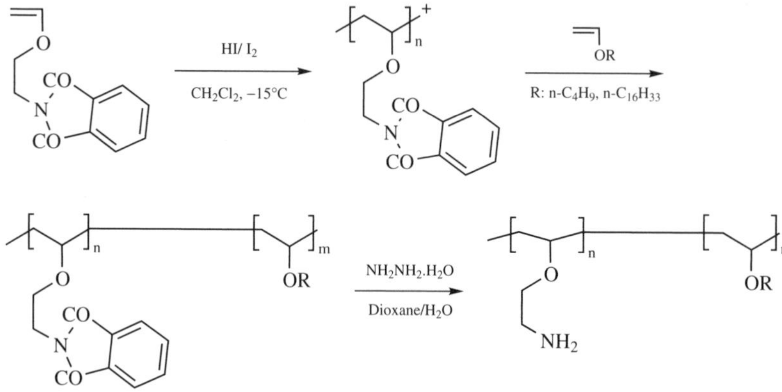

(b)

Scheme 2.8

and additional amounts of ZnX_2 must be added to the system containing living chains before the addition of AcOVE (Minoda 1987).

Amphiphilic block copolymers comprised of vinyl ethers were synthesized by sequential living cationic polymerization of isobutyl vinyl ether (IBVE) and a vinyl ether with a 1,2:5,6-diisopropylidene-D-glucose side group (Yamada 1997). The HCl adduct of isobutyl vinyl ether, $CH_3CH(OiBu)Cl$ was used as the initiator in the presence of ZnI_2. The molecular weight distribution of the final diblock copolymers was narrow ($M_w/M_n < 1.1$), and the experimentally determined molecular weights and compositions agreed satisfactorily with the theoretical ones. Deprotection of the glucose groups with a trifluoroacetic acid/water mixture led to the preparation of the desired copolymers having pendant glucose residues in their chain (Scheme 2.9).

Double hydrophilic block copolymers containing methyl vinyl ether and vinyl alcohol were synthesized by analogous reactions on well-defined diblock copolymer precursors (Forder 1996a). The synthesis of the precursors was achieved by cationic polymerization of methyl vinyl ether first in CH_2Cl_2 at $-78°C$ using the initiating system IBVE-HCl/$SnCl_4$/nBu_4NCl. After completion of the polymerization, the second monomer benzyl vinyl ether was introduced, and its polymerization was carried out at $-40°C$, leading to diblock copolymers with relatively narrow molecular weight distributions and the expected molecular weights. Catalytic

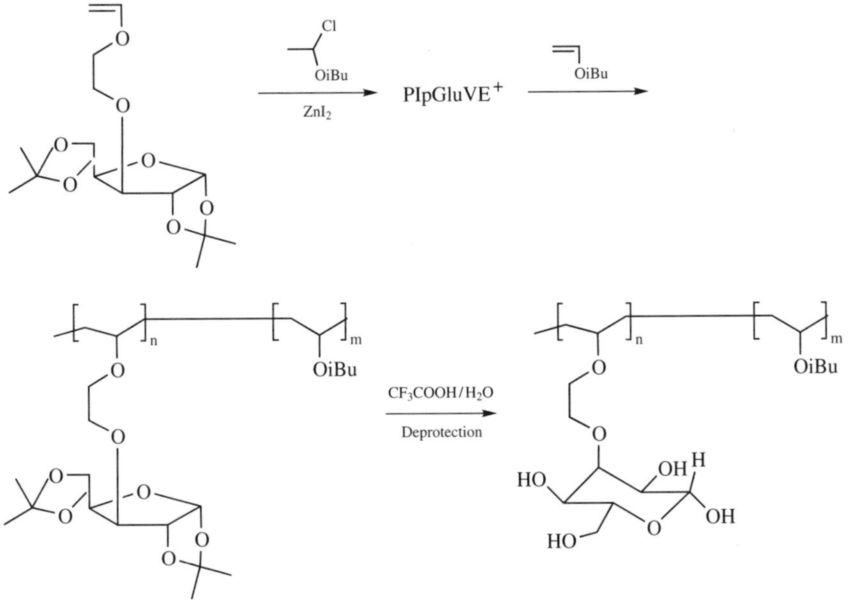

Cl
OiBu
ZnI_2
$PIpGluVE^+$
OiBu
O
O
O
O
O
O
O
n
m
OiBu
CF_3COOH/H_2O
Deprotection
HO
OH H
OH
HO

Scheme 2.9

hydrogenolysis of the benzyl groups under mild conditions gave the aforementioned doubly hydrophilic block copolymers (Scheme 2.10).

Other types of nonionic amphiphilic block copolymers were synthesized by sequential polymerization of 2-acetoxyethyl vinyl ether with different alkyl substituted (n-butyl, isooctyl, n-hexadecyl) vinyl ethers (Minoda 1987). In these cases the best results were obtained when AcOVE was the first monomer to be polymerized giving well-defined block copolymers with narrow molecular weight distributions and no homopolymer impurities. Hydrolysis of the AcOVE block gave diblock copolymers with a poly(2-hydroxyethyl vinyl ether) hydrophilic block and a poly alkyl substituted vinyl ether hydrophobic block (Scheme 2.11). In an

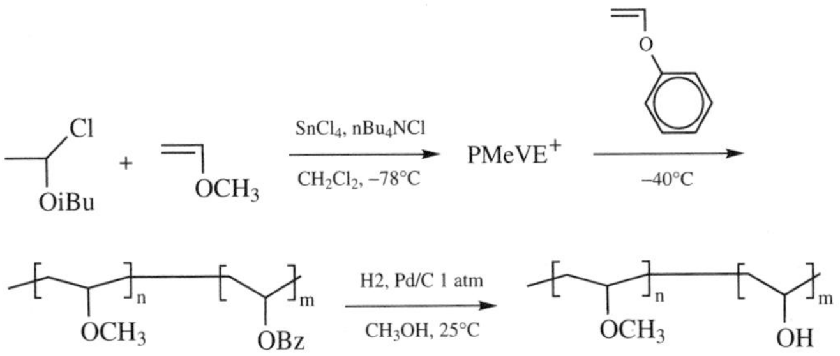

Cl
OiBu
+
OCH_3
$SnCl_4$, nBu_4NCl
CH_2Cl_2, −78°C
$PMeVE^+$
O
−40°C
n
m
OCH_3
OBz
H2, Pd/C 1 atm
CH_3OH, 25°C
OCH_3
OH

Scheme 2.10

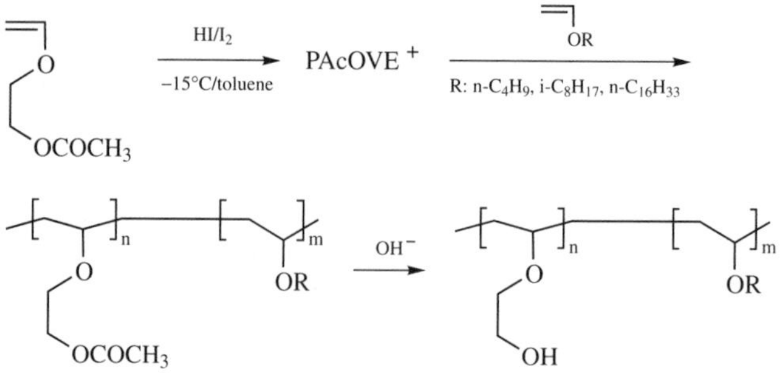

Scheme 2.11

analogous synthetic scheme, amphiphilic block copolymers were obtained by sequential polymerization of methyl vinyl ether and methyl triethylene glycol vinyl ether (Forder 1996b). The polymerizations were carried in that order, using the initiating system IBVE-HCl/$SnCl_4$/nBu_4NCl, giving polymers with different lengths of the MTEGVE block and fairly narrow molecular weight distributions (Scheme 2.12).

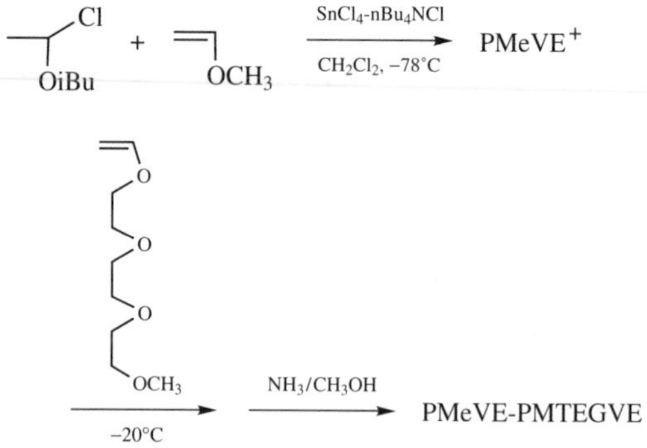

Scheme 2.12

Amphiphilic block copolymers containing carboxyl pendant groups were obtained by sequential living cationic polymerization of a vinyl ether monomer containing a malonic ester [$CH_2{=}CHOCH_2CH_2CH(COOEt)_2$, the precursor of the carboxyl-group-containing block] and various alkyl-substituted vinyl ethers (Minoda 1990). The initiating system HI/I_2 was used in toluene at low temperatures (Scheme 2.13). The final products of the polymerization reaction were essentially free of homopolymer. Their subsequent hydrolysis and decarboxylation gave the targeted amphiphilic macromolecules. Block copolymers containing pendant-protected amino groups were synthesized (Kanaoka 1990). Water soluble amphiphilic block copolymers with one block consisting of 2-hydroxyethylvinyl ether units and

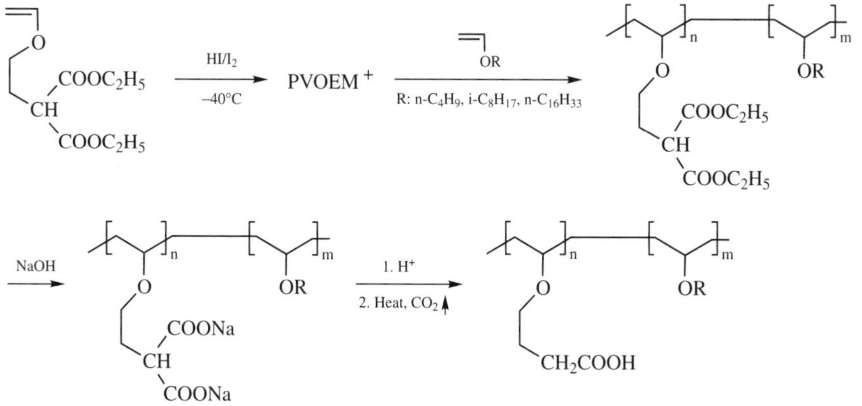

Scheme 2.13

the other of 2-(2,2,2-trifluoroethoxy)ethyl vinyl ether units were prepared by cationic polymerization using the nBVE/HCl/$ZnCl_2$ initiating system (Matsumoto 1999).

Diblock copolymers containing vinyl ethers and styrenic monomers can be synthesized by sequential addition of monomers using the vinyl ether monomer as the first block (Sawamoto 1991). The reverse process cannot give block copolymers because vinylethers are more reactive than styrenic monomers. The opposite is true for the synthesis of block copolymers comprised of vinyl ether and isobutylene blocks. Adjusting the reaction temperature is another parameter that should be taken into account in the synthesis of block copolymers.

Diblock copolymers of methyl vinyl ether and styrene were obtained by polymerizing the MVE first in CH_2Cl_2 at $-78°C$ using $CH_3CH(OiBu)Cl$ as the initiator in the presence of $SnCl_4$ as the activator and nBu_4NCl. After complete consumption of MVE, a new aliquot of the $SnCl_4/nBu_4NCl$ mixture together with the styrene monomer was added, and the polymerization temperature was raised to $-15°C$. Complete and controlled polymerization of the second monomer was observed (Scheme 2.14). The same procedure was applied for the synthesis of block copolymers containing 2-chloroethyl vinyl ether (CVE) and styrene (Scheme 2.15). In both cases well-defined diblock copolymers of various compositions were obtained (Ohmura 1994).

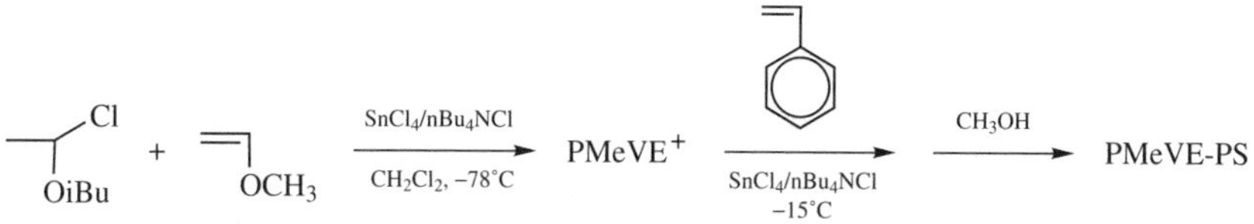

Scheme 2.14

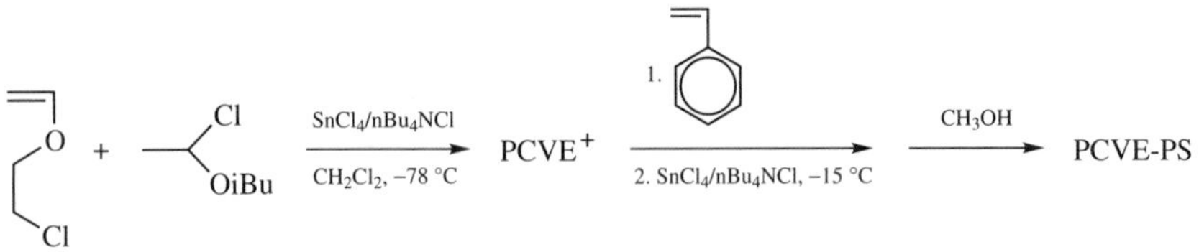

Scheme 2.15

Block copolymers of isobutyl vinyl ether and p-tert-butoxystyrene were synthesized by polymerizing the vinyl ether monomer first in toluene at $-15^{\circ}C$ using the HI/ZnX_2 systems as initiator (Kojima 1991). The final copolymers were obtained by polymerization of the p-tert-butoxystyrene monomer in the second step (Scheme 2.16). These copolymers had low polydispersities, and their molecular weights

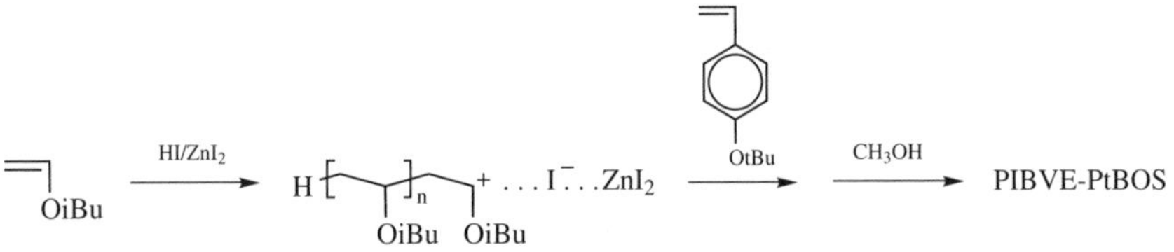

Scheme 2.16

were in agreement with the expected ones from stoichiometry. No homopolymer impurities were observed by SEC analysis.

In the synthesis of isobutylene-b-methyl vinyl ether copolymers, isobutylene is polymerized first using the initiating system 2-chloro-2,4,4-trimethylpentane/$TiCl_4$ in the presence of nBu_4NCl in CH_2Cl_2/n-hexane mixture at $-78^{\circ}C$ (Pernecker 1992). After consumption of the first monomer, the polymerization of the second, methyl vinyl ether, is conducted in the presence of excess $TiCl_4$. Relatively well-defined block copolymers are obtained (Scheme 2.17).

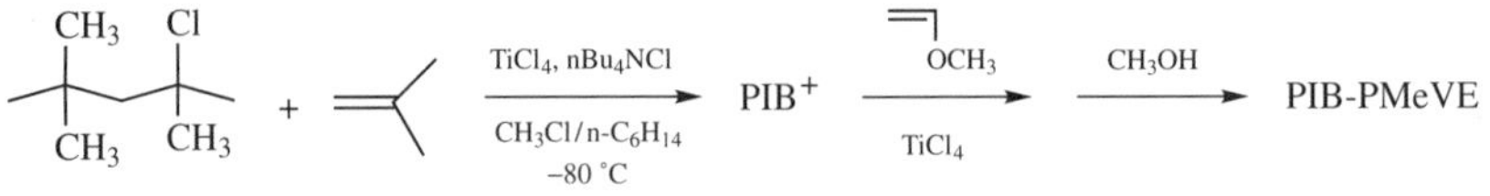

Scheme 2.17

Cationic block copolymerization of ε−caprolactone (CL) and δ−valerolactone (VL) with 1,3-dioxepan-2-one (7CC), a seven-membered cyclic carbonate, has been reported (Shibasaki 2000). The synthesis starts with the polymerization of 7CC in CH_2Cl_2 at room temperature using the initiating system n-butyl alcohol/$HCl.Et_2O$.

After quantitative conversion of 7CC, the second monomer (CL or VL) was added. The polymerization of the second monomer was conducted at 25°C for CL and at −40°C for VL. Quantitative conversion of both monomers was observed, and the final copolymers had narrow molecular weight distributions ($M_w/M_n < 1.16$). Their compositions agreed with the monomer-feed ratio in the reaction mixture (Scheme 2.18).

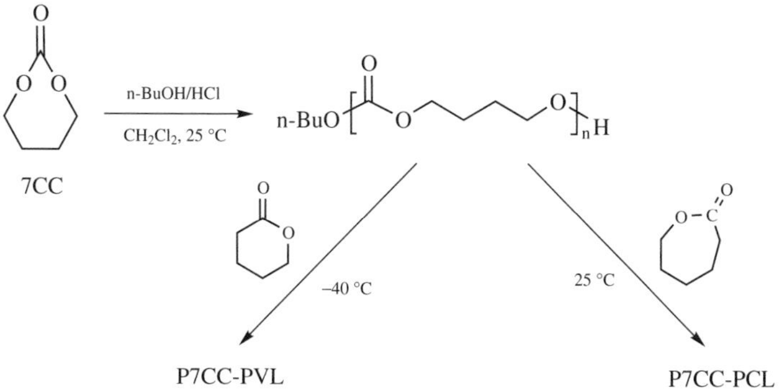

7CC —(n-BuOH/HCl, CH_2Cl_2, 25 °C)→ n-BuO[C(=O)O(CH₂)₃O]ₙH

−40 °C → P7CC-PVL

25 °C → P7CC-PCL

Scheme 2.18

The cationic block copolymerization of phosphoranimines for the preparation of well-defined phosphazene block copolymers has been reported (Allcock 2000). Diblock copolymers of the general type $[N{=}PCl_2]_n[N{=}PR(R')]_m$ were prepared using a variety of phosphoranimines ($PhCl_2P{=}NSiMe_3$, Me-(Et)ClP=$NSiMe_3$, $Me_2ClP{=}NSiMe_3$, $Ph_2ClP{=}NSiMe_3$, $PhF_2P{=}NSiMe_3$). The first block was poly (dichlorophosphazene), which was synthesized by polymerization of $Cl_3P{=}NSiMe_3$ using PCl_5 as the initiator in CH_2Cl_2 at 35°C. Halogen replacement reactions of the precursor polymers gave fully organosubstituted block copolymers (Scheme 2.19).

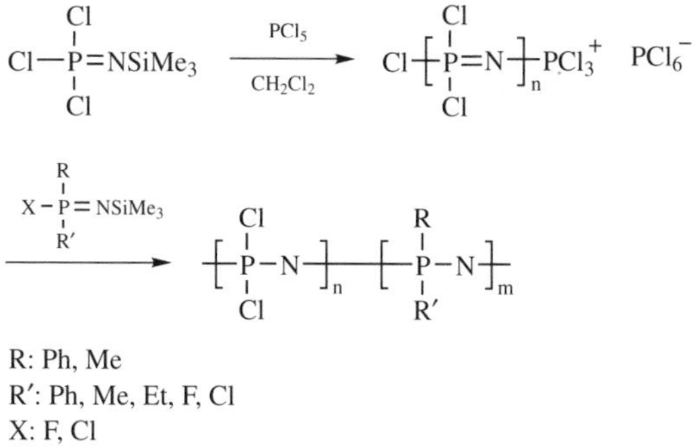

$$Cl_3P{=}NSiMe_3 \xrightarrow[CH_2Cl_2]{PCl_5} Cl{-}[PCl_2{=}N]_n{-}PCl_3^+\ PCl_6^-$$

$$\xrightarrow{X{-}PRR'{=}NSiMe_3} {-}[PCl_2{-}N]_n{-}[PRR'{-}N]_m{-}$$

R: Ph, Me
R′: Ph, Me, Et, F, Cl
X: F, Cl

Scheme 2.19

A few examples of diblock copolymers prepared by cationic polymerization and sequential addition of monomers are given in Table 2.1.

TABLE 2.1. AB Diblock Copolymers Synthesized by Cationic Polymerization

1st Monomer	2nd Monomer	Reference
Isobutylene	Styrene	Kennedy 1980
	α-Methyl styrene	Kennedy 1982
	Methyl vinyl ether	Hadjikyriacou 1999
	Isobutyl vinyl ether	Hadjikyriacou 1995
Alkyl vinyl ethers	p-Methoxy styrene	Miyamoto 1984
Cetyl vinyl ether	Methyl, ethyl vinyl ether	Miyamoto 1984
Methyl, ethyl vinyl ether	Cetyl vinyl ether	Miyamoto 1984
	Styrene	Ohmura 1991
	Methyl triethylene glycol vinyl ether	Forder 1996
Isobutyl vinyl ether	Ethyl propenyl ether	Sawamoto 1986
Alkyl vinyl ethers	Functional vinyl ethers	Kanaoka 1990, Minoda 1987, Minoda 1990
p-Methoxy styrene	p-Hydroxy styrene	Satoh 2000
Isobutyl vinyl ether	p-*tert*-butoxy styrene	Kojima 1991
Ethyl vinyl ether	N-vinyl carbazole	Rooney 1976

2. SYNTHESIS OF TRIBLOCK COPOLYMERS

2.1. Symmetric ABA Triblock Copolymers

Due to remarkable differences in reactivity between monomers belonging to different chemical groups towards cationic polymerization, preparation of symmetric linear ABA block copolymers by sequential monomer addition is very difficult.

TABLE 2.2. ABA Triblock Copolymers Synthesized by Cationic Polymerization

Monomer A	Monomer B	Reference
Styrene	Isobutylene	Fodor 1984
p-Methyl styrene	Isobutylene	Kennedy 1990a
Tert-butyl styrene	Isobutylene	Kennedy 1990a
Indene	Isobutylene	Kennedy 1990a
p-Fluoro styrene	Isobutylene	Kennedy 1990b
p-Chlorostyrene	Isobutylene	Kennedy 1990b
α-Methyl styrene	Isobutylene	Kennedy 1980
α-Methyl styrene	Isobutyl vinyl ether	Sawamoto 1982a
Methyl vinyl ether	Isobutyl vinyl ether	Sawamoto 1982b
Methyl vinyl ether	Cetyl vinyl ether	Miyamoto 1985
Cetyl vinyl ether	Methyl, ethyl vinyl ether	Miyamoto 1985
2-methyl-2-oxazoline	Tetrahydrofuran	Wang 2000
Phosphoranimine derivative	Phosphoranimine derivative	Allcock 2000

In most cases ABA triblock copolymers have been synthesized by using a difunctional initiator and two monomer additions (Table 2.2). A variety of soluble difunctional cationic initiators are available (Sawamoto 1991). The synthetic strategy developed by Faust et al. (Faust 2000, Hadjikyriacou 2000) can also be used in the synthesis of ABA block copolymers when crossover from B to A is not straightforward due to reactivity differences. PαMeS-PIB-PαMeS and PpMeS-PIB-PpMeS triblock copolymers were synthesized in this way (Li 1995, Fodor 1995). However, one should always have in mind that capping introduces another step in the synthesis of the triblock copolymer, increasing the possibility of premature chain termination due to incorporation of impurities in the system (Scheme 2.20).

Other triblock copolymers having PIB as the central block have been synthesized by Kennedy et al. The reaction scheme involves the production of a difunctional PIB chain using the difunctional initiating system p-dicumyl methyl ether(DCE)/$TiCl_4$. Then p-tert-butylstyrene is added in the polymerization mixture, keeping the temperature low (−78°C) (Scheme 2.21). The reaction is terminated with methanol (Kennedy 1991). It has been observed that, in the case of α-methylstyrene as the second monomer, the reaction temperature must be increased

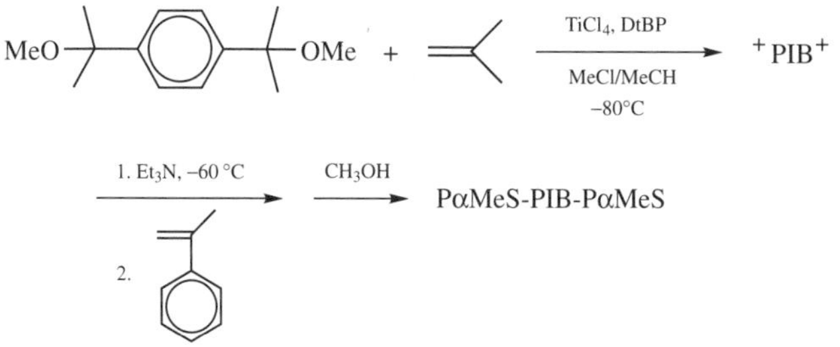

MeO OMe + $\xrightarrow[\text{MeCl/MeCH, −80°C}]{TiCl_4\text{, DtBP}}$ $^+$PIB$^+$

$\xrightarrow[\text{2.}]{\text{1. }Et_3N\text{, −60 °C}}$ $\xrightarrow{CH_3OH}$ PαMeS-PIB-PαMeS

Scheme 2.20

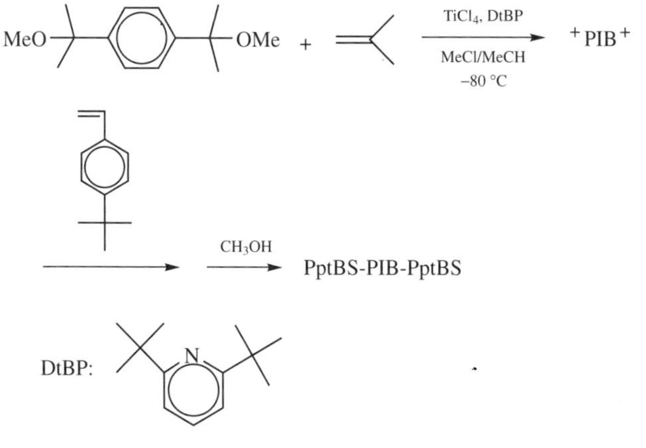

MeO OMe + $\xrightarrow[\text{MeCl/MeCH, −80 °C}]{TiCl_4\text{, DtBP}}$ $^+$PIB$^+$

$\longrightarrow$ $\xrightarrow{CH_3OH}$ PptBS-PIB-PptBS

DtBP: N

Scheme 2.21

slightly (to −60°C) in order to accelerate the crossover reaction and to minimize the PIB homopolymer impurities in the final triblock (Tsunogae 1994). Addition of Et_3N in the polymerization mixture prior to α-MeSt addition also proved beneficial to the properties of the final triblock.

PS-PIB-PS triblock copolymers were synthesized by cationic polymerization, and they were evaluated as thermoplastic elastomers in their initial form and after sulfonation of the PS end blocks (Storey 1999 and 1997).

ABA triblocks with a central PIB block and poly(indene) outer blocks have been prepared and tested as possible thermoplastic elastomers (Kennedy 1993). Difunctional PIB was synthesized by using DCE as the difunctional initiator in the presence of $TiCl_4$ as the activator, dimethyl acetamide as the electron donor, and 2,6-di-tert-butylpyridine as the proton trap. After polymerization of isobutylene, indene was introduced to the system, and, finally, the reaction was terminated with methanol (Scheme 2.22).

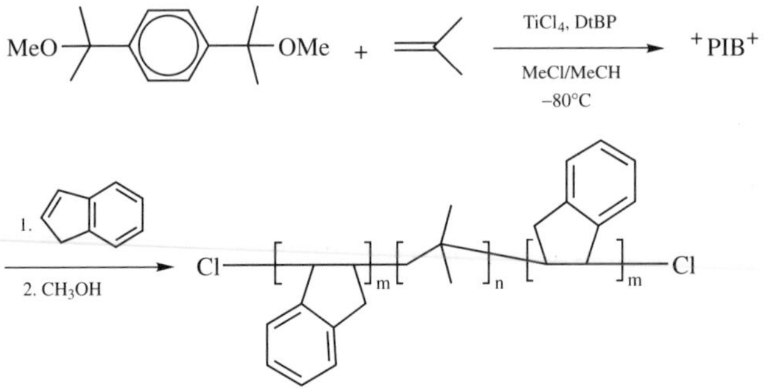

Scheme 2.22

Symmetric triblock copolymers containing poly(tetrahydrofuran) as the central block and poly(2-methyl-2-oxazoline) as the terminal blocks were synthesized by cationic polymerization (Wang 2000). The difunctional initiator trifluoromethanesulfonic acid anhydride was used to polymerize THF in the first step. 2-methyl-2-oxazoline (MeOx) was added then, and the reaction temperature was raised to 70°C, after evaporation of the solvent and replacing it with CH_3CN (Scheme 2.23). The final copolymers exhibited relatively narrow molecular weight distributions. Hydrolysis of the ABA triblock copolymers with concentrated HCl in a methanol/water mixture resulted in the formation of linear triblock copolymer with polyTHF central block and polyethylenimine outer blocks.

An alternative strategy for the synthesis of symmetric ABA linear triblock copolymers involves the formation of living AB chains and subsequent coupling with an appropriate coupling agent. Coupling agents, like the chlorosilanes routinely used in anionic polymerization, are not available in cationic polymerization. However, a number of nonhomopolymerizable compounds can be used for the

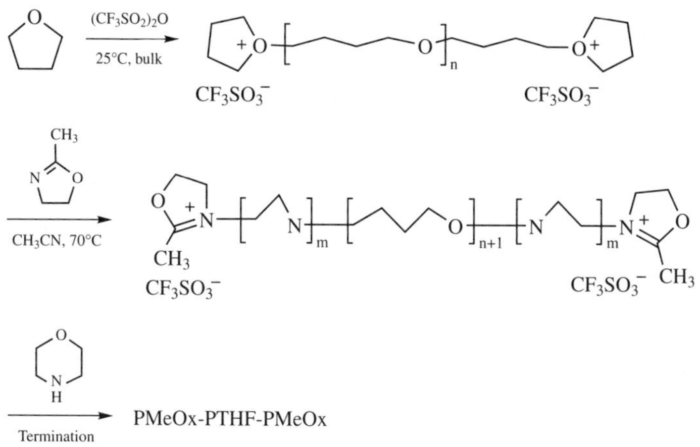

Scheme 2.23

efficient coupling of cationic living chains. DPE derivatives, like 2,2-bis-[4-(1-phenylethenyl)phenyl]propane (BDPEP) and 2,2-bis[4-(1-tolylethenyl)phenyl]propane (BDTEP), and furanyl compounds, like 2,5-bis[1-(2-furanyl)-1-methylethyl]-furan (BFPF), were successfully utilized as coupling agents (Cao 2000). BDPEP was used for the synthesis of PS-PIB-PS and PαMeS-PIB-PαMeS triblock copolymers by coupling the respective diblock copolymers. Coupling is effected by consecutive addition reactions to the double bonds of the coupling agent. Kinetic investigations have shown that the second addition is faster than the first, a result that is corroborated by the fact that high coupling efficiency was observed even when excess of BDPEP was used. Of course, some adjustment of Lewis acidity and choice of solvent and reaction temperature must be made. For example, 1,3-bis((1-trimethylsilylethyl)vinyl)benzene was found to behave as an efficient coupling

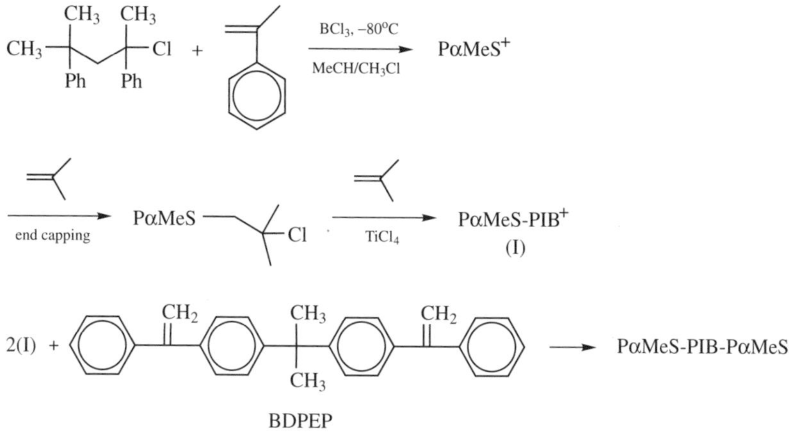

Scheme 2.24

agent for living PIB chains in hexane/CH_3Cl (60/40 v/v) solvent mixtures at $-80°C$, using $TiCl_4$ as the Lewis acid. Other bis-allylsilanes show limited coupling efficiency in conjuction with BCl_3 in CH_3Cl at $-40°C$ (Scheme 2.24).

2.2. ABC Triblock Terpolymers

Linear triblock copolymers of the ABC type can, in principle, be prepared by cationic polymerization using the sequential three-step addition method. Patrikios et al. (Patrickios 1997) reported the synthesis of an ABC triblock copolymer where A is methyl vinyl ether (MVE), B is ethyl vinyl ether (EVE), and C is methyl tri(ethylene glycol) vinyl ether (MTEGVE). The synthesis started by the polymerization of MVE in CH_2Cl_2 at $-78°C$, in the presence of HCl/nBu_4NCl initiating system. After allowing for complete polymerization of MVE, $SnCl_4$ was added to the system, followed by the addition of EVE. When consumption of the second monomer was complete, the desired amount of MTEGVE was added, and the temperature was raised to $-20°C$ (Scheme 2.25). Other triblock copolymers

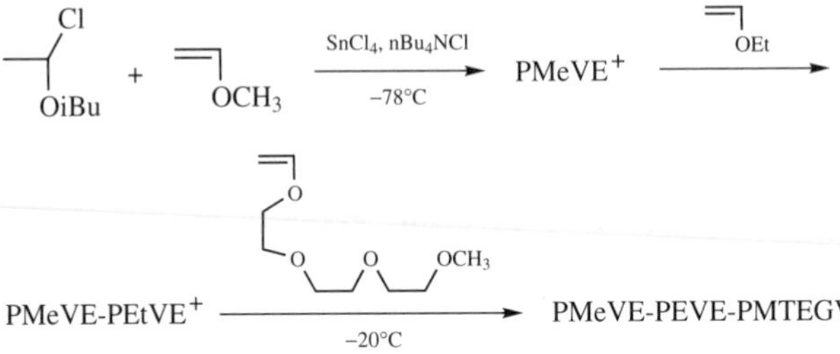

Cl, OiBu + OCH$_3$ —($SnCl_4$, nBu_4NCl; −78°C)→ PMeVE$^+$ —(OEt)→

PMeVE-PEtVE$^+$ —(O, O, O, OCH$_3$; −20°C)→ PMeVE-PEVE-PMTEGVE

Scheme 2.25

having the sequences MVE-MTEGVE-EVE, MTEGVE-MEV-EVE, EVE-MEV-MTEGVE were also synthesized with the same initiating system in the same solvent but with different addition order. MEV and EVE polymerization were accomplished at $-78°C$, whereas the polymerization of MTEGVE was always performed at $-20°C$. Almost quantitative yields of all monomers were observed. The final triblock copolymers had low molecular weights and relatively narrow molecular weight distributions if one takes into account that these polymers were synthesized by three monomer additions.

REFERENCES

Allcock H. R., Reeves S. D., Nelson J. M., Manners I. (2000) Macromolecules 33, 3999.

Cao X., Sipos L., Faust R. (2000) Polym. Bull. 45, 121.

Faust R. (1999) Polym. Prepr. 40(2), 960.

Faust R. (2000) Macromol. Symp. 157, 101.

Faust R., Shaffer T. D. (Eds) (1997) Cationic Polymerization: Fundamentals and Applications, ACS Symposium Series, Vol 665, Washington, DC.

Fodor Z., Faust R. (1994) J. Macromol. Sci. A31(12), 1983.

Fodor Z., Faust R. (1995) J. Macromol. Sci. A32(3), 575.

Forder C., Patrickios C. S., Armes S. P., Billingham N. C. (1996) Macromolecules 29, 8160.

Forder C., Patrikios C. S., Billingham N. C., Armes S. P. (1996) Chem. Commun. 883.

Hadjikyriacou S., Faust R. (1995) Macromolecules 28, 7893.

Hadjikyriacou S., Faust R. (1996) Macromolecules 29, 5261.

Hadjikyriacou S., Faust R. (1999) Macromolecules 32, 6393.

Hadjikyriacou S., Faust R. (2000) Macromolecules 33, 730.

Higashimura T., Aoshima S., Sawamoto M. (1988) Makromol. Chem. Macromol. Symp. 13/14, 457.

Kanaoka S., Minoda M., Sawamoto M., Higashimura T. (1990) J. Polym. Sci. Part A: Polym. Chem. 28, 1127.

Kennedy J. P., Guhaniyogi S. C., Ross L. R. (1982) J. Macromol. Sci.-Chem. A18, 119.

Kennedy J. P., Huang S. Y., Smith R. A. (1980) J. Macromol. Sci.-Chem. A14, 729.

Kennedy J. P., Ivan B. (1992) Designed Polymers by Carbocationic Macromolecular Engineering: Theory and Practice, Hanser Publishers, New York.

Kennedy J. P., Kurian J. (1990a) J. Polym. Sci. Part A: Polym. Chem. 28, 3725.

Kennedy J. P., Meguriya N., Keszler B. (1991) Macromolecules 24, 6572.

Kennedy J. P., Midha S., Tsunogae Y. (1993) Macromolecules 26, 429.

Kennedy J. P., Puskas J. E., Kaszas G., Hager W. G. (1990b) U.S. Patent 4,946,899.

Kennedy J. P., Smith R. A. (1980) J. Polym. Sci. Chem. Ed. 18, 1539.

Kojima K., Sawamoto M., Higashimura T. (1991) Macromolecules 24, 2658.

Li D., Faust R. (1995) Macromolecules 28, 1383.

Li D., Faust R. (1995) Macromolecules 28, 4893.

Li D., Hadjikyriacou S., Faust R. (1996) Macromolecules 29, 6061.

Matsumoto K., Kubota M., Matsuoka H., Yamaoka H. (1999) Macromolecules 32, 7122.

Matyjaszewski K. (Ed) (1996) Cationic Polymerization: Mechanism, Synthesis and Application, Marcel & Dekker, New York.

Minoda M., Sawamoto M., Higashimura T. (1987) Macromolecules 20, 2045.

Minoda M., Sawamoto M., Higashimura T. (1990) Macromolecules 23, 1897.

Miyamoto M., Sawamoto M., Higashimura T. (1984) Macromolecules 17, 265.

Miyamoto M., Sawamoto M., Higashimura T. (1984) Macromolecules 17, 2228.

Ohmura T., Sawamoto M., Higashimura T. (1994) Macromolecules 27, 3794.

Patrickios C. S., Forder C., Armes S. P., Billingham N. C. (1997) J. Polym. Sci. Part A: Polym. Chem. 35, 1181.

Pernecker T., Kennedy J. P., Ivan B. (1992) Macromolecules 25, 1642.

Rooney J. M., Squire D. R., Stannett V. M. (1976) J. Polym. Sci. Polym. Chem. Ed. 14, 1877.

Satoh K., Kamigaito M., Sawamoto M. (2000) Macromolecules 33, 5830.

Sawamoto M. (1991) Prog. Polym. Sci. 16, 111.

Sawamoto M., Ebara K., Tanizaki A., Higashimura T. (1986) J. Polym. Sci. Polym. Chem. Ed. 24, 2919.

Sawamoto M., Kennedy J. P. (1982a) J. Macromol. Sci.-Chem. A18, 1293.

Sawamoto M., Kennedy J. P. (1982b) J. Macromol. Sci.-Chem. A18, 1301.

Shibasaki Y., Sanada H., Yokoi M., Sanda F., Endo T. (2000) Macromolecules 33, 4316.

Storey R. F., Baugh D. W., Choate K. R. (1999) Polymer 40, 3083.

Storey R. F., Chisholm B. J., Lee Y. (1997) Polymer Eng. Sci. 37, 73.

Tsunogae Y., Kennedy J. P. (1994) J. Polym. Sci. Part A: Polym. Chem. 32, 403.

Wang Y., Goethals E. J. (2000) Macromolecules 33, 808.

Yamada K., Yamaoka K., Minoda M., Miyamoto T. (1997) J. Polym. Sci. Part A: Polym. Chem. 35, 255.

CHAPTER 3

BLOCK COPOLYMERS BY LIVING FREE RADICAL POLYMERIZATION

Free radical polymerization is the oldest mechanism for polymerization of vinyl monomers (Odian 1981). This kind of polymerization is widely used for the industrial preparation of a large number of polymeric materials (e.g., LDPE, PVC, etc.). A large range of monomers can be polymerized and copolymerized by free radical polymerization, under less rigorous experimental conditions compared with ionic polymerizations. Free radical polymerization processes are tolerant of protic and aqueous solvent media and certain functional monomers. However, the disadvantage of the free radical mechanism is the preparation of polydisperse polymers with little control over their molecular characteristics due to automatic termination and chain transfer reactions.

Recent advances in free radical polymerization have led to the development of synthetic methods for eliminating or suppressing the undesired termination and chain transfer reactions (Matyjaszewski 1998). The two most important methods include the use of stable free radicals, such as nitroxides, as reversible terminating agents to control the polymerization process (Moad 1982, Georges 1993) and the use of transition metal complexes, such as CuX/bipyridine and other metal complexes with Ru, Fe, Ni, Rh, and Pd, which, through a reversible catalytic action that involves atom transfer, stabilize radical intermediates (atom transfer radical polymerization) (Matyjaszewski 1995, Wang 1995, Kato 1995, Percec 1995, Wayland 1994, Granel 1996). Other procedures include reversible degenerative transfer (Chiefan 1998, Chong 1999), use of iniferters and organometallic compounds complexed with ligands and activated by stable radicals (Qiu 1997, Sawamoto 1996).

In nitroxide-mediated controlled free radical polymerization, reversible deactivation of the growing chains is accomplished by covalent bond formation

(Malmstrom 1998). At low temperatures the C-ON bond is stable, whereas, at higher temperatures, homolytic cleavage is possible, giving the macroradical and the nitroxide radical. The macroradical can then grow through the addition of new monomer units. Covalent bond formation with the nitroxide radical gives the dormant form of the propagating radical. This cycle can be performed many times until almost complete consumption of monomer is achieved. In this way the concentration of free radical remains very low, decreasing the possibility for termination reactions. Rapid exchange between dormant and active species ensures narrow molecular weight distributions. However, some termination takes place as the molecular weight increases, and this kind of polymerization should be thought of as controlled, rather than true, living polymerization. Nitroxide-mediated free radical polymerization has found success in the polymerization of mainly styrenic monomers although the polymerization of dienic and acrylic monomers (Keoshkerian 1998, Benoit 2000) has been recently reported using appropriate nitroxides. The synthesis of unimolecular initiators, i.e., molecules that can be cleaved in two parts, thus providing the necessary free radicals and the nitroxide ones, has improved many aspects of living free radical polymerization processes (Hawker 1994).

In the case of atom transfer free radical polymerization, a transition metal compound acts as halogen atom carrier in a reversible redox reaction (Matyjaszewski 1998). Transition metal atoms are usually complexed with an appropriate ligand that acts in various ways during the reaction. A continuous shift of the metal atom between different oxidation states leads to the formation of free radicals, which can then react with an alkene monomer forming the propagating macroradicals. The propagation step consists of a series of monomer addition to the radicals through halogen exchange. The polymeric halides are the dormant form of the propagating radicals. This is another way of keeping the concentration of active centers low, suppressing possible termination reactions.

The discovery of techniques for making the free radical polymerization a controlled polymerization mechanism opened the way for the preparation of well-defined block copolymers. Because free radical polymerization can be applied for the preparation of polymers consisting of a wide variety of monomers, many of which cannot be polymerized with another type of polymerization mechanism, the possibilities for synthesizing novel block copolymers are enormous.

1. SYNTHESIS OF AB DIBLOCK COPOLYMERS

Diblock copolymers can be synthesized by living free radical polymerization with the technique of sequential monomer addition. Because the mechanism involves propagation through radicals, considerations involving monomer purity and reactivity in conjunction with relative rates of crossover and propagation reactions for the second monomer still exist, but they are more relaxed than in the case of ionic mechanisms. This is a consequence of the fact that, in both nitroxide-mediated and atom-transfer variations of the method, the intermediate products can actually be isolated and further purified (purification mainly involves elimination of traces of the first monomer). The polymerization of the second monomer can be initiated by the macromolecular initiator already formed, in more or less the same way as it would be done in a normal homopolymerization.

Nitroxide-mediated stable free radical polymerization (SFRP) has been used for the preparation of diblock copolymers of styrene and other styrenic derivatives (Keoshkerian 1995, Listingovers 1997). For the synthesis of poly[(4-acetoxystyrene)-b-styrene] copolymers, styrene was polymerized first using benzoylperoxide as initiator, 2,2,6,6-tetramethylpiperidinoxy (TEMPO) as the nitroxide stabilizer, and camphorsulphonic acid as the accelarator at 130°C (Scheme 3.1). The TEMPO

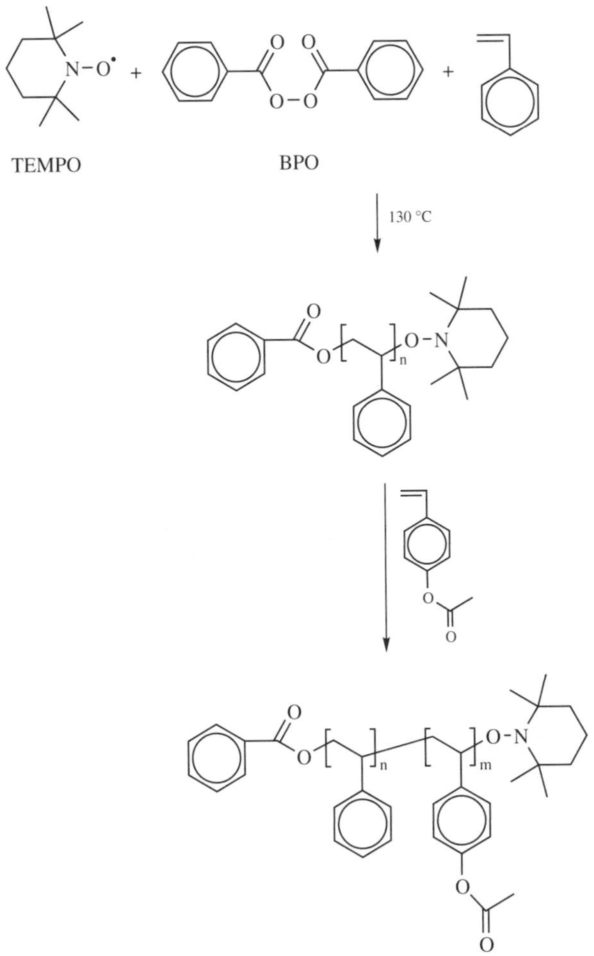

Scheme 3.1

end-capped polystyrene formed was precipitated in methanol and thoroughly dried. It was used as a macromolecular initiator for the subsequent polymerization of 4-acetoxystyrene, resulting in well-defined diblock copolymer. Block copolymers of vinylbenzyl chloride (VBC) and styrene were synthesized by first polymerizing VBC by BPO/TEMPO initiating system. The conversions were kept low in an attempt to increase chain-end activity. The polymerization of VBC was shown to proceed in a controlled manner. The TEMPO-terminated PVBC chains were used for the polymerization of styrene, resulting in the desired block copolymers (Scheme 3.2). Block copolymers containing PS as the first block and a random

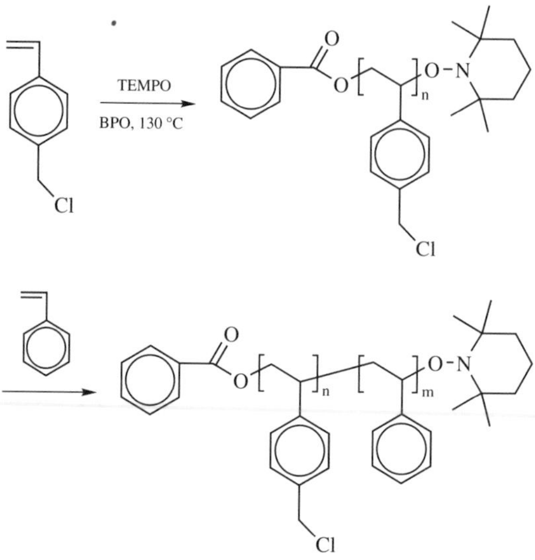

TEMPO
BPO, 130 °C

Scheme 3.2

copolymer of styrene (St) and acrylonitrile (AN) as the second block were also prepared (Fukuda 1996). Styrene was polymerized first using BPO/TEMPO at low conversion to give TEMPO-terminated PS of high functionality. This macromolecule was isolated and purified. It was used for the polymerization of an azeotropic mixture of St and AN (S:AN = 63:100). The final copolymer had rather low polydispersity and composition close to the expected one, with no detectable homo-PS impurity (Scheme 3.3).

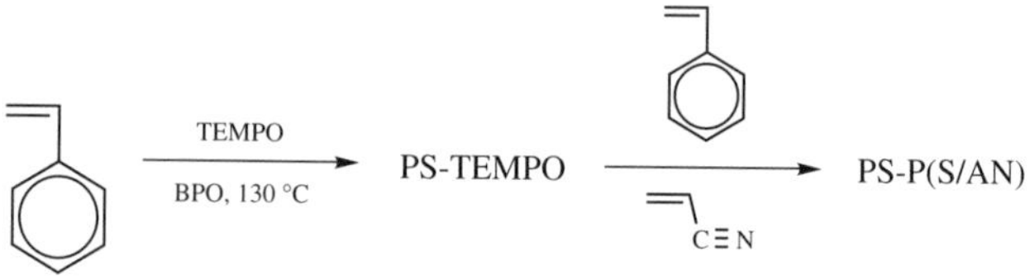

TEMPO
BPO, 130 °C
PS-TEMPO
PS-P(S/AN)

Scheme 3.3

Gnanou et al. reported the synthesis of poly(n-butylacrylate-b-styrene) diblock copolymers (Robin 1999). n-Butylacrylate (n-BuA) was polymerized first in the presence of AIBN and an acyclic β-phosphonylated nitroxide, i.e., N-tert-butyl-N-(1-diethylphosphono-2,2-dimethyl)propyl nitroxide (DEPN), which was found to promote controlled polymerization of alkyl acrylates in a ratio AIBN:DEPN = 1:2.5, at 120°C. After elimination of remaining n-BuA, styrene was introduced to the reaction mixture giving the desired diblock copolymer, which had a short styrene block (Scheme 3.4). The inverse order of monomer addition resulted in a

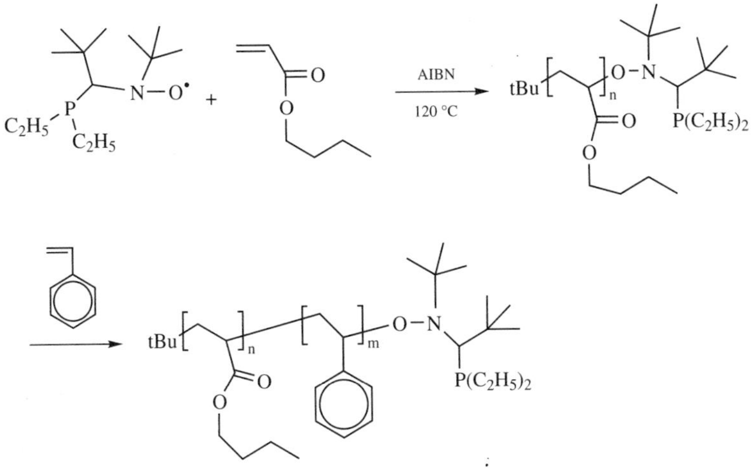

Scheme 3.4

PS-PnBuA block copolymer of high polydispersity, while about 20% of PS homopolymer was present in the crude reaction product. This behavior was attributed to a faster cross-initiation reaction from PnBuA living ends to PS chain-ends, than vice versa.

Hawker et al. (Benoit 2000) developed a synthetic route for the preparation of block copolymers containing dienes with styrene, acrylate, or methacrylate derivatives. They synthesized a unimolecular initiator derived from 2,2,5-trimethyl-4-phenyl-3-azahexane-3-nitroxide and styrene, which was subsequently employed in the polymerizations. The synthesis was realized by polymerizing first tert-butylacrylate or styrene at 120°C. The first block was isolated and dried. Then it was used for the polymerization of isoprene, under similar conditions, giving PtBuA-PI block copolymers with narrow molecular weight distributions and relatively high molecular weights (Scheme 3.5a). In the case of PS-PI diblocks, the reverse order of block formation was equally successful (Scheme 3.5b). Narrow molecular weight distribution PIs, with predictable molecular weights and having the nitroxide group at one end were prepared and used for the polymerization of styrene. Molecular weight distributions were similar to that of their PS-PI analogs.

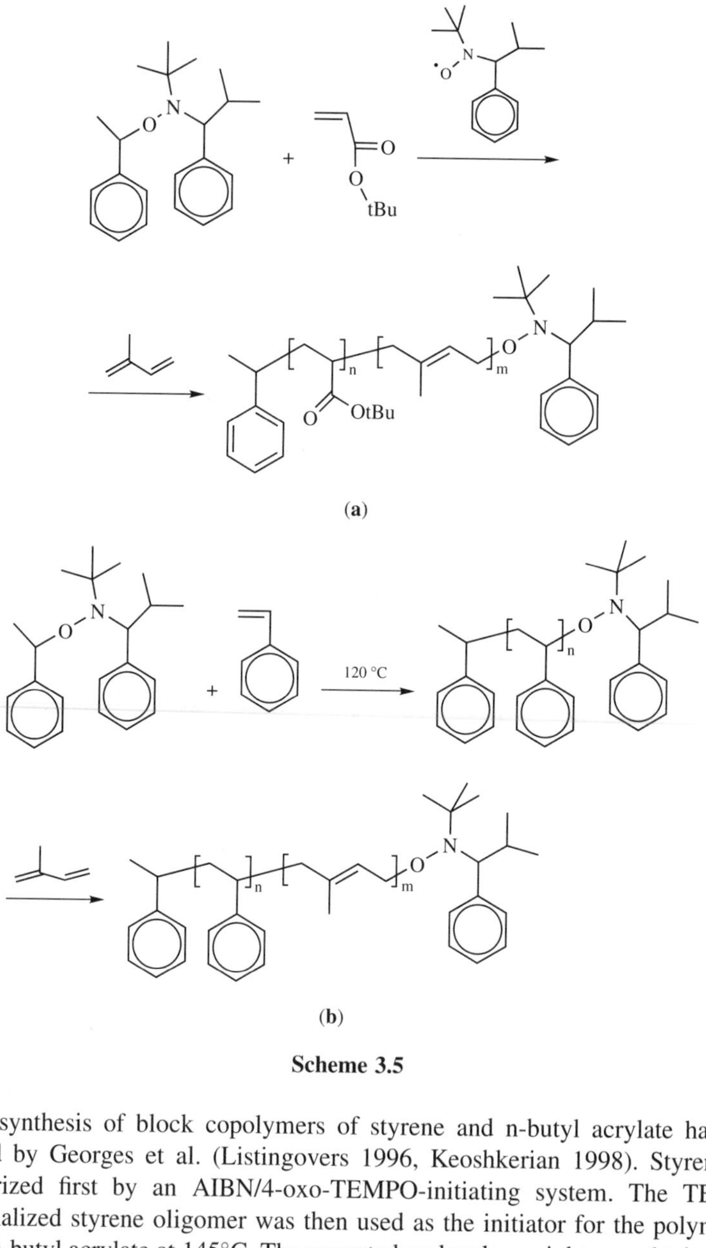

(a)

(b)

Scheme 3.5

The synthesis of block copolymers of styrene and n-butyl acrylate has been reported by Georges et al. (Listingovers 1996, Keoshkerian 1998). Styrene was polymerized first by an AIBN/4-oxo-TEMPO-initiating system. The TEMPO-functionalized styrene oligomer was then used as the initiator for the polymerization of n-butyl acrylate at 145°C. The expected molecular weight was obtained, and the resulting diblock copolymer had a molecular weight distribution typical for the living free radical polymerization mechanism (Scheme 3.6). Preliminary results from n-butyl acrylate homopolymerization, with AIBN/4-oxo-TEMPO, revealed that the polymerization of this monomer was a controlled process, although the molecular weight distributions were found to broaden as the molecular weight was

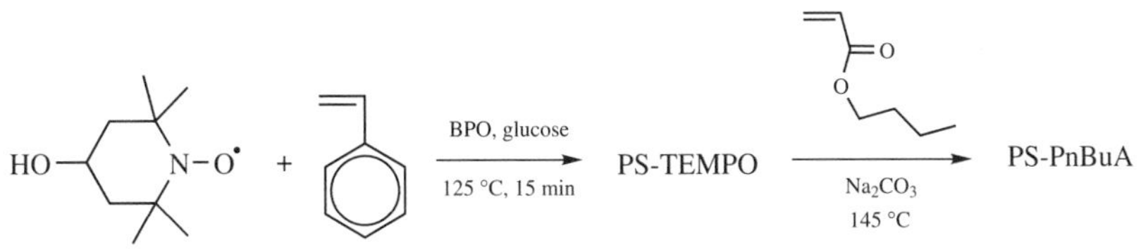

HO–(TEMPO)–N–O• + styrene —BPO, glucose; 125 °C, 15 min→ PS-TEMPO —n-butyl acrylate; Na_2CO_3, 145 °C→ PS-PnBuA

Scheme 3.6

increased. Nevertheless, this was the first successful attempt to prepare block copolymers containing (meth)acrylate monomers by stable free radical polymerization. Sugars and ene-diols were found to have a beneficial effect on the livingness of acrylate polymerizations through control of the excess free-nitroxide levels in the reaction medium. The same oxo-TEMPO was also used in order to produce PS–polydiene block copolymers in considerable yields and with satisfactory control over molecular weight and molecular weight distributions.

Block copolymers of styrene and 2-vinylpyridine were prepared by using a combination of a conventional radical initiator and TEMPO (Halari 2001). The synthetic route employed led to the formation of well-defined block copolymers by polymerizing first 2-vinylpyridine and by choosing the appropriate polymerization temperature for each monomer.

The utilization of SFRP in the preparation of block copolymers containing side-chain liquid crystalline blocks has been reported in several cases. Monomers that contain liquid crystalline (LC) side groups are, in general, very difficult to polymerize by ionic polymerization mechanisms, due to the sensitivity of the side groups to active center attack. The sensitivity is increased if the side group contains heteroatoms. Additionally, even if the side groups do not interfere with the polymerization mechanism, high purity of the LC monomer is required, which is not always easy to achieve. Extreme monomer purity is not a prerequisite for SFRP, and the method is preferred by many investigators. A typical example is the synthesis of block copolymers containing p-acetoxystyrene as the coil segment and [(4′-methoxyphenyl)-4-oxybenzoate]-6-hexyl-(4-vinylbenzoate) as the liquid crystalline segments is shown in Scheme 3.7 (Bignozzi 1999).

SFRP was also utilized for the synthesis of acidic and zwitterionic block copolymers. Poly(4-sodium sulfonate styrene) was the first block in all cases. The second block was comprised from 4-dimethylamino styrene or 4-(sodium carboxylate)styrene monomers (Gabaston 1999). In the case of conversions, 4-(sodium carboxylate)styrene was low.

A larger variety of diblock copolymers have been synthesized by ATRP due to the greater flexibility of the method, in terms of polymerizable monomers. However, as in the SFRP case, two fundamental criteria must be fulfilled for the production of well-defined block copolymers. The first is chain-end functionality, i.e., the chain end of the first block must be fully functionalized with the appropriate halogen atom, so each chain must be able to initiate polymerization of the second monomer. Complete functionalization can be achieved if termination and transfer reactions are eliminated completely. Usually, functionalization efficiency can be

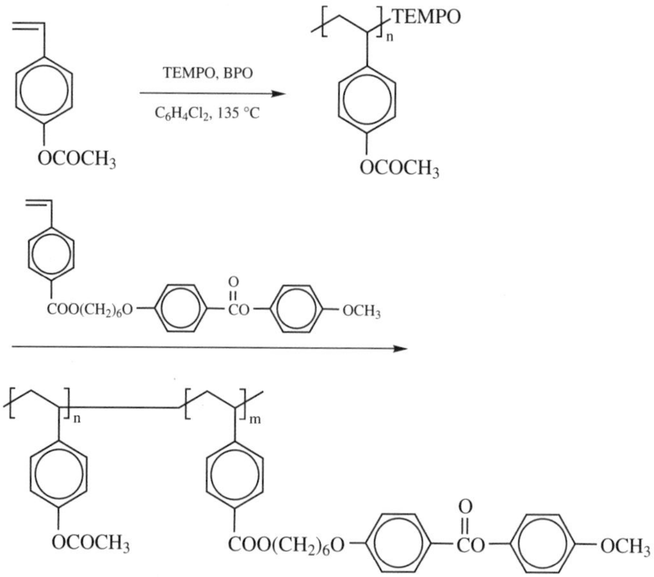

TEMPO, BPO

$C_6H_4Cl_2$, 135 °C

$OCOCH_3$ TEMPO n $OCOCH_3$

$COO(CH_2)_6O$ – CO – OCH_3

n m $OCOCH_3$ $COO(CH_2)_6O$ – CO – OCH_3

Scheme 3.7

increased by varying reaction conditions, i.e., initiating system, reaction temperature, and reaction time (Matyjaszewski 1999).

The second criterion is efficiency of the cross-propagation reaction, in relation to the propagation reaction of the second monomer, a common feature in all living polymerizations. In order to obtain block copolymers with controllable molecular weight characteristics, initiation of the second monomer must be faster than its propagation. Experimental evidence suggests that the initiation rate depends strongly on the alkyl halide structure and actually increases in the order: primary < secondary < tertiary (Matyjaszewski 1998). For example, initiation of an acrylate polymerization by a PMMA macroinitiator is expected to give well-defined block copolymers, whereas polyacrylate macroinitiators must not initiate efficiently the MMA polymerization. Experiments verified this expectation. However, it was found that the nature of the halogen atom, present at the chain end and in the transition metal compound, plays a major role. When acrylate polymerization was initiated with a PMMA having a terminal Cl atom, using CuBr/dNbpy as the catalyst, polymerization was effected, but a large amount of the macroinitiator remained in the final product (Shipp 1998). This was due to relatively slow cross-initiation. The use of a Br-terminated PMMA together with CuCl/dNbpy as catalyst gave block copolymers with no detectable amount of homopolymer (Matyjaszewski 1998). This example demonstrates, in the best way, the features that must be considered for a successful employment of ATRP in block copolymer synthesis and their relative interplay (Scheme 3.8). The nature of the ligands in the metal complex plays a significant role, especially in the case where nitrogen atoms are present in

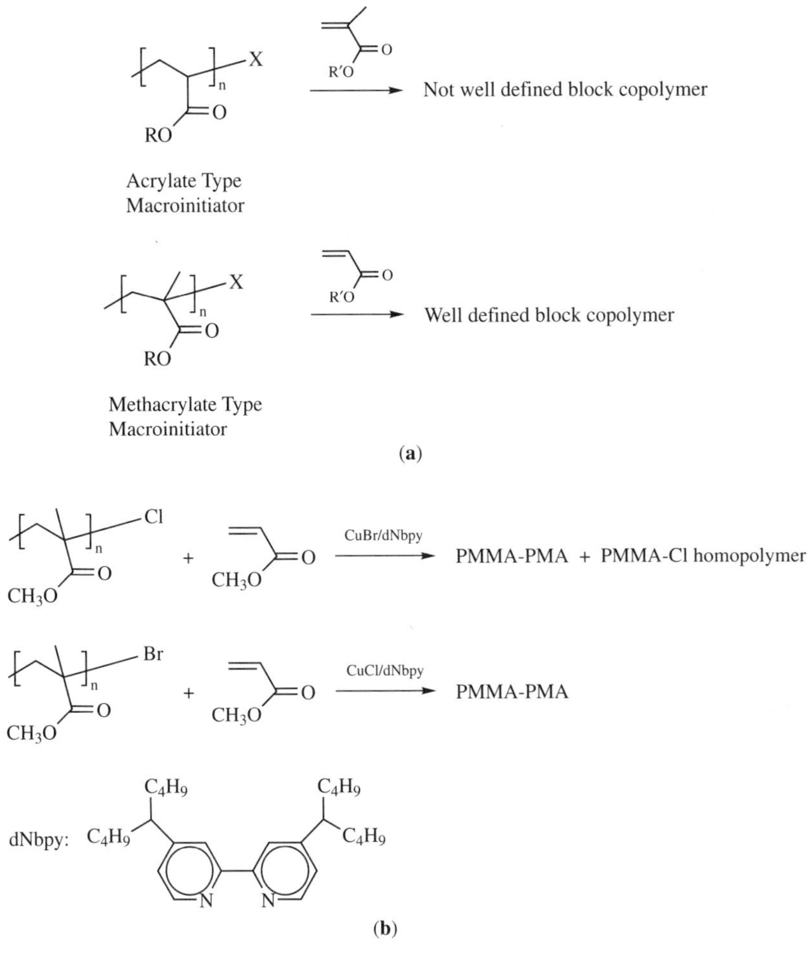

X
n
O
RO
R′O
O
Not well defined block copolymer

Acrylate Type
Macroinitiator

X
n
O
RO
R′O
O
Well defined block copolymer

Methacrylate Type
Macroinitiator

(a)

Cl
n
O
CH_3O
+
CH_3O
O
CuBr/dNbpy
PMMA-PMA + PMMA-Cl homopolymer

Br
n
O
CH_3O
+
CH_3O
O
CuCl/dNbpy
PMMA-PMA

dNbpy: C_4H_9 C_4H_9 C_4H_9 C_4H_9 N N

(b)

Scheme 3.8

the monomer to be polymerized. Because these monomers can also complex the metal used in the initiating system, a stronger complexing ligand must be used.

It is widely accepted that ATRP is mostly suited for (meth)acrylate monomers. A wide range of hydrophilic and hydrophobic monomers including 2-hydroxyethyl methacrylate (HEMA), 2-trimethylsilyloxyethyl acrylate (TMS-HEA), 2-(dimethylamino)ethyl methacrylate (DMAEMA), 4-vinylpyridine (4VP), N,N-dimethylacrylamide (DMAA), methyl acrylate, tert- and n-butyl acrylate (BuA), and styrene (St) has been polymerized by ATRP (Matyjaszewski 1998 and 1999).

Block copolymers of 2-trimethylsilyloxyethylacrylate (TMS-HEA) and n-BuA were synthesized using the system CuBr/N,N,N′,N″,N″-pentamethyldiethylenetriamine and methyl 2-bromopropionate as initiator (Muhdebach 1998). When n-BuA was polymerized, first the precursor polymers obtained had rather broad molecular weight distributions attributed to slow deactivation of the growing chains. Better

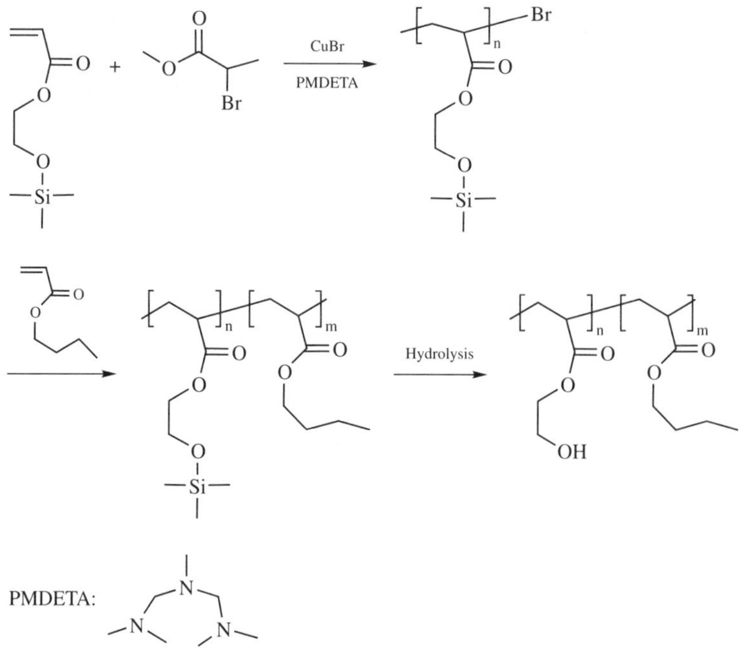

Scheme 3.9

control of the characteristics of the final copolymer was observed when TMS-HEA was used as the first monomer, probably due to a more favorable ratio of the cross-initiation and propagation reactions. The copolymers exhibited predefined molecular weights and low polydispersities. Hydrolysis of the TMS-HEA blocks resulted in the formation of HEA-BA diblocks (Scheme 3.9).

Amphiphilic block copolymers containing PS, PMA, or PMMA blocks as the hydrophobic ones and poly(dimethylaminoethyl methacrylate) (DMAEMA) as the hydrophilic one were prepared by ATRP (Zhang 1998). The 1-phenylethylbromide/CuCl/bpy initiating system was used for the preparation of the PS blocks, whereas the p-toluenesulfonyl chloride or methyl 2-bromopropionate/Cu/dNpy system was used for the synthesis of PMMA and PMA, respectively. DMAEMA was then added to the purified macroinitiators in dichlorobenzene. The reaction took place at 90°C in the presence of CuCl as the catalyst and 1,1,4,7,10,10-hexamethyltriethylenetetramine (HMTETA) as the complexing agent. Block copolymers with narrow molecular weight distributions and predictable molecular weights were obtained. Contamination of the final products with homopolymer was minimal (Scheme 3.10).

The synthesis of block copolymers with fluorine-containing blocks has been reported. Styrene or MMA were polymerized by ATRP using (1-bromoethyl)benzene or 2-bromopropionate as initiators, respectively, and the CuCl/pentamethyldiethylenetriamine system. The macromolecular initiators were isolated and purified before their employment for the polymerization of Fx-14, a methacrylic

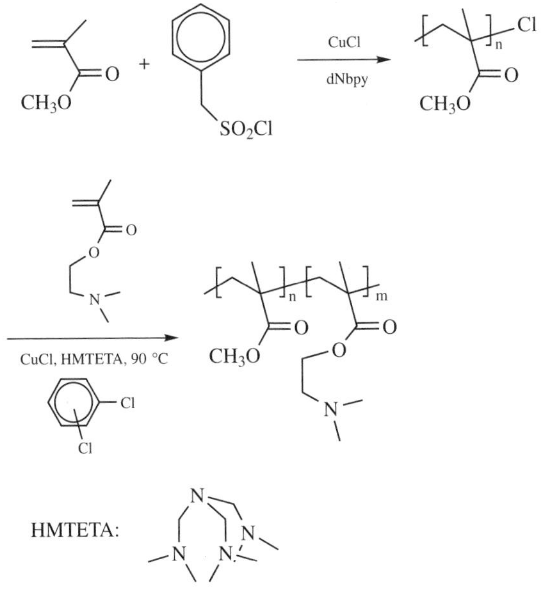

CuCl
dNbpy
CH_3O
SO_2Cl
CuCl, HMTETA, 90 °C
HMTETA:

Scheme 3.10

monomer with a long fluorine-containing side chain (Scheme 3.11) under similar experimental conditions. In this way diblock copolymers of the type PS-PFx-14 and PMMA-PFx-14 were obtained (Li 2000). Diblock copolymers with other fluorinated methacrylate monomers were also synthesized using ATRP methodology (Zhang 1999).

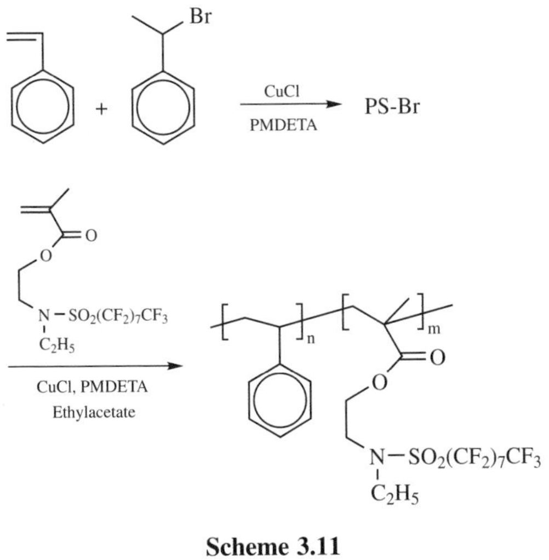

CuCl
PMDETA
PS-Br
$N-SO_2(CF_2)_7CF_3$
C_2H_5
CuCl, PMDETA
Ethylacetate

Scheme 3.11

2. SYNTHESIS OF ABA TRIBLOCK COPOLYMERS

The synthesis of symmetric linear ABA triblock copolymers, through living radical polymerization, is usually effected through the use of difunctional initiators together with sequential two-step monomer addition. A number of such initiators exist (Matyjaszewski 1998). The central B block is synthesized first and used as a macromolecular initiator for the polymerization of the outer A blocks.

Block copolymers of the type PMMA-b-PnBuA-b-PMMA were synthesized by sequential addition of n-BuA and MMA using 1,2-bis(2-bromopropionyloxy)ethane as initiator by Matyjaszewski et al. (Matyjaszewski 1999). The triblock copolymer had a narrow molecular weight distribution (Scheme 3.12). The same authors

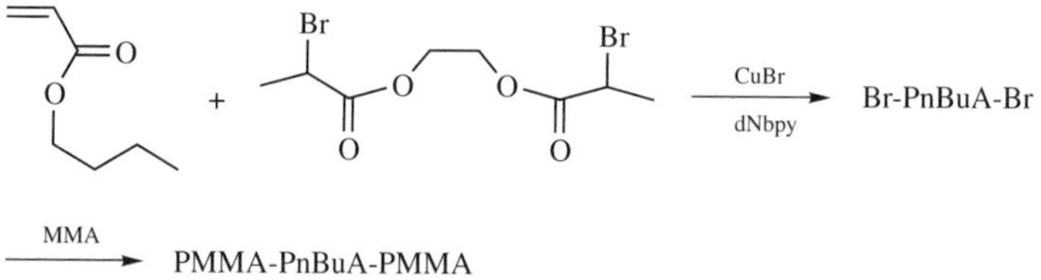

Br Br O O O O O O + CuBr dNbpy Br-PnBuA-Br

MMA PMMA-PnBuA-PMMA

Scheme 3.12

presented the synthesis of PMMA-b-PDMAEMA-b-PMMA triblock copolymers. Triblock copolymers with PMMA central blocks and PDMAEMA outer blocks were also prepared (Scheme 3.13).

Triblock copolymers of BuA and TMS-HEA also synthesized by Matyjaszewski et al. (Muhlebach 1998) using 1,2-bis(2-bromopropionyloxy)ethane and the system CuBr/N,N,N′,N″,N″-pentamethyldiethylenetriamine. Either monomer was used as the central block, but the final copolymers had rather broad molecular weight distributions (Scheme 3.14).

The same group succeeded in the preparation of a symmetric triblock copolymer with sequential three-step addition of monomers (Matyjaszewski 1999). tert-Butylacrylate was polymerized first, followed by the addition of styrene. A third addition of the appropriate amount of t-BuA resulted in the formation of the third block, which had the same molecular weight as the first block. The final triblock copolymer had a molecular weight distribution equal to 1.13, which is strong evidence of the synthesis of a well-defined triblock copolymer. The outer blocks of the copolymer were hydrolyzed, giving a poly(acrylic acid)-b-PS-b-poly(acrylic acid) symmetric triblock copolymer (Scheme 3.15).

The synthesis, by ATRP, of symmetric triblock copolymers having outer blocks comprised of perfluorinated methacrylate monomers has been reported (Zhang 1999). The inner blocks were comprised of styrene, methyl methacrylate, or n-butyl methacrylate. Dibromo or dichloromethybenzene was used as a difunctional initiator in conjuction with CuX catalysts (X = Br, Cl) and bpy as the ligand

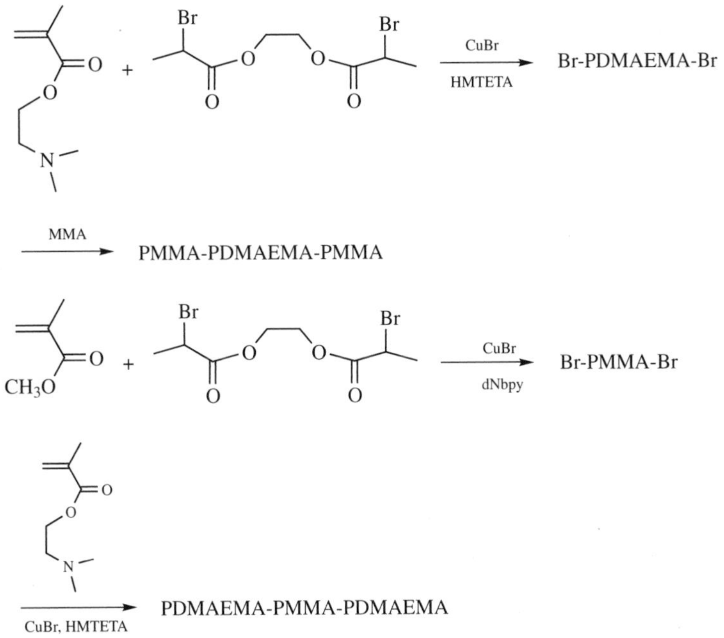

CuBr, HMTETA → Br-PDMAEMA-Br

MMA → PMMA-PDMAEMA-PMMA

CuBr, dNbpy → Br-PMMA-Br

CuBr, HMTETA → PDMAEMA-PMMA-PDMAEMA

Scheme 3.13

CuBr, PMEDTA, 80 °C → Br-PnBuA-Br

→ PTMSHEA-PnBuA-PTMSHEA

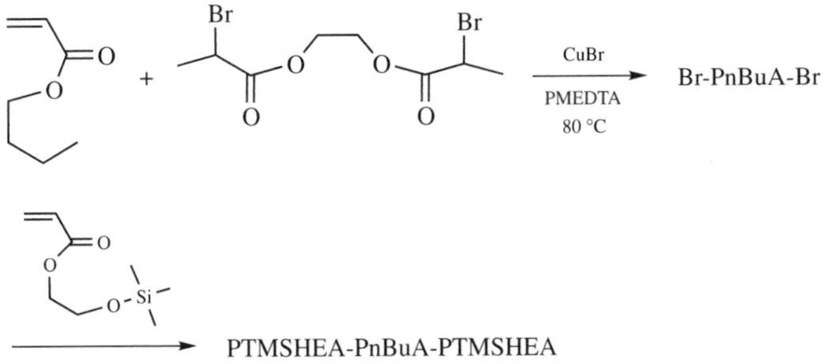

Scheme 3.14

(Scheme 3.16). The molecular weight distributions of the final triblock copolymers were in the range 1.27 to 1.49 depending on the combination of the monomers used.

ABA-type block copolymers with 4-acetoxystyrene A blocks and styrene B blocks were synthesized by stable free radical, TEMPO-mediated, polymerization (Listingovers 1997). A three-step monomer addition scheme was utilized. 4-acetoxystyrene was polymerized first using BPO as initiator and camphorsulphonic acid as accelerator, followed by the polymerization of styrene. A final aliquot of 4-acetoxystyrene was polymerized using the diblock macromolecular initiator,

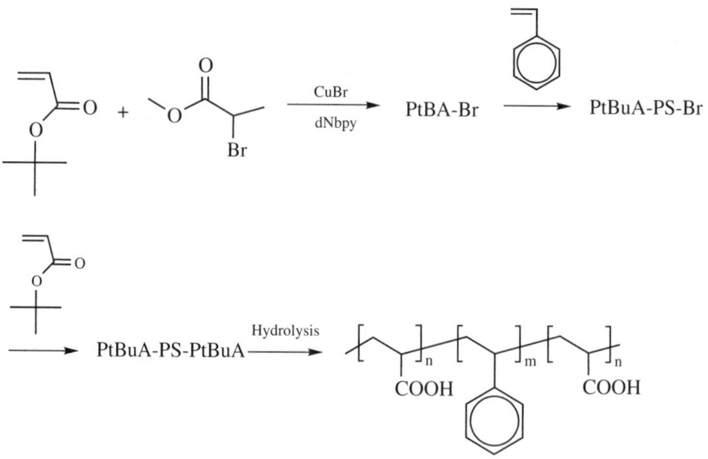

CuBr
dNbpy
PtBA-Br
PtBuA-PS-Br
Hydrolysis
PtBuA-PS-PtBuA
COOH
COOH

Scheme 3.15

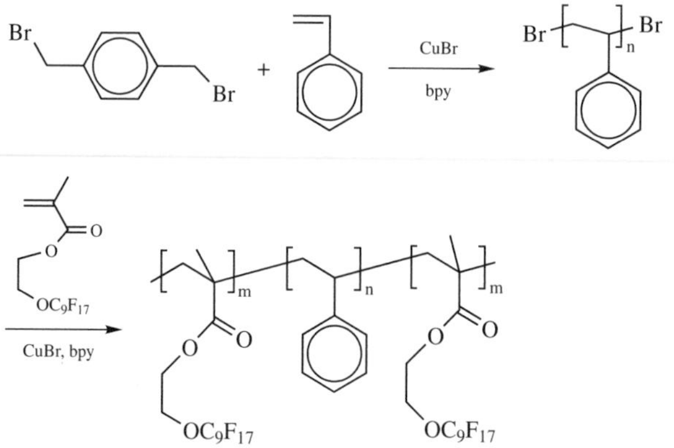

CuBr
bpy
OC_9F_{17}
CuBr, bpy
OC_9F_{17}
OC_9F_{17}

Scheme 3.16

resulting in the designed ABA block copolymers (Scheme 3.17). In each step the intermediate products were isolated and carefully purified before their use as macromolecular initiators in the polymerization of the next monomer. Molecular weights were close to the theoretical ones, and only a slight increase in the polydispersity was observed after formation of each part of the molecule.

Acrylic thermoplastic elastomers of the ABA type were synthesized by Jérôme et al. using ATRP (Moineau 2000). Using $NiBr_2(PPh_3)_2$ as the catalyst, PMMA-PnBuMA-PMMA block copolymers were synthesized with the aid of a difunctional macroinitiator (meso-2,5-dibromoadipate). CuCl and CuBr were used as catalysts in order to increase the crossover reaction in the second step (initiation of MMA outer blocks). The modification proved to be successful because the mechanical

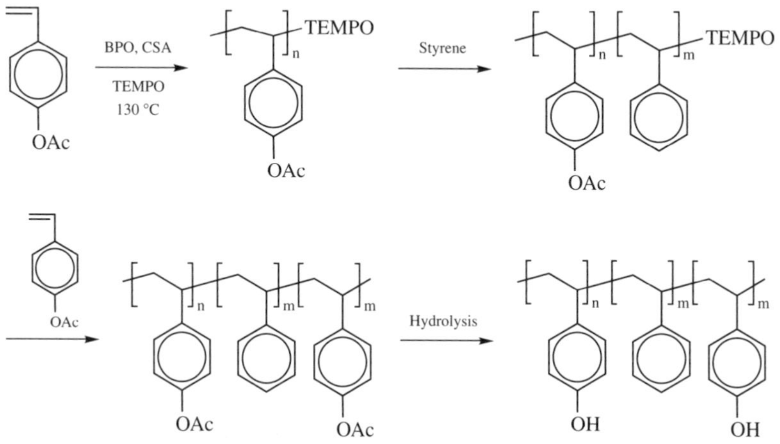

Scheme 3.17

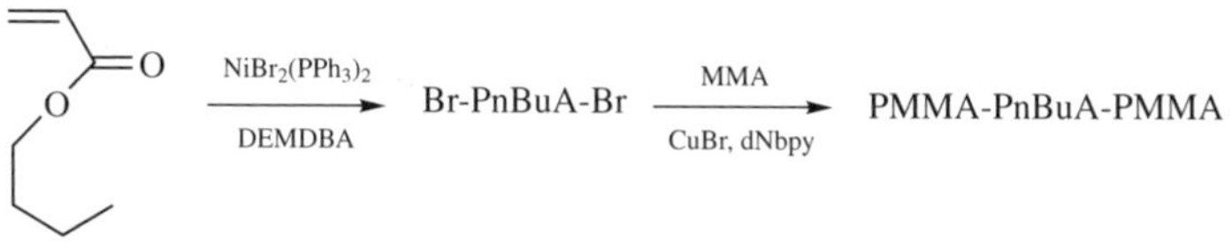

Scheme 3.18

properties of the resulting elastomers were enhanced due to a better defined chain architecture (Scheme 3.18).

3. SYNTHESIS OF ABC TRIBLOCK TERPOLYMERS AND ABCD TETRABLOCK QUARTERPOLYMERS

As discussed earlier the synthesis of triblock copolymers containing three different monomers is accomplished by a three-step sequential monomer addition. Matyjaszewski and coworkers (Matyjaszewski 1999) reported the synthesis of a PtBuA-b-PS-b-PMA triblock using this method. *tert*-BuA was polymerized first using methyl 2-bromopropionate as the initiator, and the CuBr system followed by the addition of styrene and methyl acrylate (MA). After addition and polymerization of each monomer, the intermediate product was isolated and purified, giving special attention to complete elimination of the remaining monomer. Complete elimination of the previous monomer was essential because each polymerization was allowed to proceed until 90% conversion of monomer in order to minimize side reactions, like termination. The final copolymer had a molecular weight of 24,800 and a polydispersity of 1.1, which indicates low compositional heterogeneity, if one takes into account that each intermediate polymer isolated had low molecular

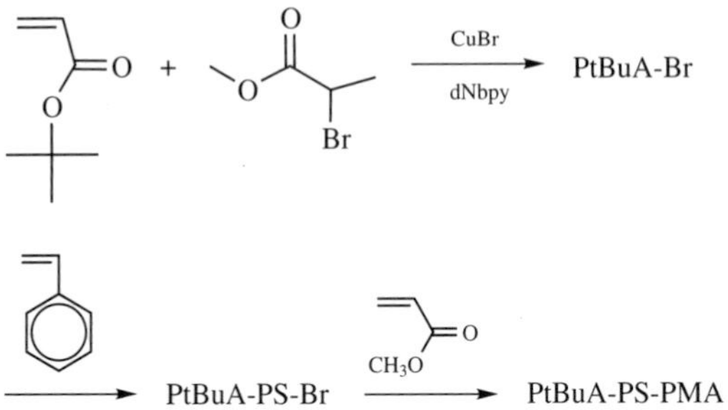

CuBr, dNbpy → PtBuA-Br → PtBuA-PS-Br → PtBuA-PS-PMA

Scheme 3.19

weight distribution ($M_w/M_n \sim 1.11$). The absence of broadening of the molecular weight distribution after each monomer addition also indicates that side reactions were minimized (Scheme 3.19).

This triblock copolymer was chain-extended by polymerization of MMA, giving an ABCD type tetrablock copolymer of total molecular weight $M_n = 33{,}300$ and relatively broader molecular weight distribution ($M_w/M_n \sim 1.3$). The last experimental result may be attributed to difficulties in the polymerization of MMA (Scheme 3.20).

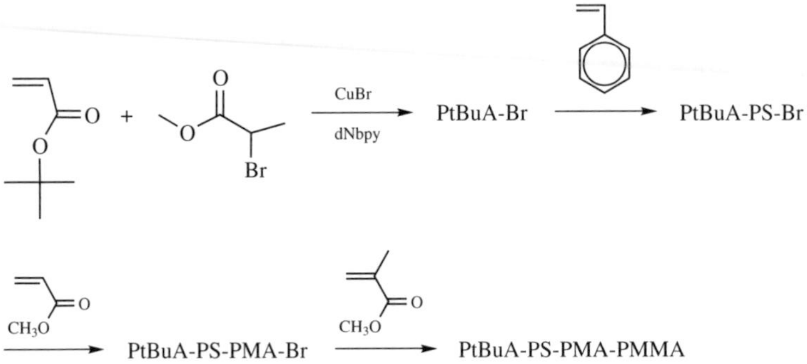

CuBr, dNbpy → PtBuA-Br → PtBuA-PS-Br → PtBuA-PS-PMA-Br → PtBuA-PS-PMA-PMMA

Scheme 3.20

The synthesis of a variety of ABC triblock copolymers was recently reported by Davis and Matyjaszewski (Davis 2001).

ABC triblock copolymers were also synthesized by SFRP. A poly(4-acetoxystyrene)-b-poly(styrene)-b-poly(methyl methacrylate) block copolymer was prepared by sequential addition of monomers (Keoshkerian 1998). Polymerization of 4-acetoxystyrene was effected first using BPO as initiator, in the presence of TEMPO, a radical trapper and camphorsulphonic acid, as the accelerator at 130°C. The first TEMPO-terminated block was isolated and, after removal of the remaining monomer, was used for the polymerization of styrene. The active diblock so formed was subsequently used for the polymerization of methyl methacrylate. Due to

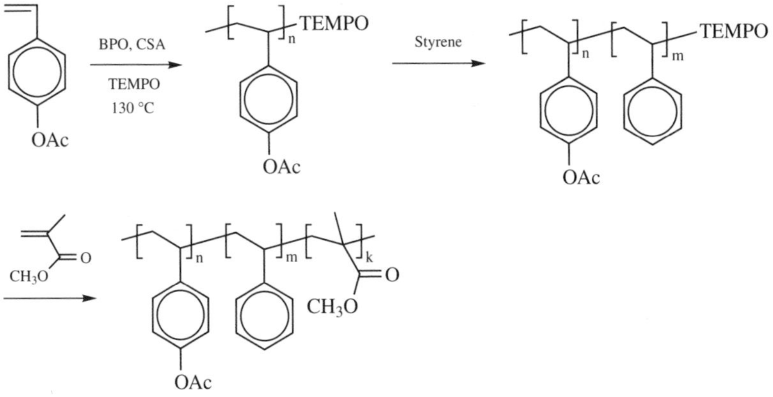

Scheme 3.21

difficulties in the polymerization of methacrylate monomers by SFRP as mentioned earlier, the conversion of the third block was low (~10%), even after 24 h of polymerization (Scheme 3.21).

REFERENCES

Benoit D., Harth E., Fox P., Waymouth R. M., Hawker C. J. (2000) Macromolecules 33, 363.

Bignozzi M. C., Ober C. K., Laus M. (1999) Macromol. Rapid Commun. 20, 622.

Chiefan J., Chong Y. K., Ercole F., Krstina J., Jeffem J., Le T. P. T., Manadunne R. T. A., Meijs G. F., Moad C. L., Moad G., Rizzardo E., Thang S. H. (1998) Macromolecules 31, 5559.

Chong B. Y. K., Le T. P. T., Moad G., Rizzardo E., Thang S. H. (1999) Macromolecules 32, 2071.

Davis K. A., Matyjaszewski K. (2001) Macromolecules 34, 2101.

Fukuda T., Terauchi T., Goto A., Tsujii Y., Miyamoto T. (1996) Macromolecules 29, 3050.

Gabaston L. I., Furlong S. A., Jackson R. A., Armes S. P. (1999) Polymer 40, 4505.

Georges M. K., Veregin R. P. N., Kazmaier P. M., Hamer G. K. (1993) Macromolecules 26, 2987.

Granel C., DuBois P., Jerome R., Teyssie P. (1996) Macromolecules 29, 8576.

Halari I., Pispas S., Hadjichristidis N. (2001) J. Polym. Sci. Part A: Polym. Chem. 39, 2889.

Hawker C. J. (1994) J. Am. Chem. Soc. 116, 11314.

Kato M., Kamigaito M., Sawamoto M., Higashimura T. (1995) Macromolecules 28, 1721.

Keoshherian B., Georges M. K., Boils-Boissier D. (1995) Macromolecules 28, 6381.

Keoshkerian B., Georges M., Quinlan M., Veregin R., Goodbrand B. (1998) Macromolecules 31, 7559.

Li Y., Zhang W., Yu Z., Huang J. (2000) Polym. Prepr. 41(1), 202.

Listigovers N. A., Georges M. K., Honeyman C. H. (1997) Polym. Prepr. 38(1), 410.

Listingovers N. A., Georges M. K., Odell P. G., Keoshkerian B. (1996) Macromolecules 29, 8992.

Malmstrom E. E., Hawker C. J. (1998) Macromol. Chem. Phys. 199, 923.

Matyjaszewski K. (1998) in Controlled Free Radical Polymerization, Matyjaszewski K. (Ed) ACS Symp. Ser. Vol, 685, p.258.

Matyjaszewski K. (Ed) (1998) Controlled Radical Polymerization, ACS Symposium Series, Washington, DC, Vol. 685. .

Matyjaszewski K., Acar M. H., et al. (1999) Polym. Prepr. 40(2), 966.

Matyjaszewski K., Gaynor S., Wang J. S. (1995) Macromolecules 28, 2093.

Matyjaszewski K., Shipp D. A., Wang J. L., Grimaud T., Patten T. (1998) Macromolecules 31, 6836.

Matyjaszewski K., Wang J-L., Grimaud T., Shipp D. (1998) Macromolecules 31, 1527.

Moad G., Rizzardo E., Solomon D. H. (1982) Macromolecules 15, 909.

Moineau G., Minet M., Teyssie P., Jérôme R. (2000) Macromol. Chem. Phys. 201, 1108.

Muhlebach A., Gaynor S. G., Matyjaszewski K. (1998) Macromolecules 31, 6046.

Odian G. (1981) Principles of Polymerization, J. Wiley & Sons, New York.

Percec V., Barboin B. (1995) Macromolecules 28, 7970.

Qiu J., Matyjaszewski K. (1997) Acta Polym. 48, 169.

Robin S., Gnanou Y. (1999) Polym. Prepr. 40(2), 387.

Sawamoto M., Kamigaito M. (1998) Trends in Polym. Sci. 4, 371.

Shipp D. A., Wang J-L., Matyjaszewski K. (1998) Macromolecules 31, 8005.

Wang J. S., Matyjaszewski K. (1995) J. Am. Chem. Soc. 117, 5614.

Wang J. S., Matyjaszewski K. (1995) Macromolecules 28, 7901.

Wayland B. B., Poszmik G., Mukerjee S. L., Fryd M. (1994) J. Am. Chem. Soc. 116, 7943.

Zhang X., Matyjaszewski K. (1998) Polym. Prepr. 39(2), 560.

Zhang Z-B., Ying S-K., Shi Z-Q. (1999) Polymer 40, 5439.

CHAPTER 4

BLOCK COPOLYMERS BY GROUP TRANSFER POLYMERIZATION

Group transfer polymerization (GTP) is a Michael-type catalyzed addition reaction (Webster 1983, Webster 1992, Brittain 1992). A silyl ketene acetal is usually used as the initiator. The silane group is transferred to the growing chain end after the addition of each monomer unit (Scheme 4.1). Thus, the chain end remains active until complete consumption of the monomer. Due to the livingness of the polymerization, the molecular weight of the polymer synthesized can be predetermined by the amount of the initiator and the monomer used. This type of polymerization reaction has been widely applied to the polymerization of (meth)acrylic monomers at room temperature, in the presence of a variety of side groups, which are sensitive to ionic or radical polymerization reactions (Webster 1990). GTP is considerably tolerant to certain functionalities such as tertiary amines, epoxides, styrenic, and allylic groups. Monomers with functionalities like -OH and -COOH groups can be polymerized after protection with appropriate protective groups. Deprotection after polymerization gives access to hydrophilic polymers. Other monomers, like acrylic esters, acrylonitrile, N,N-dimethylacrylamide, and α-methylene-γ-butyrylactones, have been polymerized by GTP. The use of anionic or Lewis acid catalysts, in conjuction with the initiator, is advantageous to the progress of polymerization because these catalysts coordinate with the silicon atom (anionic catalysts) or the monomer (Lewis acids) facilitating group transfer (Sogah 1990).

Aldol group transfer polymerization (AGTP) comprises a variation of group transfer polymerization where an aldehyde can be used as the initiator for the polymerization of a silylvinyl ether, in the presence of a catalyst (usually a nucleophile or Lewis acid) (Boettcher 1988). Aromatic aldehydes are preferred over aliphatic ones (Scheme 4.2). Electrophilic compounds like alkyl halides, acetals, and anhydrides that react with silyl enolates generating an aldehyde can be

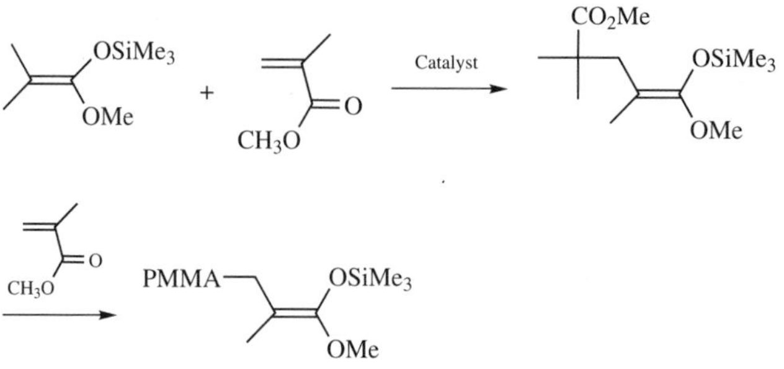

Scheme 4.1

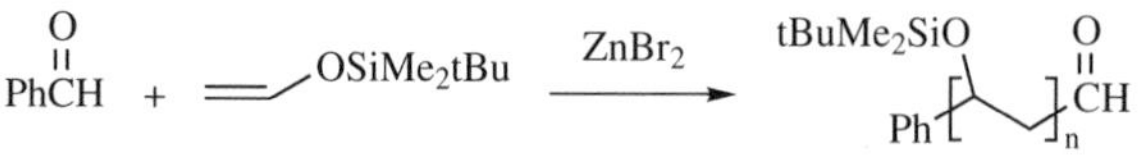

Scheme 4.2

used as aldol group transfer polymerization initiators. AGTP was originally developed for the synthesis of poly(vinylalcohol) after hydrolysis of the side silyl groups of the poly(silyl vinyl ether) (Sogah 1986 and 1987). In aldol group transfer polymerization, the silyl group is transferred from the monomer to the initiator in contrast with normal GTP where the silyl group is transferred from the initiator to the monomer.

1. SYNTHESIS OF AB DIBLOCK COPOLYMERS

Because a variety of monomers can be polymerized by GTP, and some of them cannot be polymerized by other polymerization methods, it is obvious that new classes of block copolymers can be obtained. One advantage of GTP, compared with anionic polymerization, is the relative ease in the polymerization of (meth) acrylate monomers. From a practical perspective, GTP has the advantage over anionic polymerization in allowing polymerization of acrylates at room temperature instead of the low temperatures used in anionic polymerization. Diblock copolymers can be synthesized by a two-step sequential monomer addition (Scheme 4.3).

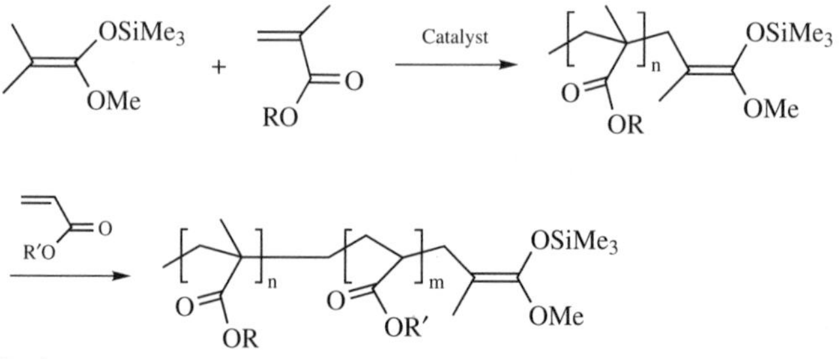

Scheme 4.3

When a diblock copolymer consisting of one methacrylate block and one acrylate block has to be synthesized, polymerization starts from the less reactive methacrylate monomer followed by the polymerization of the acrylate-type monomer (Webster 1983). This is because the dialkylsilane ketene acetal group that is formed at the end of the methacrylate block is more reactive compared with the monoalkyl silane ketene acetal groups present at the end of the acrylic block. In this way the more reactive monomer (acrylate) is polymerized with a more reactive initiator, whereas the monoalkyl silyl ketene acetal groups are less effective for the polymerization of methacrylate monomers. It has to be noted that this mode of addition is the reverse of that used in anionic polymerization.

Sogah and coworkers (Sogah 1987) reported the synthesis of block copolymers containing methyl methacrylate and n-butylmethacrylate, or allyl methacrylate as well as MMA and 2-hydroxyethyl methacrylate (HEMA) or lauryl methacrylate (LMA) with a variety of initiating systems. The molecular weight distributions of the final products were narrow, and the experimentally determined molecular weights were close to the ones calculated from the monomer-to-initiator ratio. Scheme 4.4 shows the general reaction used for the synthesis of diblock copolymers containing MMA and laurylmethacrylate.

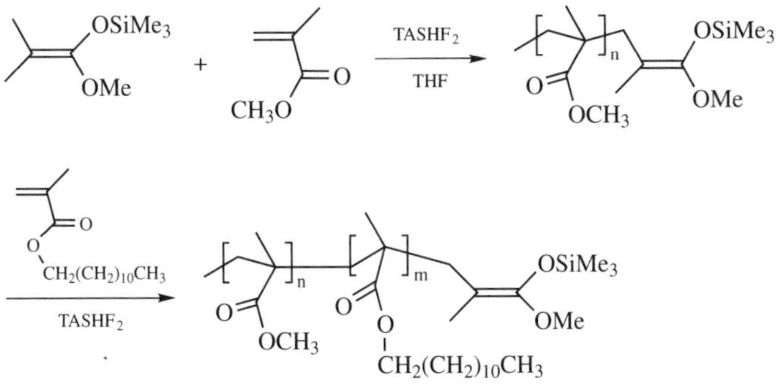

OSiMe$_3$ OMe + CH$_3$O O —TASHF$_2$ / THF→ [ ]$_n$ O OCH$_3$ OSiMe$_3$ OMe

O O CH$_2$(CH$_2$)$_{10}$CH$_3$ —TASHF$_2$→ [ ]$_n$ [ ]$_m$ O OCH$_3$ O O CH$_2$(CH$_2$)$_{10}$CH$_3$ OSiMe$_3$ OMe

Scheme 4.4

Diblock copolymers with methyl methacrylate first blocks and glycidyl methacrylate (GMA) (Simms 1987, Scheme 4.5), (dimethylamino)ethyl methacrylate (DMAEMA) (Scheme 4.6), or tetrahydropyranyl methacrylate (THPMA) (Scheme 4.7) (Dikker 1990) have been synthesized by GTP using the sequential monomer addition method. The tetrahydropyranyl groups in the last case were hydrolyzed, resulting in a poly(methylmethacrylate)-b-poly(methacrylic acid) amphiphilic diblock copolymer.

By sequential block copolymerization of MMA and methacrylic monomers with mesogenic side groups, using GTP methodology, block copolymers with an amorphous and a side chain liquid crystalline block were prepared (Scheme 4.8)

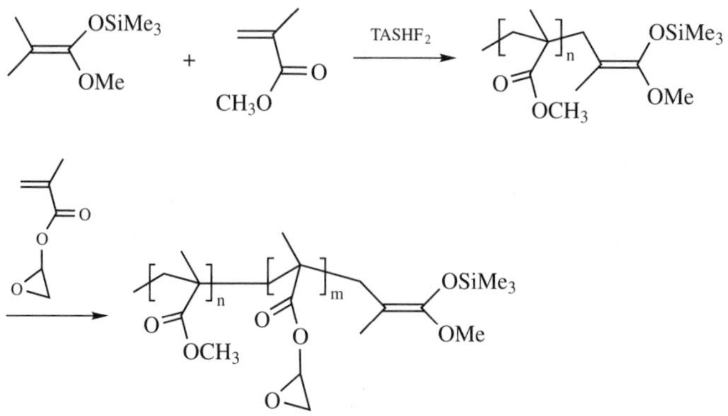

OSiMe3 OMe + CH3O O TASHF2 n O OCH3 OSiMe3 OMe

O O O n O OCH3 O m O O OSiMe3 OMe

Scheme 4.5

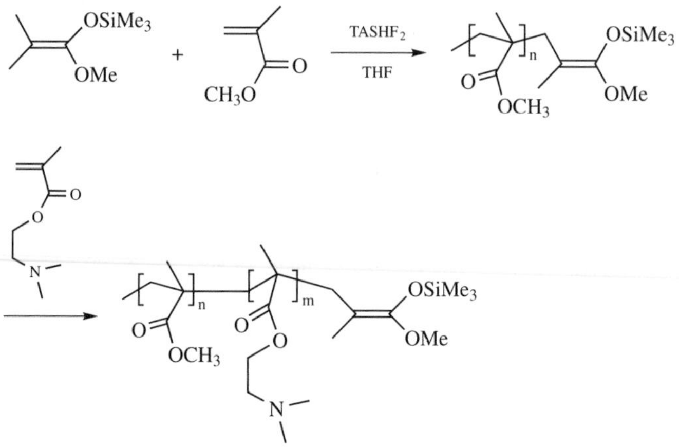

OSiMe3 OMe + CH3O O TASHF2 THF n O OCH3 OSiMe3 OMe

O O N n O OCH3 O m O N OSiMe3 OMe

Scheme 4.6

O O O OSiMe3 OMe + CH3O O TASHF2 THF n O OCH3 OSiMe3 OMe

n O OCH3 O m O O OSiMe3 OMe Hydrolysis n O OCH3 O m OH OSiMe3 OMe

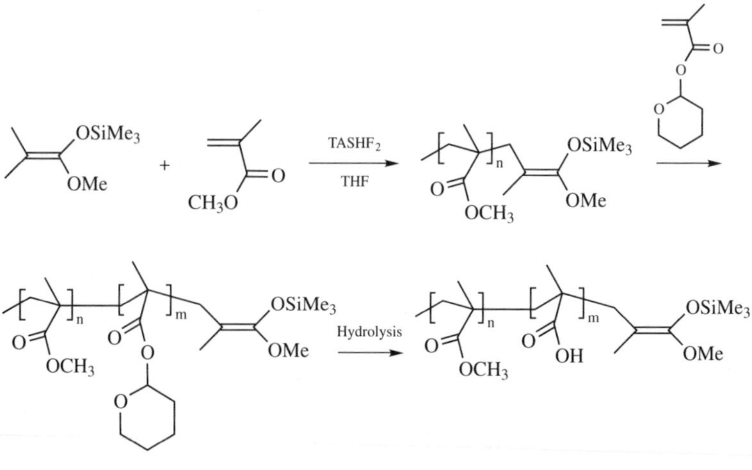

Scheme 4.7

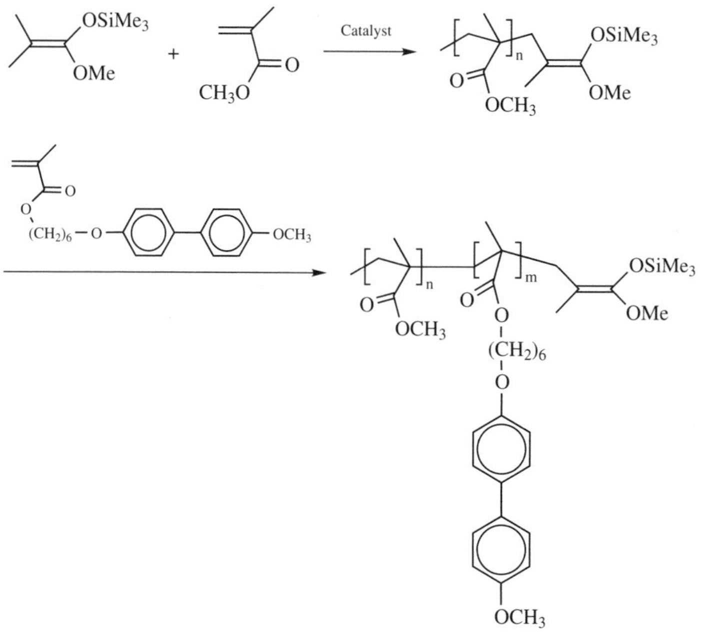

Scheme 4.8

(Hefft 1990). These mesogenic monomers could not be polymerized by anionic polymerization because their side groups react with the anionic active centers.

Amphiphilic diblock copolymers of (dimethylamino)ethyl methacrylate and alkyl methacrylates with varying hydrophobicity, including MMA, n-octyl methacrylate, n-lauryl methacrylate and allyl methacrylate, and n-BuMA were synthesized recently by Armes et al. (Lowe 1999). 1-methoxy-1-trimethylsiloxy-2-methyl-1-propene (MTS) was used as the initiator and tetra-n-butylammonium bibenzoate (TBABB) as the catalyst in all cases. Monodisperse block copolymers, with no homopolymer contamination, have been prepared. The dimethylamino groups of DMAEMA were converted to sulfopropylbetaine groups by reaction with 1,3-propane sultone, giving water-soluble block copolymers (Scheme 4.9). Block polyelectrolytes were synthesized by the same group containing 2-(dimethylamino) ethyl methacrylate and 2-(diethylamino)ethyl methacrylate block copolymers using GTP methodology (Lee 1999), and their solution properties as a function of the pH of the aqueous solution were studied (Scheme 4.10). Muller et al. (Muller 1991) have reported the synthesis of diblock copolymer containing DMAEMA and decyl methacrylate by GTP (Scheme 4.11).

Diblock copolymers of n-BuMA and benzyl methacrylate (BzMA) have been prepared using the same initiator/catalyst system (Scheme 4.12). Hydrolysis of the benzylic groups led to the formation of well-defined PnBuMA-b-poly(methacrylic acid) block copolymers (Vamvakaki 1998). Armes et al. prepared diblock copolymers containing an oligo(ethylene glycol) monomethacrylate (OEGMA) block and

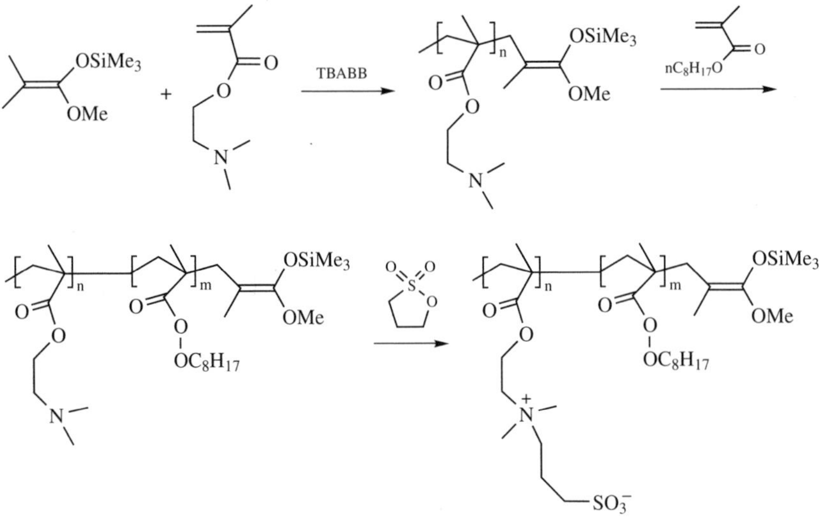

Scheme 4.9

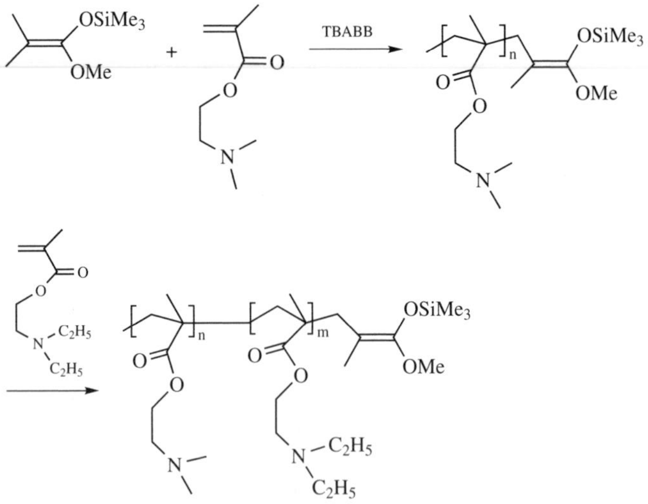

Scheme 4.10

benzyl (BzMA) (Scheme 4.13) or tetrahydropyranyl (THPMA) methacrylate blocks (Butum 2000). Hydrolysis of the BzMA and THPMA blocks gave block copolymers containing two water-soluble blocks. Amphiphilic block copolymers containing poly(methacrylic acid) blocks were also prepared by GTP using tert-butyl methacrylate (Rannard 1993) and trimethylsilyl methacrylate (Lim 1999) monomers as precursors to methacrylic acid segments.

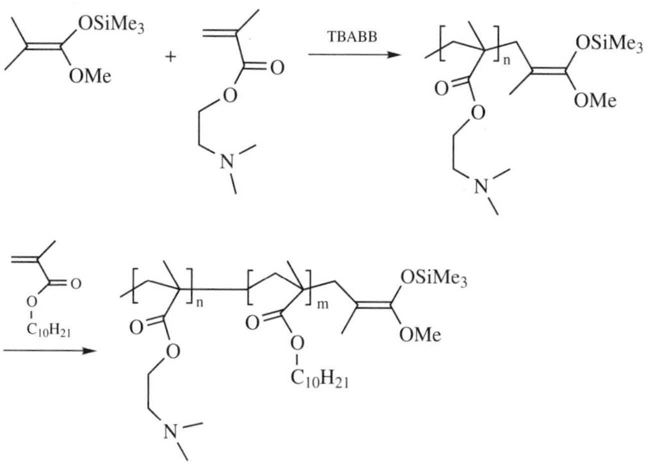

Scheme 4.11

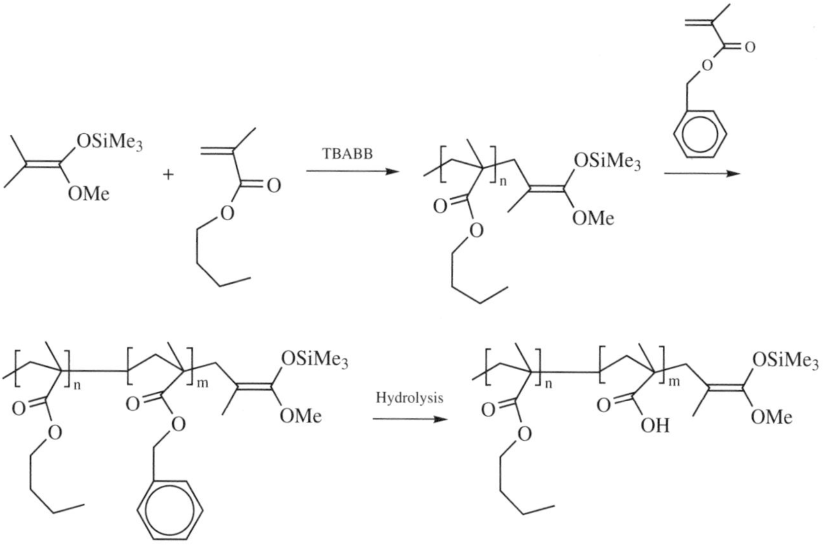

Scheme 4.12

Block copolymers containing two water-soluble blocks were also synthesized by GTP. (Diethylamino)ethyl methacrylate (DEAMA) was polymerized first in the presence of MTS/TBABB initiator/catalyst system in THF at room temperature. 2-(N-morpholino)ethyl methacrylate (MEMA) was added after complete consumption of DEAMA (Scheme 4.14). Well-defined block copolymers were produced with interesting properties in aqueous solutions due to the different solubilities of the two blocks at different p^H and temperatures (Butum 1999).

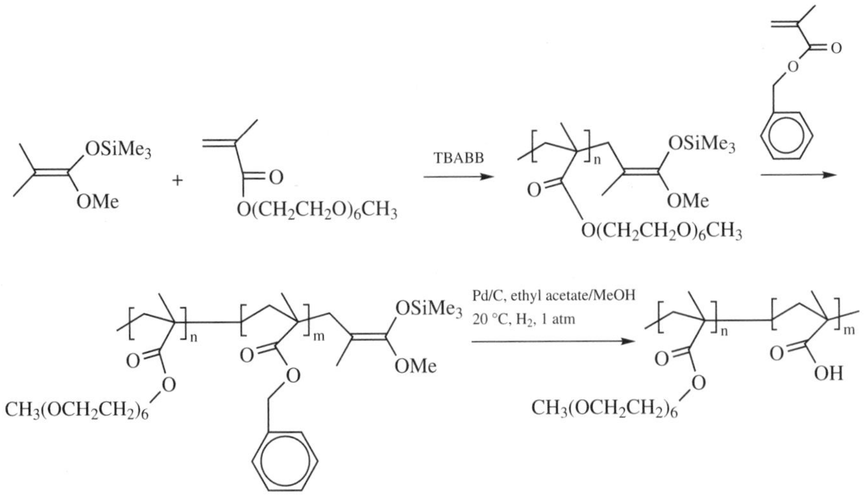

Scheme 4.13

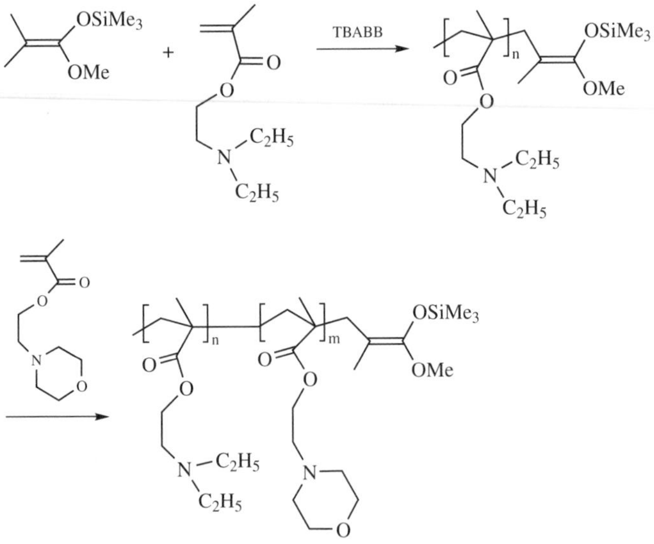

Scheme 4.14

2. SYNTHESIS OF ABA TRIBLOCK COPOLYMERS

Linear symmetric ABA triblock copolymers were synthesized by GTP, using a difunctional initiator and a two-step monomer addition scheme. A number of difunctional GTP initiators were synthesized for this purpose (Sogah 1987, Yu 1988) (Scheme 4.15).

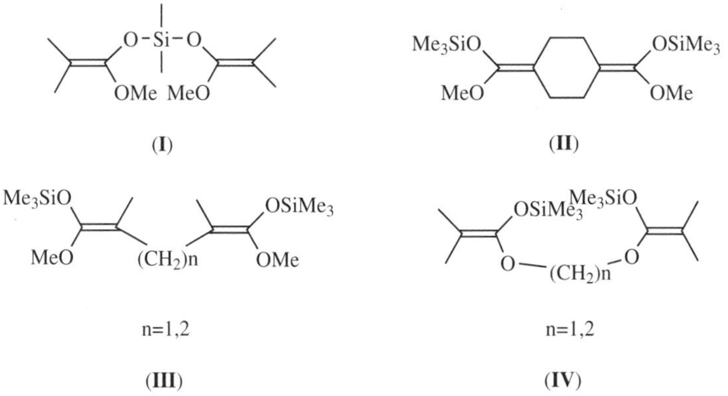

O-Si-O
OMe MeO
(I)

Me_3SiO $OSiMe_3$
MeO OMe
(II)

Me_3SiO $OSiMe_3$
MeO $(CH_2)n$ OMe
n=1,2
(III)

$OSiMe_3$ Me_3SiO
O—$(CH_2)n$—O
n=1,2
(IV)

Scheme 4.15

ABA triblock copolymers containing MMA/MA, allyl methacrylate/MMA, 2-hydroxyethyl methacrylate/allyl methacrylate (Scheme 4.16), and BuMA/MMA were synthesized using the initiators I to IV shown in Scheme 4.15. A triblock copolymer of the type poly(MMA-co-2-hydroxyethyl methacrylate)-b-poly(lauryl methacrylate)-b-poly(MMA-co-2-hydroxyethyl methacrylate) was synthesized with initiator I (Sogah 1987).

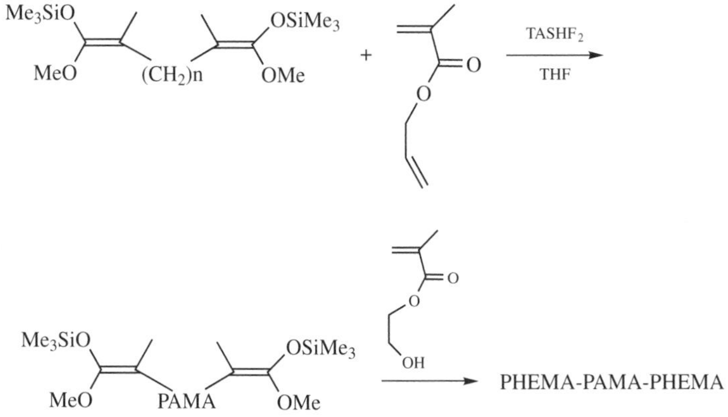

Me_3SiO $OSiMe_3$
MeO $(CH_2)n$ OMe + O O $\xrightarrow[THF]{TASHF_2}$

Me_3SiO $OSiMe_3$
MeO PAMA OMe
O O OH
$\longrightarrow$ PHEMA-PAMA-PHEMA

Scheme 4.16

Amphiphilic ABA and BAB triblock copolymers were synthesized by Armes et al., where A is a 2-(dimethylamino)ethyl methacrylate (DMAEMA) hydrophilic block and B is a 2-(diethylamino)ethyl methacrylate (DEAEMA) or methylmethacrylate (MMA) block (Scheme 4.17). The difunctional initiator bis(methoxy trimethylsiloxy) cyclohexane (CHMTS) was used in conjunction with TBABB as the catalyst. Monomer additions were made according to the block sequence required for the formation of the two types of triblock copolymers. No significant

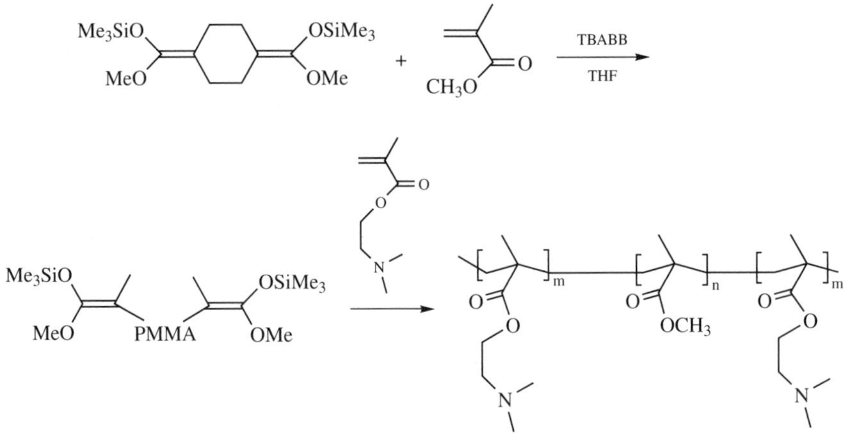

Scheme 4.17

differences were observed in the molecular characteristics of the obtained copolymers, leading to the conclusion that the preparation of the triblock copolymers can be accomplished with the same degree of chemical control whatever is the sequence of the monomers (Unali 1999).

In another case, ABA triblock copolymers having an n-laurylmethacrylate central block and benzylmethacrylate outer blocks were synthesized by GTP (Purcell 1997) using the difunctional initiator 1,5-bis(trimethylsiloxy)-1,5-dimethoxy-2,4-dimethyl-1,4-pentadiene (BDDP) (III) (Scheme 4.18). The synthetic route involved the polymerization of n-laurylmethacrylate using BDDP and TBABB in THF. Polymerization proceeded slowly. The appropriate amount of benzylMA was added after consumption of the first monomer. A number of samples were prepared in high yields, having molecular weights in the range 21,000 to 135,000. Molecular weight distributions were relatively narrow, but they tended to broaden at molecular weights higher than 130,000.

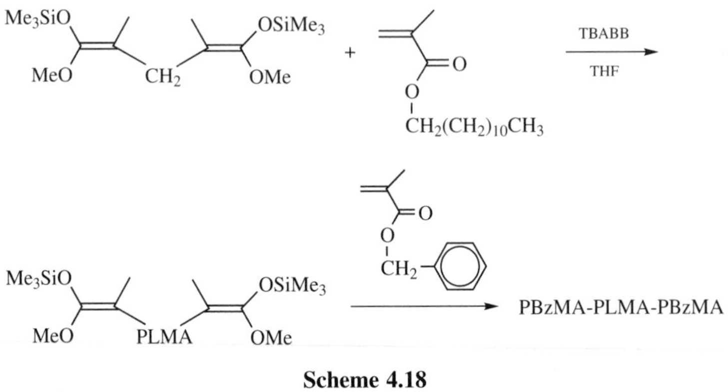

Scheme 4.18

Triblocks containing polycaprolactone (PCL) as the central block and PMMA as the outer blocks were prepared by reacting α,ω-dihydroxyl-terminated PCL with acryloyl chloride to give α,ω-diacrylopropylcaprolactone (Sogah 1987). Subsequent reaction with trimethylsilyl cyanide gave a macromolecular difunctional GTP intitiator. This initiator was used for the polymerization of MMA, in a second step, producing PMMA-PCL-PMMA triblock copolymers (Scheme 4.19).

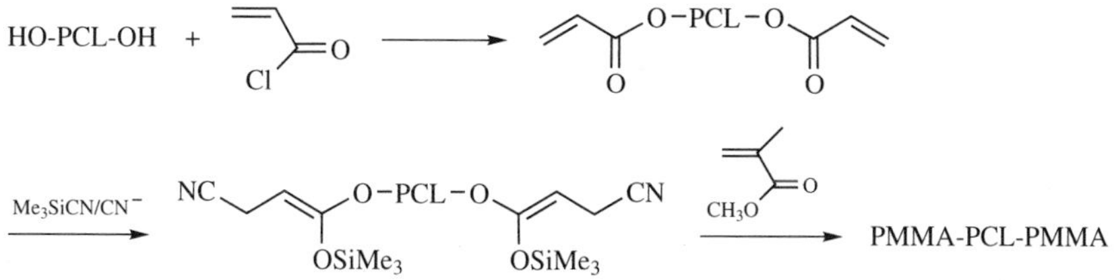

HO-PCL-OH + Cl–C(=O)–CH=CH₂ ⟶ O-PCL-O

$\xrightarrow{Me_3SiCN/CN^-}$ NC ... O-PCL-O ... CN (OSiMe₃, OSiMe₃) $\xrightarrow{MMA}$ PMMA-PCL-PMMA

Scheme 4.19

Analogous triblocks with a PEO central block and PMMA outer blocks were synthesized by α,ω-dimethacrylate-terminated PEOs (Budde 1998). Dimethylethylsilane was added to the terminal groups by use of a Wilkinson catalyst in THF under argon. The difunctional macromolecular initiator produced was used for the polymerization of MMA in the presence of tris(dimethylamino)sulfonium difluoride as catalyst (Scheme 4.20). Diblock copolymers were synthesized by a similar reaction scheme but using monofunctional PEO macroinitiators.

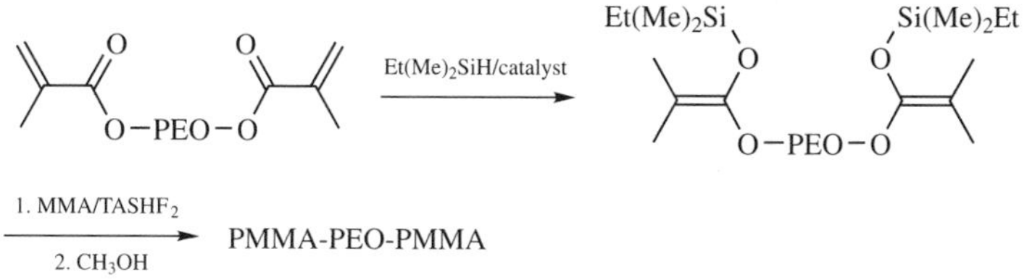

O-PEO-O $\xrightarrow{Et(Me)_2SiH/catalyst}$ Et(Me)₂Si ... O-PEO-O ... Si(Me)₂Et

$\xrightarrow[2.\ CH_3OH]{1.\ MMA/TASHF_2}$ PMMA-PEO-PMMA

Scheme 4.20

3. SYNTHESIS OF ABC TRIBLOCK TERPOLYMERS

Linear triblock terpolymers containing three chemically different blocks have been synthesized by GTP using the three-step sequential addition method.

Using this technique PMMA-b-PnBuMA-b-poly(allyl methacrylate) ABC triblock terpolymers were synthesized by Sogah et al. (Sogah 1987) (Scheme 4.21).

Patrickios et al. (Patrickios 1994) have published the synthesis of ABC triblock copolymers comprised of DMAEMA, MMA (or phenylethyl methacrylate, PEMA),

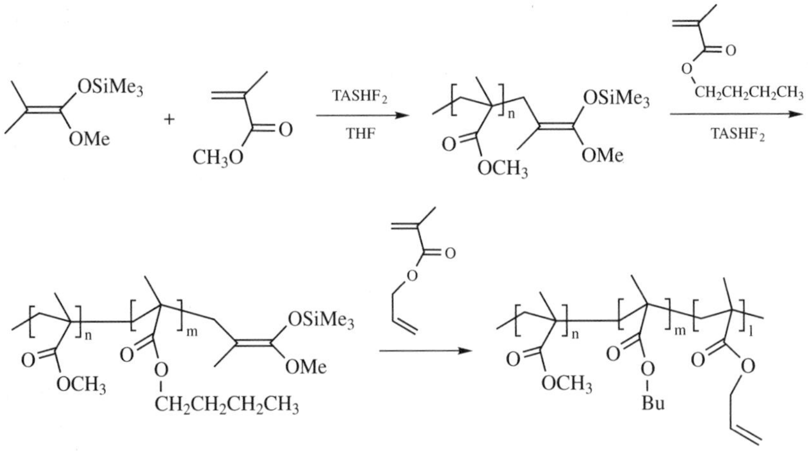

Scheme 4.21

and methacrylic acid. The last block was derived from postpolymerization deprotection of trimethylsilylmethacrylate (TMSMA) or THPMA. Sequential three-step addition of monomers was employed (Scheme 4.22). 1-methoxy-1-(trimethylsiloxy)-2-methyl-1-pentene was used as the initiator and tetrabutylammonium

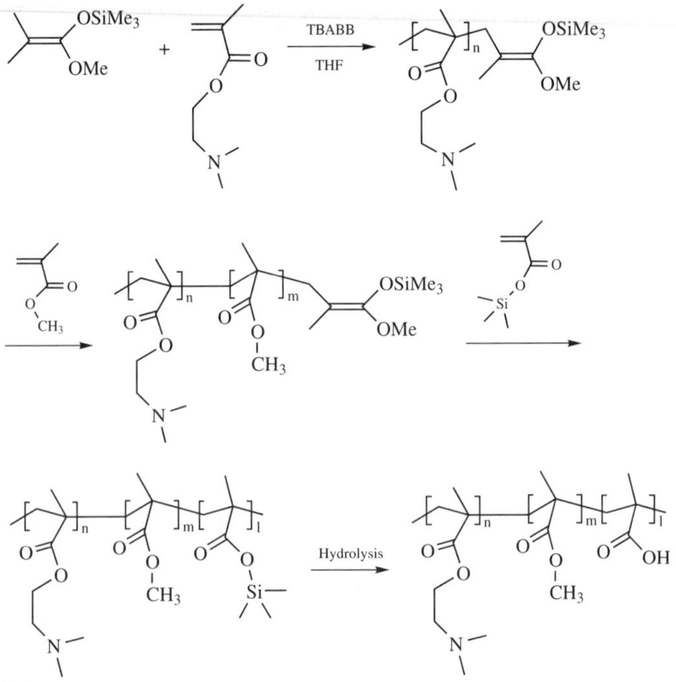

Scheme 4.22

biacetate as the catalyst. The methacrylic acid precursor block was always formed last, and the nonionic block (PMMA) was always used as the middle block. Block copolymers with relatively narrow molecular weight distributions but rather low molecular weights were obtained. Their properties as polyampholytes were investigated.

More recently, Patrickios et al. (Patrickios 1998) reported on the synthesis of ABC, ACB, and BAC triblock copolymers of methyl methacrylate, DMAEMA, and THPMA . The length of each block was chosen to be equal to 12 monomeric units in all cases, and the block sequence was varied in order to produce different topological isomers. An example is given in Scheme 4.23. The polydispersities of

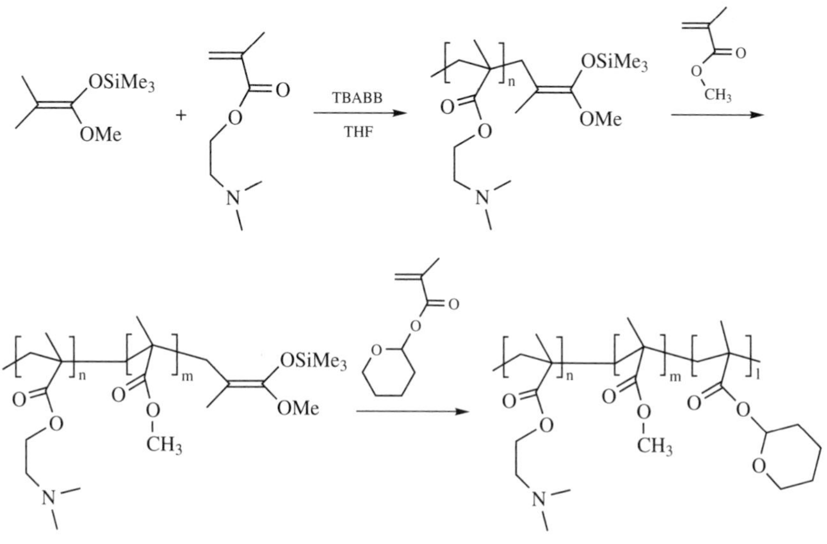

OSiMe$_3$ OMe + TBABB THF OSiMe$_3$ OMe O CH$_3$ OSiMe$_3$ OMe CH$_3$ CH$_3$

Scheme 4.23

the final copolymers were low, and the molecular weights and compositions determined experimentally were in close agreement with the theoretically expected ones.

A poly(2-ethylhexylacrylate)-b-poly(methyl methacrylate)-b-poly(acrylic acid) linear triblock terpolymer was also synthesized using GTP methodology and postpolymerization transformation reactions (Kriz 1998). 2-Ethylhexyl acrylate, MMA, and tert-butyl acrylate were sequencially polymerized, giving the PEHA-PMMA-PtBuA precursor. Hydrolysis of the PtBA block resulted in the final ABC triblock terpolymer (Scheme 4.24). This terpolymer formed three-layered micelles in water due to the incompatibility of the three chemically different blocks.

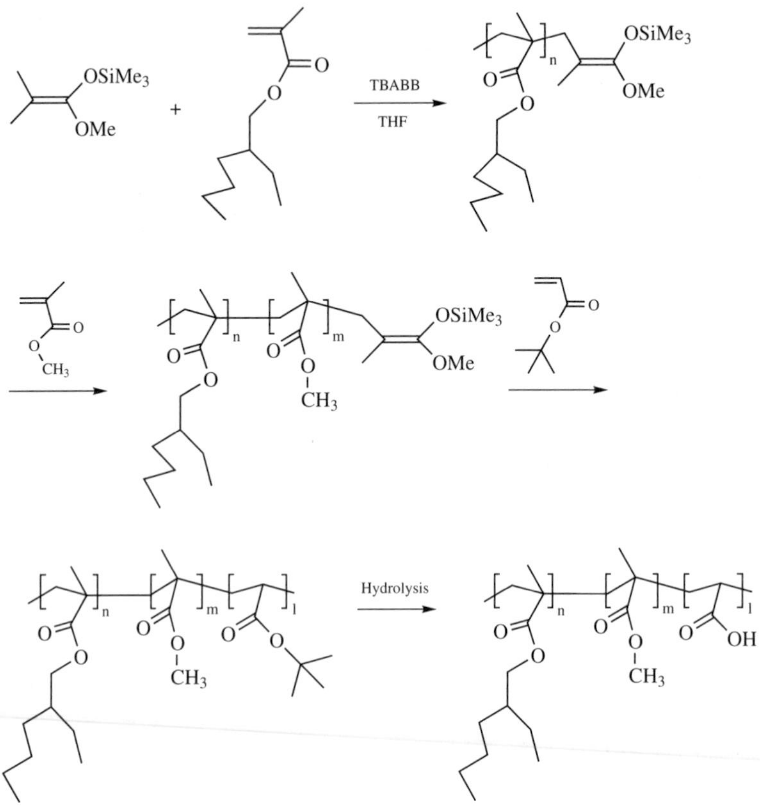

Scheme 4.24

REFERENCES

Boettcher F. P. (1988) Makromol. Chem., Macromol. Symp., 31, 128.

Brittain W. J. (1992) Rubber Chem. Technol. 65, 588.

Budde H., Horing S. (1998) Macromol. Chem. Phys. 199, 2541.

Butun V., Armes S. P., Billingham N. C., Tuzar Z. (1999) Polym. Prepr. 40(1), 261.

Butun V., Vamvakaki M., Billingham N. C., Armes S. P. (2000) Polymer 41, 3173.

Dicker I. B., Cohen G. M., Farnham W. B., Hertler W. R., Laganis E. D., Sogah D. Y. (1990) Macromolecules 23, 4034.

Hefft M., Springer J. (1990) Makromol. Chem., Rapid Commun. 11, 397.

Kriz J., Masar B., Plestil J., Tuzar Z., Pospisil H., Doskosilova D. (1998) Macromolecules 31, 41.

Lee A. S., Gast A. P., Butun V., Armes S. P. (1999) Macromolecules 32, 4302.

Lim K. T., Webber S. E., Johnston K. P. (1999) Macromolecules 32, 2811.

Lowe A. B., Billingham N. C., Armes S. P. (1999) Macromolecules 32, 2141.

Muller M. A., Augestein M., Dumont E., Pennewiss H. (1991) New Polym. Mater. 2, 315.

Patrickios C. S., Hertler W. R., Abbott N. L., Hatton T. A. (1994) Macromolecules 27, 930.

Patrickios C. S., Lowe A. B., Armes S. P., Billingham N. C. (1998) J. Polym. Sci. Part A: Polym. Chem. 36, 617.

Purcell A., Armes S. P., Billingham N. C. (1997) Polym. Prepr. 38(1), 502.

Rannard S. P., Billingham N. C., Armes S. P., Mykytiuk J. (1993) Eur. Polym. J. 29, 407.

Simms J. A., Spinelli H. J. (1987) J. Coat. Technol. 59, 125.

Sogah D. Y., Hertler W. R., Dicker I. B., DePra P. A., Butera J. R. (1990) Makromol. Chem., Macromol. Symp. 32, 75.

Sogah D. Y., Hertler W. R., Webster O. W., Cohen G. M. (1987) Macromolecules 20, 1473.

Sogah D. Y., Webster O. W. (1986) Macromolecules 19, 1775.

Sogah D. Y., Webster O. W. (1987) in Recent Advances in Mechanistic and Synthetic Aspects of Polymerization, Fontanille M., Guyot A. (Eds), D. Reidel, Dordrecht, The Netherlands, p. 61.

Unali G. F., Armes S. P., Billingham N. C., Tuzar Z., Hamley I. W. (1999) Polym. Prepr. 40(1), 259.

Vamvakaki M., Billingham N. C., Armes S. P. (1998) Polymer 39, 2331.

Webster O. W. (1990) Makromol. Chem., Macromol. Symp. 33, 133.

Webster O. W., Anderson (1992) in New Methods for Polymer Synthesis, Mijs W. J. (Ed), Plenum Press, New York, , Chapter 1.

Webster O. W., Hertler W. R., Sogah D. Y., Farnham W. B., RajanBabu T. V. (1983) J. Am. Chem. Soc. 105, 5706.

Yu H., Choi W., Lim K., Choi S. (1988) Macromolecules 21, 2893.

CHAPTER 5

BLOCK COPOLYMERS BY RING OPENING METATHESIS POLYMERIZATION

Ring opening metathesis polymerization (ROMP) has emerged in recent years as a valuable tool for the polymerization of a wide variety of strained cyclic alkene monomers (Feast 1995). ROMP, which is a transition-metal-mediated polymerization technique, has been shown to proceed in a living manner if the transition metal initiator, coinitiator, and other experimental conditions are properly chosen. A characteristic example is the polymerization of norbornene with titanacyclobutane complexes (Gilliom 1986, Grubbs 1989). The metalcyclobutane is in equilibrium with its ring-opened carbene form (Scheme 5.1), which is the actual polymerizing form of the initiator. Propagation proceeds in the absence of any deleterious side reactions until the monomer is completely consumed. Polymerization can be terminated by adding a ketone or an aldehyde in order to deactivate the metal site. Other types of initiators based on tungsten and molybdenum complexes have been reported (Kress 1983, Schrock 1990 and 1988). Ruthenium-based complexes have also been used as ROMP initiators (Nguyen 1992, Wu 1993). Some of these complexes require the presence of a Lewis acid as a cocatalyst for increased activity (e.g., $W(CH\text{-}t\text{-}Bu)(OCH_2\text{-}t\text{-}Bu)_2X_2$, where $X = \text{halide}$). Their reactivity can be controlled to a certain extent by the nature of the ligands (Feldman 1991). For example, when OAryl ligands are used, the more electron withdrawing phenoxides usually give more active W-based catalysts (Quignard 1986). The chemical nature of the ligands (and the solvent) can also play a significant role in the microstructure of the final polymer (Bazan 1990, Feast 1992).

A lot of efforts have been made in order to prepare catalysts that are more tolerant of a number of functions present in the monomer. Most of these catalysts are based on molybdenum (Bazzan 1990 and 1991). The nature of the solvent and its ability to complex to the metal center also play some role. For instance, the

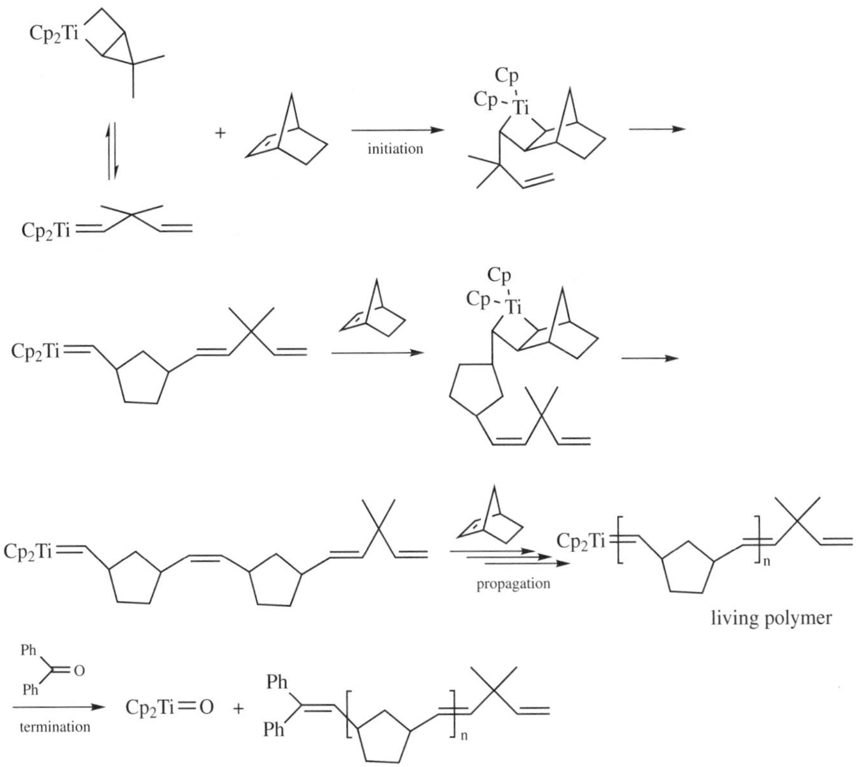

Scheme 5.1

polymerization of 5-cyanonorbornene is not possible using Mo(CH-t-Bu)(NAr)(O-t-Bu)$_2$ in toluene, a noncoordinating solvent, but polymerization proceeds in a controllable way in THF (Scheme 5.2). Polymerization of functional monomers can be affected by the position of the functional group in the monomer, by influencing the rate of formation and stability of the metallacycle intermediate (Schrock 1990). The nature of the functional group on norbornene monomers can alter the relative

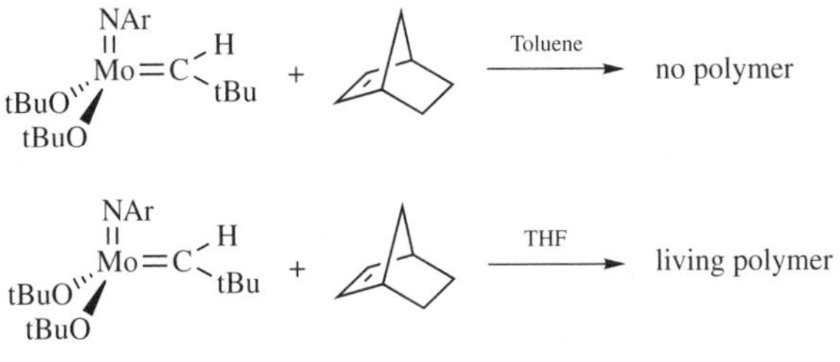

Scheme 5.2

rates of initiation and propagation reactions. It has been found that, for monosubstituted norbornenes, the rate of propagation is higher than the rate of initiation, which can cause broad molecular weight distributions. However, the rates can be modified by the addition of compounds that bind to a greater extent to the propagating alkylidene complex, like trimethylphosphine or quinuclidine, thus slowing down the propagation reaction (Schlund 1989, Wu 1992).

Because a number of substituted and nonsubstituted strained cycloalkene monomers can be polymerized in a living manner by ROMP, the synthesis of a variety of block copolymers based on these monomers can be accomplished.

1. SYNTHESIS OF AB DIBLOCK COPOLYMERS

Synthesis of diblock copolymers by ROMP is possible through two-step sequential addition of monomers. The course of the copolymerization reactions can be followed by ^{1}HNMR in many systems. Characteristic propagating alkylidene resonance peaks are seen in the ^{1}HNMR spectrum during the polymerization of the first monomer. After the complete consumption of the first monomer, the second monomer is added, and the ^{1}HNMR peaks, characteristic of the propagating alkylidene species of the second monomer, can be observed (Feast 1994).

Diblock copolymers containing norbornene, benzonorbornadiene, 6-methylbenzonorbornadiene, and endo- and exo-dicyclopentadiene were synthesized by Grubbs et al. (Cannizzo 1988). A titanacyclobutane initiator was used at temperatures ranging between 65°C and 79°C. The blocks had relatively short lengths, and the copolymers obtained exhibited narrow molecular weight distributions (Schemes 5.3a and 5.3b).

Schrock and coworkers (Sankaran 1991) reported the synthesis of diblock copolymers containing a polynorbornene block and a block comprised of an organometallic derivative of norbornane, namely Sn[2,3-trans-bis[(tert-butyl-

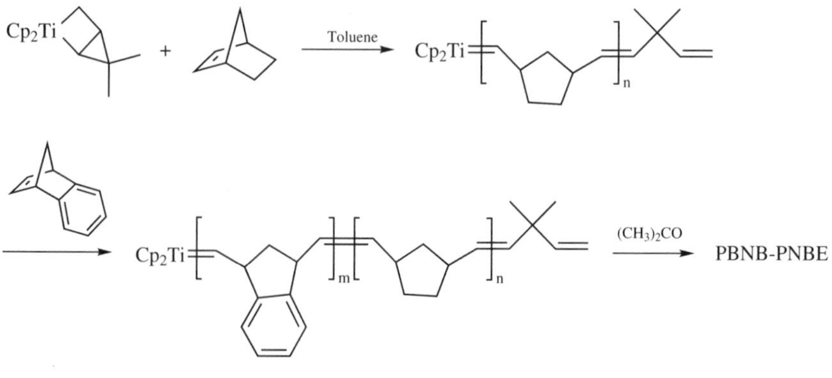

Cp_2Ti + Toluene → Cp_2Ti ... n

→ Cp_2Ti ... m ... n $\xrightarrow{(CH_3)_2CO}$ PBNB-PNBE

(*a*)

Scheme 5.3

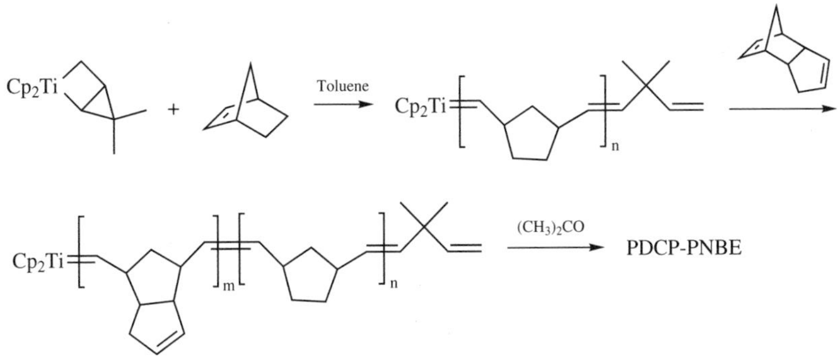

Toluene

$(CH_3)_2CO$ PDCP-PNBE

(*b*)

Scheme 5.3 (*Continued*)

amido)methyl]norborn-5-ene]Cl_2, using ROMP. The tungsten alkylidene complex, W(CH-t-Bu)(N-2,6-i-$Pr_2C_6H_3$)(O-t-Bu)$_2$ was employed as the initiator. Norbornene monomer was polymerized first followed by the polymerization of the organometallic monomer. The polymerization reaction was terminated by first adding trans-1,3-pentadiene, yielding a vinyl-terminated polymer and a new alkylidene complex, which was then terminated by benzophenone. This intermediate function before termination was necessary because the organometallic chain end was found to react rapidly with aldehydes giving a complex mixture of products (Scheme 5.4).

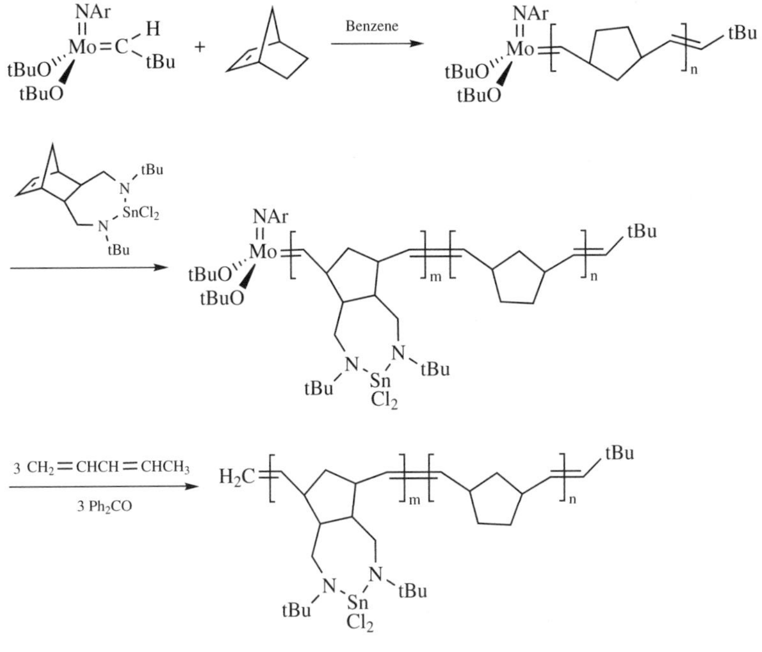

Benzene

3 CH_2=CHCH=$CHCH_3$

3 Ph_2CO

Scheme 5.4

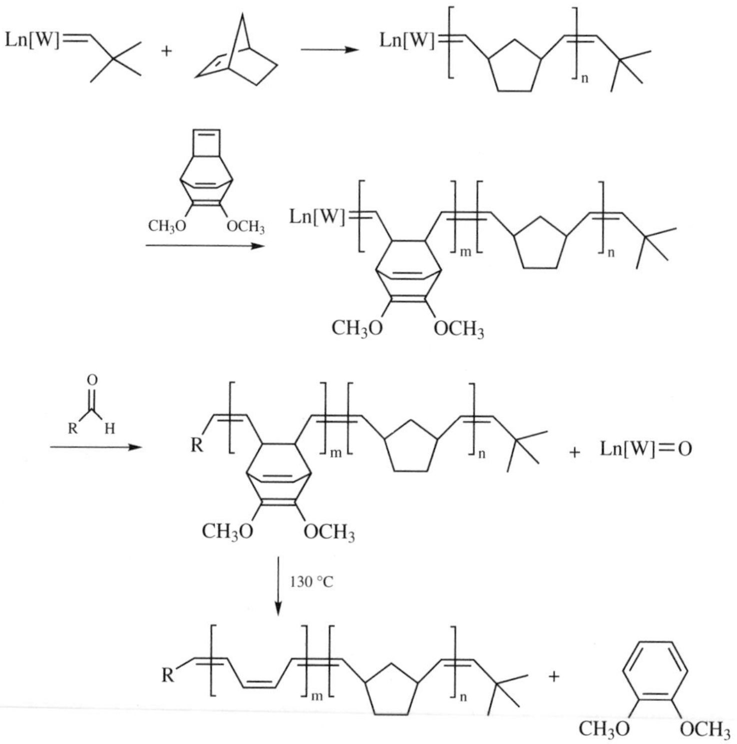

Scheme 5.5

ROMP was also used for the preparation of block copolymers containing a norbornene or a 5-((trimethylsiloxy)methyl)-norbornene first block and a second block consisting of p-dimethoxybenzotricyclo[4.2.2.0^{2,5}]deca-3,7,9-triene (Sauders 1991) (Scheme 5.5). The latter monomer is a polyacetylene precursor. A tungsten alkylidene complex was used as catalyst, and all polymerizations were performed in toluene. Well-defined polymers with narrow molecular distributions were obtained. The second block was converted to polyacetylene by heating the precursor polymer at 130°C for 10 min under inert atmosphere. The synthesis of diblock copolymers with a second block comprised of bis(trifluoromethyl)-tricyclo[4.2.2.0^{2,5}] deca-3,7,9-triene (Scheme 5.6) and benzotricyclo[4.2.2.0^{2,5}]deca-3,7,9-triene (Scheme 5.7) segments was also reported by the same group (Krause 1988).

Block copolymers having an amorphous and a side chain liquid crystalline block were prepared by Komiya and Schrock (Komiya 1993). The amorphous blocks were made of norbornene, 5-cyano-2-norbornene and methyltetracyclododecane, whereas the liquid crystalline blocks were made of n-[((4′-methoxy-4-biphenyl)yl)oxy]alkyl bicyclo[2.2.1]hept-2-ene-5-carboxylates (n = 3,6) monomers. The initiator Mo(CH-t-Bu)(N-2,6-C_6H_3-i-Pr_2)(O-t-$Bu)_2$ was used due to its tolerance of many different functional groups present in the monomer. THF was the polymerization solvent. Well-defined block copolymers with polydispersities

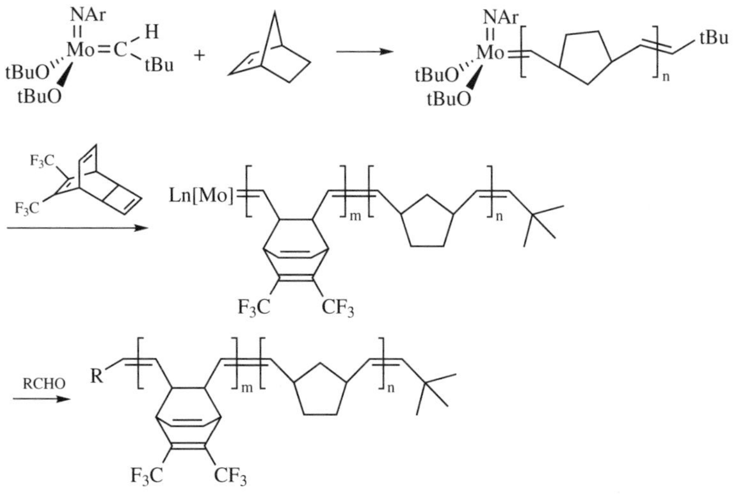

Scheme 5.6

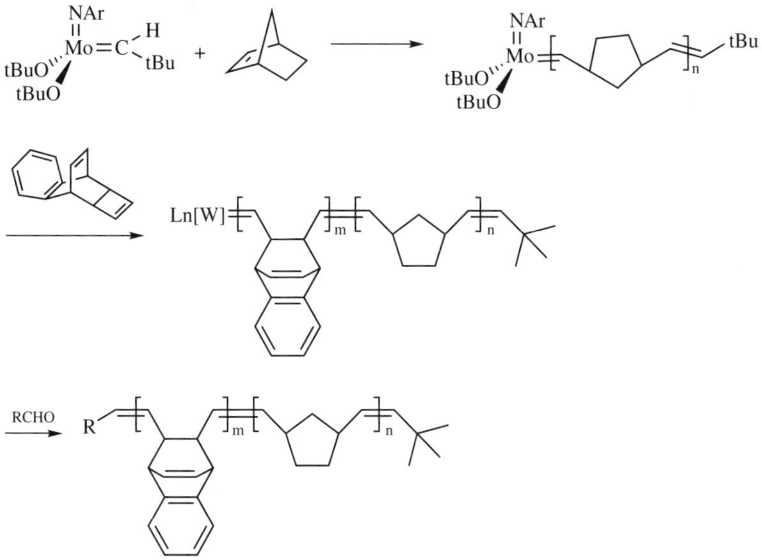

Scheme 5.7

ranging between 1.06 to 1.25 and amorphous/liquid crystalline segment ratios between 75/25 and 20/80 were obtained (Scheme 5.8).

The synthesis of block copolymers with blocks comprised of 5-(N-carbazolyl methylene)-2-norborne (CbzNB), a carbazole functionalized norbornene, and of a trimethylsilyl-protected alcohol functionalized norbornene, was recently reported

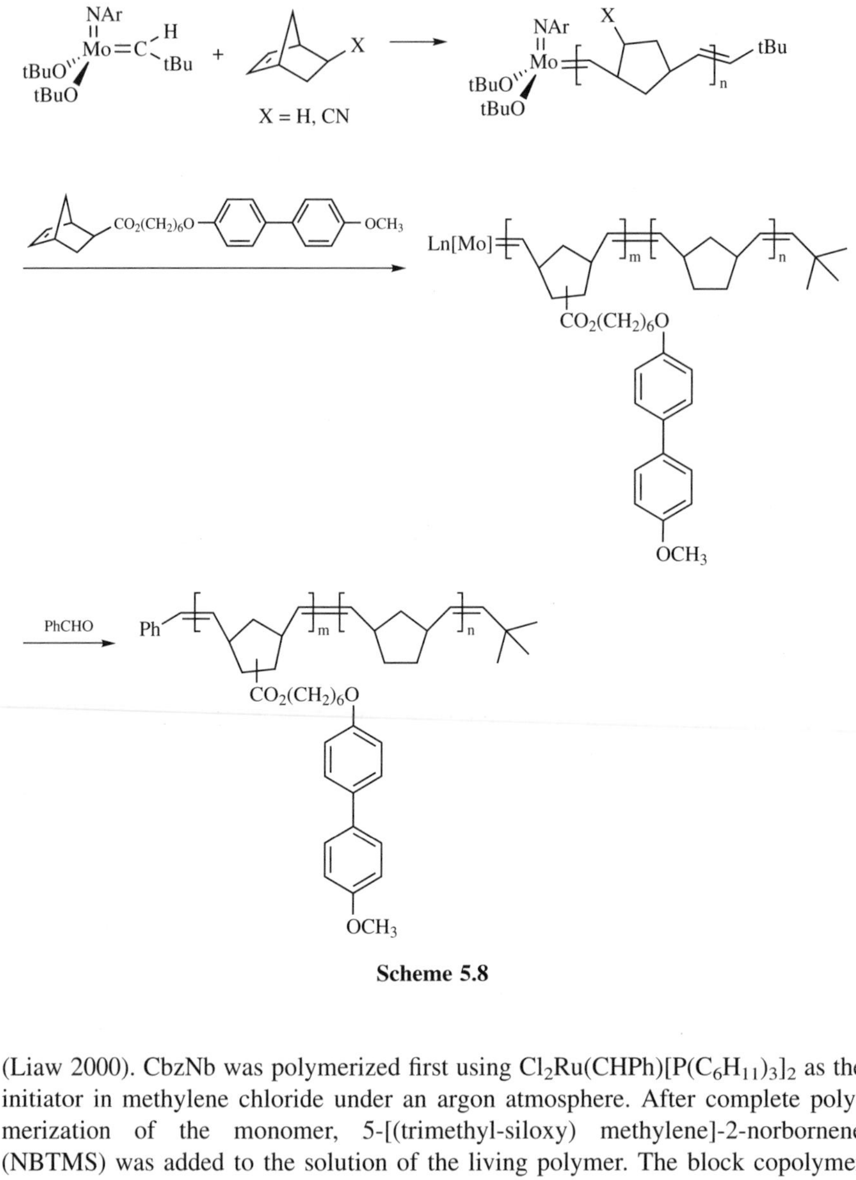

Scheme 5.8

(Liaw 2000). CbzNb was polymerized first using $Cl_2Ru(CHPh)[P(C_6H_{11})_3]_2$ as the initiator in methylene chloride under an argon atmosphere. After complete polymerization of the monomer, 5-[(trimethyl-siloxy) methylene]-2-norbornene (NBTMS) was added to the solution of the living polymer. The block copolymer obtained was hydrolyzed in THF using a small amount of HCl, giving the alcohol-functionalized norbornene block copolymer (Scheme 5.9).

A number of diblock copolymers containing fluorinated cyclic olefin monomers have been synthesized due to recent advances in the synthesis of well-defined ROMP initiators (Feast 1994). The most commonly used catalyst for the preparation of this family of materials is the Mo complex Mo(CH-t-Bu)(NAr)$(O\text{-}t\text{-}Bu)_2$. Some of the copolymers produced by block copolymerization of fluorine containing monomers are shown in Scheme 5.10.

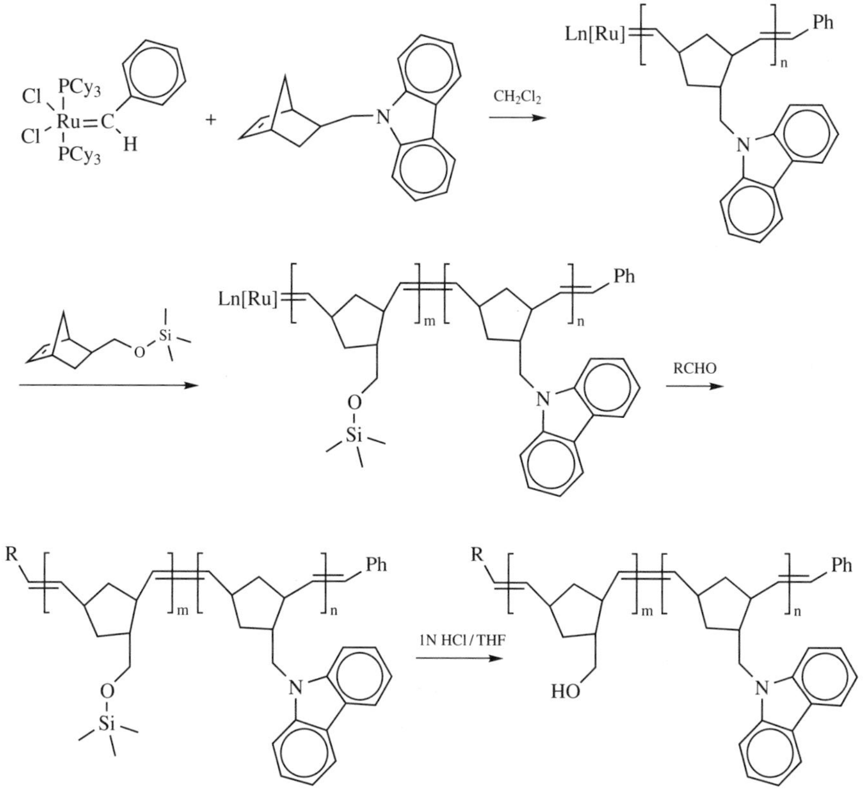

Scheme 5.9

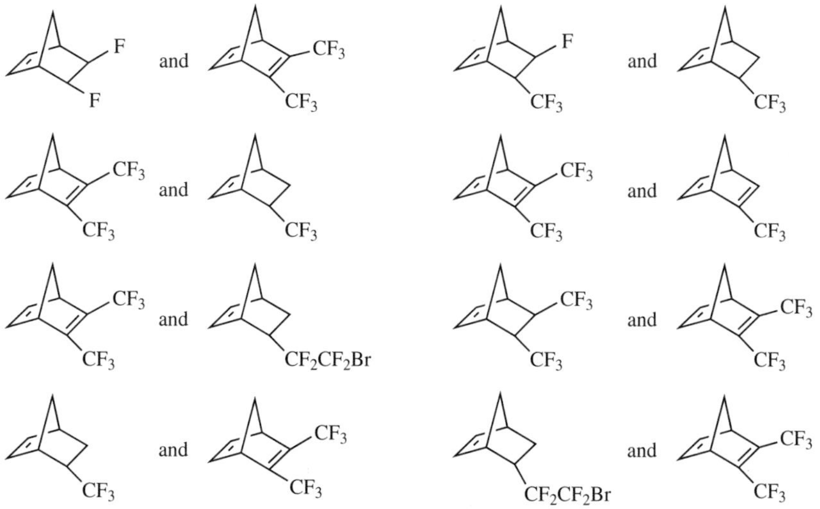

Scheme 5.10

2. SYNTHESIS OF ABA TRIBLOCK COPOLYMERS

Linear triblock copolymers have been synthesized by ring opening metathesis polymerization. The most commonly employed synthetic scheme is three-step addition of monomers. Difunctional (and tetrafunctional) initiators were also reported (Risse 1989).

Schrock et al. have synthesized ABA block copolymers, with A being polynorbornene and B being poly{7,8-bis(trifluoromethyl)-tricyclo[4.2.2.$0^{2,5}$]-deca-3,7,9-triene by sequential monomer addition (Krouse 1988). This was accomplished due to the ability of the two monomers to be copolymerized in a controllable living reaction scheme irrespective of the order of monomer addition in the polymerization mixture. $Mo(CHCMe_2)(NAr)(OCMe_3)_2$ was used as initiator in toluene, giving polymers with narrow molecular weight distributions. Heating the final block copolymers at 90°C for about 60 min resulted in polynorbornene-polyene-polynorbornene triblock copolymers. Under these conditions the $o\text{-}C_6H_4(CF_3)_2$ group is eliminated, giving a polyene chain (Scheme 5.11).

ABA block copolymers synthesized by three-step sequential monomer addition were also reported by Grubbs et al. (Cannizzo 1988). Triblock copolymers

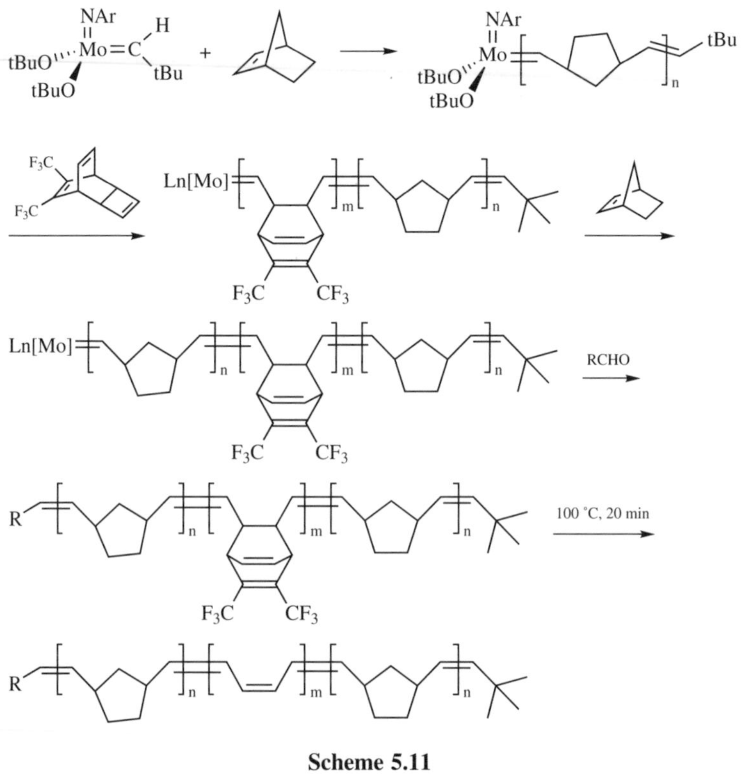

NAr
tBuO Mo=C H tBu
tBuO
+
NAr
Mo
tBuO
tBuO
tBu
n

F3C
F3C
Ln[Mo]
m
n
F3C CF3

Ln[Mo]
n
m
n
F3C CF3
RCHO

R
n
m
n
F3C CF3
100 °C, 20 min

R
n
m
n

Scheme 5.11

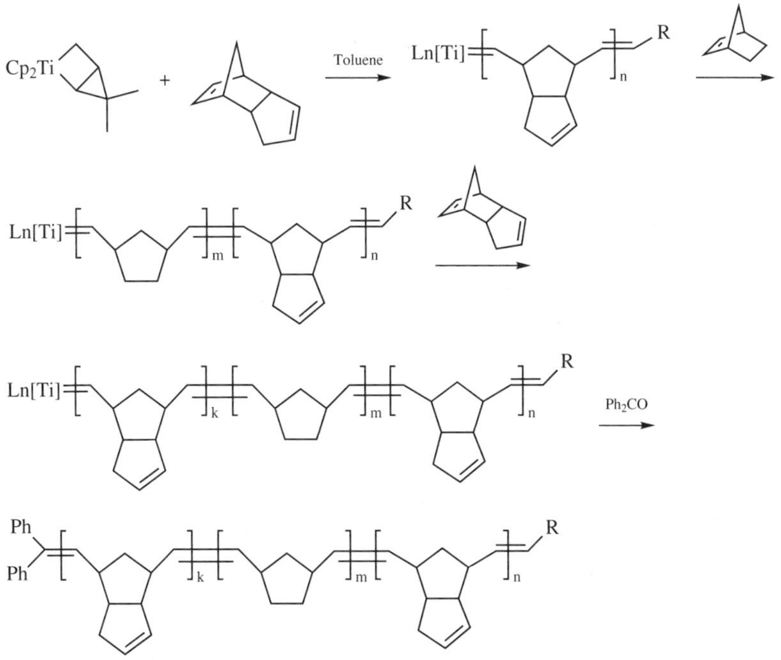

Cp_2Ti + Toluene → Ln[Ti] ... n R →

Ln[Ti] ... m ... n R →

Ln[Ti] ... k ... m ... n R Ph_2CO →

Ph Ph ... k ... m ... n R

(*a*)

Cp_2Ti + Toluene → Ln[Ti] ... n R →

Ln[Ti] ... m ... n R →

Ln[Ti] ... k ... m ... n R Ph_2CO →

Ph Ph ... k ... m ... n R

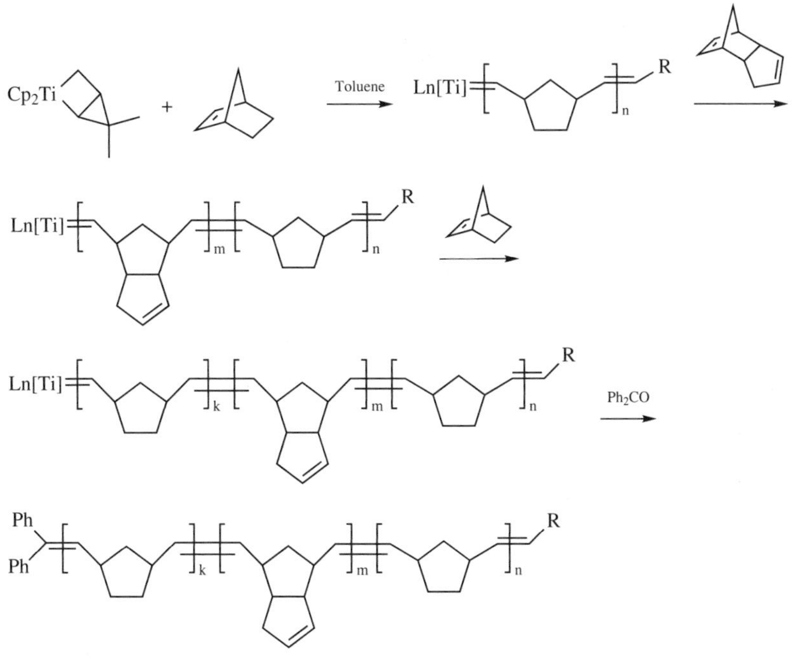

(*b*)

Scheme 5.12

containing norbornene, dicyclopentadiene, benzonorbornadiene, and 6-methylbenzonorbornadiene were synthesized using titanacyclobutane complexes as initiators (Scheme 5.12).

REFERENCES

Bazan G., Cho N-H., Schrock R. R., Gibson V. C. (1991) Macromolecules 24, 495.

Bazan G., Schrock R. R., Khosrani E., Feast W. J., Gibson V. C., O'Reagan M. B., Thomas J. K., Davis W. M. (1990) J. Am. Chem. Soc. 112, 8387.

Cannizzo L. F., Grubbs R. H. (1988) Macromolecules 21, 1961.

Feast W. J., Gibson V. C., Khosravi E., Marshall E. L. (1994) J. Chem. Soc. Chem. Commun. 9.

Feast W. J., Gibson V. C., Marshall E. L. (1992) J. Chem. Soc., Chem. Commun. 1157.

Feast W. J., Khosravi E. (1995) in New Methods in Polymer Synthesis, Chapman & Hall, Vol. 2, Chapter 3.

Feldman J., Schrock R. R. (1991) Prog. Inorg. Chem. 1, 39.

Gilliom L. R., Grubbs R. H. (1986) J. Am. Chem. Soc. 108, 733.

Grubbs R. H., Tumas W. (1989) Science 243, 907.

Komiya Z., Schrock R. R. (1993) Macromolecules 26, 1387.

Kress J., Osborn J. A. (1983) J. Am. Chem. Soc. 105, 6346.

Krouse S. A., Schrock R. R. (1988) Macromolecules 21, 1885.

Liaw D-J., Tsai C-H. (2000) Polymer 41, 2773.

Nguyen S-B. T., Jonhson L. K., Grubbs R. H. (1992) J. Am. Chem. Soc. 114, 3974.

Quignard F., Leconte M., Basset J-M. (1986) Mol. Catal. 36, 13.

Risse W., Grubbs R. H. (1991) J. Mol. Cat. 65, 211.

Risse W., Wheeler D. R., Cannizzo L. F., Grubbs R. H. (1989) Macromolecules 22, 3205.

Sankaran V., Cohen R. E., Cummins C. C., Schrock R. R. (1991) Macromolecules 24, 6664.

Saunders R. S., Cohen R. E., Schrock R. R. (1991) Macromolecules 24, 5599.

Schlund R., Schrock R. R., Crowe W. E. (1989) J. Am. Chem. Soc. 111, 8004.

Schrock R. R. (1990) Acc. Chem. Res. 23, 158.

Schrock R. R., DePue R. T., Feldman J., Schaverien C. J., Dewan J. C., Liu A. H. (1988) J. Am. Chem. Soc. 110, 1423.

Schrock R. R., Murdzek J. S., Bazan G. C., Robbins J., DiMare M., O'Regan M. (1990) J. Am. Chem. Soc. 112, 3875.

Wu Z., Benedicto A. D., Grubbs R. H. (1993) Macromolecules 26, 4975.

Wu Z., Wheeler D. R., Grubbs R. H. (1992) J. Am. Chem. Soc. 114, 146.

CHAPTER 6

SYNTHESIS OF BLOCK COPOLYMERS BY A COMBINATION OF DIFFERENT POLYMERIZATION METHODS

The well-documented fact that not all monomers can be polymerized by every available polymerization mechanism limits the possible combinations of monomers that can be employed in a block copolymer chain, if only one polymerization method is used. However, synthetic schemes have been devised that allow the polymerization mechanism to be changed appropriately in order to suit the monomers that are going to be incorporated in the same macromolecule. These polymerization processes widen considerably the variety of block copolymers that can be produced.

Active center transformation reactions usually involve the efficient end functionalization of a polymer chain that is going to comprise one of the blocks of the final copolymer. This concept can be also described as the use of a suitable macromolecular initiator for the polymerization of the second monomer. Transformation of one chain-end followed by the polymerization of the second monomer produces a diblock copolymer, whereas the transformation of both ends of a linear polymer chain results in the formation of an ABA triblock copolymer (Scheme 6.1). Transformation reactions must be 100% efficient because any unfunctionalized macromolecular chains will not initiate the polymerization of the second monomer and will be present in the final product as a homopolymer impurity. Transformation of chain-ends may involve several steps. The preparation of well-defined block copolymers requires that every individual step, including the separate polymerization reactions, must be highly efficient and finely controlled. It is obvious that these synthetic strategies involve special care in the design and the application of the chosen synthetic methodology.

Several transformation mechanisms have been employed so far for the preparation of block copolymers. Some of them are outlined in this chapter.

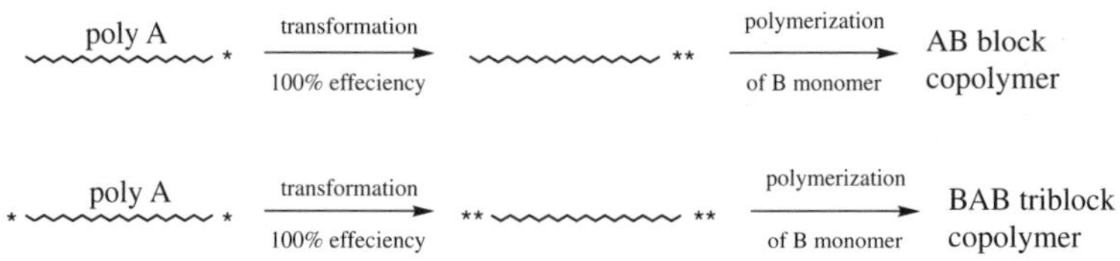

poly A ~~~~ * —transformation, 100% effeciency→ ~~~~ ** —polymerization of B monomer→ AB block copolymer

* ~~~~ poly A ~~~~ * —transformation, 100% effeciency→ ** ~~~~ ** —polymerization of B monomer→ BAB triblock copolymer

Scheme 6.1

1. SYNTHESIS OF BLOCK COPOLYMERS BY ANIONIC TO CATIONIC MECHANISM TRANSFORMATION

The synthesis of polystyrene-b-polytetrahydrofuran block copolymers is a typical example of transformation of active anionic centers to cationic ones. Living PSLi is transformed to ω-bromopolystyrene. The best way to produce this functional polymer is intermediate formation of the macromolecular Grignard reagent and subsequent reaction with Br_2, according to Scheme 6.2 (Burgess 1977). An

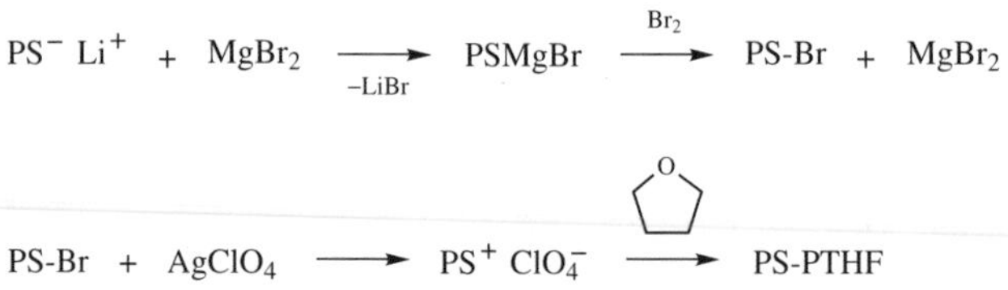

$$PS^- Li^+ + MgBr_2 \xrightarrow[-LiBr]{} PSMgBr \xrightarrow{Br_2} PS\text{-}Br + MgBr_2$$

$$PS\text{-}Br + AgClO_4 \longrightarrow PS^+ ClO_4^- \xrightarrow{\text{THF}} PS\text{-}PTHF$$

Scheme 6.2

alternative route is functionalization of PSLi with excess phosgene (Franta 1976). The terminal halogen atom can be activated using silver salts, e.g., $AgSbF_6$, $AgClO_4$, $AgPF_6$, generating a carbocation. This cationic active center is used for the polymerization of THF, forming the poly(tetramethylene oxide) block.

In another case, difunctional living polybutadiene, produced by the use of a difunctional initiator, is reacted with excess EO giving an α,ω-polybutadiene diol (which is now commercially available) (Scheme 6.3). The OH groups are converted

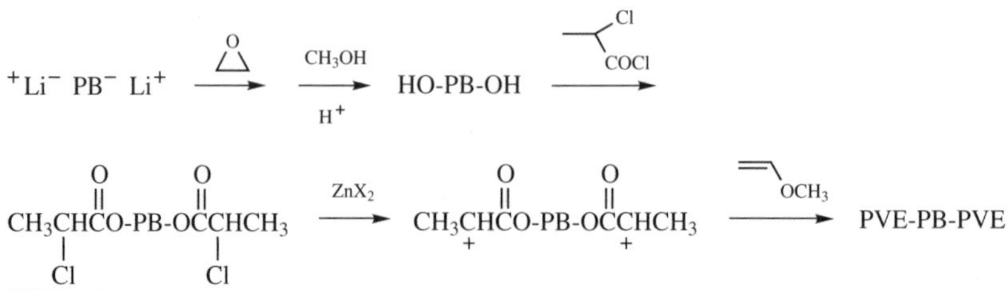

$$^+Li^- PB^- Li^+ \xrightarrow{\text{EO}} \xrightarrow[H^+]{CH_3OH} HO\text{-}PB\text{-}OH \xrightarrow{CH_3CH(Cl)COCl}$$

$$CH_3CH(Cl)CO\text{-}PB\text{-}OCCH(Cl)CH_3 \xrightarrow{ZnX_2} CH_3\overset{+}{C}HCO\text{-}PB\text{-}OC\overset{+}{C}HCH_3 \xrightarrow{CH_2=CHOCH_3} PVE\text{-}PB\text{-}PVE$$

Scheme 6.3

to haloether groups, and the new difunctional polymer is used as a macromolecular initiator for the cationic polymerization of vinyl ethers in the presence of ZnX_2 (Cramail 1992).

In another case, poly(dimethylsiloxane-b-2-ethyl-2-oxazoline) block copolymer was synthesized in two steps (Liu 1993) (Scheme 6.4). The first step involved the

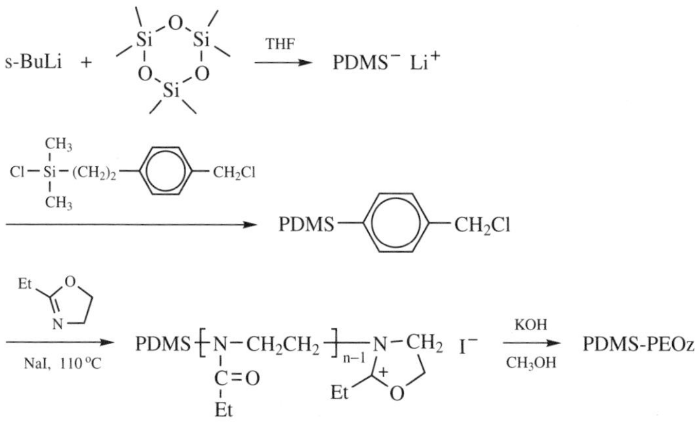

s-BuLi + (hexamethylcyclotrisiloxane) $\xrightarrow{\text{THF}}$ $PDMS^- \ Li^+$

$\xrightarrow{Cl-Si(CH_3)_2-(CH_2)_2-C_6H_4-CH_2Cl}$ PDMS–C_6H_4–CH_2Cl

$\xrightarrow[\text{NaI, 110 °C}]{\text{Et-oxazoline}}$ PDMS$\left[N(C(=O)Et)-CH_2CH_2\right]_{n-1}$–$N^+$(oxazolinium, Et)–$CH_2$ I^- $\xrightarrow[CH_3OH]{KOH}$ PDMS-PEOz

Scheme 6.4

ring opening anionic polymerization of dimethylsiloxane in THF with s-BuLi as the initiator. The living PDMS was subsequently reacted with a benzyl chloride containing chlorosilane-terminating agent, giving the monofunctional-benzylchloride-terminated PDMS. After purification, the end-functionalized PDMS was used as the macroinitiator for the ring-opening-cationic polymerization of 2-ethyl-2-oxazoline (EOz) in the presence of NaI in chlorobenzene at 110°C. A number of copolymers having narrow molecular weight distributions and a range of compositions were synthesized.

Poly(ethylene glycol-b-ethylenimine) copolymers were prepared by the anionic-ring-opening polymerization of ethylene oxide in THF with potassium 3,3-diethoxypropanolate (Akiyama 2000). The living chains were end-capped with methanesulfonyl chloride, giving the heterotelechelic acetal-PEG-SO_2CH_3. This was used as a macroinitiator for the cationic-ring-opening polymerization of 2-methyl-2-oxazoline in nitromethane at 60°C. Alkaline hydrolysis of the amide group in the POZ block to a secondary amino group, by addition of NaOH, resulted in the desired block copolymers (Scheme 6.5).

PS-PMMA-PS triblock copolymers were synthesized by transformation of polymerization mechanism from anionic to cationic (Neubauer 1997). MMA was polymerized first using a difunctional initiator. The anionic sites were converted to cationic by reaction with 1,4-bis-(1-bromoethyl)benzene, and the PS blocks were formed by cationic polymerization of styrene (Scheme 6.6). In an alternative way, a PS-PMMA diblock copolymer was synthesized first by anionic polymerization, the living ends were transformed to cationic using the same functionalization reaction as above, followed by the cationic polymerization of styrene.

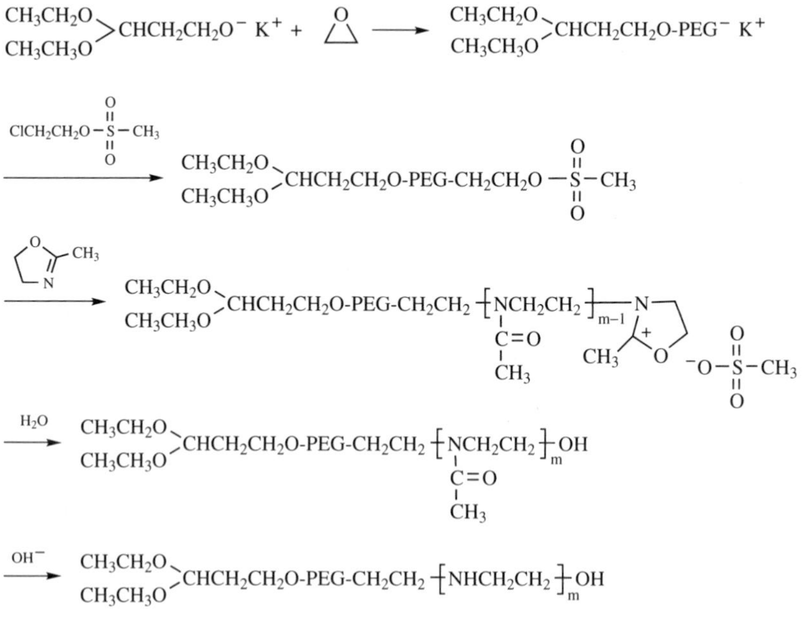

$$(CH_3CH_2O)(CH_3CH_3O)CHCH_2CH_2O^- K^+ + \text{(ethylene oxide)} \longrightarrow (CH_3CH_2O)(CH_3CH_3O)CHCH_2CH_2O\text{-PEG}^- K^+$$

$$\xrightarrow{ClCH_2CH_2O-SO_2-CH_3} (CH_3CH_2O)(CH_3CH_3O)CHCH_2CH_2O\text{-PEG-}CH_2CH_2O-SO_2-CH_3$$

$$\xrightarrow{\text{2-methyl-2-oxazoline}} (CH_3CH_2O)(CH_3CH_3O)CHCH_2CH_2O\text{-PEG-}CH_2CH_2\text{-}[N(C{=}O)CH_3\text{-}CH_2CH_2]_{m-1}\text{-}N^+(\text{oxazolinium, }CH_3)\ \ {}^-O-SO_2-CH_3$$

$$\xrightarrow{H_2O} (CH_3CH_2O)(CH_3CH_3O)CHCH_2CH_2O\text{-PEG-}CH_2CH_2\text{-}[N(C{=}O)CH_3\text{-}CH_2CH_2]_m\text{-}OH$$

$$\xrightarrow{OH^-} (CH_3CH_2O)(CH_3CH_3O)CHCH_2CH_2O\text{-PEG-}CH_2CH_2\text{-}[NHCH_2CH_2]_m\text{-}OH$$

Scheme 6.5

$$Li^+\ {}^-D^-\ Li^+ + CH_2{=}C(CH_3)COOCH_3 \longrightarrow Li^+\ {}^-PMMA^-\ Li^+ \xrightarrow{CH_3CH_2CH(Br)\text{-}C_6H_4\text{-}CH(Br)CH_2CH_3}$$

$$CH_2CH_2(Br)\text{-}C_6H_4\text{-}CH_2(CH_3)\text{-PMMA-}CH_2(CH_3)\text{-}C_6H_4\text{-}CH(Br)CH_2CH_3 \xrightarrow[\text{styrene}]{ZnBr_2} \xrightarrow{CH_3OH} \text{PS-PMMA-PS}$$

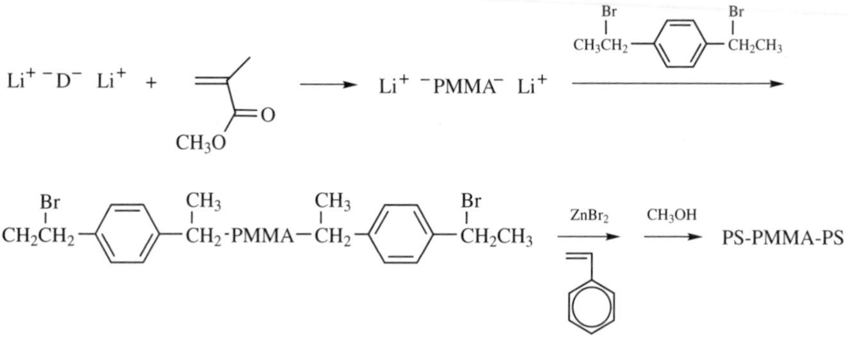

Scheme 6.6

2. SYNTHESIS OF BLOCK COPOLYMERS BY ANIONIC TO LIVING FREE RADICAL MECHANISM TRANSFORMATION

Recently, the transformation of a living anionic center to a functional group containing TEMPO was reported (Kobatake 1997 and 1999). The reaction scheme involved the reaction of a suitably modified TEMPO derivative with a living polymer prepared by anionic polymerization. The functionalization reaction has been shown to be highly efficient, and the formation of block copolymers was accomplished by controlled free radical polymerization of styrene in the second step (Scheme 6.7a). In a similar way (Yoshida 1994), PSLi was reacted with 1-oxo-4-methoxy-2,2,6,6-tetramethylpiperidine at −78°C in THF in order to produce a TEMPO-terminated PS macroinitiator, which was used for the polymerization of methyl, ethyl, and butyl acrylates to give the corresponding block copolymers (Scheme 6.7b).

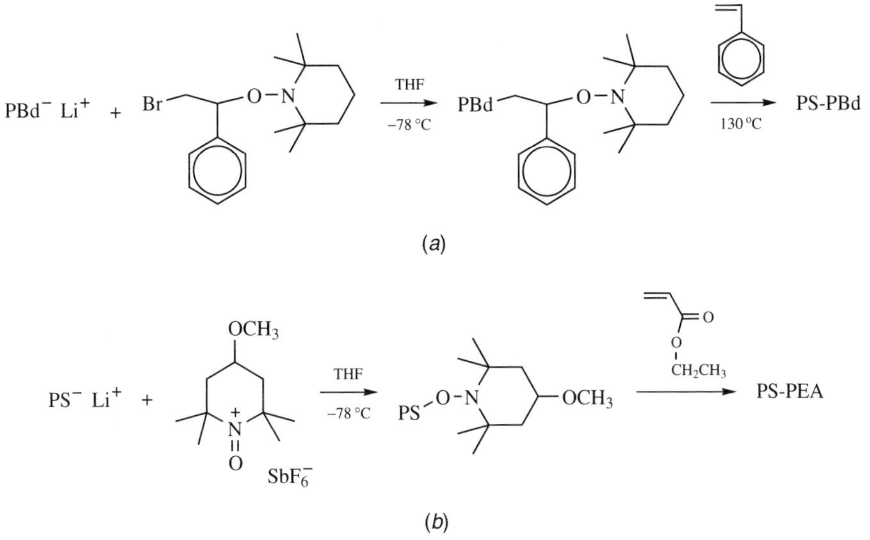

PBd⁻ Li⁺ + Br–CH₂–CH(Ph)–O–N(TEMP) —THF, −78 °C→ PBd–CH₂–CH(Ph)–O–N(TEMP) —styrene, 130 °C→ PS-PBd

(*a*)

PS⁻ Li⁺ + 4-methoxy-2,2,6,6-tetramethylpiperidine-1-oxoammonium SbF_6^- —THF, −78 °C→ PS–O–N(TEMP)–OCH₃ —$CH_2{=}CHCOOCH_2CH_3$→ PS-PEA

(*b*)

Scheme 6.7

Anionically polymerized styrene- and diene-based living polymers were terminated with ethylene oxide, giving hydroxyl end-functionalized homopolymers. The terminal OH groups were converted to 2-bromoisobutyroxy groups, which were subsequently used as initiating sites for the ATRP polymerization of (meth)acrylates (Scheme 6.8). In another case living poly(dimethyl siloxane) prepared by

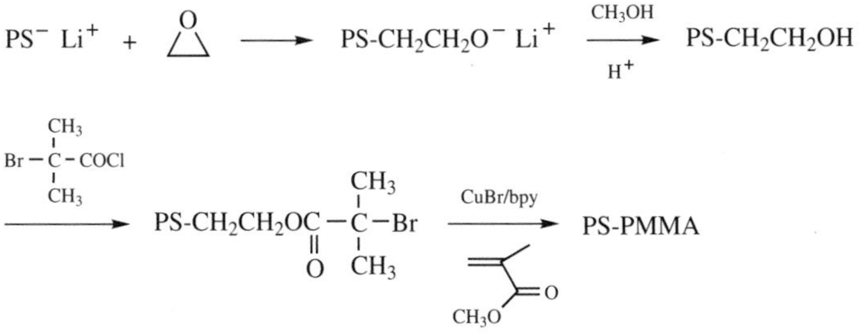

$$PS^- Li^+ + \text{ethylene oxide} \longrightarrow PS\text{-}CH_2CH_2O^- Li^+ \xrightarrow[H^+]{CH_3OH} PS\text{-}CH_2CH_2OH$$

$$\xrightarrow{Br-C(CH_3)_2-COCl} PS\text{-}CH_2CH_2OC(=O)-C(CH_3)_2-Br \xrightarrow[CH_2=C(CH_3)COOCH_3]{CuBr/bpy} PS\text{-}PMMA$$

Scheme 6.8

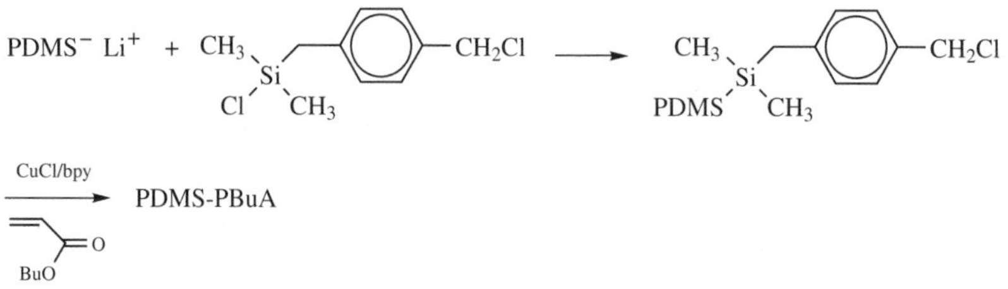

Scheme 6.9

anionic polymerization was terminated with $ClSiMe_2CH_2PhCH_2Cl$, producing the chlorobenzyl-terminated analogue. The terminal group was used for the polymerization of styrene and acrylate monomers, resulting in well-defined block copolymers (Scheme 6.9) (Matyjaszewski 1998b).

Block copolymers of isoprene and styrene bearing a fluorescent probe at the junction point were synthesized by an anionic-to-living free radical polymerization mechanism transformation (Tong 2000). First, the polymerization of isoprene was realized anionically in cyclohexane using s-BuLi as the initiator. Reaction of the living chains with the dye derivative 1-(9-phenanthryl)-1-phenylethylene resulted in the dye end-functionalized PI block, which was subsequently reacted with excess dibromo-p-xylene to afford a Br terminated PI chain. This end-functionalized polymer was used, in conjuction with CuBr/bpy, to give the second PS block through atom transfer radical polymerization (Scheme 6.10).

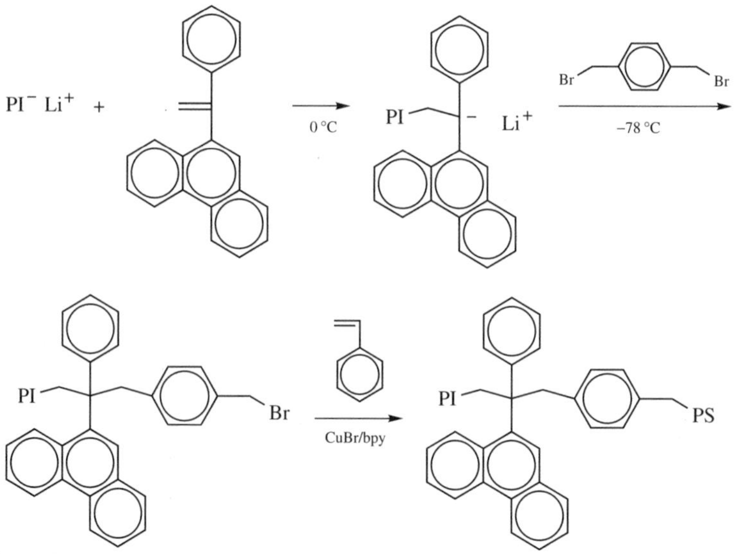

Scheme 6.10

3. SYNTHESIS OF BLOCK COPOLYMERS BY CATIONIC TO ANIONIC MECHANISM TRANSFORMATION

Hydroxyl terminated poly(isobutyl vinyl ether) (PIBVE) prepared by terminating the living cations with K_2CO_3/H_2O followed by $NaBH_4$ was used as an anionic macromolecular initiator, in the presence of stannous octanoate as the catalyst, for the polymerization of ε-caprolactone (Verma 1992). In this way well-defined block copolymers of poly(vinyl ether)-b-poly(ε-caprolactone) were synthesized (Scheme 6.11).

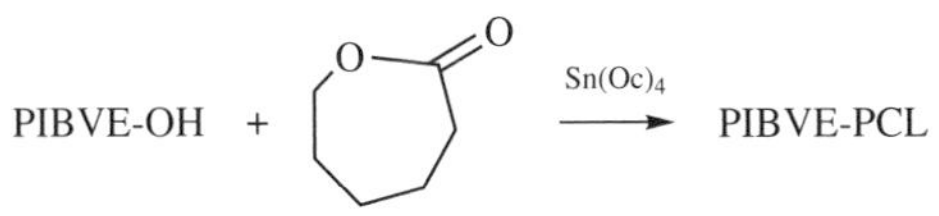

PIBVE-OH + O O $\xrightarrow{Sn(Oc)_4}$ PIBVE-PCL

Scheme 6.11

In another example the terminal chlorine atom of a polyisobutylene chain, prepared by a typical cationic polymerization procedure, was converted to the benzyl anion, which, in turn, was able to polymerize methyl methacrylate anionically (Kennedy 1991). Thus, polyisobutylene-b-poly(methyl methacrylate) diblock copolymers were produced. The living cationic polymerization of isobutylene results, even after quenching with methanol, in chlorine-terminated polymers. Therefore, no special end-functionalization reactions are needed in order to produce the terminal Cl functionality (Scheme 6.12).

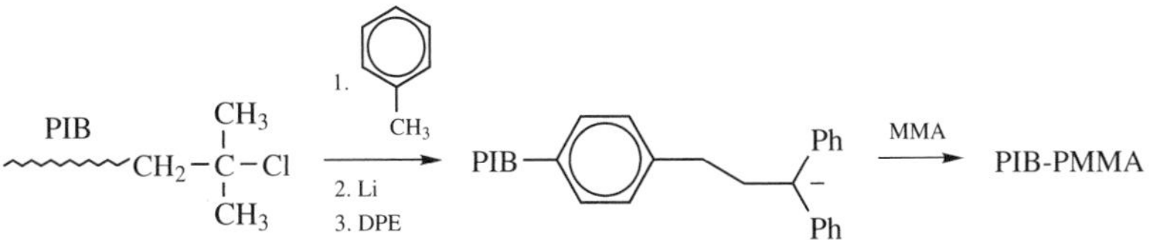

PIB $CH_2-C(CH_3)_2-Cl$ — 1. CH_3 2. Li 3. DPE → PIB — Ph Ph — MMA → PIB-PMMA

Scheme 6.12

Living PIB chains, produced by cationic polymerization of isobutylene, were end-capped with 1,1-diphenyl-1-methoxy (DPOMe) or 2,2-diphenylvinyl (DPV) terminal groups (Feldthusen 1998). These end groups were metallated quantitatively with K/Na alloy in THF. The resulting macroanionic initiators were used for the polymerization of tert-butyl methacrylate, producing well-defined block copolymers. Metallation of the DPOMe or DPV termini with Li dispersion and subsequent polymerization of MMA gave PIB-b-PMMA block copolymers. The use of difunctional linear PIB macroinitiators resulted in the formation of PMMA-b-PIB-b-PMMA triblock copolymers (Scheme 6.13).

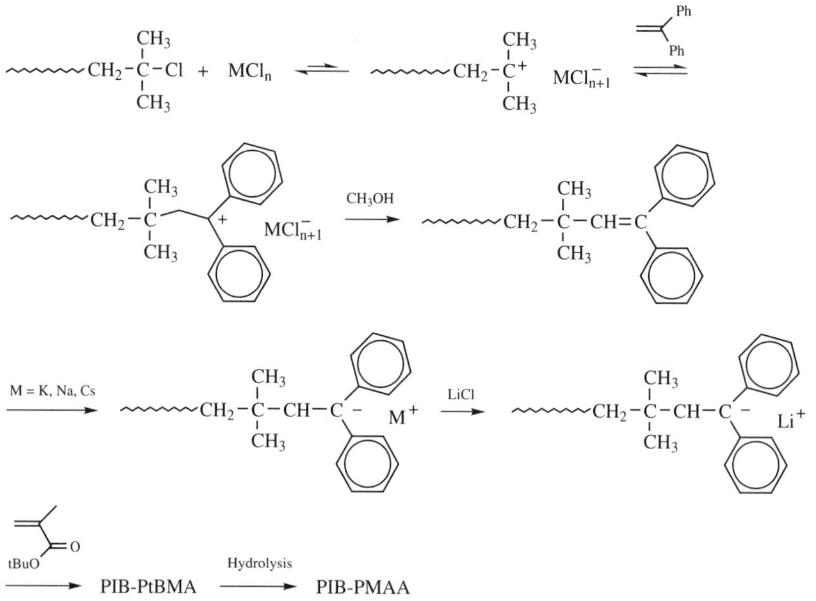

$$\sim\sim CH_2-C(CH_3)_2-Cl + MCl_n \rightleftharpoons \sim\sim CH_2-C^+(CH_3)_2 \; MCl_{n+1}^-$$

Ph, Ph

CH₃OH

M = K, Na, Cs

LiCl

M^+ Li^+

tBuO, O

$\longrightarrow$ PIB-PtBMA $\xrightarrow{\text{Hydrolysis}}$ PIB-PMAA

Scheme 6.13

4. SYNTHESIS OF BLOCK COPOLYMERS BY CATIONIC TO ONIUM MECHANISM TRANSFORMATION

The cationic-ring-opening polymerization of heterocyclic monomers is a well-established polymerization process. Direct sequential polymerization of vinyl and cyclic monomers is not possible due to the different initiating systems required in the two cases. However, block copolymers containing blocks composed of vinyl and cyclic monomers can be produced by cationic to onium transformation processes.

Poly(isobutyl vinyl ether), produced by cationic polymerization, can be end-functionalized with a chlorine atom by end-capping with 2-chloroethyl vinyl ether segment. The Cl can be converted into the more reactive iodide by reaction with sodium iodide. The terminal iodide can be appropriately activated to initiate the ring-opening cationic polymerization of 2-ethyl-2-oxazoline, giving PIBVE-b-poly(oxazoline) block copolymers (Liu 1993, Scheme 6.14).

The tertiary or benzylic-terminal chlorine atoms of polyisobutylene or poly(p-chlorostyrene), respectively, synthesized by cationic polymerization, using tertiary alkyl chlorides as the initiators, can be activated by silver salts in order to polymerize THF by a ring-opening cationic mechanism (Scheme 6.15a). Thus, PIB-PTHF (Kennedy 1990) and PpMeS-PTHF (Gadkari 1989) block copolymers were prepared.

The preparation of poly(oxetane-b-ε-caprolactone) via a cationic to anionic polymerization mechanism transformation has also been reported (Takeuchi 2000). The methodology involved the use of triflate complexes of bulky titanium bisphenolates as the moieties for initiating polymerization (Scheme 6.15b).

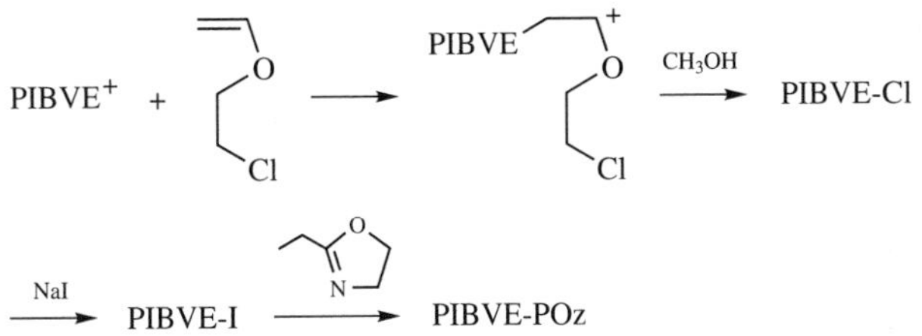

$PIBVE^+$ + O, Cl → PIBVE, +, O, Cl → (CH_3OH) PIBVE-Cl

→ (NaI) PIBVE-I → (O, N) PIBVE-POz

Scheme 6.14

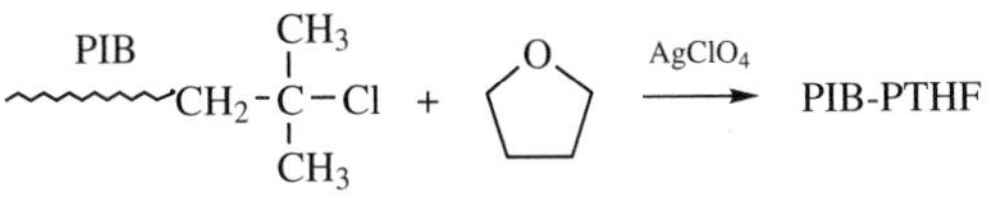

PIB ～CH_2-C(CH_3)$_2$-Cl + O → ($AgClO_4$) PIB-PTHF

(*a*)

R_1, R_2, O, Ti, Cl, Cl → (AgOTf) R_1, R_2, O, Ti, OTf, Cl → (O)

R_1 = tBu, Ph, H
R_2 = Me, H

R_1, R_2, O, Ti, Cl, $[O\sim\sim]_{n-1}$ O^+, ^-OTf → ($Q^+ X^-$) R_1, R_2, O, Ti, Cl, $[O\sim\sim]_n$ X

→ (O, O) R_1, R_2, O, Ti, Cl, $[O\sim\sim C(=O)]_m [O\sim\sim]_n$ X

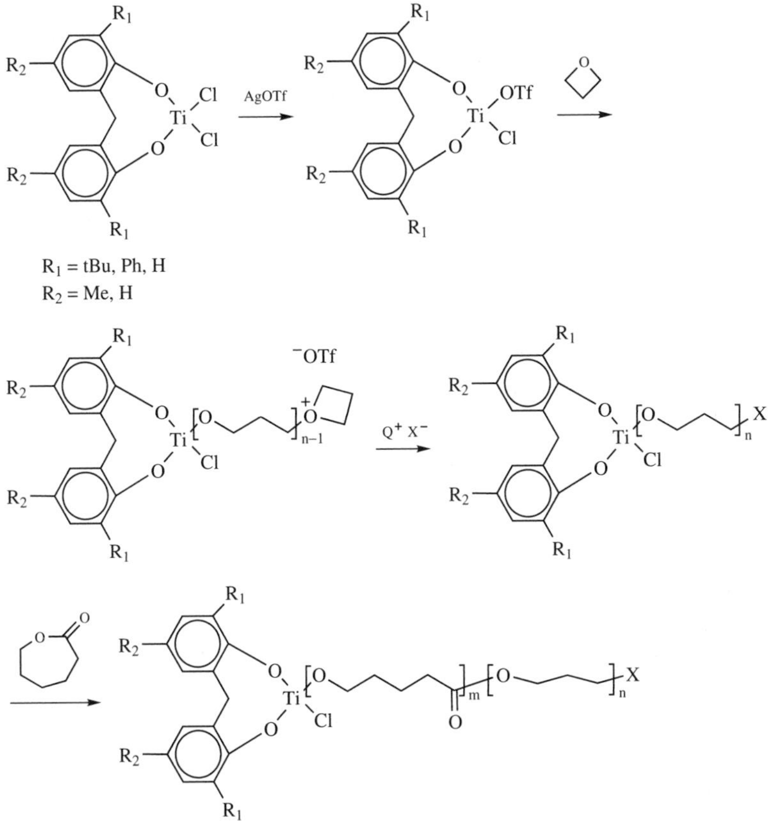

(*b*)

Scheme 6.15

5. SYNTHESIS OF BLOCK COPOLYMERS BY CATIONIC TO LIVING FREE RADICAL MECHANISM TRANSFORMATION

Polystyrene prepared by the 1-PhEtCl/$SnCl_4$ system in the presence of n-Bu_4NCl possesses a Cl-terminal atom. This end-functionalized polymer can be used directly for the atom transfer radical polymerization of methyl acrylate and methyl methacrylate in the presence of the catalytic complex CuCl/4,4′-di-(5-nonyl) 2,2′-bipyridine (dNbpy) (Coca 1997a). Well-defined PS-PMA and PS-PMMA block copolymers were obtained (Scheme 6.16).

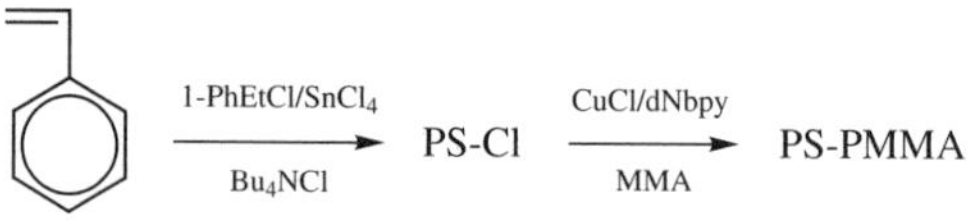

$$\text{styrene} \xrightarrow[\text{Bu}_4\text{NCl}]{\text{1-PhEtCl/SnCl}_4} \text{PS-Cl} \xrightarrow[\text{MMA}]{\text{CuCl/dNbpy}} \text{PS-PMMA}$$

Scheme 6.16

Polyisobutylene prepared by cationic polymerization, using a difunctional initiator, was end-capped with a few units of styrene. The resulting macromolecule had two terminal chlorine atoms that were used as initiating sites for the polymerization of styrene, methyl acrylate, isobornyl acrylate, and methyl methacrylate by an ATRP mechanism (Coca 1997b). Symmetric ABA triblock copolymers of the types PS-PIB-PS, PMA-PIB-PMA, PIBA-PIB-PIBA, and PMMA-PIB-PMMA were, thus, synthesized, in the presence of a CuCl /bipyridyl complex (Scheme 6.17). Molecular weight distributions were unimodal and narrow.

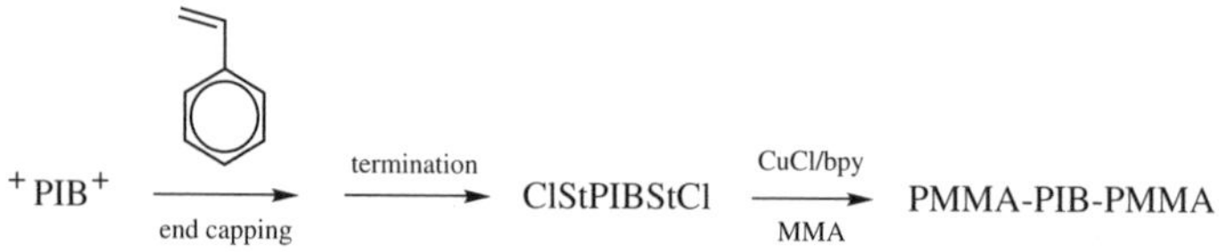

$$^{+}\text{PIB}^{+} \xrightarrow[\text{end capping}]{\text{styrene}} \xrightarrow{\text{termination}} \text{ClStPIBStCl} \xrightarrow[\text{MMA}]{\text{CuCl/bpy}} \text{PMMA-PIB-PMMA}$$

Scheme 6.17

The final triblock contained no homopolymer impurities from the starting macromolecular initiator, indicating high functionalization and polymerization efficiency.

Living cationic ring-opening polymerization of THF with bromopropionyl/silver triflate system, leads, after termination of the polymerization with H_2O, to the formation of a bifunctional polytetrahydrofuran chain having bromo and hydroxyl end groups. The terminal bromine atom of this macromolecule was used for the ATRP of styrene and methyl methacrylate to give PS-PTHF and PMMA-PTHF diblock copolymers with a hydroxyl end group (Matyjaszewski 1998, Scheme 6.18).

When THF polymerization was carried out with the difunctional initiator $(Tf)_2O$, a difunctional PTHF chain was formed. The living ends were reacted with sodium

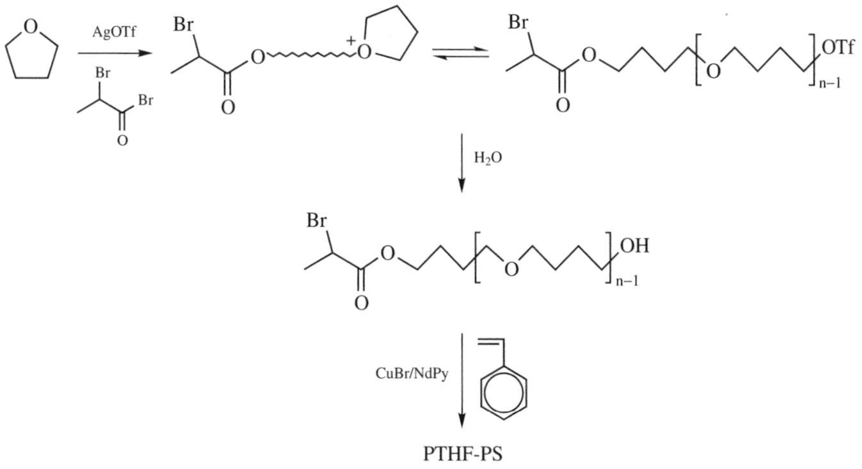

PTHF-PS

Scheme 6.18

bromopropionate in order to generate a bromodifunctional chain. These functions served as initiating sites for the ATRP polymerization of MMA, giving PMMA-PTHF-PMMA triblock copolymers (Scheme 6.19). Similar attempts were made for the ATRP of styrene and methyl acrylate with less satisfactory results.

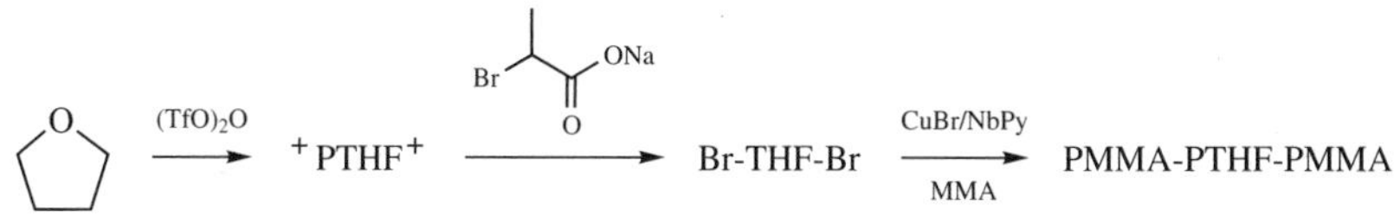

$(TfO)_2O$ → $^+PTHF^+$ → Br-THF-Br → (CuBr/NbPy, MMA) PMMA-PTHF-PMMA

Scheme 6.19

PS-b-PTHF-b-PS triblock (and diblock) copolymers were also synthesized by a cationic to stable free radical polymerization (SFRP) transformation (Yoshida 1998 and 1996). Difunctional living PTHF chains, prepared by cationic polymerization using $(Tf)_2O$ as initiator, were reacted with sodium 4-oxy-TEMPO giving PTHF chains with two terminal TEMPO groups quantitatively. These groups served as initiating sites, in the presence of BPO, for the stable free radical polymerization of styrene, resulting in the aforementioned triblock copolymers (Scheme 6.20). Molecular weight distributions were rather broad, but consumption of the second monomer was nearly quantitative.

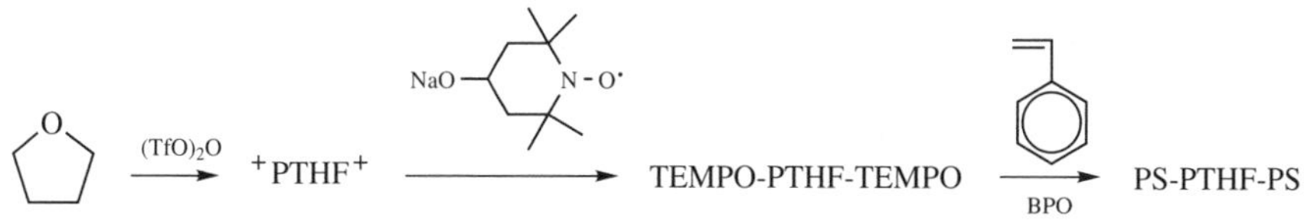

$(TfO)_2O$ → $^+PTHF^+$ → TEMPO-PTHF-TEMPO → (BPO) PS-PTHF-PS

Scheme 6.20

6. SYNTHESIS OF BLOCK COPOLYMERS BY LIVING FREE RADICAL TO CATIONIC MECHANISM TRANSFORMATION

Linear PS prepared by ATRP using the $PhCH_2X/CuX/Bpy$ initiator/catalyst systems possess terminal halogen atoms that can be activated by $AgClO_4$ in order to initiate the cationic ring-opening polymerization of THF, giving PS-PTHF copolymers (Scheme 6.21).

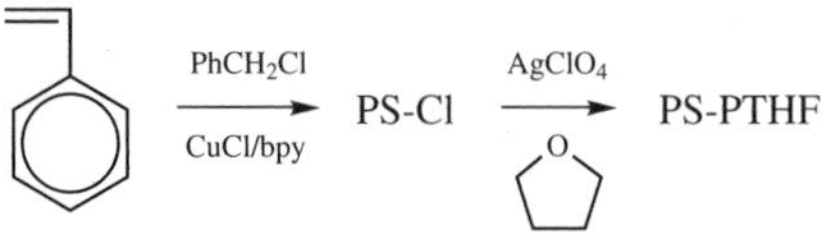

$PhCH_2Cl$, CuCl/bpy → PS-Cl → $AgClO_4$, THF → PS-PTHF

Scheme 6.21

Difunctional PS terminated by bromine atoms at both ends was synthesized by ATRP using the difunctional initiator 1,2-bis(2′-bromobutyryloxy)ethane in the presence of CuBr/bipyridine complex (Xu 2000). Reaction of the terminal bromine atoms with silver perchlorate resulted in a macromolecular initiator that was subsequently used in the cationic ring-opening polymerization of THF, giving PTHF-b-PS-b-PTHF triblock copolymers (Scheme 6.22).

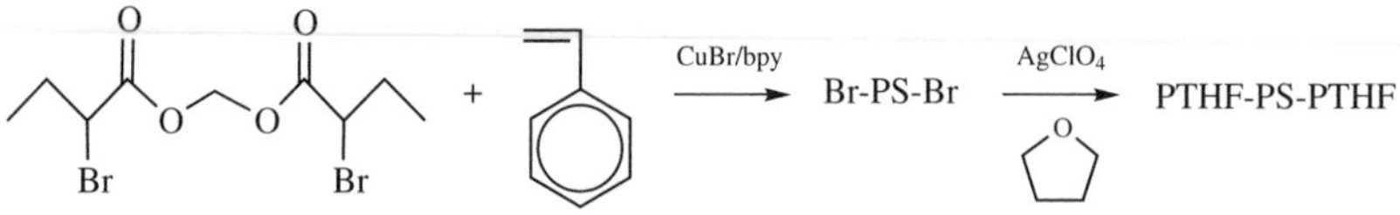

+ CuBr/bpy → Br-PS-Br → $AgClO_4$, THF → PTHF-PS-PTHF

Scheme 6.22

Diblock copolymers of p-methoxy styrene and styrene with cyclohexene oxide (CHO) were synthesized by a combination of ATRP and cationic polymerization (Duz 1999). First, the poly(p-methoxy styrene) and PS, bromine-terminated, blocks

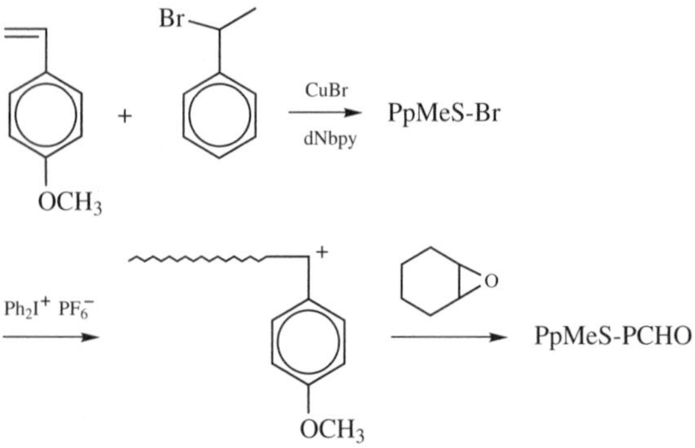

OCH_3 + Br → CuBr, dNbpy → PpMeS-Br

$Ph_2I^+ PF_6^-$ → + OCH_3 → PpMeS-PCHO

Scheme 6.23

were prepared by ATRP using (1-bromoethyl)benzene as the initiator in the presence of CuBr as the catalyst and 4-4′-di-(5-nonyl)-2,2′-bipyridine as the complexing agent. The terminal Br atoms were activated by $Ph_2I^+PF_6^-$, and CHO was polymerized cationically to give the desired block copolymers (Scheme 6.23).

7. SYNTHESIS OF BLOCK COPOLYMERS BY RING OPENING METATHESIS TO LIVING FREE RADICAL MECHANISM TRANSFORMATION

Living polynorbornene chains prepared using molybdenum or tungsten complexes were reacted with an excess of 4-formylbenzyl bromide in a Witting type reaction, giving a polynorbornene chain with a terminal benzylic bromide function (Coca 1997c). This chain-end was used to initiate ATRP of styrene and methyl acrylate. The resulting polynorbornene-b-PS and polynorbornene-b-PMA diblock copolymers showed unimodal molecular weight distributions. No homopolymer impurity coming from the macromolecular initiator was detected by SEC (Scheme 6.24). In a similar reaction scheme, polydicyclopentadiene (PDCPD) benzylic bromide end-functionalized chains were used as macroinitiators for styrene and methylacrylate atom transfer radical polymerization, resulting in well-defined PDCPD-b-PS and PDCPD-b-PMA diblock copolymers.

N
O–Mo
O
+
RT
Mo–O
n N
O
Br
n
Br
CuBr/dNbpy
PNB-PS

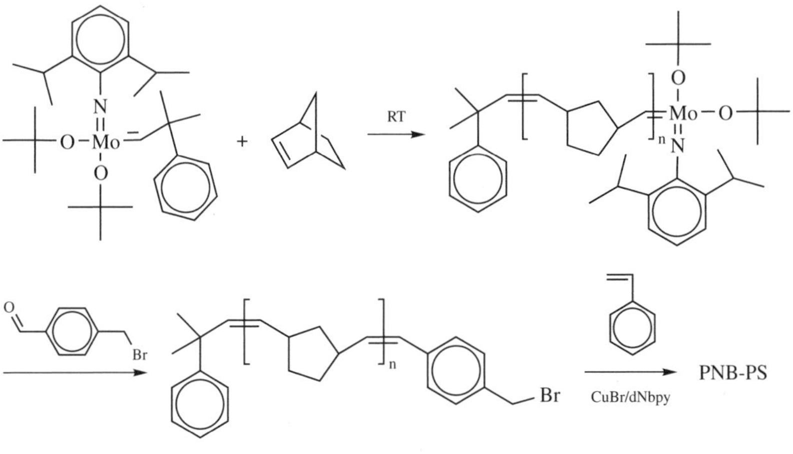

Scheme 6.24

Triblock copolymers with polybutadiene middle blocks and polystyrene or poly(methyl methacrylate) end blocks were synthesized by a combination of ROMP and ATRP (Bielawski 2000). The middle blocks were prepared by ROMP polymerization of cyclooctadiene in the presence of 1,4-chloro-2-butene or cis-2-butene-1,4-diol bis(2-bromo)propionate using an Ru complex as catalyst. The resulting polymers were end-capped at both ends with allyl chloride or 2-bromo propionyl ester groups. These end groups were subsequently used for the polymerization of styrene or methyl methacrylate under ATRP conditions, using CuX/

bipyridine catalytic systems (X = Cl or Br) (Scheme 6.25). The final block copolymers had relatively narrow monomodal molecular weight distributions.

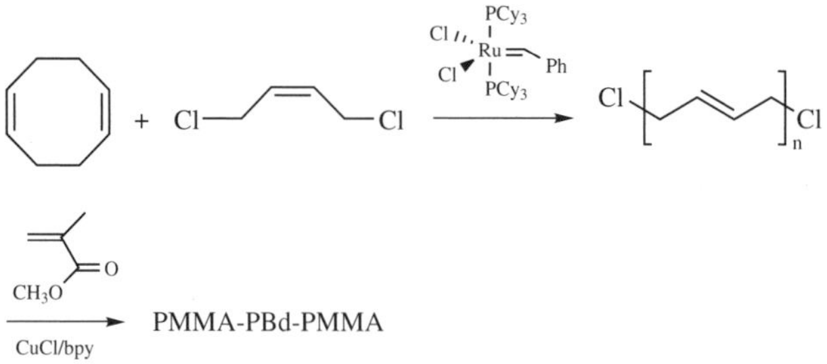

PCy$_3$ Cl Ru Ph Cl PCy$_3$

+ Cl—Cl → Cl–[ ]$_n$–Cl

CH$_3$O O

→ PMMA-PBd-PMMA

CuCl/bpy

8. SYNTHESIS OF BLOCK COPOLYMERS BY RING OPENING METATHESIS TO GROUP TRANSFER MECHANISM TRANSFORMATION

Living polynorbornene was synthesized by ring-opening metathesis polymerization and was terminated using p-$CHOC_6H_4CHO$. Thus, an aldehyde group was

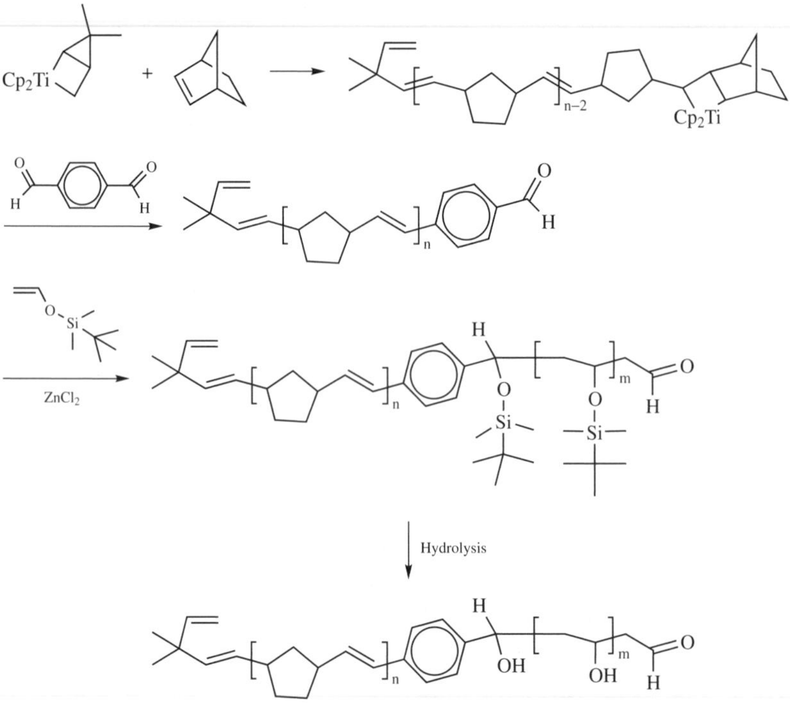

Cp$_2$Ti + → [ ]$_{n-2}$ Cp$_2$Ti

O H O H → [ ]$_n$ O H

O Si → ZnCl$_2$ [ ]$_n$ H [ ]$_m$ O O O Si Si H

Hydrolysis

[ ]$_n$ H [ ]$_m$ OH OH O H

Scheme 6.26

introduced at the chain terminus. This group was used as the initiating site for the aldol group transfer polymerization of t-butyldimethylsilyl vinyl ether, producing a polynorbornene-b-poly(silyl vinyl ether) diblock copolymer (Risse 1989). The final copolymers had the predetermined molecular weights and compositions. Hydrolysis of the silyl groups in the presence of tetrabutylammonium fluoride and methanol resulted in amphiphilic block copolymers of poly(vinyl alcohol) and polynorbornene (Scheme 6.26).

9. OTHER COMBINATIONS

Poly(ethylene)-b-poly(ethylene oxide) block copolymers (Hillmyer 1996) were synthesized by first preparing a PBd high 1,4 precursor by anionic polymerization and functionalization of the living end with one unit of ethylene oxide. Termination gave the hydroxyl-functionalized PBd, which was subsequently hydrogenated in order to obtain a polyethylene block. The hydroxyl groups were then reactivated with potassium naphalenide to yield the corresponding alkoxide, which was used as the initiator for the anionic polymerization of ethylene oxide, resulting in the targeted diblock copolymers (Scheme 6.27). A series of block copolymers with narrow molecular weight distribution and a range of compositions was prepared.

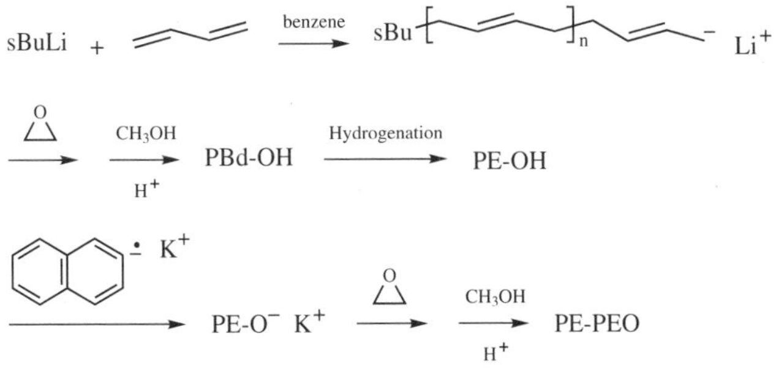

sBuLi + (butadiene) $\xrightarrow{\text{benzene}}$ sBu-[polybutadiene]$_n$-(butenyl)$^-$ Li^+

$\xrightarrow{\text{ethylene oxide}}$ $\xrightarrow[H^+]{CH_3OH}$ PBd-OH $\xrightarrow{\text{Hydrogenation}}$ PE-OH

$\xrightarrow{\text{naphthalene}^{\cdot -}\ K^+}$ PE-O^- K^+ $\xrightarrow{\text{ethylene oxide}}$ $\xrightarrow[H^+]{CH_3OH}$ PE-PEO

Scheme 6.27

Hydroxyl difunctional poly(ethylene glycol) (PEG) was used as the precursor in order to synthesize PS-PEG-PS amphiphilic triblock copolymers (Jankova 1998). The terminal hydroxyl groups of PEG were reacted with 2-bromopropionyl chloride, yielding a dibromo-functionalized PEG, which was used in the next step as a difunctional macroinitiator for the ATRP polymerization of styrene. Complete conversions of the macroinitiator to the triblock copolymers were observed (Scheme 6.28).

Amphiphilic block copolymers containing 2-ethyl-2-oxazoline and L-lactide (Lee 1999) were synthesized by first polymerizing 2-ethyl-2-oxazoline cationically

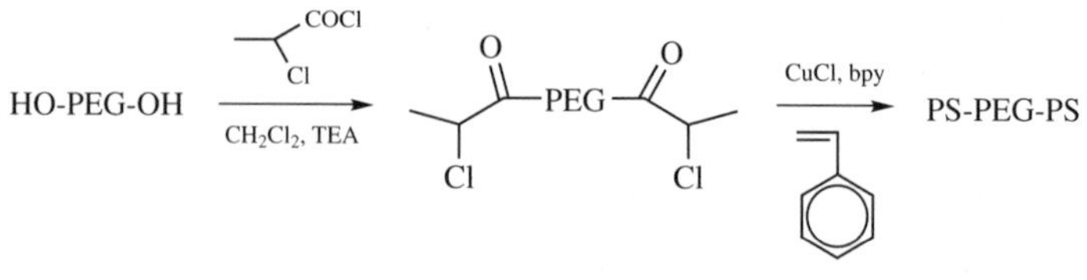

Scheme 6.28

in the presence of methyl p-toluenesulfonate in acetonitrile. By addition of KOH, hydroxyl groups were introduced at the PEtOz chain-ends and were used for the polymerization of L-lactide in the presence of stannous octoate in chlorobenzene (Scheme 6.29). In a similar manner, PEtOz-PCL block copolymers were formed.

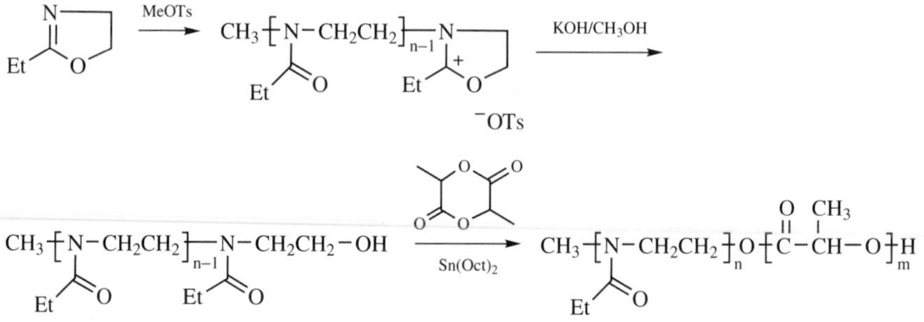

Scheme 6.29

In the same context, anionically prepared hydroxyfunctionalized PBd (or PI) has been used as a macromolecular initiator for the ring-opening polymerization of DL-lactide in the presence of $AlEt_3$ (Scheme 6.30). By keeping the lactide conversion lower than 90%, narrow distribution diblock copolymers were obtained, having a range of compositions (Wang 2000).

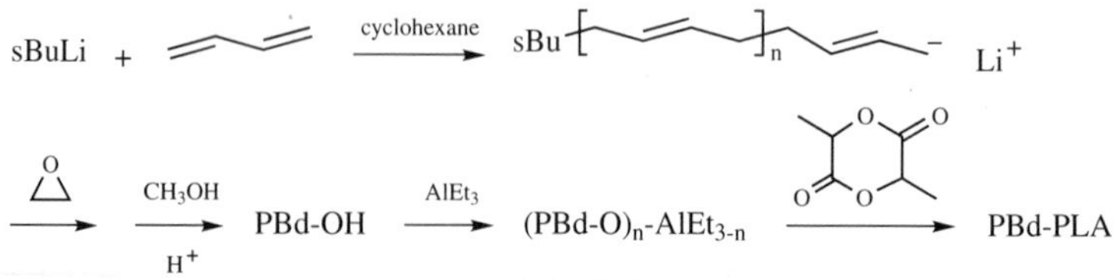

Scheme 6.30

PS-polyphosphazene block copolymers were synthesized by utilizing a PS phosphoranimine terminator for the cationic polymerization of $Cl_3P = NSiMe_3$ (Prange 2000).

10. BIFUNCTIONAL (DUAL) INITIATORS

Recently, the concept of using a single initiator (bifunctional, dual, or double-head initiator) in order to perform two mechanistically distinct polymerizations, without the need of intermediate transformation or activation steps, has been introduced. The potential of this synthetic scheme to produce block copolymers comprised of monomers that are polymerized by different polymerization mechanisms in a simple and efficient way is obvious.

Sogah et al. first reported the synthesis of multifunctional initiators having in their molecule initiating sites for different types of polymerizations (Puts 1997, Sogah 1997) and their use in the synthesis of block (and graft) copolymers of vinylpyridine, styrene, and 2-oxazoline. Later (Weimer 1998), the same group reported the synthesis of PS-poly(2-phenyl-2-oxazoline) by a bifunctional initiator capable of polymerizing the oxazoline monomer by a cationic ring-opening mechanism and PS by a living free radical polymerization.

Hawker, Hedrick, Jérôme, and coworkers (Hawker 1998) synthesized a hydroxy β-functionalized alkoxyamine and used it for the living free radical polymerization of styrene and the living ring-opening polymerization of ε-caprolactone in two steps. The diblock copolymers obtained had narrow molecular weight distributions. Similarly, hydroxylfunctionalized ATRP initiators were used for the polymerization of ε-caprolactone and other vinyl monomers to afford block copolymers having low polydispersities and predictable molecular weights in both blocks (Scheme 6.31).

Arndtsen and Lim (Lim 2000) used a palladium complex for the cationic polymerization of tetrahydrofuran and the ring-opening metathesis polymerization of norbornene (Scheme 6.32).

11. SYNTHESIS OF BLOCK COPOLYMERS BY DIRECT COUPLING OF PREFORMED LIVING BLOCKS

AB and ABA block copolymers of styrene (B) and poly(ethyl vinyl ether) (A) were synthesized by direct coupling of living mono- or difunctional polystyrene anions with living poly(ethyl vinyl ether) cations (Creutz 1994). Monofunctional living polystyrene was polymerized in THF at $-78°C$ using the s-BuLi adduct with α-methylstyrene as the initiator. Monofunctional poly(ethyl vinyl ether) was polymerized in toluene using the HI/ZnI_2 initiating system. The orange solution of polystyrene was added rapidly to the solution of the living PEVE. Instantaneous elimination of the color was observed immediately after mixing of the solutions, indicating coupling between the opositively charged macroions. Addition of PSLi was stopped when a slight yellowish color was observed to persist in the solution

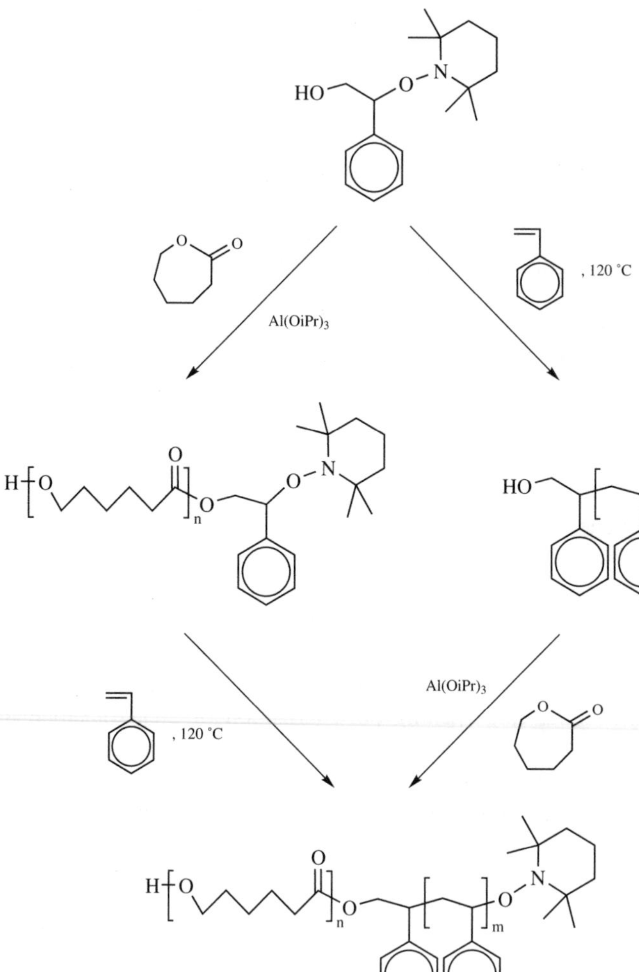

HO O N

Al(OiPr)$_3$

, 120 °C

O O

H O O n O O N

HO m O N

, 120 °C

Al(OiPr)$_3$

H O O n O m O N

Scheme 6.31

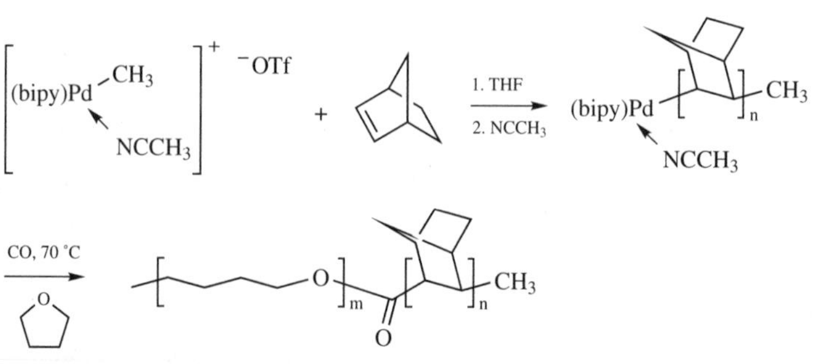

(bipy)Pd CH$_3$ NCCH$_3$ $^+$ $^-$OTf + 1. THF 2. NCCH$_3$ (bipy)Pd n CH$_3$ NCCH$_3$

CO, 70 °C O m O n CH$_3$ O

Scheme 6.32

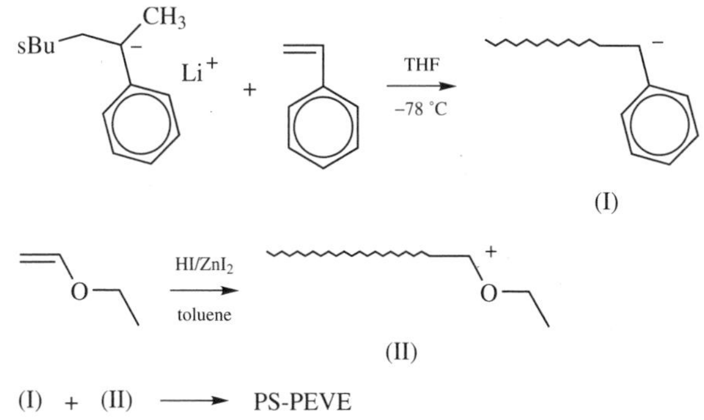

(I) + (II) → PS-PEVE

Scheme 6.33

(Scheme 6.33). Characterization by SEC indicated the formation of the desired diblock copolymers, which possessed narrow molecular weight distributions and the expected molecular weights and compositions. Triblocks of the ABA type were obtained by reaction of difunctional polystyryllithium, resulting from the initiation of styrene polymerization with lithium naphthalene with the monofunctional poly(ethyl vinyl ether) cation in a similar way. In a similar fashion, PS-PTHF block copolymers were synthesized by direct coupling of PSLi anions and polytetrahydrofuran cations (Richards 1978).

In another study living poly(butyl vinyl ether) chains, prepared by cationic polymerization, were coupled with living poly(methyl methacrylate) prepared by group-transfer polymerization. Both diblock and triblock copolymers were synthesized in a controllable and predictable way (Verma 1990).

Diblock copolymers comprised of PIB and PMMA blocks were synthesized by a combination of cationic and group-transfer polymerizations (Takacs 1995). The terminal C-Cl bonds of a PIB precursor prepared by cationic polymerization were converted to diphenylcarbenium ions by reaction with DPE. Living PMMA, from GTP, was subsequently added, resulting in the aforementioned PIB-PMMA block copolymers (Scheme 6.34).

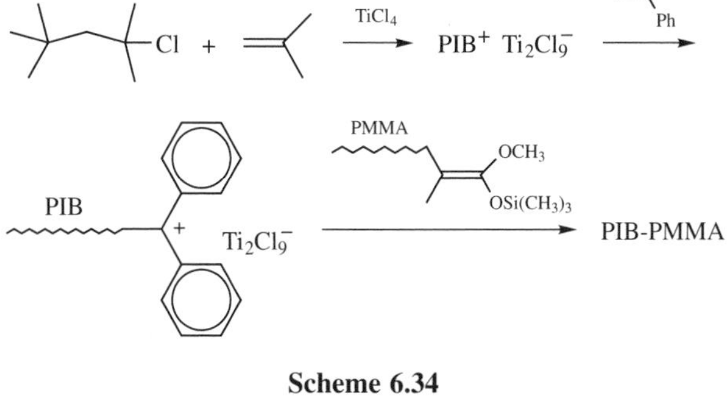

Scheme 6.34

12. SYNTHESIS OF BLOCK COPOLYMERS BY COUPLING OF END-FUNCTIONALIZED PREPOLYMERS

The synthesis of an P4VP-b-PNLO-b-PS-b-PVPh tetrablock copolymer, where PNLO is a nonlinear optical block, was achieved by reaction of aniline-terminated diblock P4VP-PNLO-Ph-NH_2 with poly(4-tert-butyldimethylsiloxy)styrene-PS-COOH diblock copolymer (Pan 2000). The synthetic strategy involved living anionic polymerization of 4VP and end-functionalization with the phenol group. These phenol end groups were used for the polymerization of the NLO monomer, 4-(4-(4-fluorophenylsulfonyl)phenyl)-sulfonyl-4′-N-ethyl-N-2-(4-hydroxyphenolic) ethylazobenzene, catalyzed by potassium carbonate. The PS-PBDMSS block copolymer was synthesized by anionic polymerization and was end-functionalized by reaction with CO_2. Coupling of the two preformed block copolymers resulted in P4VP-PNLO-PS-PBDMSS, which, after hydrolysis of the sililoxy groups, formed the desired tetrablock copolymer (Schemes 6.35a and 6.35b).

A PEO-PDMS-PEO triblock copolymer was prepared by hydrosilylation of α,ω-dihydropoly(dimethylsiloxane) with a vinyl end-functional poly(ethylene oxide) (Yang 1992, Scheme 6.36).

Poly(p-phenyleneethylene)-b-poly(ethylene oxide) (PPE-PEO) rod-coil block copolymers were also synthesized by direct coupling of carboxy-terminated PPE and hydroxyl-terminated PEO (Franke 1998, Scheme 6.37).

Li$^+$ + [4-vinylpyridine] $\xrightarrow[-78\ °C]{THF}$ P4VP$^-$ Li$^+$

Br–CH$_2$–C$_6$H$_4$–OSi(CH$_3$)$_2$tBu $\longrightarrow$ P4VP-Ph-OSiMe$_2$tBu $\xrightarrow[THF]{Bu_4NF}$ P4VP-Ph-OH

F–C$_6$H$_4$–O–[C$_6$H$_4$–SO$_2$]$_2$–C$_6$H$_4$–N=N–C$_6$H$_4$–N(Et)(CH$_2$)$_n$O–C$_6$H$_4$–OH (NLO)

$\xrightarrow{}$ DMAc, K$_2$CO$_3$, 165 °C

P4VP-PhO-PNLO-PhOH $\xrightarrow{DMAc,\ K_2CO_3,\ 165\ °C}$ P4VP-PhO-PNLO-NH$_2$

(*a*)

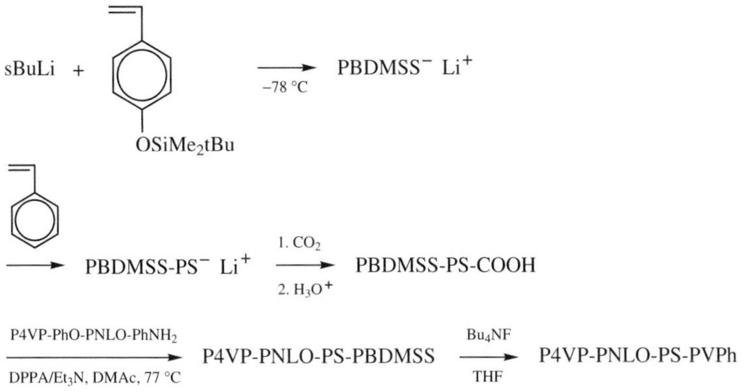

sBuLi + $\xrightarrow{-78\ °C}$ $PBDMSS^- Li^+$

$OSiMe_2tBu$

$\longrightarrow$ $PBDMSS\text{-}PS^- Li^+$ $\xrightarrow[2.\ H_3O^+]{1.\ CO_2}$ PBDMSS-PS-COOH

$\xrightarrow[DPPA/Et_3N,\ DMAc,\ 77\ °C]{P4VP\text{-}PhO\text{-}PNLO\text{-}PhNH_2}$ P4VP-PNLO-PS-PBDMSS $\xrightarrow[THF]{Bu_4NF}$ P4VP-PNLO-PS-PVPh

(*b*)

Scheme 6.35

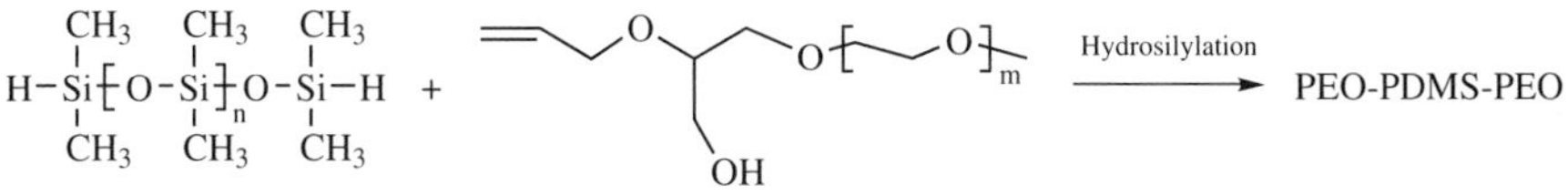

$H\text{-}Si(CH_3)_2\text{-}[O\text{-}Si(CH_3)_2]_n\text{-}O\text{-}Si(CH_3)_2\text{-}H$ + OH ... $[\ O\ ]_m$ $\xrightarrow{Hydrosilylation}$ PEO-PDMS-PEO

Scheme 6.36

C_6H_{13} ... $[\ ]_n$ ... O, OH + $CH_3[OCH_2CH_2]_mOH$

Diisopropylcarbodiimide

CH_2Cl_2

$N(CH_3)_2$ ... N , CH_3 ... SO_3H

C_6H_{13} ... $[\ ]_n$ ... O, $O[CH_2CH_2O]_mCH_3$

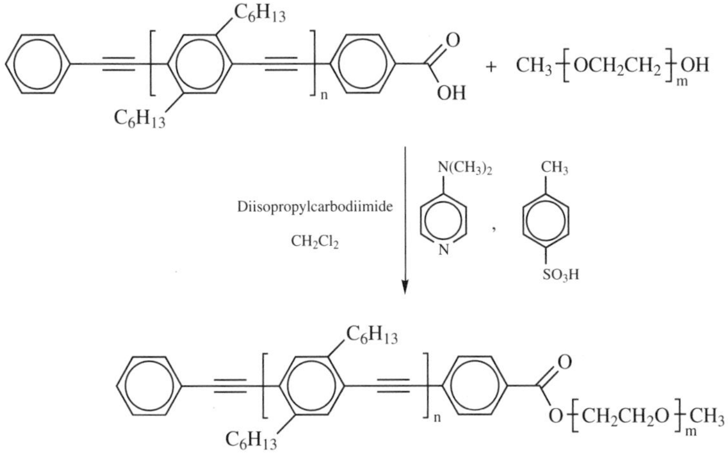

Scheme 6.37

REFERENCES

Akiyama Y., Harada A., Nagasaki Y., Kataoka K. (2000) Macromolecules 33, 5841.

Bielawski C. W., Morita T., Grubbs R. H. (2000) Macromolecules 33, 678.

Burgess F. J., Cunliffe A. V., MacCallum J. R., Richards D. H. (1977) Polymer 18, 719.

Coca S., Matyjaszewski K. (1997a) Macromolecules 30, 2808.

Coca S., Matyjaszewski K. (1997b) J. Polym. Sci. Part A: Polym. Chem. 35, 3595.

Coca S., Paik H-J., Matyjaszewski K. (1997c) Macromolecules 30, 6513.

Cramail H., Deffieux A. (1992) Makromol. Chem. 193, 2793.

Creutz S., Vanooren C., Jérôme R., Teyssié P. (1994) Polym. Bull. 33, 21.

Duz A. B., Yagci Y. (1999) Eur. Polym. J. 35, 2031.

Feldthusen J., Ivan B., Müller A. H. E. (1997) Macromolecules 30, 6989.

Feldthusen J., Ivan B., Müller A. H. E. (1998) Macromolecules 31, 578.

Francke V., Rader H. J., Geerts Y., Müllen K. (1998) Macromol. Chem. Phys. 19, 275.

Franta E., Lehmann J., Reibel L. C., Penczek S. (1976) J. Polym. Sci. Chem. Ed. 56, 139.

Gadkari A., Kennedy J. P. (1989) J. Appl. Polym. Sci. Appl. Polym. Symp. 44, 19.

Hawker C. J., Hedrick J. L., Malmstrom E. E., Trollsas M., Mecerreyes D., Moineau G., Dubois Ph., Jérôme R. (1998) Macromolecules 31, 213.

Hillmyer M. A., Bates F. S. (1996) Macromolecules 29, 6994.

Jankova K., Chen X., Kops J., Batsberg W. (1998) Macromolecules 31, 538.

Kennedy J. P., Kurian J. (1990) Polym. Bull. 23, 259.

Kennedy J. P., Price J. L., Koshimura K. (1991) Macromolecules 24, 6567.

Kobatake S., Harwood H. J., Quirk R. P., Priddy D. B. (1997) Macromolecules 30, 4238.

Kobatake S., Harwood H. J., Quirk R. P., Priddy D. B. (1999) Macromolecules 32, 10.

Lee S. C., Chang Y., Yoon J-S., Kim C., Kwon I. C., Kim Y-H., Jeong S. Y. (1999) Macromolecules 32, 1847.

Lim N. K., Arndtsen B. A. (2000) Macromolecules 33, 2305.

Liu Q., Konas M., Davis R. M., Riffle J.S. (1993) J. Polym. Sci. Part A: Polym. Chem. 31, 2825.

Liu Q., Wilson G. R., Davis R. M., Riffle J. S. (1993) Polymer 34, 3030.

Matyjaszewski K. (1998a) Makromol. Symp. 132, 85.

Matyjaszewslki K. (1998b) Controlled Free Radical Polymerization, ACS Symp. Ser., Washington DC, Vol. 685.

Neubauer A., Poser S., Arnold M. (1997) J. Macromol. Sci.-Pure Appl. Chem. A34(9), 1715.

Pan J., Chen M., Warner W., He M., Dalton L., Hogen-Esch T. E. (2000) Macromolecules 33, 4673.

Prange R., Reeves S. D., Allcock H. R. (2000) Macromolecules 33, 5763.

Puts R. D., Sogah D. Y. (1997) Macromolecules 30, 7050.

Richards D. H., Kingston S. B., Souel T. (1978) Polymer 19, 68.

Risse W., Grubbs R. H. (1989) Macromolecules 22, 1558.

Sogah D. Y., Puts R. D., Trimble A., Sherman O. (1997) Polym. Prepr. 38(1), 731.

Takacs A., Faust R. (1995) Macromolecules 28, 7266.

Takeuchi D., Aida T. (2000) Macromolecules 33, 4607.

Tong J-D., Ni S., Winnik M. A. (2000) Macromolecules 33, 1482.

Verma A., Glagola M, Prasad A., Marand H., Riffle J. S. (1992) Makromol. Chem. Macromol. Symp. 54/55, 95.

Verma A., Nielsen A., McGrath J. E., Riffle J. S. (1990) Polym. Bull. 23, 563.

Wang Y., Hillmeyer M. A. (2000) Macromolecules 33, 7395.

Weimer M. W., Scherman O. A., Sogah D. Y. (1998) Macromolecules 31, 8425.

Xu Y., Pan C. (2000) J. Polym. Sci. Part A: Polym. Chem. 38, 337.

Yang J., Wegner G. (1992) Macromolecules 25, 1791.

Yoshida E., Ishizone T., Hirao A., Nakahama S., Takata T., Endo T. (1994) Macromolecules 27, 3119.

Yoshida E., Sugita A. (1996) Macromolecules 29, 6422.

Yoshida E., Sugita A. (1998) J. Polym. Sci. Part A: Polym. Chem. 36, 2059.

CHAPTER 7

SYNTHESIS OF BLOCK COPOLYMERS BY CHEMICAL MODIFICATION

In the previous chapters, the different living polymerization mechanisms available for the synthesis of well-defined block copolymers were discussed. Although a wide variety of monomers can be polymerized using one or a combination of these mechanisms, the requirement of a living polymerization process, in order to produce well-defined block copolymers, puts some limitations on the nature of monomers that can actually be used. This is especially true in the case of monomers that contain functional groups. The problem can be partially solved by the use of monomers with protected functionality, the protecting group being appropriately chosen with respect to the functional group and the polymerization mechanism that is going to be used. Even in this case, the requirement of most living polymerizations of a high monomer purity is sometimes incompatible with the nature of the monomer or the synthetic scheme employed for its preparation.

Another way of tackling the problem of synthesizing block copolymers, consisting of monomers that are difficult to polymerize by a living mechanism, is to employ polymer-analogous reactions. In this general methodology, a well-defined and suitably chosen precursor polymer is chemically transformed into another polymer by making use of well-known organic reactions. Because the chemical nature of the polymer will be altered, its properties will also differ from those of the precursor. In the case of block copolymers, chemical modification can be performed on one of the blocks selectively, using the appropriate chemical route, or on the whole copolymer. Sequential modification of each block is also possible if the order of chemical transformation of each block and the reaction conditions employed each time have been chosen judiciously.

The ultimate goal of this synthetic methodology is the development of relatively mild, but effective, reaction conditions. The transformation of the precursor block to a new one, having the desired chemical nature (and properties), must be accomplished without giving rise to any degradation, crosslinking, and, in general, any side reactions that can deteriorate the other molecular characteristics of the copolymer. When the precursor block copolymer has been chosen, the number of segments in the new copolymer and in each block are automatically chosen, as well as its molecular architecture. These molecular characteristics, together with the molecular weight distribution, should normally remain unaltered.

The synthesis of block copolymers by chemical transformation of preformed precursor copolymers must be viewed as another route for extending the available range of block copolymers. Several polymer-analogous reactions have appeared in the literature. Some of them are presented and discussed in this chapter.

1. HYDROGENATION

Hydrogenation of unsaturated polymers is a typical example of a polymer chemical modification reaction (McGrath 1995). Hydrogenation of the polydiene part of PS-PBd-PS and PS-PI-PS triblock copolymers is used for the commercial production of thermoplastic elastomers (Kratons) with enhanced thermal and oxidative stability (Chapter 21). Quantitative hydrogenation can be accomplished by heterogeneous (Rosedale 1998), homogeneous (Mohammadi 1987), or chemical methodology (Mango 1973, Hahn 1992).

Heterogeneous catalysts, like the $Pd/CaCO_3$ system (Gehlsen 1993, Gotro 1984), are powerful hydrogenation catalysts but usually show low selectivity. This particular heterogeneous catalyst is efficient for the hydrogenation of aliphatic double bonds of polydienes (Gotro 1984) as well as for the hydrogenation of aromatic rings under more severe conditions (Gehlsen 1993). Thus, the $Pd/CaCO_3$ catalytic system has been used for the preparation of poly(ethylene-co-propylene)-b-polyethylene block copolymers from anionically prepared PI-PBd-1,4 block copolymers (Rangarajan 1993) and poly(ethylene-co-propylene)-b-poly(vinylcyclohexane) copolymers from PI-PS precursors (Gehlsen 1993) (Scheme 7.1a).

Homogeneous catalysts show more selectivity and require relatively milder reaction conditions. Thus, it is possible to prepare polyethylene-b-PS block copolymers from PBd-high 1,4-b-PS diblocks synthesized by anionic polymeriza-

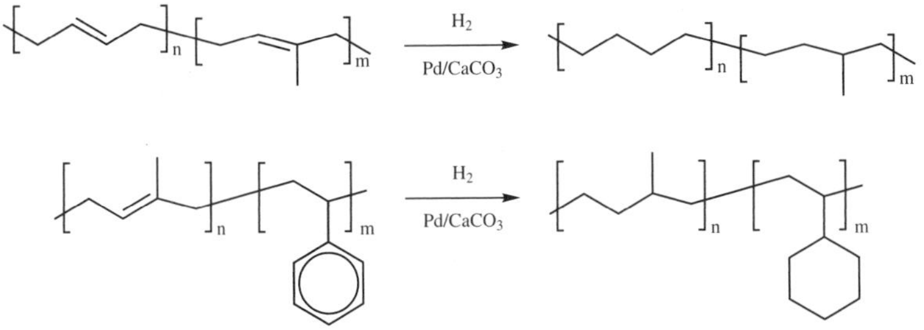

H_2, $Pd/CaCO_3$ (n, m)

H_2, $Pd/CaCO_3$ (n, m)

(a)

Scheme 7.1

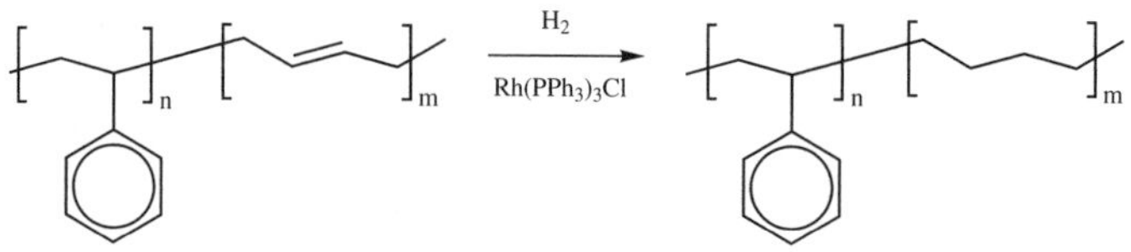

H_2

$Rh(PPh_3)_3Cl$

(*b*)

Scheme 7.1 (*Continued*)

tion, by selective hydrogenation of the PBd block using the Wilkinson catalyst (Schulz 1987). Other catalysts based on Cr or Ni have been employed for the selective hydrogenation of PBd (Yu 1997) in PBd-b-poly(alkylmethacrylate) block copolymers, due to their increased tolerance to the presence of polar, i.e., ester, groups (Scheme 7.1b).

Hydrogenation of dienes using the tosylhydrazide method, is an example of a noncatalytic (chemical) hydrogenation method (Hahn 1992). It requires relative mild reaction conditions and has been used in several cases for the selective hydrogenation of diene blocks in the presence of blocks bearing reactive (polar) functionalities, such as the ester functionality of alkylmethacrylates. However, caution should be exercised in its use because there is always the possibility of the formation of substituted products. Another disadvantage is the use of large quantities of the hydrogenation reagents and, thus, their elimination from the final product, deserves special attention and extra effort. On the other hand, hydrogenation using catalytic systems requires small amounts of catalyst that, under the appropriate conditions, do not react with the substrate and can be easily removed from the reaction product, especially in the case of heterogeneous catalysts.

2. HYDROLYSIS

This simple polymer-analogous reaction is applied for the preparation of blocks carrying carboxy, hydroxy, and amine groups. These groups, present in the precursor block in the appropriate protected form, can be deprotected under mild acidic or basic conditions, depending on the desired functionality.

Thus, poly(methacrylic or acrylic acid) blocks can be formed by acidic hydrolysis of poly(tert-butyl methacrylate or acrylate) blocks. For instance, PS-b-poly(methacrylic acid) block copolymers can be derived from PS-P(tert-butyl methacrylate) precursors by hydrolysis (Ramireddy 1992, Scheme 7.2a).

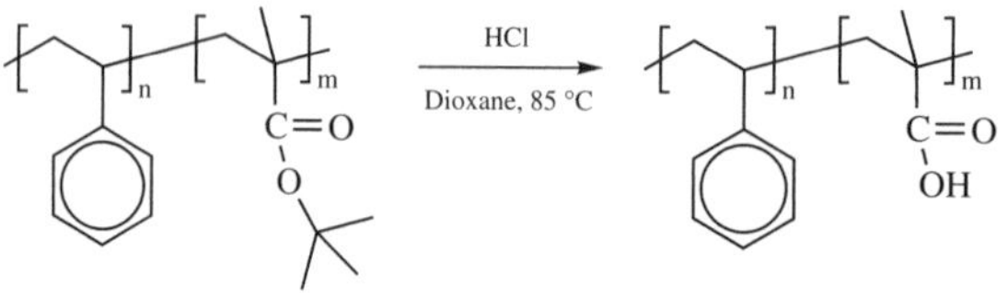

HCl

Dioxane, 85 °C

C=O

O

C=O

OH

(*a*)

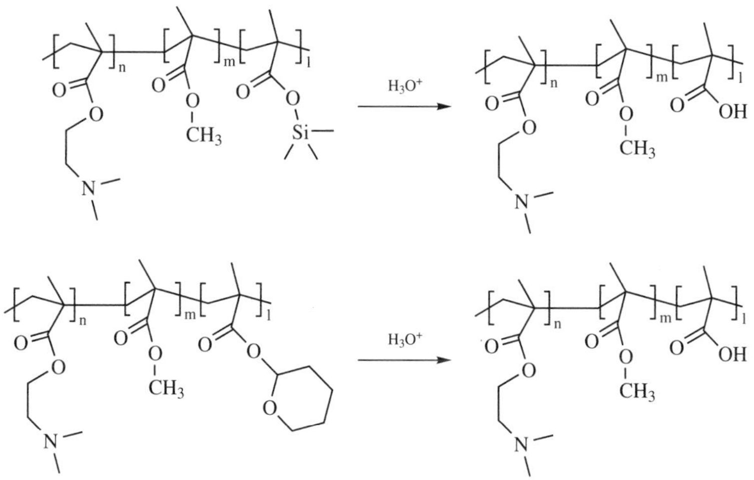

(b)

Scheme 7.2

Other protecting groups like silyl or pyranyl groups (Dicker 1990, Butun 2000) can serve as precursors to the COOH group (Scheme 7.2b).

Ester protected vinyl ether monodisperse blocks give well-defined poly(vinyl alcohol) blocks after hydrolysis (Forder 1995, Aoshima 1994).

3. QUATERNIZATION

Basic tertiary amine moieties present in one of the blocks of a block copolymer can be converted to quaternary ammonium salts by reaction with acids, like HCl, or alkyl halides (Selb 1985, 1980a,b, and 1981a,b). In this way the poly(2-vinyl pyridine) or poly(4-vinylpyridine) blocks of PS-P2VP or PS-P4VP diblocks have been transformed to quaternary ammonium salts in the presence of HCl, CH_3I, or benzylchloride, producing cationic block polyelectrolytes. The new copolymers could be dispersed in water in contrast with the precursors that are insoluble in aqueous media (Scheme 7.3a).

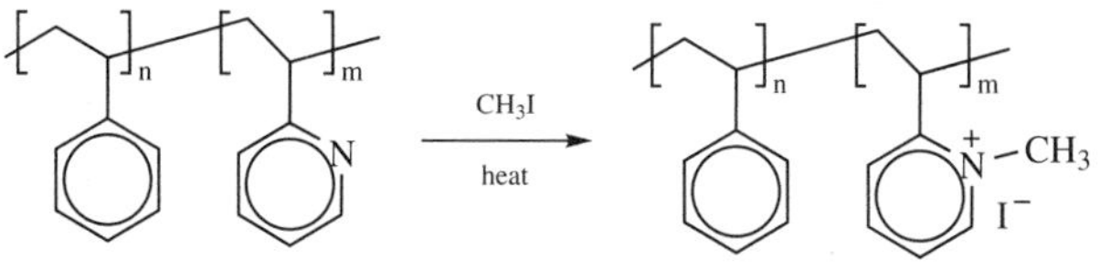

(a)

Scheme 7.3

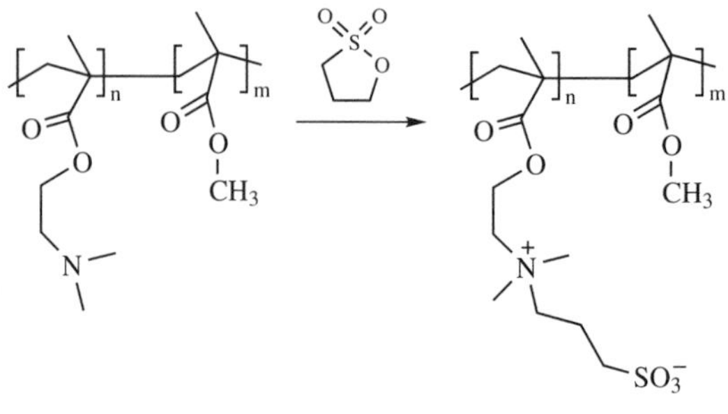

(*b*)

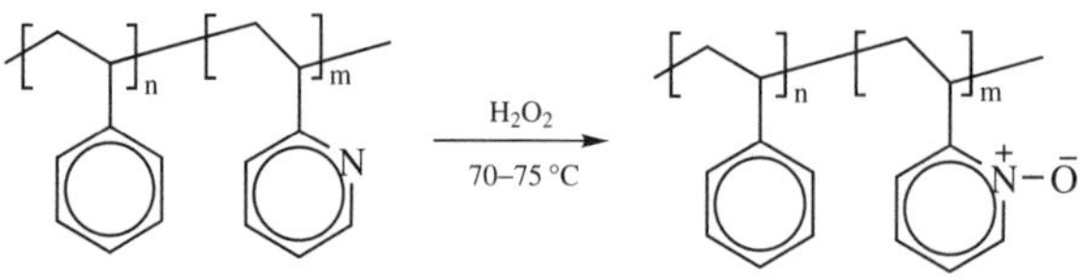

H_2O_2
70–75 °C

(*c*)

Scheme 7.3 (*Continued*)

When the quaternizing reagent is a cycloalkylsultone, or lactone, the conversion of the tertiary amine group to a zwitterionic group (sulfobetaine or carboxy betaine) is accomplished. Block copolymers containing a hydrophilic block with zwiiter-ionic side groups and a hydrophobic block were produced by reaction of poly (dimethylaminoethylmethacrylate)-b-poly(methacrylate) diblocks, prepared by GTP, with 1,3-propane sultone (Lowe 1999, Scheme 7.3b).

Tertiary amine groups can also be transformed to N-O oxide groups, resulting in water-soluble blocks from hydrophobic block precursors (Scheme 7.3c).

4. SULFONATION

Polystyrene blocks can be converted to polyelectrolyte blocks by sulfonation. Almost complete sulfonation of the phenyl rings, primarily at the para position, is accomplished via a H_2SO_4/P_2O_5 complex or sulfur trioxide complex with triethyl phosphate (TEP) in organic (CH_2Cl_2, CH_2ClCH_2Cl) media (Vink 1970, Valint 1988). The sulfonated PS homopolymer precipitates as the reaction proceeds, due to changes in the solubility, and can be easily recovered. Poly(styrene sulfonate)-b-poly(tert-butyl styrene) block copolymers have been synthesized from PS-PtBuS precursors (Valint 1988). Due to the bulky tert-butyl group, sulfonation of the PtBuS blocks is highly improbable. Thus, only the PS block can be functionalized, resulting in hydrophobic-hydrophilic block copolymers (Scheme 7.4a).

Poly(ethylene-alt-propylene)-b-poly(styrene sulfonate) block copolymers were synthesized from PI-1,4-b-PS anionically prepared precursors (Guenoun 1998). The PI block was hydrogenated first, giving poly(ethylene-alt-propylene)-b-poly-

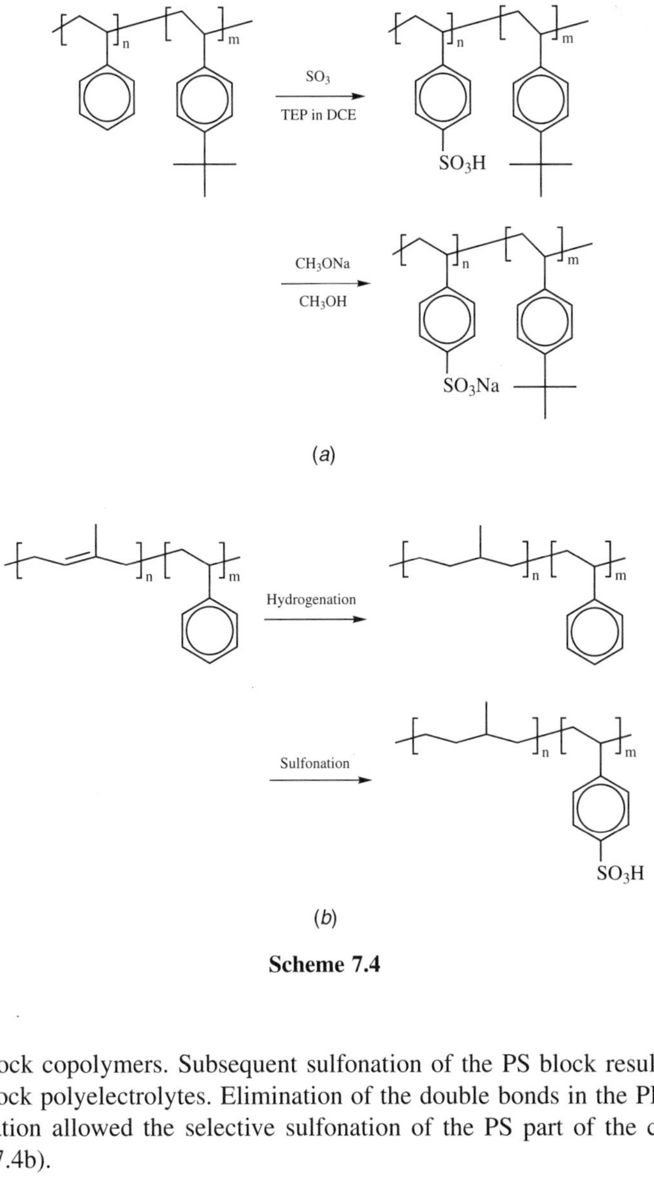

(*a*)

(*b*)

Scheme 7.4

styrene block copolymers. Subsequent sulfonation of the PS block resulted in the desired block polyelectrolytes. Elimination of the double bonds in the PI block by hydrogenation allowed the selective sulfonation of the PS part of the copolymer (Scheme 7.4b).

5. HYDROBORATION/OXIDATION

Another chemical modification reaction that leads to functionalized intermediates is the hydroboration/oxidation reaction that introduces hydroxyl groups into the double bonds of a diene block. Initial hydroboration of the double bonds is accomplished by using 9-borabicyclo[3.3.1]nonane (9-BBN), followed by

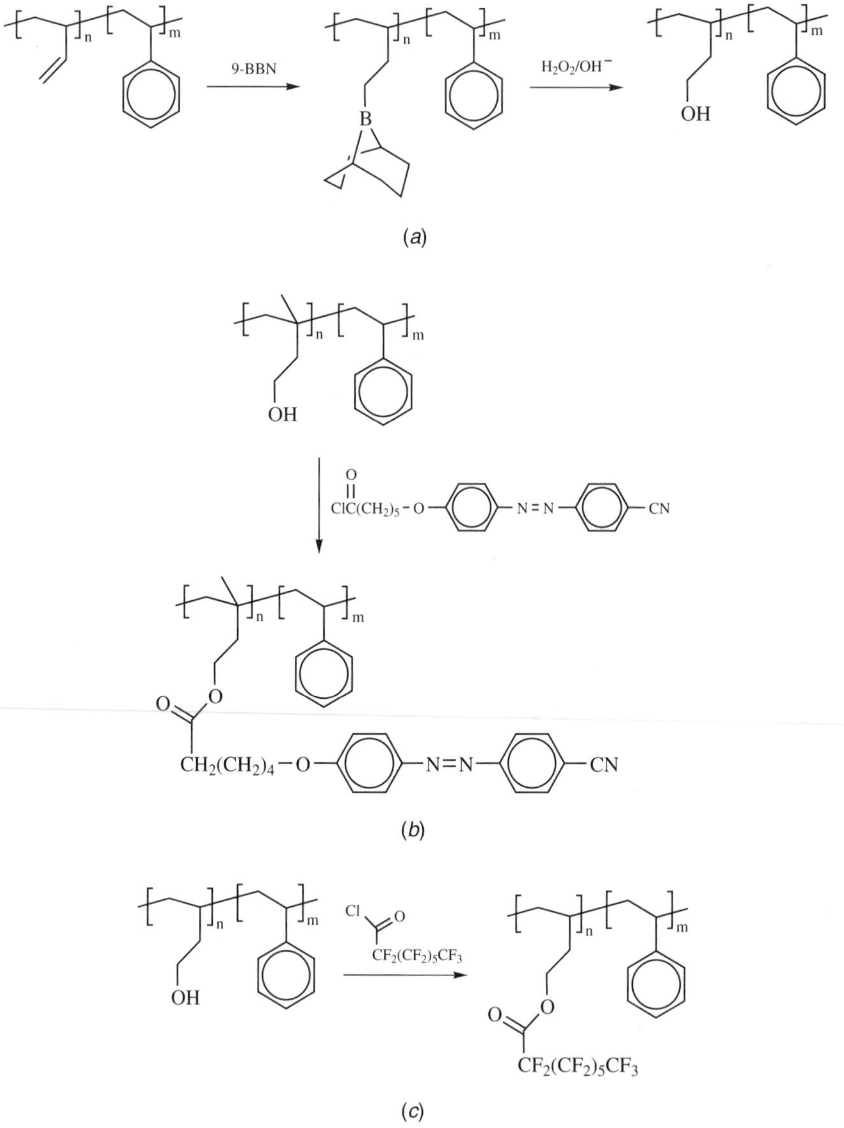

9-BBN

H_2O_2/OH^-

B

OH

(*a*)

OH

$ClC(CH_2)_5-O-$ ... $-N=N-$ ... $-CN$

O O

$CH_2(CH_2)_4-O-$ ... $-N=N-$ ... $-CN$

(*b*)

OH

Cl O

$CF_2(CF_2)_5CF_3$

O O

$CF_2(CF_2)_5CF_3$

(*c*)

Scheme 7.5

oxidation with H_2O_2/NaOH. The hydroxyl groups introduced by this reaction scheme can be further used for the introduction of other functional groups into the copolymer due to the wide range of possible reactions associated with a hydroxyl functionality (Scheme 7.5a). Thus, OH groups can be esterified with an acid chloride or an acid anhydride moiety of another compound, making possible the linking of other functional side groups on the main block copolymer chain. For example, liquid crystalline groups can be attached to the diene block of a PBd-PS precursor (Scheme 7.5b) (Iyengar 1996, Gronski 1997), or perfluorinated alkyls can be linked, producing blocks of highly hydrophobic character and poor adhesion properties (Scheme 7.5c) (Antonietti 1997). Cholesteric groups have also been attached to a precursor block copolymer through OH-modified intermediates.

6. EPOXIDATION

Epoxidation is another example of a polymer-analogous reaction that leads to the formation of reactive intermediates. A number of epoxidation agents has been reported for the epoxidation of the double bonds of polybutadiene (Jian 1991, Huang 1988, Udipi 1979). Complete conversion of the double bonds can be effected with almost no undesirable side reactions (Scheme 7.6a). The oxirane rings thus formed can participate in several ring-opening reactions for introducing various functional side groups in the diene block of the copolymer. For instance,

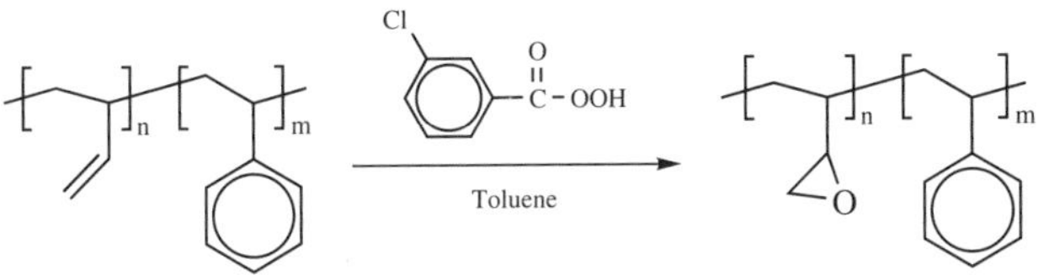

Cl

O

C - OOH

n m

Toluene

n m

O

(*a*)

Scheme 7.6

oxirane rings can be reacted with acid chlorides to give the respective ester functionality. Nucleophilic ring-opening reaction can couple a number of functional side groups, like 2-mercaptobenzothiazole or 2-mercaptopyridine groups (Scheme 7.6b) (Antonietti 1996).

7. CHLORO/BROMOMETHYLATION

Introduction of halomethyl groups (e.g., chloromethyl and bromomethyl) onto an aromatic ring of, i.e., polystyrene block of a block copolymer, can be achieved through classical chloromethyl ether reaction schemes or with halogen methylating agents that are prepared in situ during the halogen methylation reaction (Pepper 1953, Itsuno 1990) (Scheme 7.7). The polymer analogues thus produced can be viewed as reactive intermediates because the halomethyl groups can be converted to a variety of other side groups making use of known organic reactions. Furthermore, the halomethyl groups can serve as linking or initiating sites for the coupling of

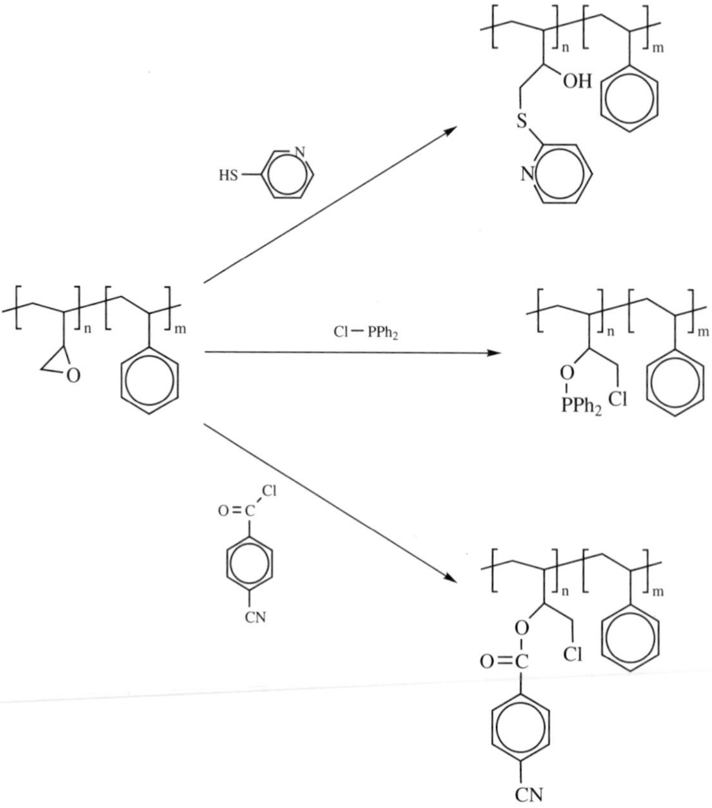

HS– (pyridine, N)

Cl–PPh_2

O=C–Cl (4-CN-phenyl)

n m O

n m OH S N

n m O PPh_2 Cl

n m O O=C Cl CN

(b)

Scheme 7.6 (*Continued*)

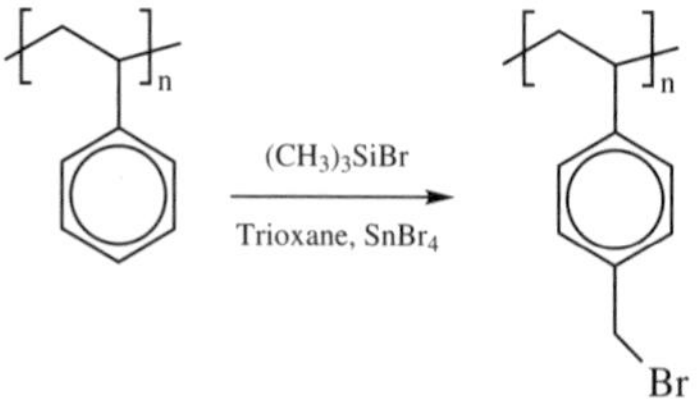

n

$(CH_3)_3SiBr$

Trioxane, $SnBr_4$

n Br

Scheme 7.7

living chains or the polymerization of another monomer (Rahlwes 1977, George 1987, Selb 1979a,b, Pitsikalis 1998), through the appropriate mechanism, giving rise to complex macromolecular architectures (grafts, block grafts, arboresent, etc., block copolymers).

8. HYDROSILYLATION

Hydrosilylation is a versatile route for the preparation of functional copolymers from polymeric materials containing double bonds. This type of reaction has been extensively used for the incorporation of different silane groups into the double bonds of polybutadiene blocks. In most cases the reaction takes place in the presence of platinum (e.g., H_2PtCl_6) (Cameron 1981, Iraqi 1992) or rhodium [e.g., $Rh(PPh_3)_3Cl$] (Guo 1990) based catalysts. The second catalytic system seems to be selective for the 1,2 double bonds, whereas addition to the 1,4 double bonds can be achieved with Pd catalyst. The extent of hydrosilylation depends on the microstructure of the diene, the nature of the silane used, and the nature of the functional groups in the copolymer. The addition of the SiH to the double bond occurs in an anti-Markownikoff fashion in most cases, but Markownikoff addition has been also observed (Guo 1990). The reaction is characterized by a slow induction period, and traces of oxygen were found to be necessary in order for the catalyst Pt(0)-divinyltetramethyldisiloxane (Pt,DVDS) (Karstedt's catalyst) to be activated. The method has been used in order to incorporate SiCl bonds into PBd chains (Scheme 7.8a). These SiCl bonds were subsequently used as branching sites (Cameron 1981)

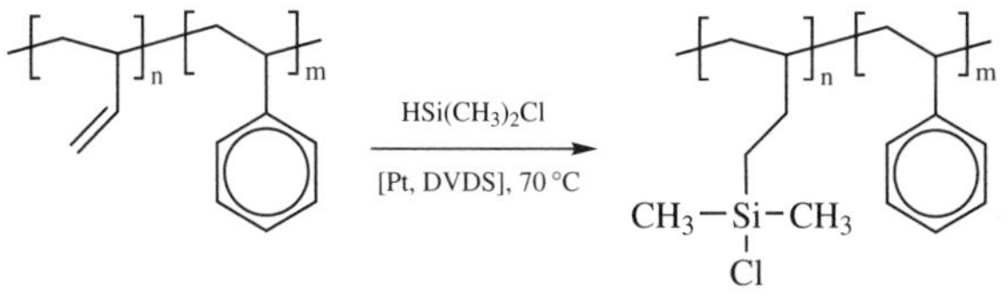

$HSi(CH_3)_2Cl$

[Pt, DVDS], 70 °C

$CH_3-Si-CH_3$

Cl

(*a*)

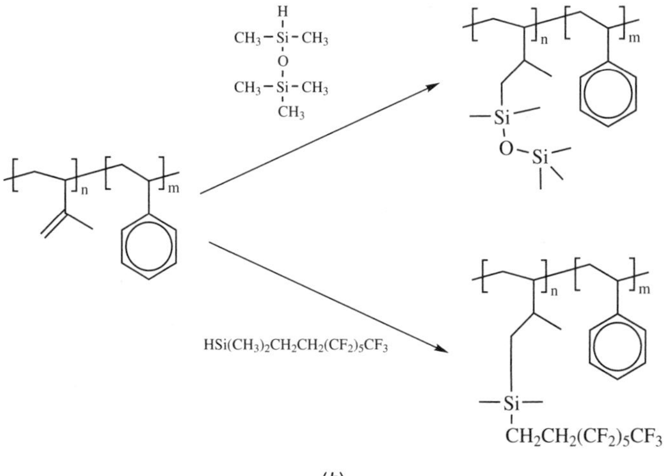

H

$CH_3-Si-CH_3$

O

$CH_3-Si-CH_3$

CH_3

$HSi(CH_3)_2CH_2CH_2(CF_2)_5CF_3$

Si

O–Si

Si

$CH_2CH_2(CF_2)_5CF_3$

(*b*)

Scheme 7.8

where living anionic chains (of homopolymer or diblock copolymer nature) were attached, giving graft or block graft copolymers with trifunctional or tetrafunctional branch points (Xenidou 1998).

With the appropriate choice of the silane, tertiary amine groups can be attached to the polydiene backbone. Hazziza-Lascar et al. (Hazziza-Lascar 1993) used hexachloroplatinic acid as a catalyst to hydrosilylate a polybutadiene with [(3-N,N-dimethylamino)-propyl]tetramethyldisiloxane.

PS-PI block copolymers were hydrosilylated with pentamethyldisiloxane via Karstedt's catalyst in toluene at 110°C (Scheme 7.8b). The reaction was completed in several days (Gabor 1994).

The pendant double bonds of another PS-PI block copolymer were hydrosilylated with tridecafluorooctyldimethylchlorosilane in anhydrous toluene (Scheme 7.8b). The reaction time reported was 10 days. Ober et al. (Jeyaprakash 2000) reported the hydrosilylation of anionically PS-PBd block copolymers of various microstructures with 1H, 1H, 2H, 2H-perfluorooctyldimethylhydrosilane in the presence of Karstedt's catalyst in the absence of solvent. Nearly quantitative yields were obtained after 24 h to 26 h. These kind of materials are useful in low-surface-energy applications.

REFERENCES

Antonietti M., Forster S., Hartmann J., Oestreich S. (1996) Macromolecules 29, 3800.

Antonietti M., Forster S., Micha M. A., Oestreich S. (1997) Acta Polymer. 48, 262.

Aoshima S., Iwasawa S., Kobayashi E. (1994) Polymer J. 26, 912.

Butun V., Vamvakaki M., Billingham N. C., Armes S. P. (2000) Polymer 41, 3173.

Cameron G. G., Qureshi M. Y. (1981) Makromol. Chem. 2, 287.

Candau F., Afchar-Taromi F., Rempp P. (1977) Polymer 18, 1253.

Dicker I. B., Cohen G. M., Farnham W. B., Hertler W. R., Laganis E. D., Sogah D. Y. (1990) Macromolecules 23, 4034.

Forder C., Armes S. P., Billingham N. C. (1995) Polym. Bull. 35, 291.

Gabor A. H., Lehner E. A., Mao G., Schneggenburger L. A., Ober C. K. (1994) Chem. Mater. 6, 927.

Gehlsen M. D., Bates F. S. (1993) Macromolecules 26, 4122.

George M. H., Majid M. A., Barrie J. A., Rezanian I. (1987) Polymer 28, 1217.

Gotro J. T., Graessley W. W. (1984) Macromolecules 17, 2767.

Gronski W., Sanger J. (1997) Makromol. Chem. Rapid Commun. 18, 59.

Guenoun P., Delsanti M., Gazeau D., Mays J. W., Cook D. C., Tirrell M., Auvray L. (1998) Eur. Phys. J. Bull.1, 77.

Guo X., Rajeev F., Rempel G. L. (1990) Macromolecules 23, 5047.

Hahn S. F. (1992) J. Polym. Sci. Part A: Polym. Chem. 30, 397.

Hazziza-Laskar J., Nurdin N., Helary G., Sauvet G. (1993). J. Appl. Polym. Sci. 50, 651.

Huang W. K., Hsiue G. H., Hou W. H. (1988) J. Polym. Sci. Part A: Polym. Chem. 26, 1867.

Iraqi A., Seth S., Vincent V. A., Cole-Hamilton D. J., Watkinson M. D., Graham I. M., Jeffrey D. J. (1992) Mater. Chem. 2, 1057.

Itsuno S., Uchikoshi K., Ito K. (1990) J. Am. Chem. Soc. 112, 8187.

Iyengar D. R., Perutz S. M., Dai C. A., Ober C. K., Kramer E. J. (1996) Macromolecules 29, 1229.

Jeyaprakash J. D., Samuel S., Dhamodharan R., Ober C. K. (2000) J. Polym. Sci. Part A: Polym. Chem. 38, 1179.

Jian X., Hay A. S. (1991) J. Polym. Sci. Part A: Polym. Chem. 29, 1183.

Lowe A. B., Billingham N. C., Armes S. P. (1999) Macromolecules 32, 2141.

Mango L. A., Lentz R. W. (1973) Makromol. Chem. 163, 13.

McGrath M. P., Sall E. D., Tremont S. J. (1995) Chem. Rev. 95, 381.

Mohammadi N. A., Rempel G. L. (1987) Macromolecules 20, 2362.

Pepper K. W., Paisley H. M., Young M. A. (1953) J. Am. Chem. Soc. 4097.

Pitsikalis M., Sioula S., Pispas S., Hadjichristidis N., Cook D. C., Li J., Mays J. W. (1999) J. Polym. Sci. Part A: Polym. Chem. 37, 4337.

Rahlwes D., Roovers J. E. L., Bywater S. (1977) Macromolecules 10, 604.

Ramireddy C., Tuzar Z., Prochazka K., Webber S. E., Munk P. (1992) Macromolecules 25, 2541.

Rangarajan P., Register R. A., Fetters L. J. (1993) Macromolecules 26, 4640.

Rosedale J. H., Bates F. S. (1988) J. Am. Chem. Soc. 110, 3542.

Selb J., Gallot Y. (1979a) Polymer 20, 1259.

Selb J., Gallot Y. (1979b) Polymer 20, 1273.

Selb J., Gallot Y. (1980a) Makromol. Chem. 181, 809.

Selb J., Gallot Y. (1980b) Makromol. Chem. 181, 2605.

Selb J., Gallot Y. (1981a) Makromol. Chem. 182, 1491.

Selb J., Gallot Y. (1981b) Makromol. Chem. 182, 1513.

Selb J., Gallot Y. (1985) in Developments in Block Copolymers, Goodman I. (Ed), Elsevier, London, Vol. 2, p. 27.

Shulz D. N. (1987) in Encyclopedia of Polymer Science and Engineering, Mark H. F. (Ed) Wiley-Interscience, New York, Vol. 7, p. 807.

Udipi K. (1979) J. Appl. Polym. Sci. 23, 3301.

Valint P. L., Bock J. (1988) Macromolecules 21, 175.

Vink H. (1970) Makromol. Chem. 131, 133.

Wang S-M., Tsiang R. C-C. (1996) J. Polym. Sci. Part A: Polym. Chem. 34, 1483.

Xenidou M., Hadjichristidis N. (1998) Macromolecules 31, 5690.

Yu J. M., Yu Y., Dubois P., Teyssie P., Jerome R. (1997) Polymer 38, 3091.

CHAPTER 8

NONLINEAR BLOCK COPOLYMERS

So far our discussion has been focused on the synthesis of linear block copolymers by the various polymerization methods available. However, nonlinear block copolymers have also attracted the interest of polymer scientists in an attempt to answer the fundamental question of how macromolecular architecture can affect block copolymer properties. Nonlinear block copolymers include star block copolymers, graft copolymers, miktoarm star copolymers, cyclic block copolymers, and a variety of other complex architectures (Pitsikalis 1998). Thanks the imagination of polymer chemists, a collection of new macromolecules has been synthesized so far, and many of them have proven to have interesting properties, in many cases distinctly different from their linear counterparts.

1. STAR BLOCK COPOLYMERS

Star block copolymers are actually star-shaped macromolecules where each arm is a block copolymer. The number of branches can vary from a few to several tens. The topological difference of this kind of macromolecules, with respect to linear block copolymers, is focused on the existence of a central branching point, which, by itself, brings a certain symmetry in the macromolecule and sometimes defines a certain amount of intramolecular ordering.

Many different synthetic approaches have been employed in order to synthesize well-defined star block copolymers with anionic, cationic, radical, and even condensation polymerization methods. They can be classified into two major categories: the method of linking agents and the method of difunctional polymerizable monomers.

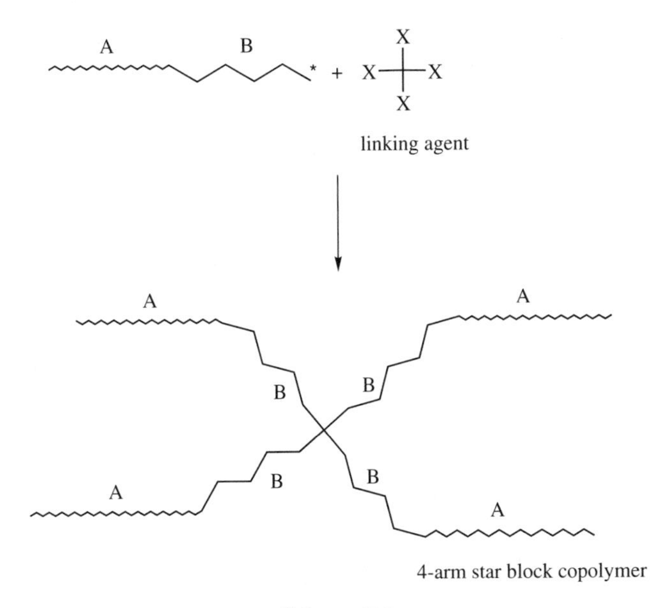

Scheme 8.1

In the method of the linking agents, the block copolymer arms are synthesized first by a living polymerization mechanism. Then they are deactivated by a multifunctional compound having an appropriate number of functional groups (Scheme 8.1) equal to the number of arms of the desired final star block copolymer. The functional groups must be able to react with the living centers in a quantitative, fast, and controlled way, giving no undesirable byproducts. The living active centers are usually employed in excess of the available functional groups in order to ensure complete substitution of the reactive groups of the linking agent by the block copolymer chains. The excess arms can be eliminated, after neutralization, from the crude reaction product by fractionation. The major advantage of this synthetic methodology is that the number of arms present on each star macromolecule is well-defined and predetermined.

Following this methodology, star block copolymers of styrene and isoprene having 3 to 18 diblock arms were synthesized by anionic polymerization using chlorosilanes as linking agents (Alward 1981). The final materials had well-defined molecular weights and functionality as well as narrow molecular weight distributions. Chlorosilane chemistry gives the advantage of choosing the way of block arm connection to the central point as it was shown by the preparation of $(PS\text{-}PI)_nSi$ and $(PI\text{-}PS)_nSi$ stars (Scheme 8.2) as well as inverse star block copolymers (Tselikas

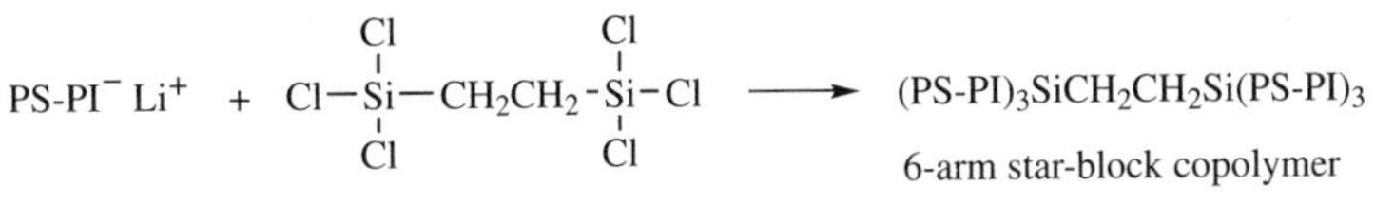

$$PS\text{-}PI^-\,Li^+ + Cl{-}SiCl_2{-}CH_2CH_2{-}SiCl_2{-}Cl \longrightarrow (PS\text{-}PI)_3SiCH_2CH_2Si(PS\text{-}PI)_3$$

6-arm star-block copolymer

Scheme 8.2

$$2\ \text{PI-PS}^{-}\text{Li}^{+} + \text{SiCl}_4 \xrightarrow{\text{titration}} (\text{PI-PS})_2\text{SiCl}_2 + 2\ \text{LiCl}$$

$$(\text{PI-PS})_2\text{SiCl}_2 + \text{PS-PI}^{-}\text{Li}^{+}\ (\text{excess}) \longrightarrow (\text{PI-PS})_2\text{Si(PI-PS)}_2$$

Scheme 8.3

1996). These latter materials had four diblock arms with the same molecular weight and same PS and PI content but two of them were connected to the Si central atom by the PI end of the diblock, whereas the other two by the PS part (Scheme 8.3).

Four-arm star block copolymers of poly(vinyl ethers) were synthesized using a tetrafunctional linking agent, i.e., $C[CH_2OC_6H_4\text{-p-}C(OSiMe_3)=CH_2]_4$ (Fukui 1995). The linking reaction was found to be well-controlled, as evidenced by the narrow molecular weight distributions and functionalities of the final products (Scheme 8.4). Four-arm star block copolymers having P(α-MeSt-b-2-hydroxyethyl

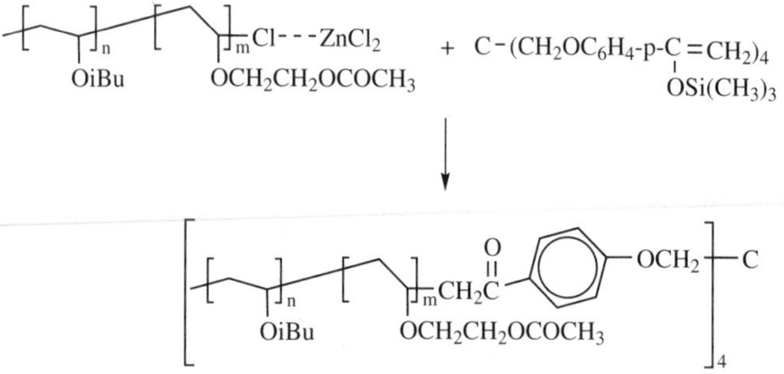

OiBu OCH2CH2OCOCH3 Cl---ZnCl2 + C-(CH2OC6H4-p-C=CH2)4 OSi(CH3)3

OiBu OCH2CH2OCOCH3 CH2C(=O) OCH2 C 4

Scheme 8.4

vinyl ether) arms were also synthesized (Fukui 1996). Using a similar approach, six-arm star block copolymers with poly(indene-b-isobutylene) arms and cyclosiloxane cores were synthesized by cationic polymerization and hydrosilylation chemistry (Shim 2000). The arms were prepared first by sequential polymerization of indene and IB with cumylchloride or cumylmethoxide/$TiCl_4$ systems and quenching the living ends with allyltrimethylsilane. The allyl end-capped arms were then reacted with hexamethylcyclohexasiloxane, giving the desired star block copolymers (Scheme 8.5). PS-PIB star block copolymers were also synthesized by using the same approach (Shim 1998).

According to the second method of synthesis of star block copolymers, a difunctional polymerizable monomer is used for the formation of the stars core (Eschway 1975). Two different procedures can be employed. In the first one, the living block copolymer arms react with a predetermined amount of the difunctional monomer (Scheme 8.6). Due to the existence of two reactive sites, the polymerization results in the formation of a network of small dimensions, which serves as a connecting point for the arms. By this procedure star molecules with a relatively

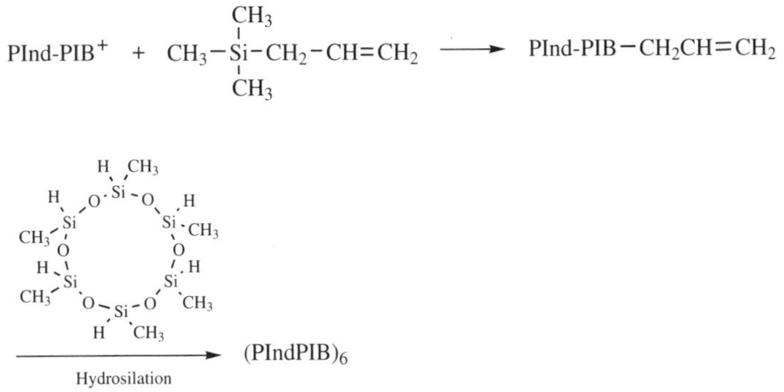

$$\text{PInd-PIB}^+ + CH_3-\underset{CH_3}{\overset{CH_3}{Si}}-CH_2-CH=CH_2 \longrightarrow \text{PInd-PIB}-CH_2CH=CH_2$$

$\xrightarrow[\text{Hydrosilation}]{\text{cyclic }[Si(H)(CH_3)O]_6}$ $(PIndPIB)_6$

Scheme 8.5

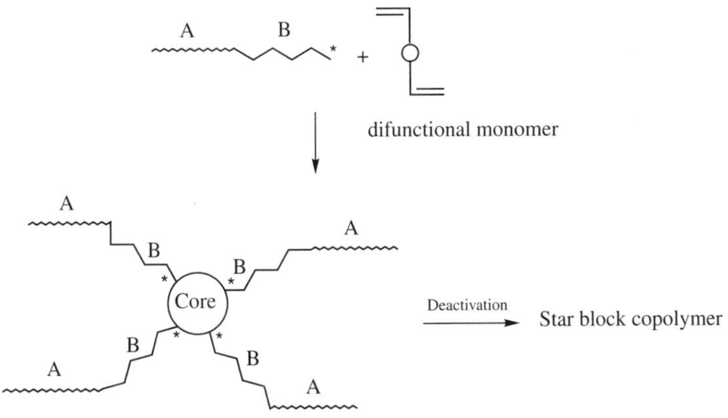

A B * + difunctional monomer

Core, Deactivation, Star block copolymer

Scheme 8.6

broad distribution in the number of arms, due to the statistical nature of the last step are prepared. Therefore, the number of arms in the final star block copolymer is not well-defined although some control can be exercised, mainly through the ratio of the concentration of the difunctional monomer to the concentration of the active centers (Bi 1975). However, other factors, including the molecular weight of the arms, overall concentration of the active centers, reaction time, and temperature can influence the functionality of the product (Tsitsilianis 1991). A small amount of unlinked arm is always present in the final crude reaction product, and sometimes presence of gel has been detected. An advantage of this procedure is that the core of the star bears a number of active sites, in principle, equal to the number of the chains that have been connected on it. These active sites can be used for the initiation of the polymerization of a second monomer, resulting in the production of $(AB)_nC_n$ stars having an equal number of branches of different chemical constitution and molecular weight (Scheme 8.7). In the second procedure, the core is

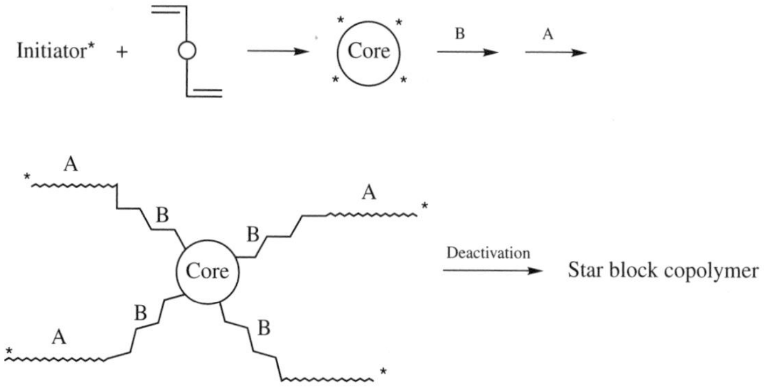

Initiator* + → Core → B → A →

Deactivation → Star block copolymer

Scheme 8.7

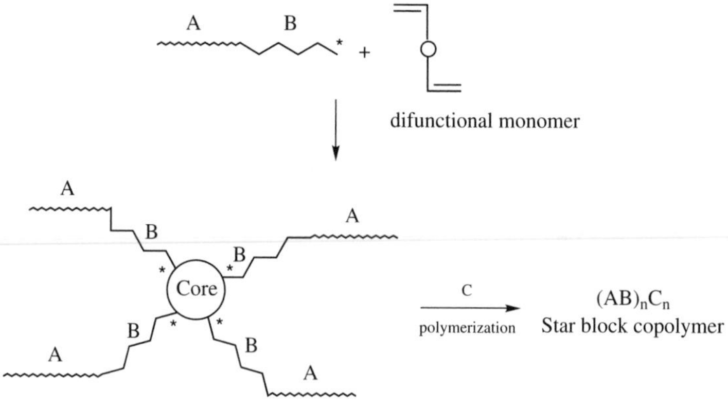

difunctional monomer

C polymerization → $(AB)_nC_n$ Star block copolymer

Scheme 8.8

synthesized first by polymerization of the difunctional monomer, followed by addition of the monomers comprising the block copolymer arms in the desired sequence (Scheme 8.8).

Through the use of the difunctional monomer approach, a number of star block copolymers has been synthesized by both anionic and cationic polymerization using divinylbenzene as the difunctional monomer due to its ability for polymerization by both ionic mechanisms. In cationic polymerization other types of difunctional monomers have been also used in order to produce star block copolymers of monomers like vinylethers. Because the formation of the core is a polymerization procedure the difunctional monomer must be able to be polymerized or its active species to initiate the polymerization of the monomers forming the blocks. Thus, living diblock copolymers of vinyl ethers and ester-containing vinyl ethers were reacted with a difunctional vinylether to produce star-shaped block copolymers (Scheme 8.9). Also, in this case the mode of diblock connection to the central core

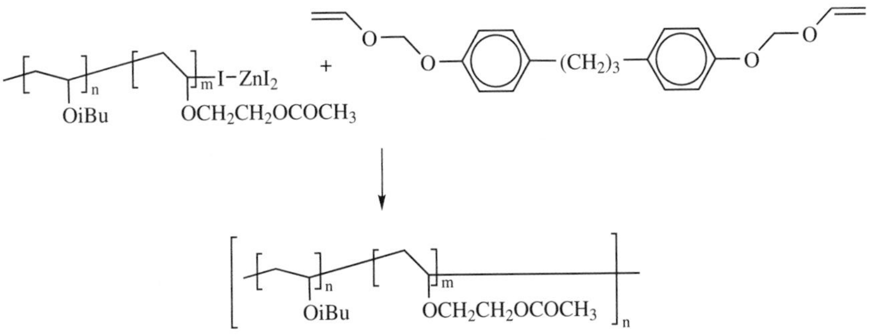

Scheme 8.9

could be chosen due to the interchangeability of the sequential polymerization of the two monomers (Kanaoka 1991).

Amphiphilic star block copolymers were synthesized by ring-opening metathesis polymerization (Saunders 1992). Sequential polymerization of norbornene-type, functionalized and nonfunctionalized monomers, using Mo(CH-tBu)(NAr)(O-t-$Bu)_2$ as initiator, produced narrow polydispersity diblocks. The living arms were reacted with the difunctional monomer endo-cis-endo-hexacyclo-$[10.2.1.1.^{3,4}1^{5,8}.0^{2,11}.0^{4,9}]$ heptadeca-6,13-diene, to produce the star block copolymers. The functionalized monomers, bearing carboxylic groups protected by trimethylsilyl groups, were deprotected after the linking reaction in order to give the amphiphilic star block copolymers (Scheme 8.10).

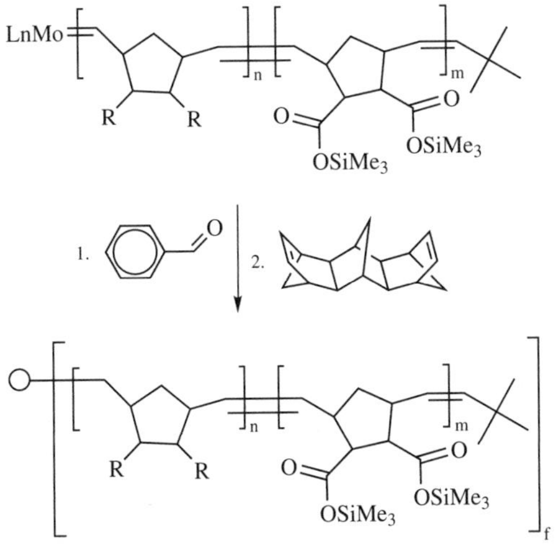

$(AB)_f$ star-block copolymer

Scheme 8.10

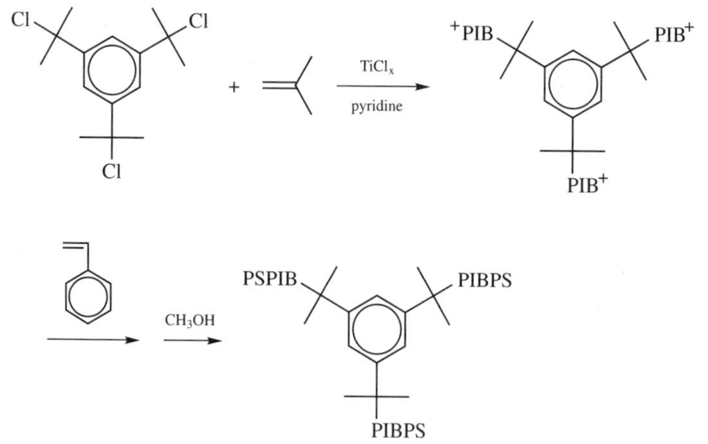

Scheme 8.11

Multifunctional initiators can be used in the preparation of star block copolymers, usually with a small number of branches. A trifunctional cationic polymerization initiator, derived from tricumylchloride, has been used for the preparation of three-arm star block copolymers of polyisobutylene and polystyrene (Scheme 8.11) (Storey 1993). Another trifunctional initiator system composed of tris(trifluoroacetate) and ethyaluminum chloride has been used for the preparation of poly-(isobutyl vinyl ether)-b-poly(2-hydroxyethyl vinyl ether) star block copolymers having narrow molecular weight distributions (Shohl 1991). Using the same initiator as Storey et al., Kennedy and coworkers prepared three-arm star polyisobutylenes by cationic polymerization. The terminal Cl groups were converted to isobutyryl bromide groups, and they were used for initiation of the free radical polymerization of methylmethacrylate, resulting in three-arm star block copolymers with PIB inner blocks and PMMA outer blocks (Keszler 2000). The same authors reported the synthesis of well-defined eight-arm star block copolymers using calix[8]arene derivatives as initiators and $BCl_3/TiCl_4$ as coinitiators. By this methodology star block copolymers containing isobutylene and styrene, as well as isobutylene and p-chlorostyrene, were synthesized and characterized (Jacob 1998, Scheme 8.12).

Sawamoto et al. reported the synthesis of a three-arm star block copolymer of 2-chloroethyl vinyl ether (CEVE) and p-(phthalimidomethyl)styrene (ImSt) using a trifunctional initiator, CH_3-C-[Ph-OCH_2CH_2O-CH(CH_3)Cl$)_3$ (Takahashi 1999). Polymerization started with the addition of CEVE in a mixture of initiator, $SnCl_4$, nBu_4NCl, and 2,6-di-tert-butyl-4-methylpyridine in CH_2Cl_2 at −78°C. ImSt was subsequently polymerized. The copolymers had relatively narrow molecular weight distributions. Postpolymerization hydrazinolysis of the imide functions of the ImSt block resulted in the formation of an amphiphilic three-arm

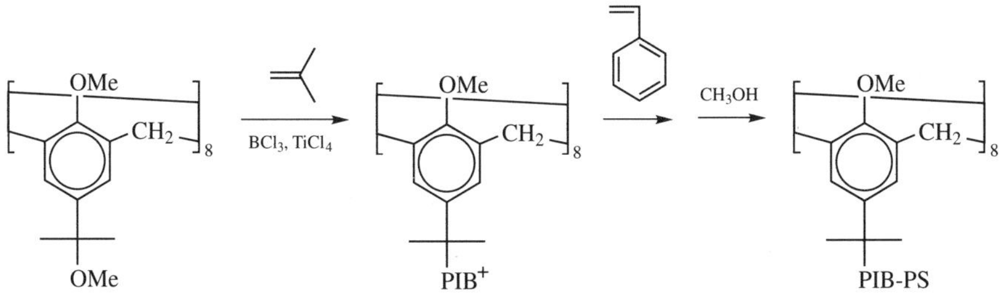

OMe
CH_2
8
BCl_3, $TiCl_4$
OMe
CH_2
8
CH_3OH
OMe
CH_2
8
OMe
PIB^+
PIB-PS

Scheme 8.12

star block copolymer consisting of CEVE hydrophobic segments and p-(amino-methyl)styrene hydrophilic segments (Scheme 8.13).

Star block copolymers of MMA and nBuMA were prepared by ATRP methodology, using a calyx[8]arene based multifunctional initiator in the presence of $RuCl_2(PPh_3)_3$ and $Al(OiPr)_3$ (Ueda 1998).

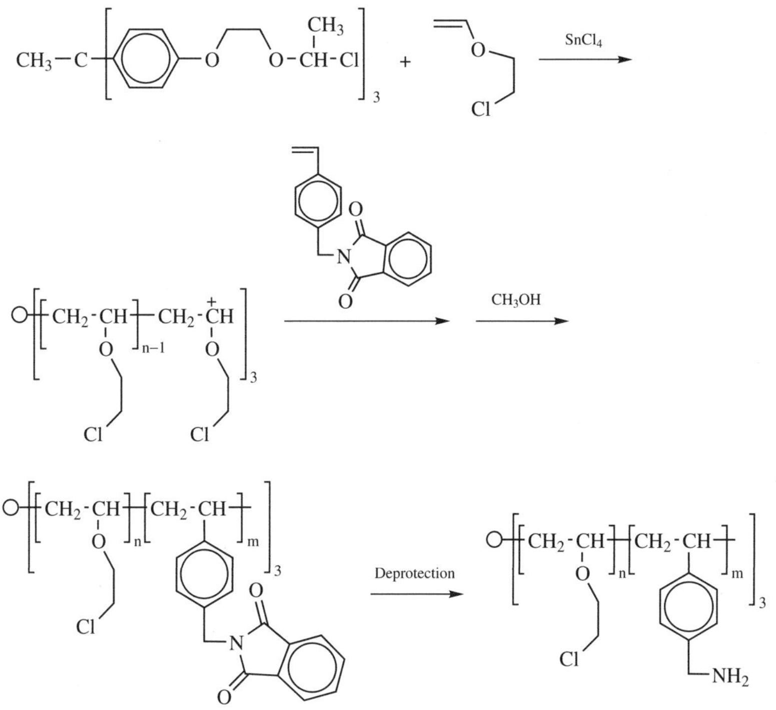

CH_3
CH_3-C
O
O-CH-Cl
3
+
O
Cl
$SnCl_4$

O
N
O
CH_2-CH CH_2-$\overset{+}{C}$H
O n−1 O
3
Cl Cl
CH_3OH

CH_2-CH CH_2-CH
O n m
3
Cl
O
N
O
Deprotection
CH_2-CH CH_2-CH
O n m
3
Cl
NH_2

Scheme 8.13

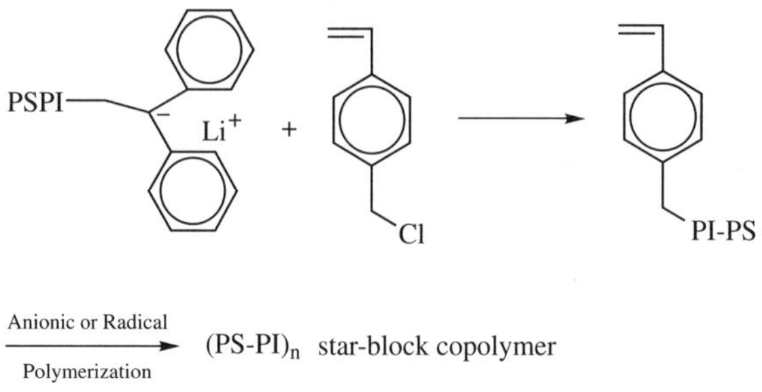

PSPI— (1,1-diphenylalkyl)$^{-}$ Li^+ + p-chloromethylstyrene (Cl) ⟶ styrene–PI-PS

$\xrightarrow[\text{Polymerization}]{\text{Anionic or Radical}}$ $(PS\text{-}PI)_n$ star-block copolymer

Scheme 8.14

Polymerization of diblock macromonomers can also produce star block copolymers (Ishizu 1994 and 1991). A diblock copolymer synthesized by anionic polymerization can be end-functionalized by p-chloromethylstyrene or DVB. The polymerizable double bond at the end of the diblock chain can be polymerized by anionic or radical polymerization mechanisms to produce polymacromonomers of low degree of polymerization, which are essentially star block copolymers (Scheme 8.14).

2. GRAFT COPOLYMERS

Graft copolymers are comprised of a main polymer chain, the backbone, having one or more side polymer chains attached to it through covalent bonds, the branches. The chemical nature and composition of the backbone and the branches differ in most cases. Branches are usually distributed randomly along the backbone although, recently, advances in synthetic methods allowed the preparation of more well-defined structures.

Randomly branched graft copolymers can be prepared by three general synthetic methods: 1) the "grafting to", 2) the "grafting from", and 3) the "grafting through" or macromonomer method (Pitsikalis 1998, Cowie 1989) (Scheme 8.15).

In the "grafting to" method, the backbone and the arms are prepared separately by a living polymerization mechanism. The backbone bears reactive groups distributed along the chain that can react with the living branches. Mixing the backbone and the living branches in the desired proportion and under the appropriate experimental conditions, a coupling reaction takes place resulting in the final graft copolymer. By the use of a living mechanism, the molecular weight, molecular weight polydispersity, and the chemical composition of the backbone and branches can be controlled. Additionally, both backbone and branches can be isolated, before coupling reaction, and characterized separately. The average number of branches can be controlled primarily by the number of the functional groups (branching sites) present in the backbone and sometimes by the ratio of the

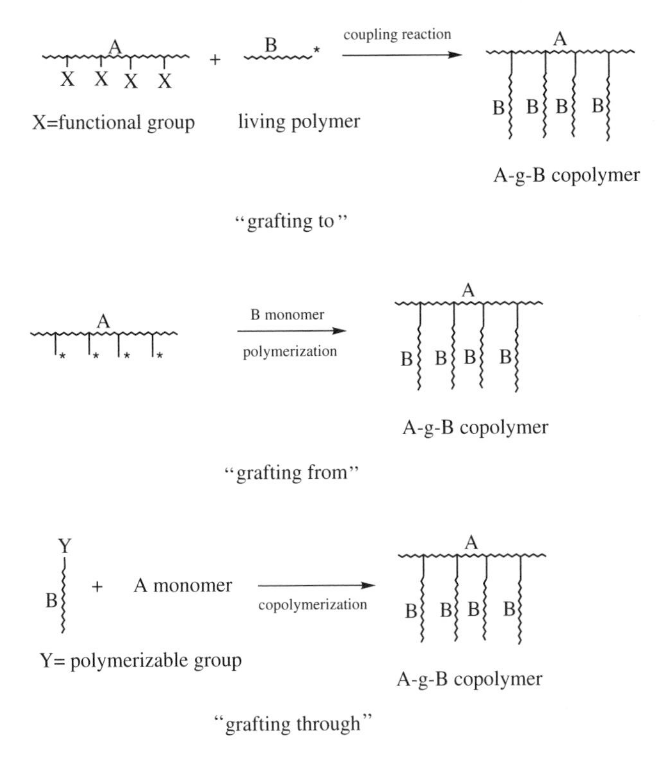

Scheme 8.15

functional groups to the active chain-end concentration of the branches. The branching sites can be introduced into the backbone either by postpolymerization reactions or by copolymerization of the main backbone monomer(s) with a suitable comonomer, bearing the desired functional group (unprotected or in a protected form if this functional group may interfere with the polymerization reaction). Branches of graft copolymers are commonly prepared by anionic polymerization, and backbones with electrophilic functionalities, such as anhydrides, esters, pyridine, chlorosilane, or benzylic halide groups, are employed. The actual average number of branches in the final copolymer can be determined from the overall molecular weight of the copolymer and the molecular weights of the backbone and the branches.

A representative example is the preparation of poly(butadiene-g-styrene) and poly(butadiene-g-styrene$_2$) copolymers (Xenidou 1999), where the PBd backbone is synthesized by anionic polymerization, followed by introduction of chlorosilane groups, via postpolymerization hydrosilylation, and, finally, linking with living polystyrene anions (Scheme 8.16).

In the "grafting from" method after the preparation of the backbone, active sites are produced along the main chain, which are able to polymerize the monomer(s)

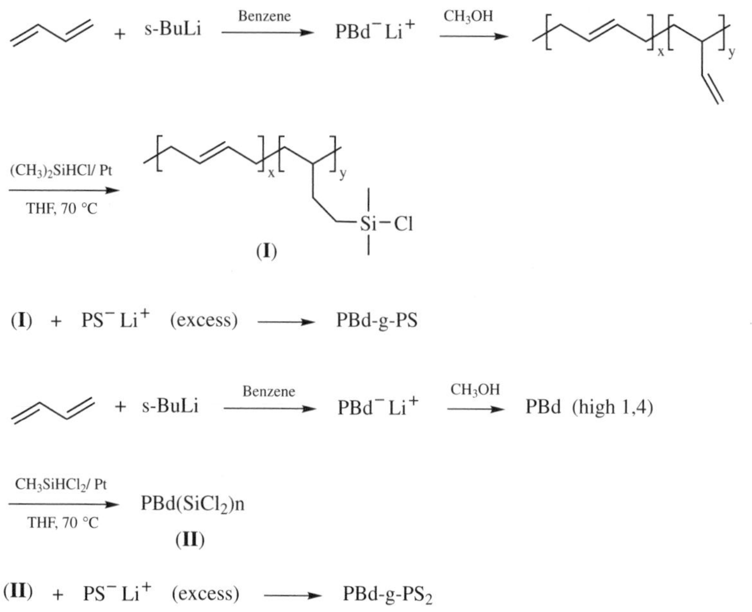

$$\text{butadiene} + \text{s-BuLi} \xrightarrow{\text{Benzene}} \text{PBd}^-\text{Li}^+ \xrightarrow{\text{CH}_3\text{OH}} [\text{1,4-PBd}]_x[\text{1,2-PBd}]_y$$

$$\xrightarrow[\text{THF, 70 °C}]{(\text{CH}_3)_2\text{SiHCl/ Pt}} [\text{1,4-PBd}]_x[\text{CH}_2\text{CH}(\text{CH}_2\text{CH}_2\text{Si}(\text{CH}_3)_2\text{Cl})]_y \quad \textbf{(I)}$$

$$\textbf{(I)} + \text{PS}^-\text{Li}^+ \text{ (excess)} \longrightarrow \text{PBd-g-PS}$$

$$\text{butadiene} + \text{s-BuLi} \xrightarrow{\text{Benzene}} \text{PBd}^-\text{Li}^+ \xrightarrow{\text{CH}_3\text{OH}} \text{PBd (high 1,4)}$$

$$\xrightarrow[\text{THF, 70 °C}]{\text{CH}_3\text{SiHCl}_2\text{/ Pt}} \text{PBd(SiCl}_2\text{)n} \quad \textbf{(II)}$$

$$\textbf{(II)} + \text{PS}^-\text{Li}^+ \text{ (excess)} \longrightarrow \text{PBd-g-PS}_2$$

Scheme 8.16

that will comprise the branches. Polymerization of the second monomer results in the formation of branches and the graft copolymer. The number of branches can be controlled by the concentration of active sites generated along the backbone, assuming that each one of them participates in the formation of one branch. Obviously, the isolation and characterization of each part of the graft copolymer in this case is almost impossible. Knowledge of precursor molecular characteristics is limited only to the case of the backbone. Isolation of the branches can be achieved only in some cases and usually involves selective chemical decomposition of the backbone, e.g., ozonolysis of polydiene backbone in poly(diene-g-styrene) graft copolymers (Hadjichristidis 1978). Following "grafting from" methodology several graft copolymers can be synthesized by the use of a number of different polymerization techniques because radicals, anions, and cations can be generated along a polymer chain. Generation of radicals on a preformed backbone has proven to be a very versatile and commercially attractive way of producing graft copolymers due mainly to its simplicity in performing the various reaction steps. The new advances in the living (controlled) radical polymerization techniques led to the preparation of several well-defined graft copolymers (Matyjaszewski 1999, Grubbs 1997, Moschogianni 2001). A typical example is the use of chloromethylated polystyrene, produced by controlled radical copolymerization of styrene and chloromethylstyrene, as a multifunctional ATRP initiator for the formation of graft copolymers with polystyrene backbones and branches comprised of different (meth)acrylate monomers (Scheme 8.17a).

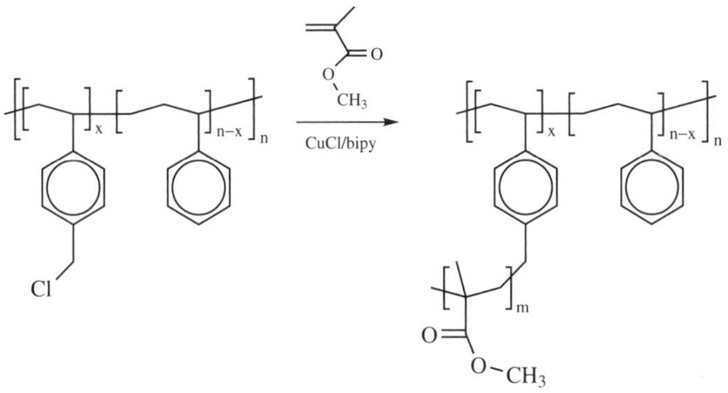

(*a*)

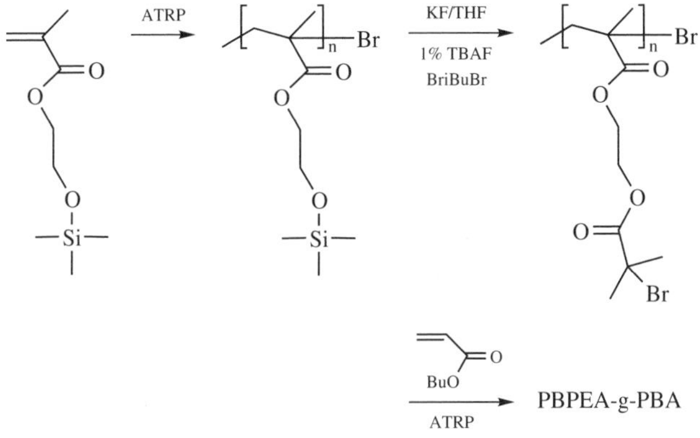

(*b*)

Scheme 8.17

Densely grafted copolymers, having almost one grafted chain per monomeric unit of the backbone, were synthesized by Matyjaszewski et al. (Beers 1998). The backbone was produced by ATRP polymerization of a functional methacrylate monomer (trimethylsilane protected 2-hydroxylethyl methacrylate), which was converted by postpolymerization reactions to poly[2-(2-bromoisobutyryloxy)ethyl methacrylate]. The bromine moieties of this polymer were used as ATRP-initiating sites for the polymerization of different monomers like styrene and butylacrylate. An alternative route using poly[2-(2-bromopropionyloxy)-ethyl acrylate) as the backbone was also described by the same authors (Scheme 8.17b).

Anionic active sites can be generated by metallation of allylic, benzylic, or aromatic C-H bonds, present in the backbone, by organometallic compounds, such as s-BuLi, in the presence of strong chelating agents that facilitate the reaction. The metallation of polydienes with s-BuLi in the presence of

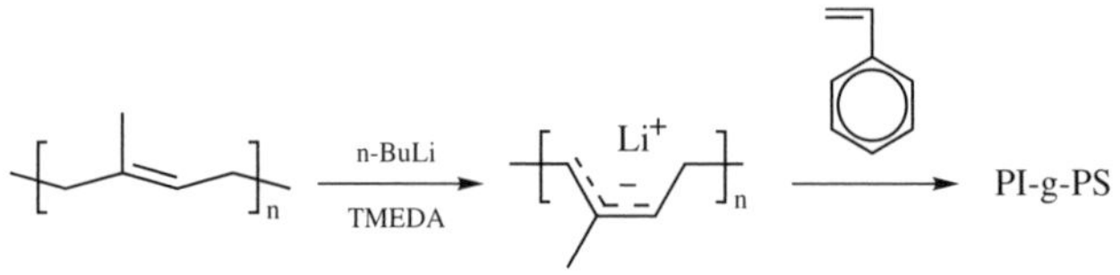

n-BuLi
TMEDA
Li^+
PI-g-PS

Scheme 8.18

N,N,N′,N′-tetramethylethylenediamine (TMEDA) constitutes a representative example (Hadjichristidis 1978). Metallation of PI and PBd by this procedure and subsequent polymerization of styrene led to the formation of PI-g-PS and PBd-g-PS copolymers with well-defined molecular characteristics (Scheme 8.18).

Cationic active sites can be formed in chains bearing labile halogen atoms. Polymers of this category include polychloroprene, poly(vinyl chloride), chlorinated styrene-butadiene rubber, etc. These groups can be used as initiating sites, in the presence of Lewis acids, for the polymerization of cationically polymerized monomers like isobutylene, styrene and derivatives, alkylvinyl ethers, tetrahydrofurane, etc., generating graft copolymers (Kennedy 1992).

In the "grafting through" method, preformed macromonomers are copolymerized with another monomer in order to produce the graft copolymer (Rempp 1984). Macromonomers are oligomeric or polymeric chains that have a polymerizable end group. In this case the macromonomer comprises the branch of the copolymer, and the backbone is formed in situ. The number of branches per backbone can be generally controlled by the ratio of the molar concentrations of the macromonomer and the comonomer (Hawker 1997, Wang 1998). Several other factors have to be considered. Among them the most important one is the copolymerization behavior of the macromonomer and the comonomer forming the backbone. Depending on the reactivity ratios, r_1 and r_2, of the reacting species, different degrees of randomness can be achieved, with respect to the placement of the branches. Because macromonomer and comonomer incorporation in the graft copolymer can vary in the course of the copolymerization reaction due to changes in the concentration of the two compounds in the mixture, different kinds of graft copolymers are formed as a function of time. Phase separation can also occur in these systems due to the formation of the copolymers, leading to increased compositional and molecular weight heterogeneity of the final product. PS macromonomer formation by anionic polymerization and its subsequent use for preparation of graft copolymer with polymethyl methacrylate backbone are given as a representative example of the use of macromonomers in the synthesis of graft copolymers (Arnold 1990, Scheme 8.19). A more detailed discussion of the use of macromonomers in block copolymer synthesis has been presented elsewhere (Pitsikalis 1998).

Graft copolymers of St and Is, with trifunctional or tetrafunctional branching points, situated equidistantly on the backbone, were synthesized by a combination of living anionic and condensation polymerization methodologies (Iatrou 1999).

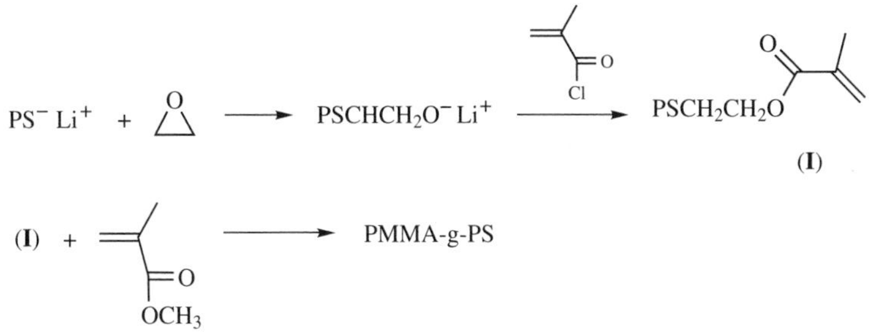

$PS^- Li^+$ + [ethylene oxide] → $PSCHCH_2O^- Li^+$ → (methacryloyl chloride) $PSCH_2CH_2O$–methacrylate **(I)**

(I) + methyl methacrylate ($C{=}O$, OCH_3) → PMMA-g-PS

Scheme 8.19

In the case of graft copolymers with the trifunctional branching points, a living PSLi was reacted with excess $MeSiCl_3$ to give a macromolecule with two terminal SiCl bonds. This was reacted, after removal of excess silane, with a difunctional PI produced by the difunctional initiator derived from PEB and s-BuLi, following a polycondensation reaction scheme, giving a well-defined graft copolymer with PI backbone and PS branches. In this way the length of the branches and the distance between them on the backbone could be controlled. Some control over the average number of branches could also be exercised through the ratio of the difunctional macromonomers used. In the case of the grafts with tetrafunctional branching points, the PS branch was reacted with $SiCl_4$ in a controllable way to give $(PS)_2SiCl_2$. The first branch was introduced by reaction with excess $SiCl_4$, which was subsequently removed, and the second PS was introduced slowly in order to avoid substitution of the third Cl. The macromolecular linking agent was reacted with a predetermined amount of a difunctional PI to give the desired graft copolymer (Scheme 8.20). In both cases crude reaction products had relatively wide molecular weight distributions due to the condensation mechanism used in the second synthetic step, but, after fractionation, narrow molecular weight distribution products were obtained.

π-shaped graft copolymers, i.e., graft copolymers with a PI backbone and two identical PS branches, were synthesized by anionic polymerization (Gido 1996). PILi was reacted with excess $MeSiCl_3$ in order to produce the macromolecular linking agent $PISiCl_2$. PSLi was slowly added to $PISiCl_2$ aiming to the replacement of the second Cl atom. The course of the reaction was followed by SEC. The junction point-functionalized PSSiClPI diblock was subsequently reacted with a difunctional living PI in a 2.1:1 ratio to give the pure π–shaped copolymer after fractionation. In this case the length of the PS branches, of the connecting part of the backbone, and the length of the two backbone parts between the branching points and the ends of the main chain could be controlled, leading to a well-defined graft architecture (Scheme 8.21).

Exact graft copolymers with a PI backbone and two PS branches, where the position of the branching points along the backbone and the length of each branch could be controlled exactly (Paraskeva 1999), were synthesized using anionic polymerization methodology and nonpolymerizable DPE derivatives for branching-point formation (Scheme 8.22). Living PILi was produced and reacted with PEB (p-isomer) in the presence of a small amount of THF. Taking advantage of the differences in the reactivity of the two double bonds and the appropriate stoichiometry,

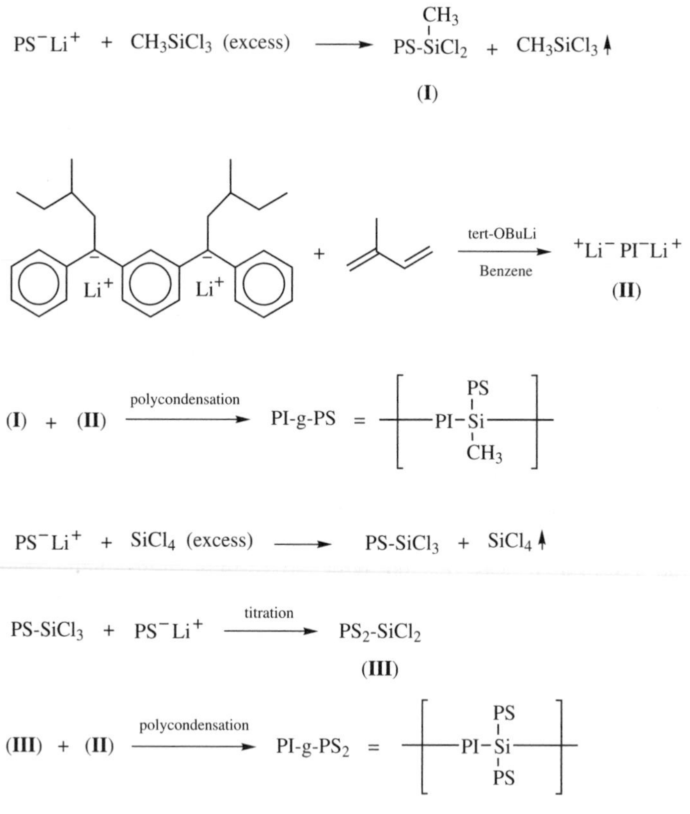

PS⁻Li⁺ + CH3SiCl3 (excess)
CH3
PS-SiCl2 + CH3SiCl3
(I)
Li⁺
Li⁺
tert-OBuLi
Benzene
⁺Li⁻ PI⁻Li⁺
(II)
(I) + (II)
polycondensation
PI-g-PS =
PS
PI-Si
CH3
PS⁻Li⁺ + SiCl4 (excess)
PS-SiCl3 + SiCl4
PS-SiCl3 + PS⁻Li⁺
titration
PS2-SiCl2
(III)
(III) + (II)
polycondensation
PI-g-PS2 =
PS
PI-Si
PS

Scheme 8.20

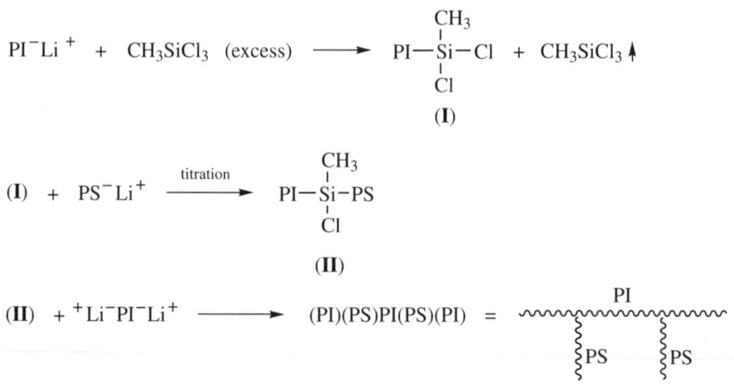

PI⁻Li⁺ + CH3SiCl3 (excess)
CH3
PI—Si—Cl + CH3SiCl3
Cl
(I)
(I) + PS⁻Li⁺
titration
CH3
PI—Si—PS
Cl
(II)
(II) + ⁺Li⁻PI⁻Li⁺
(PI)(PS)PI(PS)(PI) =
PI
PS
PS

Scheme 8.21

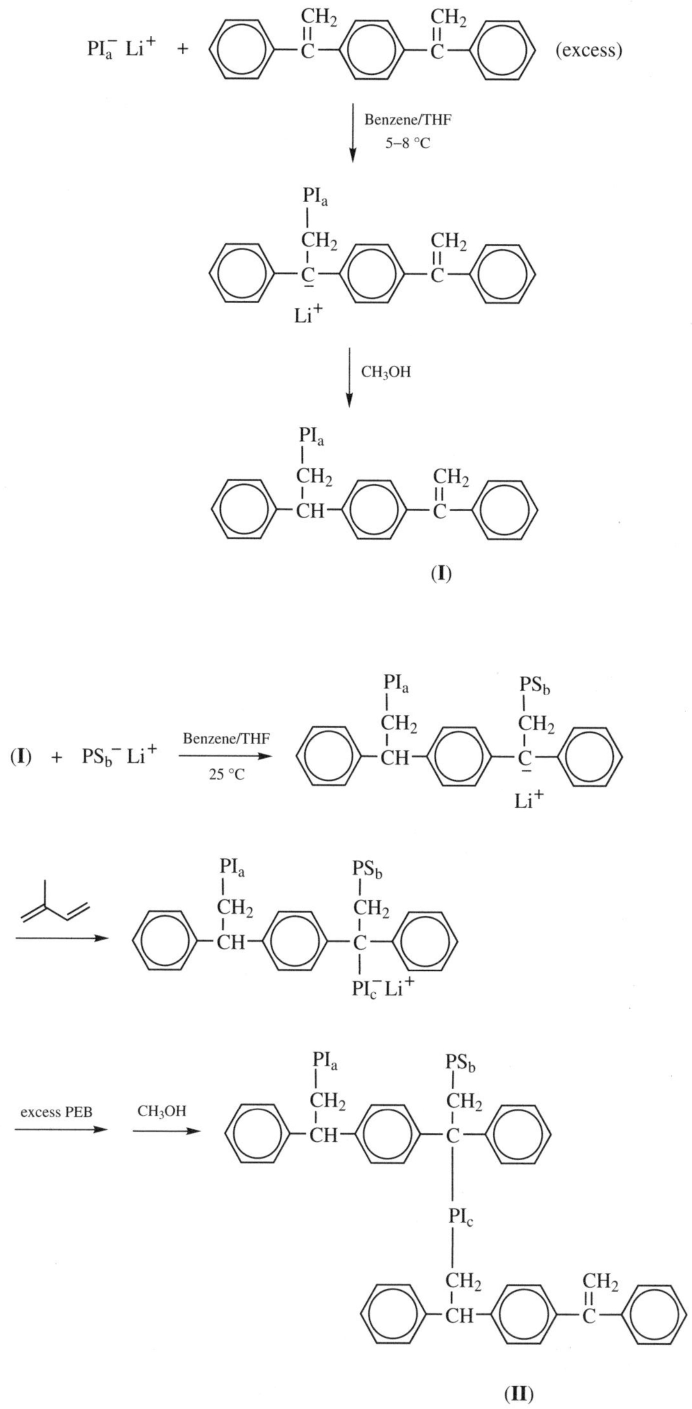

$PI_a^- Li^+$ + $CH_2=C(C_6H_5)-C_6H_4-C(C_6H_5)=CH_2$ (excess)

Benzene/THF, 5–8 °C

$PI_a-CH_2-C^-(C_6H_5)(Li^+)-C_6H_4-C(C_6H_5)=CH_2$

CH_3OH

$PI_a-CH_2-CH(C_6H_5)-C_6H_4-C(C_6H_5)=CH_2$

(**I**)

(**I**) + $PS_b^- Li^+$ → (Benzene/THF, 25 °C) $PI_a-CH_2-CH(C_6H_5)-C_6H_4-C^-(C_6H_5)(Li^+)-CH_2-PS_b$

→ (isoprene) $PI_a-CH_2-CH(C_6H_5)-C_6H_4-C(C_6H_5)(CH_2-PS_b)-PI_c^- Li^+$

→ (excess PEB) → (CH_3OH) $PI_a-CH_2-CH(C_6H_5)-C_6H_4-C(C_6H_5)(CH_2-PS_b)-PI_c-CH_2-CH(C_6H_5)-C_6H_4-C(C_6H_5)=CH_2$

(**II**)

Scheme 8.22

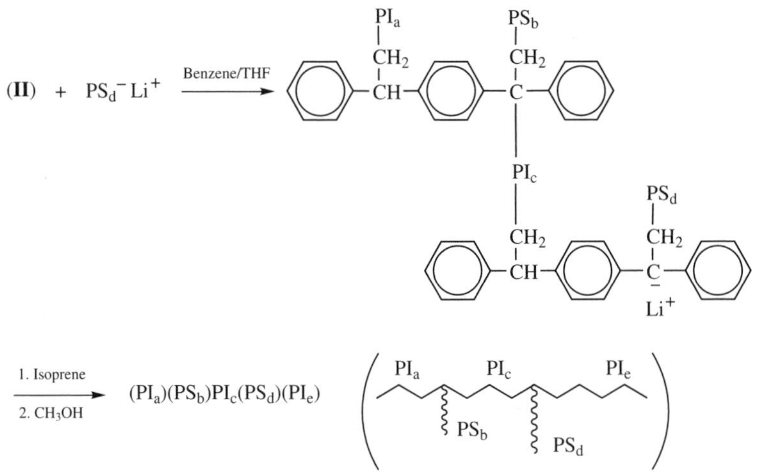

(II) + $PS_d^-Li^+$ $\xrightarrow{\text{Benzene/THF}}$

$\xrightarrow[\text{2. } CH_3OH]{\text{1. Isoprene}}$ $(PI_a)(PS_b)PI_c(PS_d)(PI_e)$

Scheme 8.22 (*Continued*)

controlled by titration, only one chain was introduced on the DPE derivative. The product was isolated and purified. Then it was reacted with living PSLi in order to introduce the second part. After complete reaction, the living centers on PEB were used to polymerize Is in the presence of THF. The resulting ABB' star had an anionic active center at the end of one of the B branches, which can react with PEB again, giving a star-shaped double bond functionalized polymer. After isolation and thorough purification of the intermediate, another PS branch was introduced by reaction with the remaining double bond. The active site was used for the formation of the last part of the molecule by polymerization of Is. Despite the numerous reaction steps and the demanding purification of the intermediate products, the method allows control over almost all molecular characteristics of the final graft copolymer and can be extended to the preparation of grafts with a higher number of branches.

3. MIKTOARM STAR COPOLYMERS

Miktoarm (mikto from the Greek word μικτός meaning mixed) star copolymers are a special class of nonlinear block copolymers where arms of different chemical nature and/or composition are linked to the same branch point (Hadjichristidis 1999). These block copolymers have been synthesized mainly by anionic polymerization methods, although some examples of synthesis by other methodologies have appeared in the literature. The number of different kind of arms can be varied as well as the total number of arms, giving rise to a variety of miktoarm stars like A_2B, A_3B, A_5B (or generally A_nB), A_nB_n, ABC, or even ABCD. Several approaches have been reported for the preparation of miktoarm stars, with each one of them having specific advantages and disadvantages.

In the so-called chlorosilane method, chlorosilane compounds of various functionality are used as linking agents where living chains are linked in a controllable way in order to produce different types of miktoarm stars. Using this method A_2B miktoarms can be synthesized by reaction of living B chains with excess of trichlorosilane, producing a macromolecule having two reactive Si-Cl bonds at its end (Mays 1990, Iatrou 1995, Pochan 1996). After elimination of excess unreacted silane under high vacuum conditions, the macromolecular linking agent can be reacted with excess living A arms to give the A_2B miktoarm star (A and B can be PS, PI, or PBd) (Scheme 8.23). Employing the same basic idea, A_3B

$$PS^-Li^+ + CH_3SiCl_3\ (excess) \longrightarrow PSSi(CH_3)Cl_2 + LiCl + CH_3SiCl_3 \uparrow$$

$$PSSi(CH_3)Cl_2 + PI^-Li^+\ (excess) \longrightarrow PS(PI)_2$$

Scheme 8.23

(Tselikas 1996) and $(BA)A_3$ miktoarm stars (Avgeropoulos 1997), with A = PI and B = PS, were prepared using $SiCl_4$ as the linking agent. For the formation of A_nB stars with $n > 3$, a stoichiometric quantity of silane of the appropriate functionality has to be used ($SiCl_n$:Li = 1:1) because silanes with functionality of five or more are nonvolatile and excess of these compounds cannot be eliminated easily from the reaction mixture by distillation or sublimation under high vacuum. Reaction of only one chlorine with living PS arm is ensured by adding a very dilute solution of PSLi dropwise to the vigorously stirred concentrated solution of silane. After completion of the capping reaction, the excess of PI arm is added, and, after several weeks, the desired miktoarm star is obtained (Velis 1999).

The synthesis of A_2B miktoarm stars with A = PS and B = P2VP involves the use of silane compounds in a different way (Khan 1992). CH_3SiHCl_2 was used as a coupling agent for the formation of a dimer, $(PS)_2Si(CH_3)H$. Living P2VP chains were end-capped with a hydrocarbon derivative having a terminal double bond. Addition of the Si-H group of the dimer to the terminal double bond of P2VP resulted in the preparation of the $(PS)_2$P2VP miktoarm stars (Scheme 8.24).

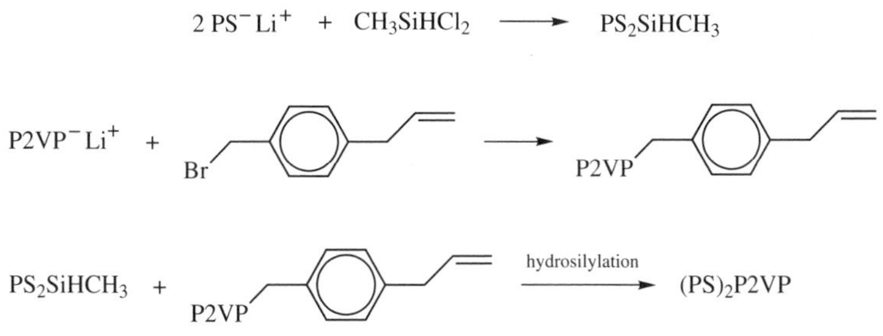

$$2\ PS^-Li^+ + CH_3SiHCl_2 \longrightarrow PS_2SiHCH_3$$

P2VP⁻Li⁺ + Br–CH₂–C₆H₄–CH₂CH=CH₂ ⟶ P2VP–CH₂–C₆H₄–CH₂CH=CH₂

PS_2SiHCH_3 + P2VP–CH₂–C₆H₄–CH₂CH=CH₂ —hydrosilylation→ $(PS)_2$P2VP

Scheme 8.24

$$PS^-Li^+ + SiCl_4\,(excess) \longrightarrow PS\text{-}SiCl_3 + LiCl + SiCl_4\uparrow$$

$$PS\text{-}SiCl_3 + PS^-Li^+ \xrightarrow{titration} (PS)_2\text{-}SiCl_2 + LiCl$$

$$(PS)_2\text{-}SiCl_2 + PBd^-Li^+\,(excess) \longrightarrow (PS)_2(PBd)_2 + 2\,LiCl$$

Scheme 8.25

A_2B_2 miktoarm stars with A = PS and B = PBd were obtained by adding the first PS arm to $SiCl_4$ in a manner analogous to the A_2B case, followed by slow addition of the second PS arm until complete substitution of the second chlorine atom (as evidenced by SEC analysis) (Iatrou 1993). The final miktoarm can be formed after complete reaction of A_2SiCl_2 with excess of the second living arms (Scheme 8.25). In a different approach, $SiCl_4$ can be reacted with living PS in a stoichiometric ratio $SiCl_4$:PSLi = 1:2, taking advantage of the steric hindrance of the phenyl ring of the PSLi chain ends, and then PS_2SiCl_2 can react with excess of second living arms (Wright 1994). In this manner $(PS)_2(PEO)_2$ miktoarm stars have been synthesized (Xie 1987). In the case of $(PI)_2(PBd)_2$ miktoarms, PI arms end-capped with a few units of styrene were added first to $SiCl_4$ followed by the addition of PBd (Algaier 1996). Alternatively, the linking of the first two PI arms was effected, without end-capping with styrene, at −40°C and using a molar ratio of $SiCl_4$:PILi = 1:2. End-capping with styrene or carrying out the linking reaction at −40°C avoids the incorporation of a third chain into the $SiCl_4$.

ABC miktoarm stars containing PS, PI, and PBd have been synthesized following the procedure outlined in Scheme 8.26 (Iatrou 1992). The first step

$$PI^-Li^+ + (CH_3)SiCl_3\,(excess) \longrightarrow PI\text{-}Si(CH_3)Cl_2 + LiCl + (CH_3)SiCl_3\uparrow$$

$$PI\text{-}Si(CH_3)Cl_2 + PS^-Li^+ \xrightarrow{titration} (PS)(PI)\text{-}Si(CH_3)Cl + LiCl$$

$$(PS)(PI)\text{-}Si(CH_3)Cl + PBd^-Li^+\,(excess) \longrightarrow (PS)(PI)(PBd) + LiCl$$

Scheme 8.26

involved addition of PI arm to excess of $SiMeCl_3$, followed by titration of PI-Si $(CH_3)Cl_2$ with continuous SEC monitoring of the linking reaction. Finally, living PBd arm was added in excess, and the (PS)(PI)(PBd) ABC miktoarm star terpolymer was produced. ABC miktoarms with PS, PI, and PDMS arms were recently prepared (Bellas 2000) as well as miktoarms with PS, PI, and P2VP arms (Zioga 2000). Using the same methodology, asymmetric AA'B miktoarm stars were synthesized with A, A′ being PI arms of different length and B being PS (Lee 1997).

In a slightly modified procedure, ABC miktoarm stars of PS, PI, and PMMA were synthesized (Sioula 1997). The synthetic scheme involved the preparation of the (PS)(PI)Si(CH_3)Cl in hydrocarbon solvent and its subsequent reaction with a dianion formed by 1,1-diphenylethylene and Li in THF. Due to the stoichiometry used, only one anion reacted with the SiCl bond, producing a reactive anionic site at the junction point of the PS-PI diblock. These anionic sites used for the polymerization of MMA at low temperatures, leading to the formation of the third arm of the miktoarm star (Scheme 8.27).

The synthesis of ABCD miktoarm star quaterpolymer where A = PI, B = P4MeS poly(4-methyl styrene), C = PS, and D = PBd was accomplished in a manner similar to the ABC case (Iatrou 1993). Advantage of differences in steric hindrance of the anions used was taken. PI was introduced first to $SiCl_4$ in the usual way. The most sterically hindered P4MeSLi anion was slowly added, and, after

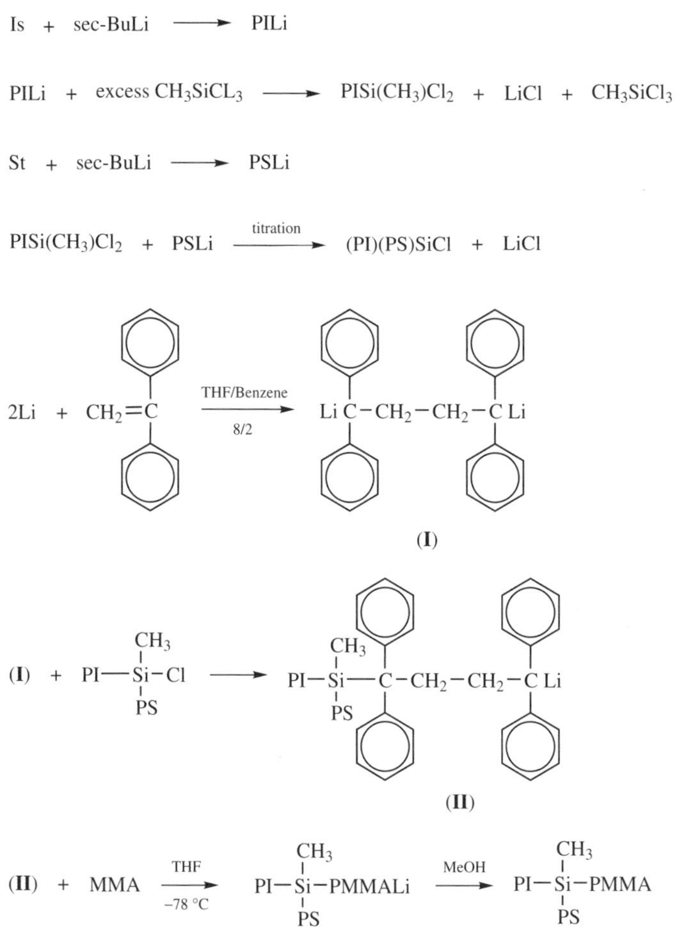

Is + sec-BuLi ⟶ PILi

PILi + excess CH_3SiCL_3 ⟶ $PISi(CH_3)Cl_2$ + LiCl + CH_3SiCl_3

St + sec-BuLi ⟶ PSLi

$PISi(CH_3)Cl_2$ + PSLi $\xrightarrow{\text{titration}}$ (PI)(PS)SiCl + LiCl

2Li + $CH_2{=}C(C_6H_5)_2$ $\xrightarrow[8/2]{\text{THF/Benzene}}$ $Li\,C(C_6H_5)_2{-}CH_2{-}CH_2{-}C(C_6H_5)_2\,Li$

(I)

(I) + $PI{-}Si(CH_3)(PS){-}Cl$ ⟶ $PI{-}Si(CH_3)(PS){-}C(C_6H_5)_2{-}CH_2{-}CH_2{-}C(C_6H_5)_2\,Li$

(II)

(II) + MMA $\xrightarrow[-78\,^{\circ}C]{\text{THF}}$ $PI{-}Si(CH_3)(PS){-}PMMALi$ $\xrightarrow{\text{MeOH}}$ $PI{-}Si(CH_3)(PS){-}PMMA$

Scheme 8.27

$$PI^-Li^+ + SiCl_4 \text{ (excess)} \longrightarrow PI\text{-}SiCl_3 + LiCl + SiCl_4 \uparrow$$

$$PI\text{-}SiCl_3 + P4MeS^-Li^+ \xrightarrow{\text{titration}} (PI)(P4MeS)\text{-}SiCl_2 + LiCl$$

$$(PI)(P4MeS)\text{-}SiCl_2 + PS^-Li^+ \xrightarrow{\text{titration}} (PI)(P4MeS)(PI)\text{-}SiCl + LiCl$$

$$(PS)(P_4MeS)(PI)\text{-}SiCl + PBd^-Li^+ \text{ (excess)} \longrightarrow (PS)(P4MeS)(PI)(PBd) + LiCl$$

Scheme 8.28

complete reaction with the second Si-Cl bond, PSLi was introduced to the macromolecular linking agent through titration. The least sterically hindered anion, PBdLi, was introduced last and in excess to ensure complete reaction with the remaining Si-Cl bond (Scheme 8.28). The final quaterpolymer was shown to have low molecular weight and compositional heterogeneity.

The synthesis of miktoarm star copolymers of the $(PS)_8(PI)_8$ type (Vergina star copolymers) was reported (Avgeropoulos 1996). The chlorosilane compound $Si[CH_2CH_2Si(CH_3)\ (CH_2CH_2Si(CH_3)Cl_2)_2]_4$ (S-Cl_{16}) was synthesized and used as the linking agent. The living PS arms were prepared first and reacted with the silane in a stoichiometry of Si-Cl:Li = 1:8, producing the $(PS)_8S\text{-}Cl_8$ star precursor. Due to the increased steric hindrance of the PSLi, addition of a second arm to the same Si atom at the periphery of the dendritic linking agent is highly improbable. Addition of the PI arm in excess gave the final $(PS)_8(PI)_8$ miktoarm star copolymer (Scheme 8.29).

$$\text{Styrene} + sec\text{-}BuLi \longrightarrow PSLi$$

$$8PSLi + S\text{-}Cl_{16} \longrightarrow (PS)_8\text{-}S\text{-}Cl_8 + 8LiCl$$

$$Is + sec\text{-}BuLi \longrightarrow PILi$$

$$(PS)_8\text{-}S\text{-}Cl_8 + \text{excess } PILi \longrightarrow (PS)_8\text{-}S\text{-}(PI)_8 + 8LiCl + PILi$$

Scheme 8.29

In the so-called divinylbenzene (DVB) method, living polymer chains are used to initiate the polymerization of DVB, a difunctional monomer, leading to the formation of a core, as in the case of star block copolymers discussed previously, which bears a number of active anionic sites equal to the number of the living arms linked to the nodule. These active centers can be used in order to polymerize a second monomer, leading to the preparation of A_nB_n miktoarm star polymers. This method has found considerable use in anionic polymerization (Tsitsilianis 1991a,b,

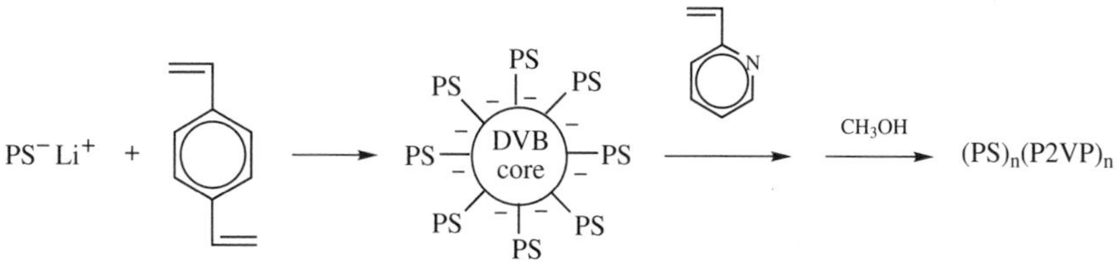

$PS^-Li^+ + \text{DVB} \rightarrow (PS)_n\text{-DVB core} \xrightarrow{\text{2-vinylpyridine}} \xrightarrow{CH_3OH} (PS)_n(P2VP)_n$

Scheme 8.30

1995, and 1997) (Scheme 8.30). Miktoarm stars have also been synthesized using the difunctional monomer technique and cationic polymerization methods (Kanaoka 1992 and 1993). Formation of A_nB_n stars containing vinyl ethers with isobutyl, acetoxy ethyl, and malonate ethyl pendant groups led to the synthesis of amphiphilic miktoarm star polymers (Scheme 8.31).

As explained before the number of arms is not well-defined, and there is a distribution of arms within the ensemble of macromolecules produced by this method. Despite its disadvantages the method can produce A_nB_n miktoarm stars where n can be varied within a relatively large number of values leading to the formation of highly branched structures in a process much easier and less demanding than the chlorosilane approach. Additionally, $(AB)_n(BA)_n$, $(AB)_nC_n$, and $(AB)_n(CD)_n$ stars can, in principle, be synthesized, expanding the variety of miktoarm star architectures available.

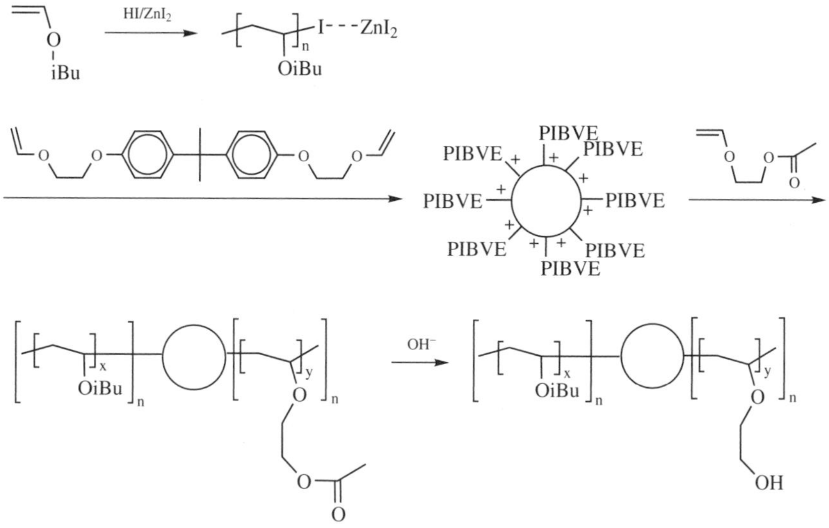

iBu vinyl ether $\xrightarrow{HI/ZnI_2}$ $[\ ]_n$–I- - -ZnI_2 (OiBu); + divinyl ether crosslinker → (PIBVE)$_n$ star core with + centers → + acetoxyethyl vinyl ether → $[(\ )_x \text{OiBu}]_n$–core–$[(\ )_y \text{O–CH}_2\text{CH}_2\text{–OAc}]_n$ $\xrightarrow{OH^-}$ $[(\ )_x \text{OiBu}]_n$–core–$[(\ )_y \text{O–CH}_2\text{CH}_2\text{–OH}]_n$

Scheme 8.31

In the 1,1-diphenylethylene-derivative method, nonpolymerizable divinylcompounds are used to transform the active chain-ends of living polymers to new active species able to polymerize other monomers. This method was used for the preparation of A_2B_2 miktoarm stars as shown in Scheme 8.32 (Quirk 1992 and 1994). The first step involved the reaction of living PSLi with 1,3-bis(1-phenylvinylbenzene) (PEB) in a ratio of Li:PEB = 2:1, leading to the formation of $(PS)_2Li_2$. Addition of the second monomer (Bd) gave the desired product (Scheme 8.32). $A_2(BA)_2$ and ABC stars were also synthesized Young et al., using similar methodology, succeeded in the synthesis of $(PI)_2(PMMA)_2$ miktoarm stars (Fernyhough 1999). The limitations of this method include: 1) possible formation of a mixture of both monoadduct and diadduct due to the similar reactivity of the two double bonds of the DPE (m-isomer). This problem can be minimized by addition of polar compounds (THF or sec-butoxides), which alter the reactivity of the double bonds; 2) maximum control over the stoichiometry of the reaction must be exercised at the first synthetic step because a mixture of star and diblock

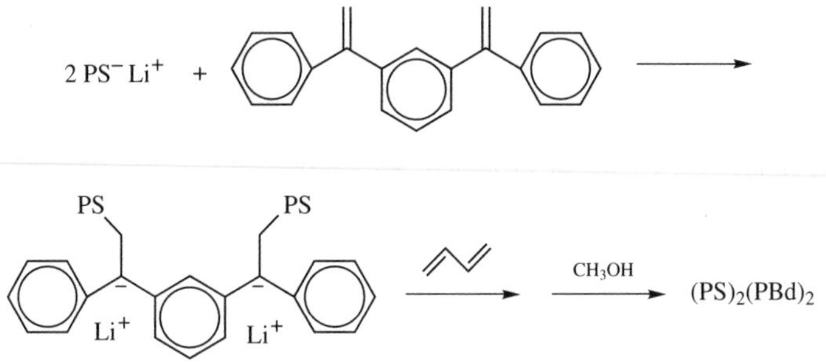

2 PS⁻Li⁺ +

PS PS

Li⁺ Li⁺

CH_3OH

$(PS)_2(PBd)_2$

Scheme 8.32

copolymers can result, i.e., if an excess of living chains is used over the double bonds; 3) the arms produced in the second step cannot be isolated and characterized independently, and the success of the reaction scheme cannot be verified.

Faust et al. used another nonpolymerizable DPE derivative, 2,2-bis(ditolylethenyl)propane (BDTEP), and living cationic polymerization in order to synthesize A_2B_2 miktoarm stars (Bae 1998). BDTEP was first reacted with living polyisobutylene chains in a 1:2 ratio to give the difunctional dimer. The two active cationic sites were used for the preparation of the other two branches by polymerization of methylvinylether, in the presence of $Ti(OEt)_4$, to provide the desired miktoarm material (Scheme 8.33).

The synthesis of (PS)(PBd)(PMMA) miktoarm star terpolymers is another representative example of the use of DPE derivatives in the synthesis of miktoarm

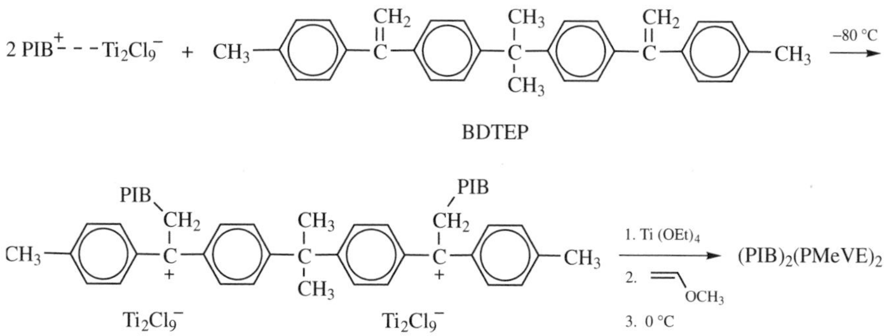

2 PIB$^{\pm}$ - - - $Ti_2Cl_9^-$ + CH_3–C_6H_4–$C(=CH_2)$–C_6H_4–$C(CH_3)_2$–C_6H_4–$C(=CH_2)$–C_6H_4–CH_3 $\xrightarrow{-80\,°C}$

BDTEP

CH_3–C_6H_4–$C^+(CH_2\text{–PIB})$–C_6H_4–$C(CH_3)_2$–C_6H_4–$C^+(CH_2\text{–PIB})$–C_6H_4–CH_3

$Ti_2Cl_9^-$ $Ti_2Cl_9^-$

1. Ti $(OEt)_4$
2. $CH_2{=}CHOCH_3$
3. 0 °C

$\longrightarrow$ $(PIB)_2(PMeVE)_2$

Scheme 8.33

stars (Huckstadt 1996). In this case PSLi was first reacted with DPE (Scheme 8.34). The DPE-capped living PS was reacted with bromomethyl-substituted DPE to produce a polymer with a diphenyl-substituted vinyl bond at one end. This bond was allowed to react with living PBdLi, and the second arm was, thus, introduced. The PBdLi addition to the double bond produced a PS-PBd diblock with an anionic

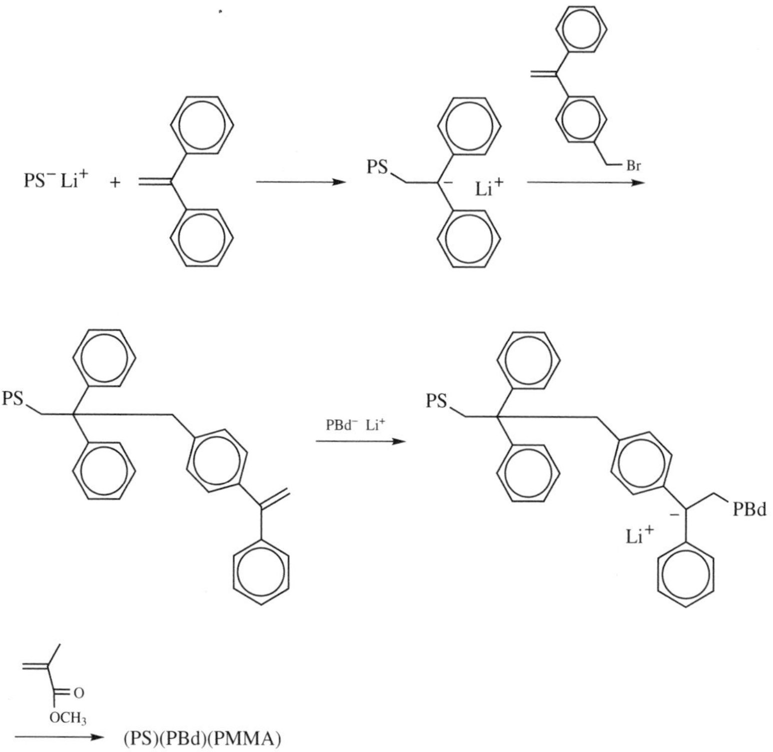

$PS^- Li^+$ + $CH_2{=}C(C_6H_5)_2$ $\longrightarrow$ PS–CH_2–$C^-(C_6H_5)_2$ Li^+ $\xrightarrow{CH_2=C(C_6H_5)\text{–}C_6H_4\text{–}CH_2Br}$

PS–CH_2–$C(C_6H_5)_2$–CH_2–C_6H_4–$C(=CH_2)C_6H_5$ $\xrightarrow{PBd^-\ Li^+}$ PS–CH_2–$C(C_6H_5)_2$–CH_2–C_6H_4–$C^-(C_6H_5)$–CH_2–PBd Li^+

$\xrightarrow{CH_2=C(CH_3)C(=O)OCH_3}$ (PS)(PBd)(PMMA)

Scheme 8.34

living center at its junction point, which was subsequently used for the polymerization of MMA, leading to the formation of the ABC miktoarm star. The same approach was used for the synthesis of ABC miktoarms of PS, PBd, and poly(2-vinylpyridine). The main disadvantages of the method outlined above lie in the fact that all capping reactions must be quantitative; intermediate products have to be purified carefully before their use in the next steps of the synthesis, and stoichiometry of the third step (PBdLi addition) must be controlled exactly, in order to minimize the amount of undesired byproducts in the final material. Characterization of the products has shown that, if considerable care is exercised, well-defined terpolymers can be produced.

Fujimoto et al. (Fujimoto 1992) reported the synthesis of (PS)(PtBuMA)(PDMS) miktoarm stars. First, the lithium salt of p-(dimethylhydroxy)silyl-α-phenyl-styrene was prepared and used for the polymerization of hexamethylcyclotrisiloxane. The vinyl terminated PDMS was reacted with PSLi, and the new macromolecular initiator formed was used for the polymerization of tert-butyl methacrylate, leading to the formation of the desired miktoarm star (Scheme 8.35).

ABC miktoarm stars containing styrene, ethylene oxide, and ε–caprolactone have been synthesized by anionic polymerization techniques (Lambert 1998). Living PSK was obtained by using cumylpotassium as the initiator and reacted with (1-[4-(2-tert-butyldimethylsiloxy)ethyl]phenyl-1-phenyethylene) to give living PS end-functionalized with a protected OH group. The active anions were used for the polymerization of EO, leading to the formation of the second arm. Deprotection of the OH group situated at the junction point of the PS-PEO diblock, with tetrabutylammonium fluoride, and thorough drying of the product followed. Diphenylmethyl sodium in THF was used for the formation of an alcholate, which

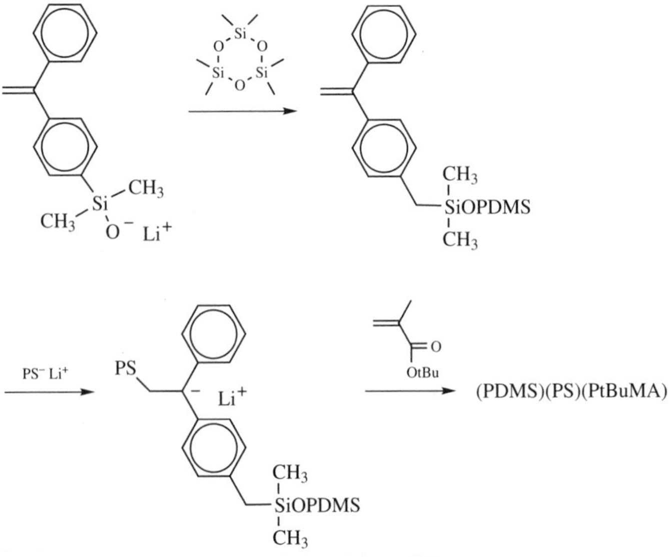

Scheme 8.35

was subsequently used for the formation of the third arm, by polymerization of ε–caprolactone (Scheme 8.36).

In an extension of the methodology involving DPE derivatives, Hirao and coworkers (Hirao 2001a,b) proposed a synthetic strategy for the preparation of chain-end and in-chain functionalized polymers with a definite number of chloromethylphenyl or bromomethylphenyl groups and their utilization in the synthesis of miktoarm star polymers. In this synthetic scheme, a macroanion is reacted with a DPE derivative having two methoxymethyl groups at the metapositions of the phenyl rings (Scheme 37a). After deactivation with methanol, the methoxymethyl groups can be converted quantitatively to chloromethyl groups by reaction with BCl_3. Reaction of this end-functional polymer with excess of polyisoprenyl or poly(tert-butyl methacrylate) anions gives A_2B miktoarm star polymers. The

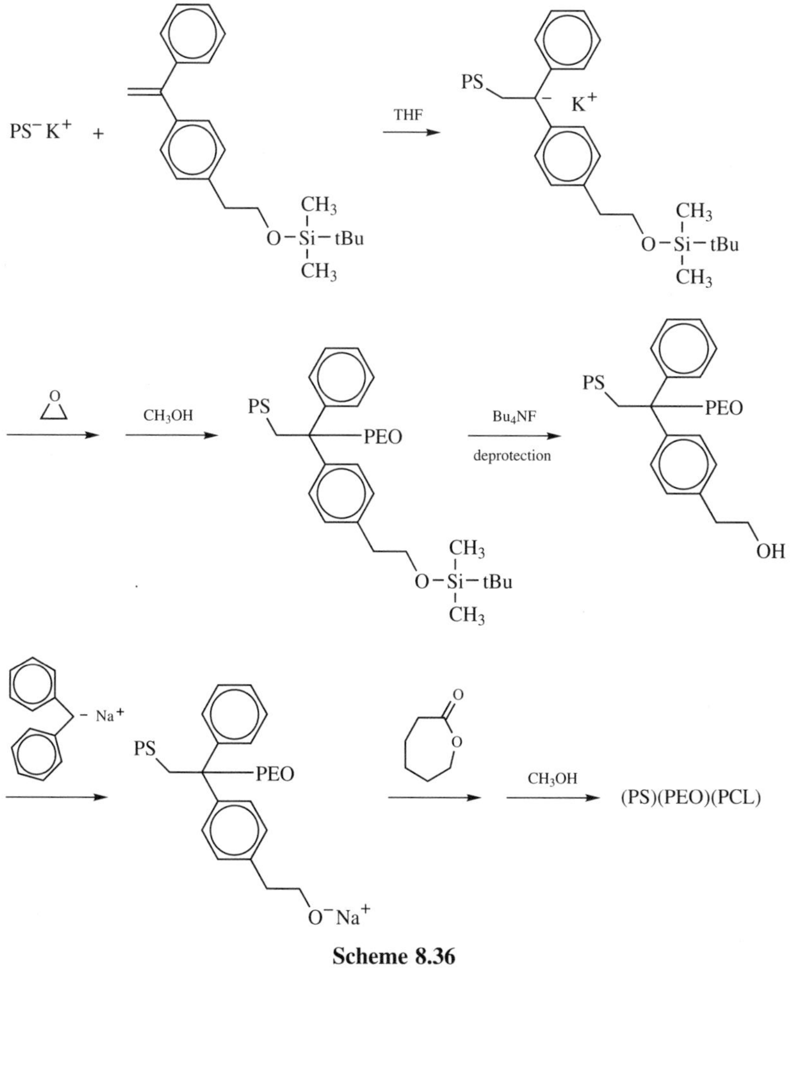

PS^-K^+ + THF → CH_3 O–Si–tBu CH_3 → PS K^+

→ CH_3OH → PS PEO → Bu_4NF deprotection → PS PEO OH

→ Na^+ PS PEO O^-Na^+ → O O → CH_3OH → (PS)(PEO)(PCL)

Scheme 8.36

CH_3O
$PS^- Li^+$ +
CH_3O
CH_3O
PS
Li^+
CH_3OH
CH_3O

CH_3O
PS
CH_3O
BCl_3
Cl
PS
Cl
$2\ PtBuMA^-\ Li^+$
PtBuMA
PS
PtBuMA

(*a*)

CH_3O
Br
sBu
CH_3O
CH_3O
PS
Li^+
CH_3O
CH_3O
OCH_3
PS
sBu
CH_3O
OCH_3
Br
OCH_3
CH_3O
PS
OCH_3
CH_3O

(*b*)

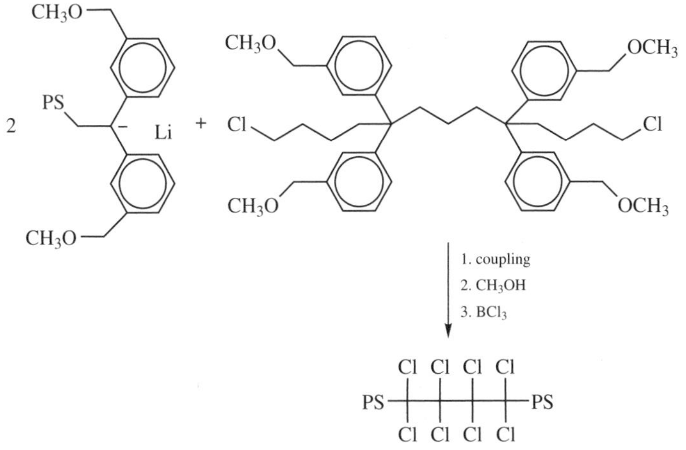

(*c*)

Scheme 8.37

number of the chloromethylphenyl (CMP) groups incorporated at the end of the macromolecular chain can be varied by reaction of the living anion with an appropriate DPE derivative, containing methoxymethyl precursor groups, in succeeding steps as shown in scheme 8.37b. Using CMP-functionalized polystyrenes, a variety of miktoarm star copolymers of the types AB_3, AB_4, A_2B_4, A_2B_{12}, ABC_2, and ABC_4, where A, B, C are polystyrene, polyisoprene, and poly(α-methylstyrene), respectively, were synthesized. In the case of in-chain functionalization, a new α,ω-dichloroalkene substituted with four chloromethylphenyl groups was used as the coupling agent (Scheme 8.37c). By using this coupling agent and end-functionalized PS with different numbers of chloromethylphenyl groups, in-chain functionalized PS with a variable number of CMP groups were synthesized.

Faust et al. recently reported the synthesis of AA′B asymmetric miktoarm stars using cationic polymerization (Yun 1999). Living polyisobutylene (PIB) was reacted with a furan derivative to produce a PIB chain bearing a terminal furanyl group. A new PIB living chain was formed having different molecular weight from the first one and was linked to the terminal furanyl group, giving a disubstituted furanyl ring. This ring was used as initiator for the polymerization of methylvinylether (MeVE), in the presence of $Ti(OiPr)_4$ as coinitiator, producing the (PIB)(PIB)′(MeVE) 3-miktoarm star (Scheme 8.38).

Other methods falling outside the aforementioned categories have appeared in the literature for the synthesis of miktoarm stars.

Naka et al. (Naka 1991) prepared noncovalent bonded A_2B miktoarm star copolymers using the complex formation between bipyridyl-terminated PEO (A) and polyoxazoline (B).

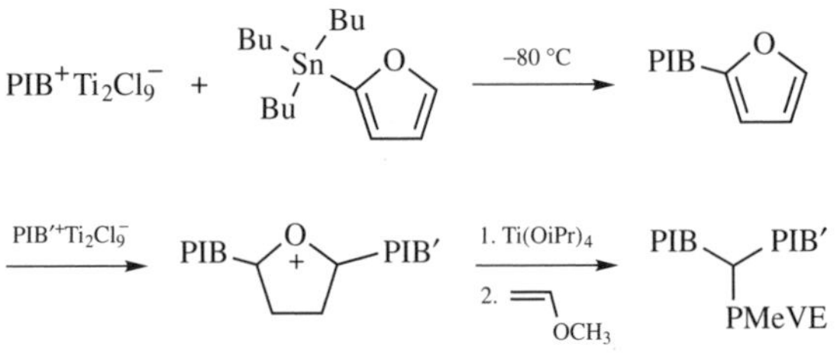

Scheme 8.38

Teyssié et al. (Ba-Gia 1980) synthesized A_2B miktoarm stars where A was PEO and B was polystyrene, polyisoprene, or poly(tert-butylstyrene) using the reaction scheme shown in Scheme 8.39. Polydispersity indexes of the final copolymers were relatively high.

Takano et al. (Takano 1992) reported the preparation of $(PS)_nPVN$ star where $n = 13$. They first synthesized a block copolymer of 2-vinylnaphthalene (VN) and (4-vinylphenyl)dimethylvinylsilane (VS). The VS block was kept short. The double bonds attached to the silicone remained unreacted during the synthesis of the diblock due to the use of a K counterion in THF at low temperature and short polymerization times and were subsequently used as branching sites for the linking of living PS chains (Scheme 8.40).

Ishizu et al. (Ishizu 1991 and 1992) used macromonomers for the preparation of miktoarm stars. PS and PI macromonomers with terminal polymerizable double bonds were copolymerized anionically to give miktoarm stars with a distribution in the total number of arms as well as in the number of each different kind of arm (Scheme 8.41). In an analogous way, diblock copolymers of styrene and tert-butylmethacrylate having polymerizable double bonds at the junction point were polymerized by anionic techniques to give the respective A_nB_n miktoarm stars. In

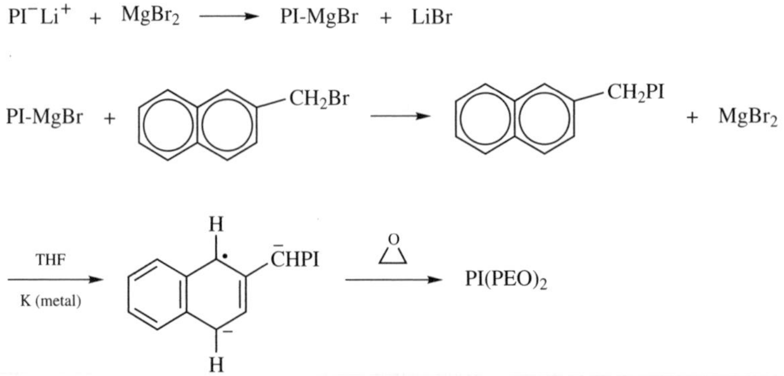

Scheme 8.39

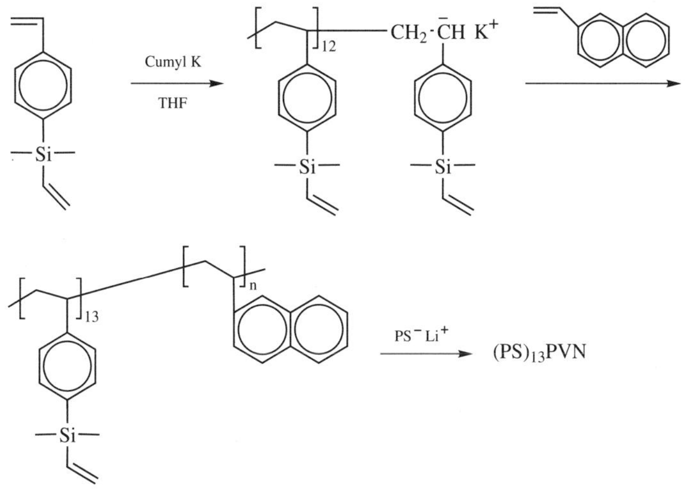

Scheme 8.40

both cases the disadvantages of the macromonomer technique are reflected in the molecular characteristics of the final products.

Pan and coworkers reported the synthesis of A_2B_2 miktoarm copolymers containing PS and PTHF or PS and poly(1,3-dioxepane) arms by using suitable initiators having two pairs of different active sites able to perform cationic ring-opening and atom transfer radical polymerization, respectively (Guo 2001a,b). Each monomer could be polymerized independently without interfering with the polymerization of the second one (Scheme 8.42).

In the same context, miktoarm stars of the type A_3B_3, where A = PMMA and B = PCL, were synthesized using a dentritic initiator having two different kinds of initiating sites (Heise 2001). MMA and caprolactone were polymerized sequentially by ATRP and living ring-opening polymerization or by simultaneous polymerization of the two monomers as shown in Scheme 8.43. The resulting star polymers exhibited molecular weights distributions around 1.22 as determined by SEC.

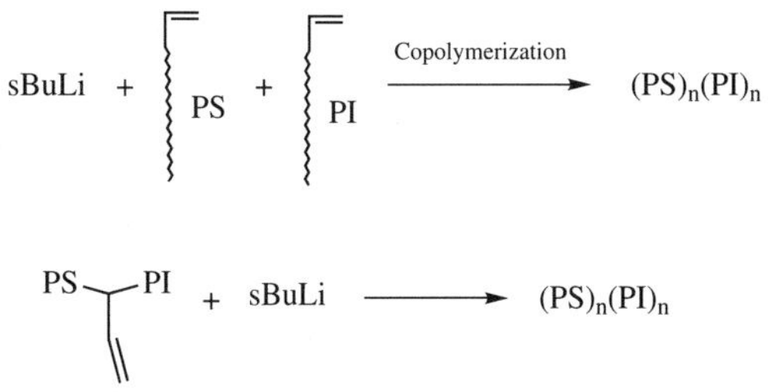

Scheme 8.41

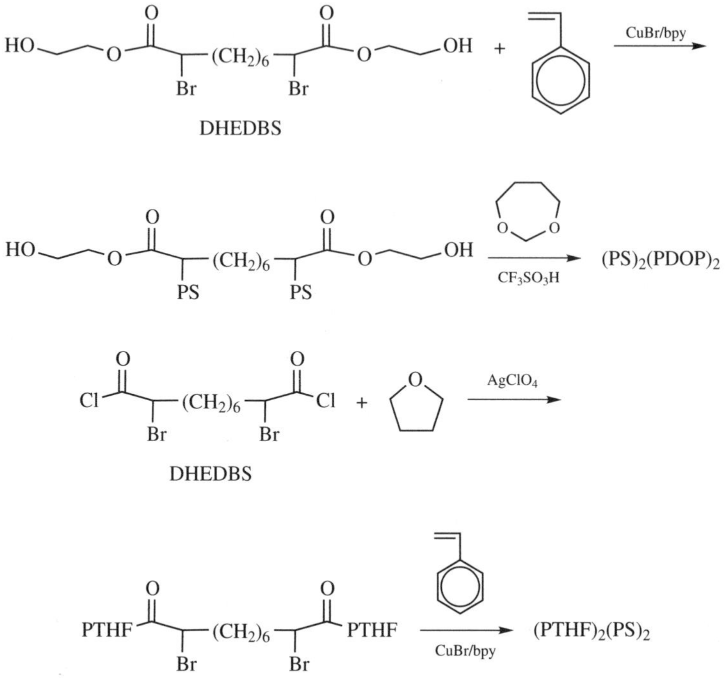

DHEDBS + styrene $\xrightarrow{CuBr/bpy}$

HO–(PS)(CH$_2$)$_6$(PS)–OH $\xrightarrow{CF_3SO_3H}$ $(PS)_2(PDOP)_2$

DHEDBS + THF $\xrightarrow{AgClO_4}$

PTHF–(Br)(CH$_2$)$_6$(Br)–PTHF + styrene $\xrightarrow{CuBr/bpy}$ $(PTHF)_2(PS)_2$

Scheme 8.42

4. OTHER COMPLEX ARCHITECTURES

Nonlinear block copolymers with even more complex architectures have been recently synthesized, making new interesting materials available for basic research studies.

Umbrella copolymers (Wang 1995a,b), having a backbone consisting of a PS-PBd diblock copolymer with a short PBd block, were synthesized by Roovers et al. The short PBd block had a high 1,2 content, due to the use of diperidinoethane as a microstructure modifier during the polymerization of butadiene. This short block was hydrosilylated, and SiCl groups were introduced. Subsequent linking of PBd and addition of P2VP living chains produced the macromolecules of the desired architecture. In a variation of the method, the PS-PBd diblocks were linked with a high functionality chlorosilane to give a star block copolymer with short PBd outer blocks. These blocks were then hydrosilylated and used as macromolecular linking agents for the incorporation of PBd and P2VP branches, resulting in the formation of umbrella star block copolymers (Scheme 8.44).

Block graft copolymers, having a PS-PBd diblock as a backbone and PS, PI, PBd, and PS-b-PI branches were prepared by a combination of anionic polymerization

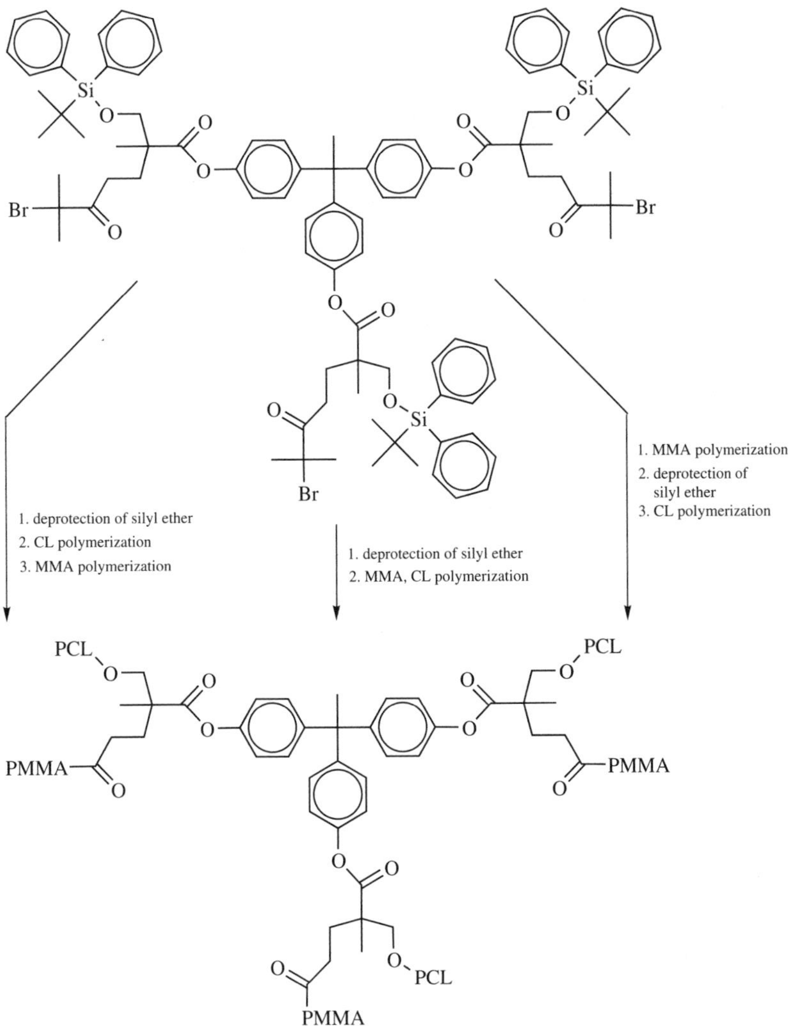

1. deprotection of silyl ether
2. CL polymerization
3. MMA polymerization

1. deprotection of silyl ether
2. MMA, CL polymerization

1. MMA polymerization
2. deprotection of silyl ether
3. CL polymerization

Scheme 8.43

and hydrosilylation reaction (Velis 2000). The diblock copolymer backbone was prepared by sequential addition of styrene and butadiene to s-BuLi. The polymerization of butadiene took place in the presence of dipyperidinoethane, resulting in high 1,2 content. The backbone was then subjected to hydrosilylation in order to incorporate the desired amount of SiCl groups on the PBd block. These groups were then used as branching sites where PSLi, PILi, PBdLi, and PSPILi living chains were linked. The use of $MeSiHCl_2$ instead of Me_2SiHCl in the hydrosilylation step produced difunctional branching sites along the PBd part of the backbone, leading to the formation of block graft copolymers with two chains grafted on each branching point (Scheme 8.45).

Se et al. (Se 1997) presented the synthesis of poly[(VS-g-I)-b-S] block graft copolymers. The diblock copolymer of 4-(dimethylvinylsilyl)styrene (VS) and styrene was prepared first by anionic polymerization. The VS monomer was polymerized selectively through the styryl double bond at low temperature in THF using cumylcesium as initiator followed by the addition of styrene. Living PILi was then allowed to react with the silylvinyl groups of the VS block, giving the final graft copolymer (Scheme 8.46).

The synthesis of block graft copolymers containing styrene, butadiene, and ethylene oxide of the types PS-b-(PBS-g-PEO) and PS-b-(PBS-g-PEO)-b-PS, where PBS is poly(p-tert-butoxystyrene), has been reported (Se 1998). The PBS blocks were converted to poly(p-hydroxystyrene), and then the OH groups, in the form of potassium alkoxides, were used for the formation of the branches by the polymerization of EO (Scheme 8.47).

Ruckenstein et al. (Ruckenstein 1998) synthesized poly[MMA-b-(AEM-g-IBVE)] block graft copolymers, where AEM is poly(2-(1-acetoxyethoxy)ethyl methacrylate) and IBVE is poly(isobutylvinylether). The MMA-AEM diblock copolymer was synthesized first. The side groups of AEM were hydrolyzed and activated by $EtAlCl_2$. These cationic sites were used for the polymerization of IBVE in order to produce the branches of the final graft copolymer (Scheme 8.48).

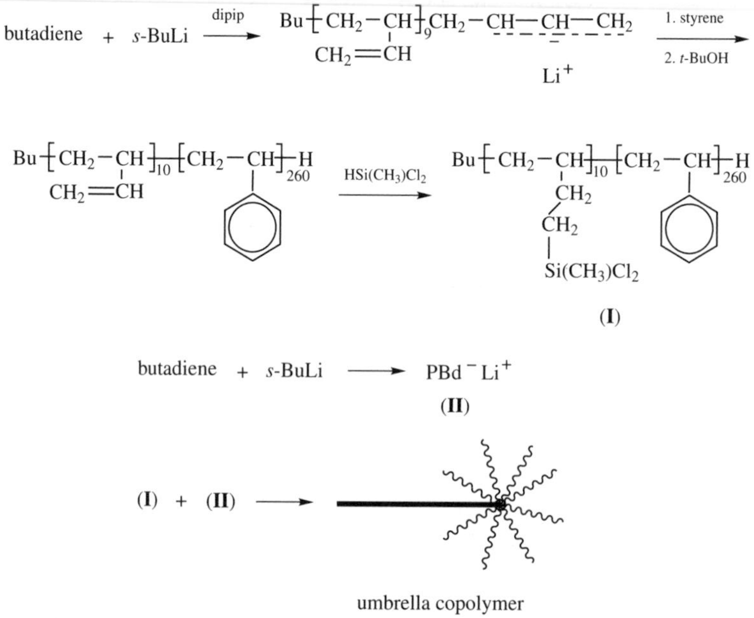

$$\text{butadiene} + s\text{-BuLi} \xrightarrow{\text{dipip}} \text{Bu}\!-\!\!\left[CH_2-\underset{\underset{CH_2=CH}{|}}{CH}\right]_9\!CH_2-CH\!=\!CH\!-\!CH_2^{-}\,Li^+ \xrightarrow[\text{2. } t\text{-BuOH}]{\text{1. styrene}}$$

$$\text{Bu}\!-\!\left[CH_2-\underset{\underset{CH_2=CH}{|}}{CH}\right]_{10}\!\left[CH_2-\underset{\underset{C_6H_5}{|}}{CH}\right]_{260}\!H \xrightarrow{HSi(CH_3)Cl_2} \text{Bu}\!-\!\left[CH_2-\underset{\underset{CH_2CH_2Si(CH_3)Cl_2}{|}}{CH}\right]_{10}\!\left[CH_2-\underset{\underset{C_6H_5}{|}}{CH}\right]_{260}\!H \quad \textbf{(I)}$$

$$\text{butadiene} + s\text{-BuLi} \longrightarrow PBd^-Li^+ \quad \textbf{(II)}$$

$$\textbf{(I)} + \textbf{(II)} \longrightarrow \text{umbrella copolymer}$$

umbrella copolymer

(*a*)

Scheme 8.44

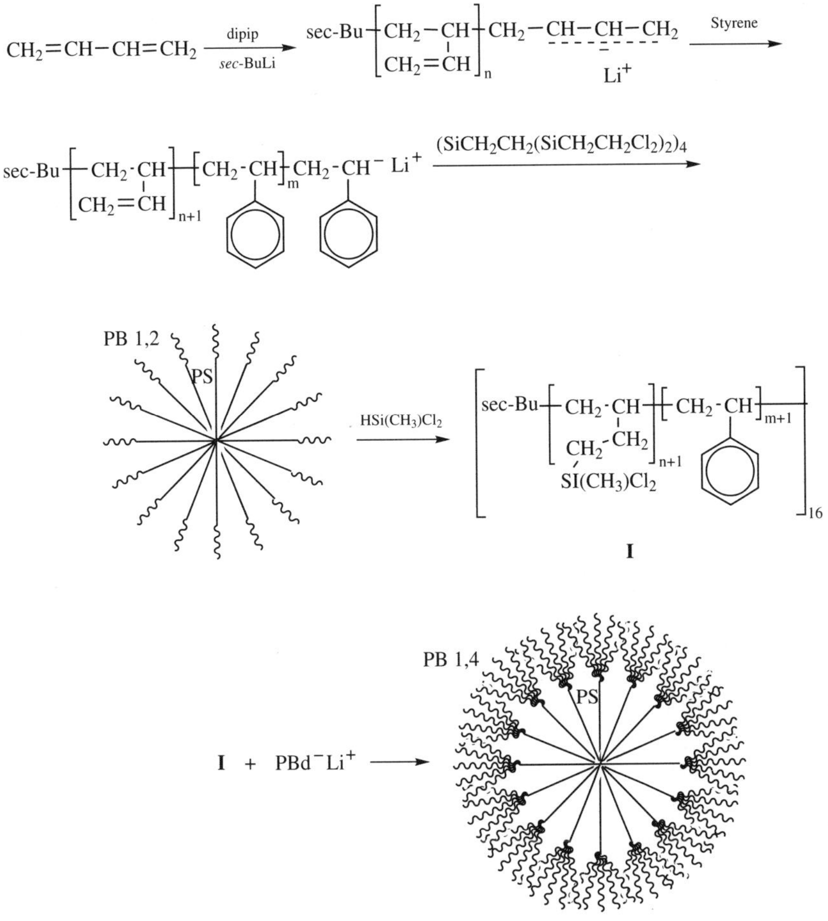

$CH_2{=}CH{-}CH{=}CH_2$ —dipip / *sec*-BuLi→ sec-Bu$[CH_2{-}CH(CH_2{=}CH)]_n CH_2{-}CH{-}CH{-}CH_2\ Li^+$ —Styrene→

sec-Bu$[CH_2{\cdot}CH(CH_2{=}CH)]_{n+1}[CH_2{\cdot}CH(C_6H_5)]_m CH_2{\cdot}CH(C_6H_5)^-\ Li^+$ —$(SiCH_2CH_2(SiCH_2CH_2Cl_2)_2)_4$→

PB 1,2

PS

—$HSi(CH_3)Cl_2$→ $[$sec-Bu$[CH_2{\cdot}CH(CH_2{\cdot}CH_2{-}SI(CH_3)Cl_2)]_{n+1}[CH_2{\cdot}CH(C_6H_5)]_{m+1}]_{16}$

I

I + PBd^-Li^+ →

PB 1,4

PS

(*b*)

Scheme 8.44

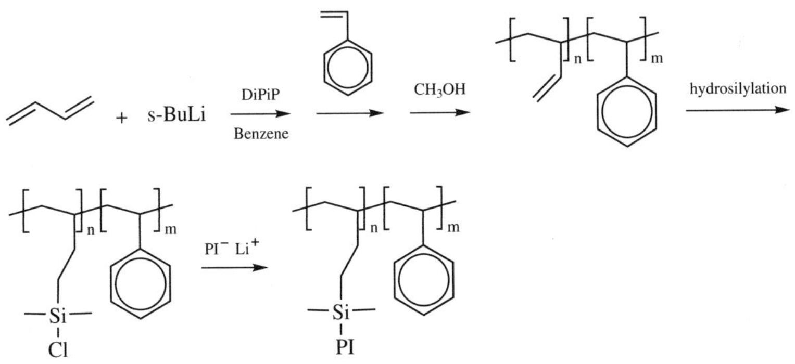

+ s-BuLi —DiPiP / Benzene→ —(styrene)→ —CH_3OH→ —hydrosilylation→

—$PI^-\ Li^+$→

Si

Cl

Si

PI

n m

Scheme 8.45

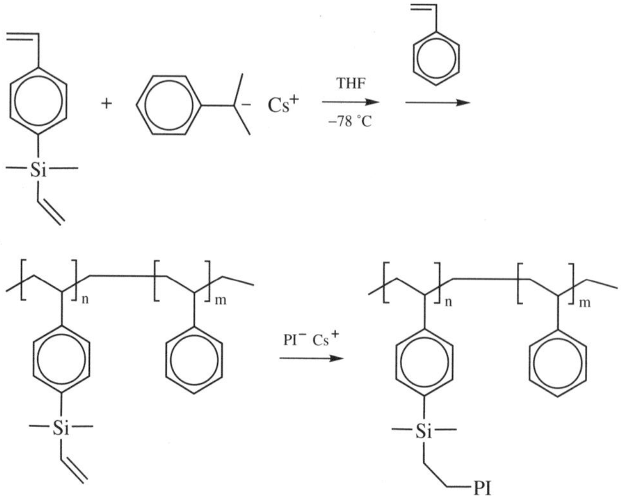

THF
−78 °C
$PI^- Cs^+$

Scheme 8.46

The same authors reported the preparation of PMMA-b-(PGMA-g-PS) and PMMA-b-(PGMA-g-PI) block graft copolymers, where PGMA is poly(glycidyl methacrylate). The PMMA-b-PGMA diblock was obtained first by anionic polymerization. Then living PSLi or PILi chains were linked to the diblock backbone by reaction of the glycidyl groups of the GMA block (Scheme 8.49).

Pan et al. (Pan 1999) selectively hydrogenated PS-PI diblock, resulting in a poly[(ethylene-co-propylene)-b-styrene] diblock. The PS block of this copolymer was chloromethylated, and the chloromethyl groups were used for the atom-tranfer radical polymerization of ethylmethacrylate, giving the desired P[(E-co-P)-b-(S-g-EMA)] block graft copolymer.

Graft, block graft, block brush, and graft-block-graft copolymers with PS backbones and PI branches were prepared by a combination of living free radical and anionic polymerization. Random and block copolymers of styrene and p-chloromerthyl styrene prepared by SFRP were used as the backbone, where anionically synthesized PI branches were tethered. In the case of graft-block-graft copolymers, anionically synthesized PS and PI branches were grafted to different parts of the backbone (Tsoukatos 2000).

H- and super-H-shaped copolymers were synthesized by anionic polymerization high-vacuum techniques (Gido 1996, Iatrou 1994). In the case of H-shaped copolymers, living PSLi and $MeSiCl_3$ were reacted in a ratio SiCl:Li = 3:2.1. Due to the sterically hindered PSLi anion, only two Cl atoms were substituted, resulting in a PS dimer having an active Si-Cl bond at the center. The macromolecular linking agent was reacted with a difunctional PI chain (the connector), synthesized using the difunctional initiator derived from PEB and s-BuLi in benzene solution and in the presence of tert-butoxide, giving the H-copolymer as shown in Scheme 8.50a. For the synthesis of the $(PI)_3PS(PI)_3$ super-H copolymers,

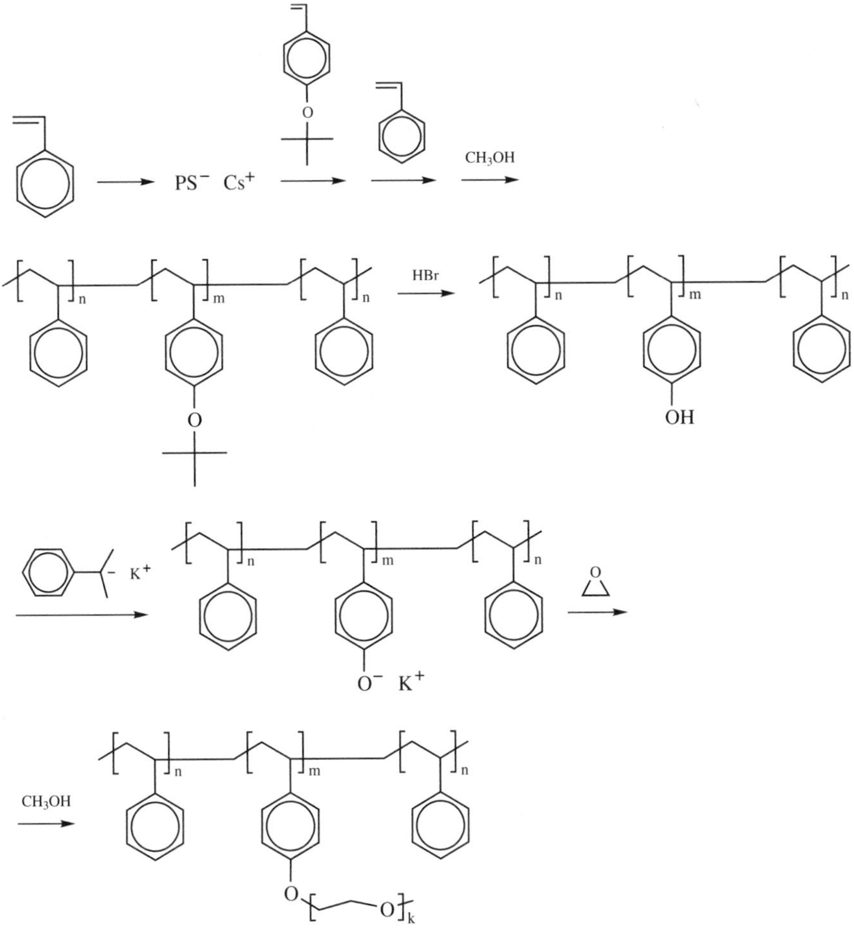

Scheme 8.47

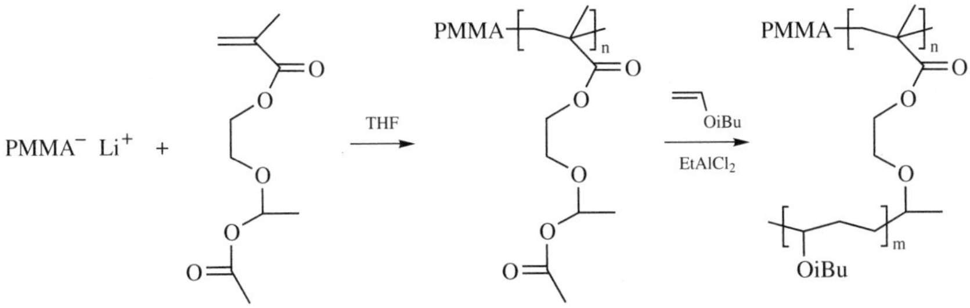

Scheme 8.48

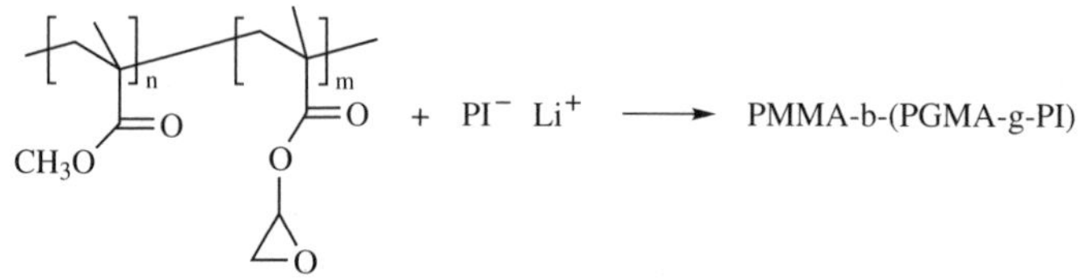

$$+ \; PI^- \; Li^+ \longrightarrow \text{PMMA-b-(PGMA-g-PI)}$$

Scheme 8.49

a difunctional PS chain, derived from the polymerization of isoprene in THF using sodium naphthalene as initiator, was reacted with a large excess of $SiCl_4$, giving a PS chain with three Si-Cl active bonds at each end. After elimination of excess $SiCl_4$ and the addition of excess of PI living arms, the $(PI)_3PS(PI)_3$ super-H-shaped copolymer was isolated (Scheme 8.50b). $(PI)_5PS(PI)_5$ copolymers (pom-pom shaped) were synthesized in a way similar to the preparation of H-shaped copolymers (Velis 1999). A hexafunctional silane was reacted with PILi in a ratio SiCl:Li = 6:5, giving the five-arm star having a SiCl group at the central point. These functional stars were reacted in a second step with a difunctional PS, yielding the desired pom-pom copolymers.

Pom-pom, or dumbbell, copolymers with a higher functionality were synthesized by first preparing a PBd-1,2-PS-PBd-1,2 triblock copolymer having short PBd blocks by anionic polymerization (Bayer 1994). The pendant double bonds in the PBd blocks were subjected to hydroboration-oxidation, producing OH groups. These groups were transformed to alkoxides and used as initiating sites for the

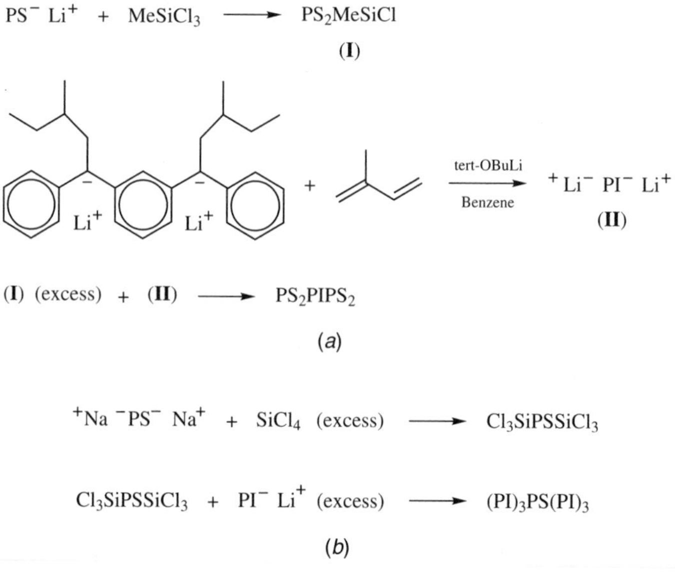

$$PS^- \; Li^+ + MeSiCl_3 \longrightarrow PS_2MeSiCl \; \textbf{(I)}$$

$$+ \xrightarrow[\text{Benzene}]{\text{tert-OBuLi}} {}^+Li^- \; PI^- \; Li^+ \; \textbf{(II)}$$

$$\textbf{(I)} \; (\text{excess}) + \textbf{(II)} \longrightarrow PS_2PIPS_2$$

(*a*)

$${}^+Na \; {}^-PS^- \; Na^+ + SiCl_4 \; (\text{excess}) \longrightarrow Cl_3SiPSSiCl_3$$

$$Cl_3SiPSSiCl_3 + PI^- \; Li^+ \; (\text{excess}) \longrightarrow (PI)_3PS(PI)_3$$

(*b*)

Scheme 8.50

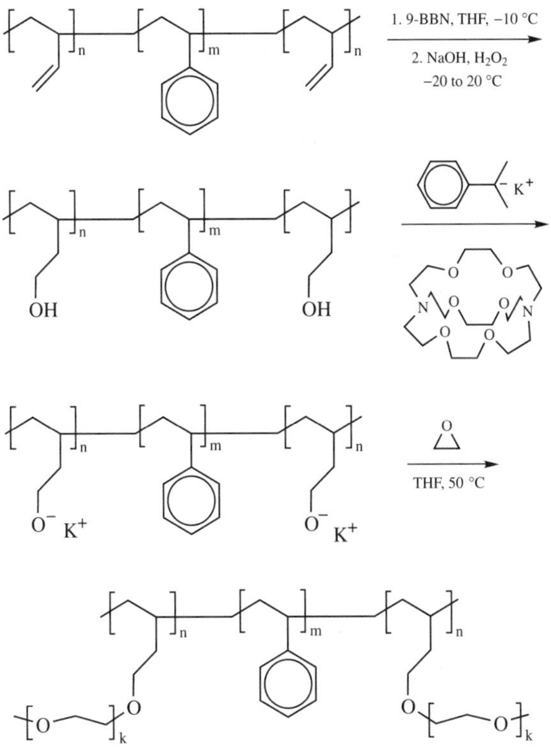

1. 9-BBN, THF, −10 °C
2. NaOH, H_2O_2
−20 to 20 °C

THF, 50 °C

Scheme 8.51

polymerization of ethylene oxide. The dumbbell-shaped $(PEO)_nPS(PEO)_n$ copolymer was, thus, prepared (Scheme 8.51).

Dumbbell copolymers were also synthesized by Fréchet et al. (Gitsov 1994). A difunctional PS living chain, prepared in THF using potassium naphthalenide as initiator, was end-capped with DPE and subsequently reacted with a fourth generation dendrimer having a bromomethyl group at its converging point {[G-4]-Br} (Scheme 8.52). In a similar fashion, Hawker et al. (Leduc 1996) prepared dendritic living free radical initiators, which were used for the synthesis of linear-dendritic copolymers. TEMPO-mediated and ATRP routes were feasible, depending on the initiator used, and polystyrene or poly(hydroxystyrene) was the linear block.

Dendrimer-like macromolecules containing styrene and ethylene were prepared by Gnanou and coworkers (Taton 1998, Angot 2000). The synthetic strategies involve the use of multifunctional initiators, carrying hydroxyl groups, for the synthesis of star-like PEOs. The end groups of the resulting polymers could be transformed to suitable derivatives for the controlled radical polymerization of styrene. Using the appropriate functionalization chemistry, each end group can

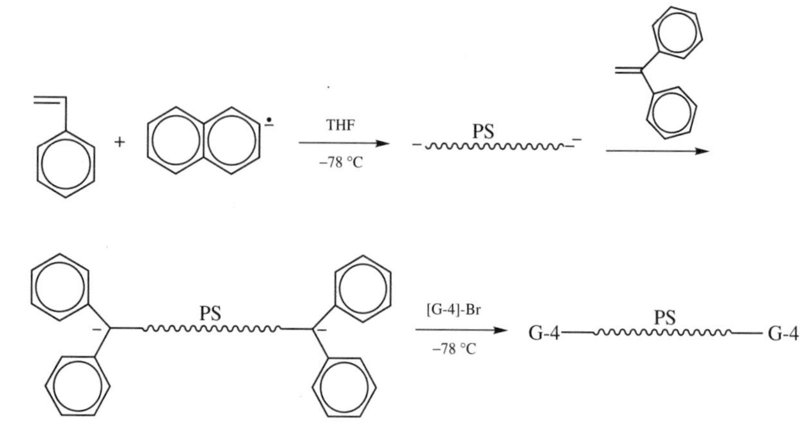

Scheme 8.52

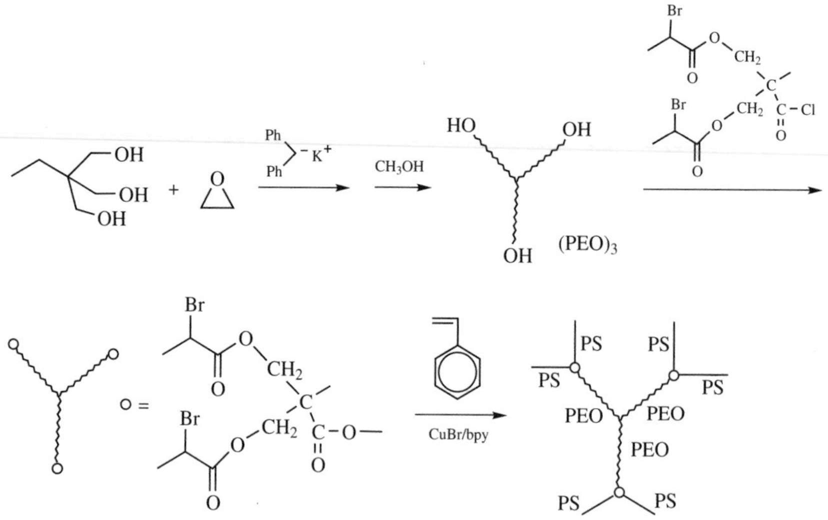

Scheme 8.53

produce two PS arms, resulting in dendritic macromolecules with amphiphilic properties (Scheme 8.53).

Arborescent graft copolymers with a highly branched PS core and a PEO shell were synthesized by a combination of anionic polymerization, repetitive postpolymerization chloromethylation reactions, followed by grafting onto procedures, and,

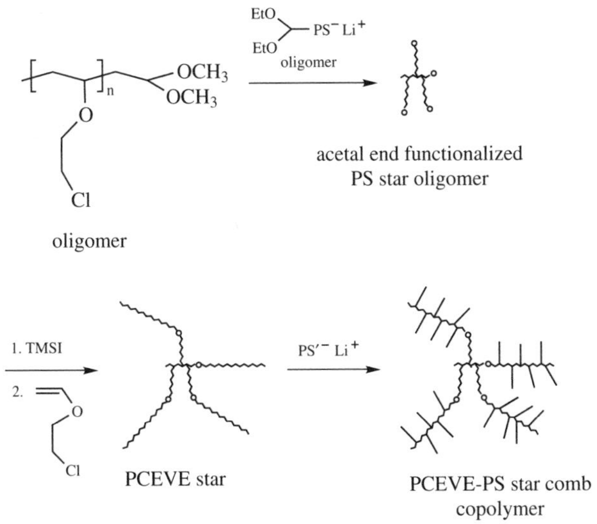

Scheme 8.54

finally, by a grafting from reaction as described below (Gauthier 1996, Kee 1999). Chloromethylated PS was reacted with living PS, end-capped with DPE, to produce a graft copolymer. This copolymer was subsequently chloromethylated, and new PS chains were grafted on it. This cycle can be performed several times, giving dendrimer-like molecules of higher generation. The last PS grafted was produced with the functional initiator (6-lithiohexyl)acetadehyde acetal, introducing protected hydroxyl groups at the periphery of the molecule. These groups were deprotected by mild acidic hydrolysis and converted to alcoholate anions by titration with potassium naphthalide. The alcoholate anions were used for the polymerization of ethylene oxide to form the PEO outer shell of the molecule.

In a similar way, Deffieux and coworkers synthesized star comb copolymers containing poly(chloroethyl vinyl ether) star cores and PS side chains (Schappacher 2000). The key point to the synthesis of these kinds of macromolecules is the high coupling efficiency of living PSLi with the chloroethyl groups of cationically synthesized poly(chloroethyl vinyl ether) (PCEVE) (Scheme 8.54).

The synthesis of cyclic block copolymers has also been reported. Using a difunctional initiator, the triblock copolymer LiP2VP-PS-P2VPLi was synthesized by sequential addition of styrene and 2-vinylpyridine (Gan 1994). Coupling of the two active ends of each chain was accomplished in very dilute solutions using 1,4-bis(bromomethyl)benzene as the coupling agent. Polymers having narrow molecular weight distributions were obtained after elimination of polycondensates by fractionation.

Employing the same difunctional initiator, end-active triblock copolymers of PS and PDMS were synthesized, and cyclization took place in dilute solution, in order

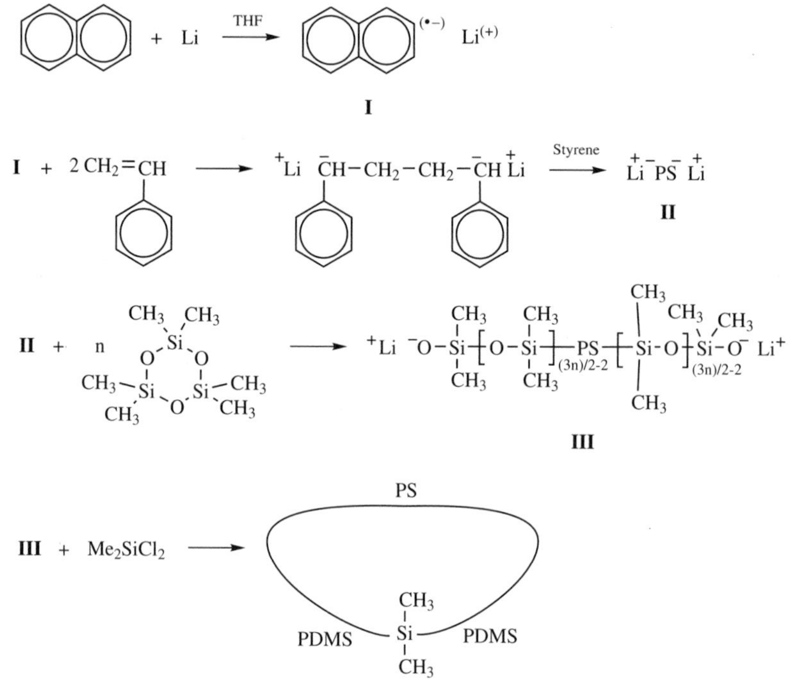

+ Li →(THF) $Li^{(+)}$ (•−)

I

I + 2 CH_2=CH → $^{+}$Li $\bar{C}$H−CH_2−CH_2−$\bar{C}$H $\overset{+}{Li}$ →(Styrene) $\overset{+}{Li}$ $^{-}$PS^{-} $\overset{+}{Li}$

II

II + n [hexamethylcyclotrisiloxane] → $^{+}$Li $^{-}$O−Si(CH_3)$_2$−[O−Si(CH_3)$_2$]$_{(3n)/2-2}$−PS−[Si(CH_3)$_2$−O]$_{(3n)/2-2}$−Si(CH_3)$_2$−O^{-} Li^{+}

III

III + Me_2SiCl_2 →

PS

PDMS Si(CH_3)$_2$ PDMS

(*a*)

+ n CH_2=CH−CH=CH_2 →(Benzene, *s*-BuOLi, 20 °C) $^{+}$Li^{-}PBd^{-}Li^{+}

I

I + n CH_2=CH → $^{+}$Li $\bar{P}$S−PBd−$\bar{P}$S Li^{+} →

→($(CH_3)_2SiCl_2$)

PBd

PS Si(CH_3)$_2$ PS

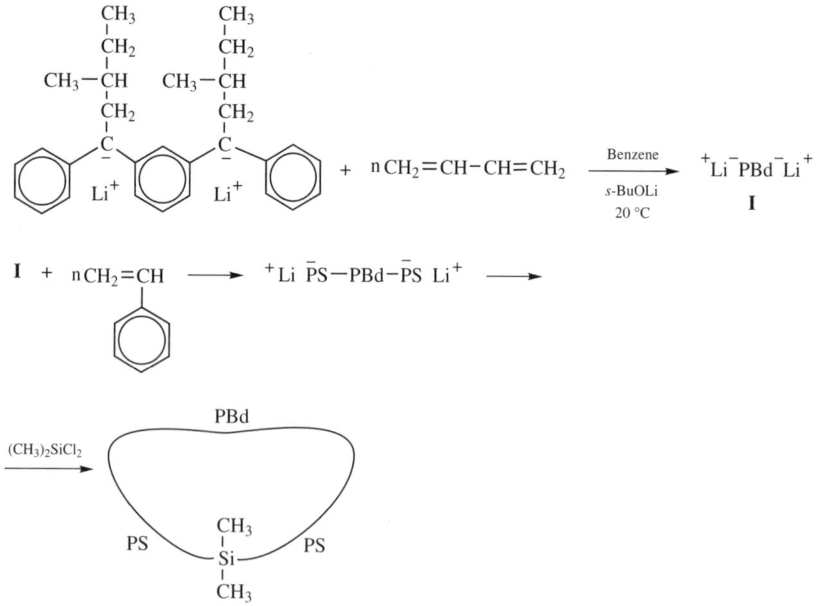

(*b*)

Scheme 8.55

to minimize the extent of polycondensation reaction, using dichlorodimethylsilane as the coupling agent (Yin 1993, Scheme 8.55a). Cyclic block copolymers of PS and PBd were synthesized using a similar synthetic strategy (Ma 1995, Scheme 8.55b). The preparation of the precursor triblock, PBd-PS-PBd, was performed in benzene using the difunctional initiator derived from s-BuLi and PEB in the presence of sec-BuOLi. Again, $(CH_3)_2SiCl_2$ was used as the coupling agent for the cyclization reaction under high dilution. Deffieux and coworkers synthesized cyclic PS-PEO block copolymers utilizing a heterodifunctional PS-PEO diblock copolymer formed by anionic polymerization and having acetal and styrenic end groups as shown in Scheme 8.56 (Cramail 2000).

Deffieux et al. also reported the preparation of cyclic PBd having two PS branches connected to the same point of the cyclic part of the molecule (Beinat 1996). The reaction sequence employed in this case is shown in Scheme 8.57. A four-arm miktoarm star of the type $(PS)_2(PBdLi)_2$ was prepared using PEB as the coupling agent, as described before. The living PBdLi ends were coupled intermolecularly with $(CH_3)_2SiCl_2$ in high dilution. The desired complex macromolecule was isolated from the polycondensates by fractionation.

Catenated copolymers were also prepared by Hogen-Esch et al. (Gan 1995). 2VP was polymerized with lithium dihydronaphthylide, a difunctional initiator, in the

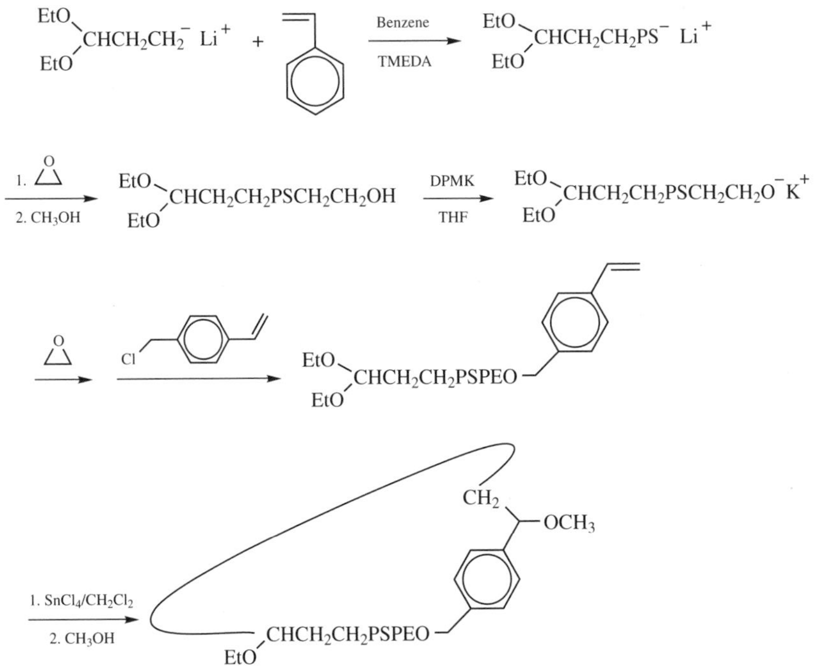

Scheme 8.56

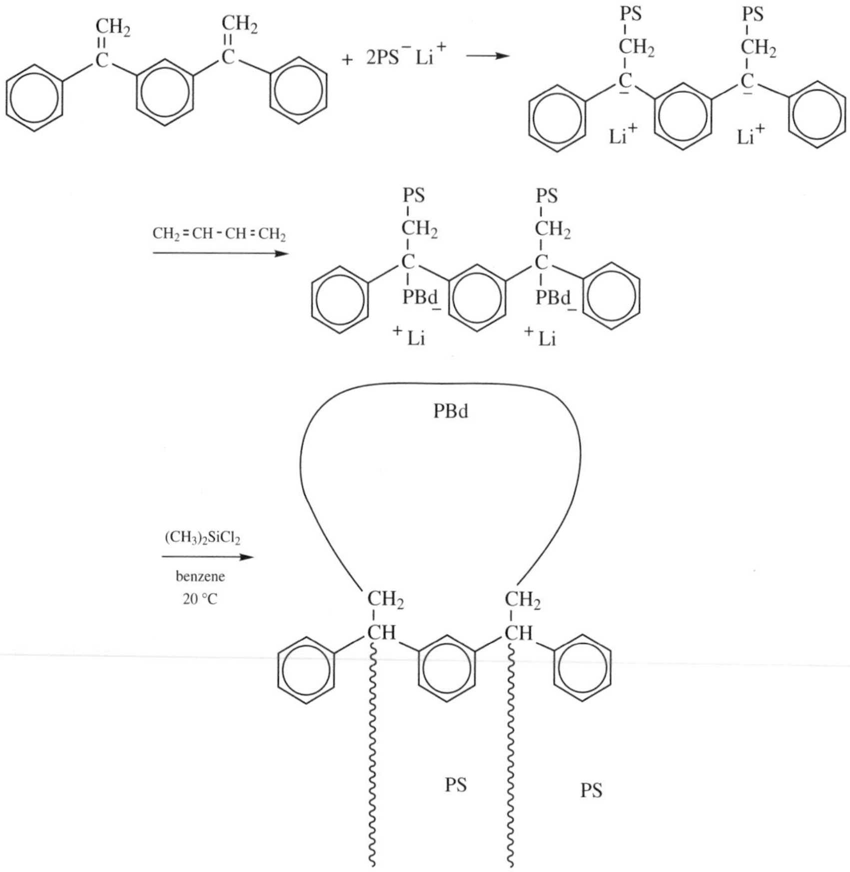

CH_2 CH_2 + $2PS^-Li^+$ → PS PS CH_2 CH_2 C C Li^+ Li^+

$CH_2=CH-CH=CH_2$ → PS PS CH_2 CH_2 C C PBd PBd ^+Li ^+Li

$(CH_3)_2SiCl_2$, benzene, 20 °C → PBd CH_2 CH_2 CH CH PS PS

Scheme 8.57

presence of cyclic PS macromolecules. The cyclization of difunctional P2VP was performed under high dilution but in the presence of a high concentration of cyclic PS (Scheme 8.58). According to the authors, catenated PS-P2VP copolymers were formed. The crude reaction product was subjected to extraction with cyclohexane and methanol in order to remove excess cyclic PS and uncatenated cyclic P2VP and P2VP polycondensates, respectively. The isolated product is not soluble in methanol or in cyclohexane, and its molecular weight was found to be equal to the sum of those of PS and P2VP.

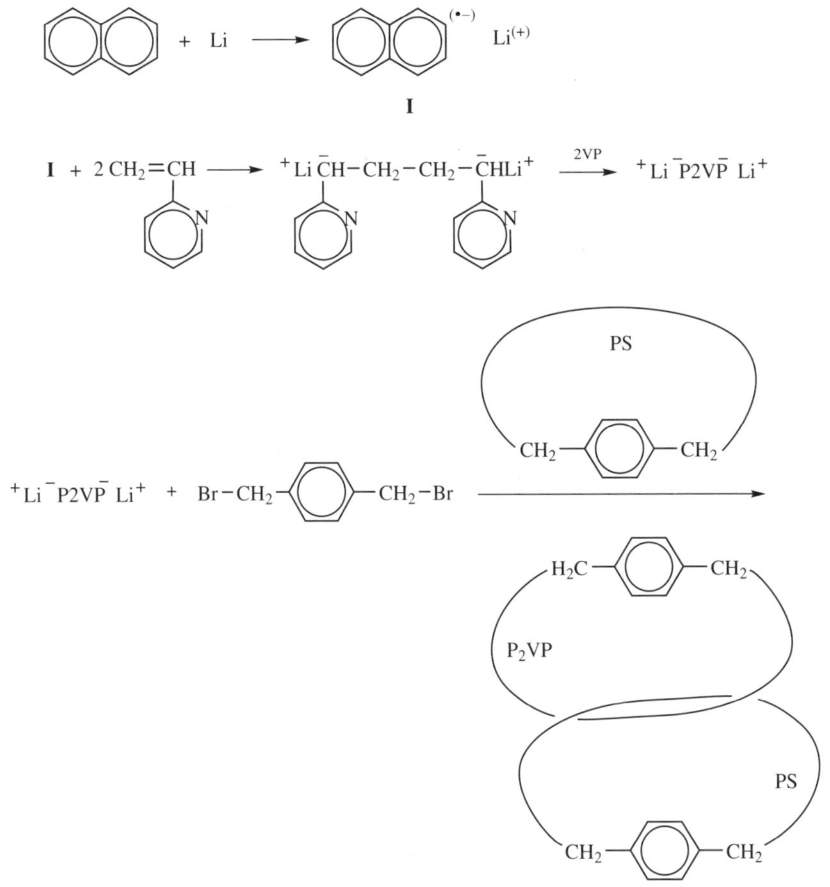

Scheme 8.58

REFERENCES

Algaier J., Young R. N., Efstratiadis V., Hadjichristidis N. (1996) Macromolecules 29, 1794.
Alward D. B., Kinning D. J., Thomas E. L., Fetters L. J. (1986) Macromolecules 19, 215.
Angot S., Taton D., Gnanou Y. (2000) Macromolecules 33, 5418.
Arnold M., Frank W., Reinhold G. (1990) Polym. Bull. 24, 1.
Avgeropoulos A., Hadjichristidis N. (1997) J. Polym. Sci. Part A: Polym. Chem. 35, 813.
Avgeropoulos A., Poulos Y., Hadjichristidis N., Roovers J. (1996) Macromolecules 29, 6076.
Bae Y. C., Faust R. (1998) Macromolecules 31, 2480.
Ba-Gia H., Jérôme R., Teyssié P. (1980) J. Polym. Sci. Polym. Chem. Ed. 18, 3483.
Bayer V., Stadler R. (1994) Macromol. Chem. Phys. 195, 2709.

Beers K. L., Gaynor S. G., Matyjaszewski K., Sheiko S. S., Möller M. (1998) Macromolecules 31, 9413.

Beinat S., Schappacher M., Deffieux A. (1996) Macromolecules 29, 6737.

Bellas V., Iatrou H., Hadjichristidis N. (2000) Macromolecules 33, 6993.

Bi L. K., Fetters L. J. (1975) Macromolecules 8, 90.

Cowie J. M. G. (1989) in Comprehensive Polymer Science, Allen G., Berington J. C. (Eds) Pergamon, Oxford, Vol. 3, p. 33.

Cramail S., Schappacher M., Deffieux A. (2000) Macromol. Chem. Phys. 201, 2328.

Eschwey H., Buchard W. (1975) Polymer 16, 180.

Fernylough C. M., Young R. N., Tack R. D. (1999) Macromolecules 32, 5760.

Fujimoto T., Zhang H., Kazama T., Isono Y., Hasegawa H., Hashimoto T (1992) Polymer 33, 2208.

Fukui H., Sawamoto M., Higashimura T. (1995) Macromolecules 28, 3756.

Fukui H., Yoshihashi S., Sawamoto M., Higashimura T. (1996) Macromolecules 29, 1862.

Gan Y., Dong D., Hogen-Esch T. E. (1995) Polym. Prepr. 36(1), 408.

Gan Y-P., Zoller J., Yin R., Hogen-Esch T. E. (1994) Macromol. Symp. 77, 93.

Gathier M., Tichagawa L., Downey J. S., Gao S. (1996) Macromolecules 29, 519.

Gido S. P., Lee C., Pochan D. J., Pispas S., Mays J. W., Hadjichristidis N. (1996) Macromolecules 29, 7022.

Gitsov I., Fréchet J. M. J. (1994) Macromolecules 27, 7309.

Grubbs R. B., Hawker C. J., Dao J., Fréchet J. M. J. (1997) Angew. Chem. Int. Ed. Eng. 36, 270.

Guo Y-M., Pan C-Y. (2001a) Polymer 34, 2863.

Guo Y-M., Xu J., Pan C-Y. (2001b) J. Polym. Sci. Part A: Polym. Chem. 39, 437.

Hadjichristidis N. (1999) J. Polym. Sci. Part A: Polym. Chem. 37, 857.

Hadjichristidis N., Roovers J. (1978) J. Polym. Sci. Polym. Phys. Ed. 16, 851.

Hawker C. J., Mecerreyes D., Elce E., Dao J., Hedrick J. L., Bakarat I., Dubois P., Jérôme R., Volksen W. (1997) Macromol. Chem. Phys. 198, 155.

Heise A., Trollsas M., Magbittang T., Hedrick J. L., Frank C. W., Miller R. D. (2001) Macromolecules 34, 2798.

Hirao A., Hayashi M., Haraguchi N. (2001a) Macromol. Rapid Commun. 21, 1171.

Hirao A., Tokuda Y., Morifuji K., Hayashi M. (2001b) Macromol. Chem. Phys. 202, 1606.

Huckstadt H., Abetz V., Stadler R. (1996) Macromol. Rapid Commun. 17, 599.

Iatrou H, Hadjichristidis N. (1993) Macromolecules 26, 2479.

Iatrou H., Avgeropoulos A., Hadjichristidis N. (1994) Macromolecules 27, 6232.

Iatrou H., Hadjichristidis N. (1992) Macromolecules 25, 4649.

Iatrou H., Mays J. W., Hadjichristidis N. (1998) Macromolecules 31, 6697.

Iatrou H., Siakali-Kioulafa E., Hadjichristidis N., Roovers J., Mays J. W. (1995) J. Polym. Sci. Part B: Polym. Phys. 33, 1925.

Ishizu K., Shimomura K., Saito R., Fukutomi T. (1991) J. Polym. Sci. Part A: Polym. Chem. 29, 607.

Ishizu K., Uchida S. (1994) Polymer 35, 4712.

Ishizu K., Yikimana S., Saito R. (1991) Polym. Commun. 32, 386.

Ishizu K., Yikimana S., Saito R. (1992) Polymer 33, 1982.

Jacob S., Majoros I., Kennedy J. P. (1998) Rubber Chem. Technol. 71, 708.

Kanaoka S., Omura T., Sawamoto M., Higashimura T. (1992) Macromolecules 25, 6407.

Kanaoka S., Sawamoto M., Higashimura T. (1991) Macromolecules 24, 5741.

Kanaoka S., Sawamoto M., Higashimura T. (1993) Macromolecules 26, 254.

Kee R. A., Gauthier M. (1999) Macromolecules 32, 6478.

Kennedy J. P., Ivan B. (1992) Designed Polymers by Carbocationic Macromolecular Engineering: Theory and Practice, Hanser, Munich.

Keszler B., Fenyvesi G., Kennedy J. P. (2000) J. Polym. Sci. Part A: Polym. Chem. 38, 706.

Khan I. M., Gao Z., Khougaz K., Eisenberg A. (1992) Macromolecules 25, 3002.

Lambert O., Reutenauer S., Hurtrez G., Riess G., Dumas P. (1998) Polym. Bull. 40, 143.

Leduc M. R., Hawker C. J., Daw J., Fréchet J. M. J. (1996) J. Am. Chem. Soc. 118, 11111.

Lee C., Gido S. P., Pitsikalis M., Mays J. W., Beck Tan N., Trevino S. F., Hadjichristidis N. (1997) Macromolecules 30, 3732.

Ma J. (1995) Macromol. Symp. 91, 41.

Matyjaszewski K., Beers K. L., Coca S., Gaynor S. G., Miller P. J., Paik H. J., Theodorescu M. (1999) Polym. Prepr. 40(1), 95.

Mays J. W. (1990) Polym. Bull. 23, 249.

Moschogianni P., Pispas S., Hadjichristidis N. (2001) J. Polym. Sci. Part A: Polym. Chem. 39, 650.

Naka A., Sada K., Chujo Y., Saegusa T. (1991) Polym. Prepr. Jap. 40(2), E116.

Pan Q., Liu S., Xie J., Jiang M. (1999) J. Polym. Sci. Part A: Polym. Chem. 37, 2699.

Paraskeva S., Hadjichristidis N. (2000) J. Polym. Sci. Part A: Polym. Chem. 38, 931.

Pitsikalis M., Pispas S., Mays J. W., Hadjichristidis N. (1998) Adv. Polym. Sci. 135, 1.

Pochan D. J., Gido S. P., Pispas S., Mays J. W., Ryan T., Fairlough P., Terrill N., Hamley I. W. (1996) Macromolecules 29, 5091.

Quirk R. P., Lee B., Schick L. E. (1992) Makromol. Chem. Macromol. Symp. 53, 201.

Quirk R. P., You T., Lee B. (1994) J. Macromol. Sci. Pure Appl. Chem. A31(8), 911.

Rempp P., Franta E. (1984) Adv. Polym. Sci. 62, 95.

Ruckenstein E., Zhang H. (1998) Macromolecules 31, 2977.

Saunders R. S., Cohen R. E., Wong S. J., Schrock R. R. (1992) Macromolecules 25, 2055.

Schappacher M., Deffieux A. (2000) Macromolecules 33, 7371.

Se K., Miyawaki K., Hirahara K., Takano A., Fujimoto T. (1998) J. Polym. Sci. Part A: Polym. Chem. 36, 3021.

Se K., Yamazaki H., Shibamoto T., Takano A., Fujimoto T. (1997) Macromolecules 30, 1570.

Shim J. S., Asthana S., Omura N., Kennedy J. P. (1998) J. Polym. Sci. Part A: Polym. Chem. 36, 2997.

Shim J. S., Kennedy J. P. (2000) J. Polym. Sci. Part A: Polym. Chem. 38, 279.

Shohl H., Sawamoto M., Higashimura T. (1991) Polym. Bull. 25, 529.

Sioula S., Tselikas Y., Hadjichristidis N. (1997) Macromolecules 30, 1518.

Storey R. F., Chisholm B. J., Lee Y. (1993) Polymer 34, 4330.

Takahashi R., Ouchi M., Satoh K., Kamogaito M., Sawamoto M. (1999) Polymer J. 31, 995.

Takano A., Okada M., Nose T., Fujimoto T. (1992) Macromolecules 25, 3596.
Taton D., Cloutet E., Gnanou Y. (1998) Macromol. Chem. Phys. 199, 2501.
Tselikas Y., Hadjichristidis N., Lescanec R. L., Honeker C. C., Wohlgemuth M., Thomas E. L. (1996) Macromolecules 29, 3390.
Tselikas Y., Iatrou H., Hadjichristidis N., Liang K. S., Mohanty K., Lohse D. J. (1996) J. Chem. Phys. 105, 2456.
Tsitsilianis C., Graff S., Rempp P. (1991) Eur. Polym. J. 27, 243.
Tsitsilianis C., Lutz P., Graff S., Lamps J. P., Rempp P. (1991) Macromolecules 24, 5897.
Tsitsilianis C., Papanagopoulos D., Lutz P. (1995) Polymer 36, 3745.
Tsitsilianis C., Voulgaris D. (1997) Macromol. Chem. Phys. 198, 997.
Tsoukatos T., Pispas S., Hadjichristidis N. (2000) Macromolecules 33, 9504.
Ueda J., Kamigaito M., Sawamoto M. (1998) Macromolecules 31, 6762.
Velis G., Hadjichristidis N. (1999) Macromolecules 32, 534.
Velis G., Hadjichristidis N. (2000) J. Polym. Sci. Part A: Polym. Chem. 38, 1136.
Wang F., Roovers J., Toporowski P. M. (1995a) Macromol. Symp. 95, 205.
Wang F., Roovers J., Toporowski P. M. (1995b) Macromol. Reports A32 (Suppls 5/6) 951.
Wang Y., Huang J. (1998) Macromolecules 31, 4057.
Wright S. J., Young R. N., Croucher T. G. (1994) Polym. Int. 33, 123.
Xenidou M., Hadjichristidis N. (1998) Macromolecules 31, 5690.
Xie H., Xia J. (1987) Makromol. Chem. 188, 2543.
Yin R., Hogen-Esch T. E. (1993) Macromolecules 26, 6952.
Yun J., Hadjikyriacou S., Faust R. (1999) Polym. Prepr. 40(1), 1041.
Zioga A., Sioula S., Hadjichristidis N. (2000) Macromol. Symp. 157, 239.

PART II

MOLECULAR CHARACTERIZATION OF BLOCK COPOLYMERS

CHAPTER 9

MOLECULAR CHARACTERIZATION OF BLOCK COPOLYMERS

Molecular and compositional characterization is an essential step in the study of block copolymers. Block copolymers present molecular weight and compositional heterogeneity. After the synthesis, the next step is to thoroughly characterize the obtained samples in order to determine their average molecular weight and composition, as well as molecular and compositional homogeneity. Additionally, information about the size of the macromolecule may be needed. Due to the imperfections present in the methodology used for the preparation of nearly every synthetic macromolecule, the resulting material is a very complex system whose identity can only be revealed by the employment of a combination of analytical techniques (Barth 1991). Sometimes a number of purification procedures must be followed before the desired materials can be isolated and characterized.

1. PURIFICATION OF BLOCK COPOLYMERS BY FRACTIONATION

In linear block copolymer samples, apart from molecular weight, there may be also differences in chemical composition. Complex copolymer structures like graft, star blocks, miktoarm stars, etc., may contain varying amount of impurities, like homopolymers or copolymers of different architecture, as a result of the synthetic strategy followed for their preparation (e.g., incomplete coupling during the intermediate reaction steps). These heterogeneities in all cases affect the final properties of the material. Fractionation is a powerful tool used to separate heterogeneous mixtures of polymers before the final characterization of each fraction (Tung 1977). Copolymers can be fractionated according to their molecular weight or chemical composition (Riess 1977). The most widely used methods are batch fractionation and column-elution fractionation (Tung 1985).

1.1. Batch Fractionation

Batch fractionation is based on the partition of polymer molecules between two liquid phases. Usually a solvent/nonsolvent system is used. The polymer is first dissolved in a good solvent for all parts of the copolymer in order to give a dilute solution (~1% w/v). By addition of a nonsolvent, and/or by lowering the temperature, the solution is then separated into two phases, the polymer rich phase and the polymer poor phase. The partition of the polymer into the two phases depends on molecular weight and chemical composition.

Phase equilibria in dilute solutions of polymers can be described by the Flory-Huggins theory (Flory 1953). The partition of an AB copolymer with a degree of polymerization x and a volume fraction of monomers A, a, can be given by:

$$Q''_{x,a}/Q'_{x,a} = e^{x(s+ka)} \tag{9.1}$$

with $k = (Q''_p - Q'_p)(\chi_A - \chi_B)$ and $s = \ln(Q''_o/Q'_o) + 2\chi(Q'_o - Q''_o)$, where $Q''_{x,a}$ and $Q'_{x,a}$ are the volume fractions of a polymer species with degree of polymerization x and chemical composition a in the polymer-rich and solvent-rich phase, respectively; Q''_p and Q'_p are the volume fractions of the total polymer in the respective phases, χ_A, χ_B are the interaction parameters of the solvent with A and B monomeric units, Q'_o is the volume fraction of solvent (or solvent mixture) in the solvent-rich phase, and Q''_o is the volume fraction of the solvent in the polymer-rich phase. If $\chi_A = \chi_B$ then $k = 0$ and fractionation is made according to molecular weight only. It is obvious that fractionation cannot be made totally without the influence of the molecular weight. The effect of chemical composition in the fractionation procedure can be regulated by choosing a solvent with an appropriate χ_A and χ_B difference. Two variations of the batch fractionation technique can be used: partial precipitation or successive precipitation fractionation and coacervate extraction or successive solution fractionation.

In partial precipitation (or successive precipitation fractionation), the copolymer is first dissolved in a good solvent at a dilute concentration. Addition of nonsolvent, which is miscible with the solvent, under stirring, results in the separation of two phases. The polymer-rich precipitated phase is isolated as a fraction. The procedure can be repeated many times with the supernatant solution or with the polymer-rich phase until the desired separation is achieved. Occlusions of the polymer solution in the precipitated phase can be avoided by heating the mixture until phase separation disappears and then by allowing it to cool slowly to the original temperature.

In coacervate extraction (or successive solution fractionation), a large amount of nonsolvent is added to the dilute polymer solution in order to drive a large portion of the polymer into the polymer-rich phase. The solvent-rich phase is recovered, and the polymer is isolated as the first fraction. The remaining polymer-rich phase is diluted with more solvent, and the procedure can be repeated. In this way the lower molecular weight fractions are collected first, in contrast with partial precipitation where the higher molecular weight fractions are collected first.

1.2. Column Elution Fractionation

As in the batch fractionation, the separation is based on the partitioning of polymers within the two phases. The polymer-rich phase is deposited on the packing material in the column as a thin layer. Deposition occurs selectively, and slow cooling of the column during deposition enhances efficiency of fractionation. The waiting time for the phases to coalesce and settle into layers is eliminated, and the fractions can be collected within a few hours by elution of the column with the solvent.

1.3. Preparative Size Exclusion Chromatography

Fractionation of copolymers can be effected by using preparative scale size exclusion chromatography. The normal instrumentation used in conventional SEC can be used with some modifications that include a high-volume injector, a set of columns with larger dimensions (e.g., 122 cm in length and 55 mm inner diameter), and an automatic sample collector. Flow rates of 10 mL/min to 100 mL/min are frequently used. Separation is done according to the hydrodynamic volume of the molecules. The use of two different concentration-sensitive detectors, e.g., RI and UV detectors, is preferred in order to have maximum information on the composition of the solute at each elution volume.

2. MOLECULAR CHARACTERIZATION

2.1. Primary Molecular Structure and Chemical Composition Determination

Many analytical techniques are available for the determination of the primary molecular structure of a block copolymer and its average chemical composition. Among them, spectroscopic techniques used for low molecular weight compounds are the most powerful and most widely employed.

2.1.1. Nuclear Magnetic Resonance Spectroscopy (NMR). NMR can provide both qualitative and quantitative information with respect to comonomer composition and stereochemical configuration of polymeric molecules (Cheng 1991, Bovey 1972, Ivin 1983). This is due to the fact that there is a proportional relation between the observed peak intensity in the NMR spectrum and the number of nuclei that produce this signal (Figure 9.1). Both conventional solution and solid-state (particularly for nonsoluble materials) NMR techniques can be used in the characterization of block copolymers. Many types of nuclei can be observed, but the most frequently used for polymers are proton (^{1}H) NMR and carbon-13 (^{13}C) NMR.

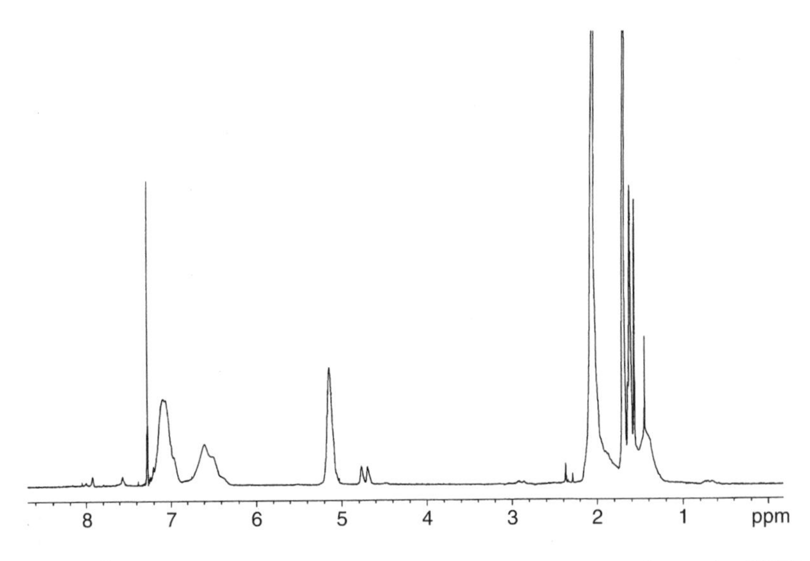

Figure 9.1. ^{1}H-NMR spectrum of a poly(styrene-b-isoprene) copolymer in $CDCl_3$ at 250 MHz. The peaks at 4.6 ppm to 5.2 ppm correspond to the olefinic protons of the isoprene segments, whereas those at 6.2 ppm to 7.2 ppm correspond to the aromatic protons of the styrene segments. The composition of the copolymers can be calculated from the areas under the peaks. The ratio of the peaks at 4.7 ppm (doublet) and at 5.1 ppm (singlet) correspond to the 3,4 and 1,4 incorporated isoprene units. The sharp peak at 7.25 ppm is due to $CDCl_3$.

Proton NMR has been widely used for many years in order to provide information on the monomeric species used in the preparation of polymers, the average composition, tacticity, and configuration of polymeric chains (Senn 1963). These studies are usually done in solution. A disadvantage is that polymer spectra are frequently poorly resolved with broad overlapping lines. This situation has been improved by the use of high magnetic field equipment.

Carbon-13 NMR is more revealing than ^{1}H-NMR in polymer work because of the inherently wider spectral separation of carbon chemical shifts that makes ^{13}C-NMR spectra more readily interpretable. Furthermore, the development of special decoupling techniques and pulse sequences have enhanced the utility of NMR spectroscopy in the field of polymers (Bovey 1982).

2.1.2. Infrared Spectroscopy (IR). In infrared absorption spectroscopy, absorption of energies corresponding to transitions between vibrational or rotational energy states gives rise to characteristic patterns. These patterns can be translated into qualitative and quantitative information regarding the presence of functional groups, which can lead to the identification of the monomer types and their concentration in a block copolymer. The technique provides information on chemical, structural, and conformational aspects of polymeric chains (Holland-Moritz 1976). Advances in instrumentation and data analysis techniques such as difference spectroscopy (spectral substraction), factor analysis, spectral deconvolution, and least square curve fitting of calibration plots for quantitative analysis, together with its ability for studying solid polymer samples, make IR a method of choice in the molecular characterization of complex block copolymers.

2.1.3. UV Spectroscopy. Due to the inherently high sensitivity of UV spectroscopy, the technique is often utilized for the identification and quantitative determination of comonomers in block copolymers (Meehan 1946). The coupling of UV detectors with SEC enhances the detection capabilities and applicability of both techniques.

2.2. Determination of Molecular Weight and Molecular Weight Distribution

The average molecular weight of a block copolymer is a very important parameter strongly related to the properties of the polymeric material. Therefore, its determination and knowledge is imperative in the study of these materials. A variety of techniques are available for the determination of different average molecular weights of block copolymers. Some of them are discussed in the following text.

2.2.1. Membrane Osmometry. Membrane osmometry is a classical method for determining the number of average molecular weights of block copolymers and polymers in general (Mays 1991). It is based on the measurement of the osmotic pressure that develops a polymer solution when it is brought into thermodynamic equlibrium with the pure solvent through a semipermeable membrane, permeable only to solvent molecules. The osmotic pressure is correlated to the concentration of the solution through eq. (9.2):

$$\pi/c = RT(1/M_n + A_2c + A_3c^2 + \cdots) \tag{9.2}$$

where A_2, A_3 are the second and third virial coefficients, a measure of solution nonideality (or, in other words, a measure of solvent-solute interactions). When the concentration is sufficient low (in the dilute solution range) A_3 can be ignored, and eq. 9.2 becomes linear in concentration. Measuring a series of polymer solutions with different concentration, the M_n of the solute can be determined from the intercept at $c = 0$ and the second virial coefficient from the slope. Sometimes, especially in the case of high molecular weight copolymers, higher concentrations must be used, and eq. 9.2 is no longer linear in c. Using the relation between A_2 and A_3 given by Stockmayer and Casassa (Stockmayer 1952), eq. 9.2 can be transformed to:

$$(\pi/c)^{1/2} = (RT/M_n)^{1/2}[1 + 1/2(A_2M_nc)] \tag{9.3}$$

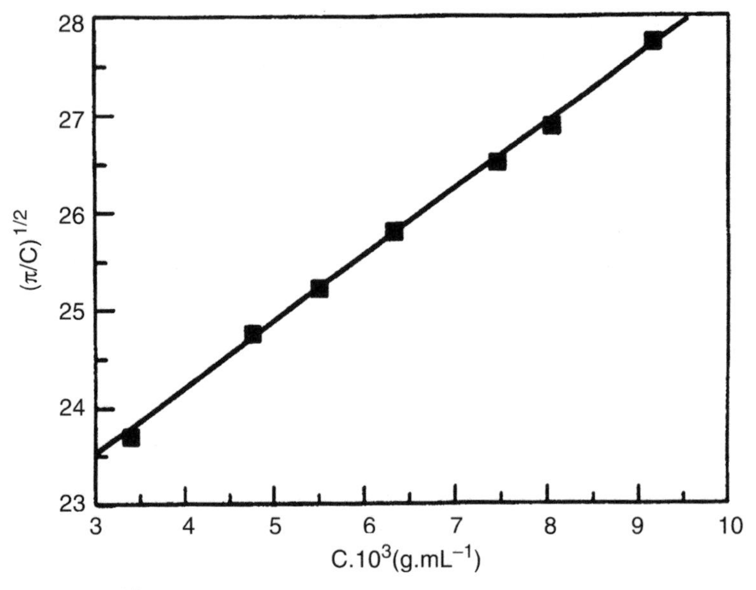

Figure 9.2. $(\pi/c)^{1/2}$ versus concentration plot for an PS-PI-P4MeS-PBd 4-miktoarm star quarterpolymer in toluene resulting from membrane osmometry measurements. Extrapolation to zero concentrations results in the determination of the true number average molecular weight, M_n, of this complex macromolecule. The second virial coefficient, A_2, can be deduced from the slope of such a plot (Iatrou 1992).

which is linear in c at a larger concentration range. Osmotic pressure, as a colligative property of the solution, depends on the number of solute molecules present. Therefore, even for the case of block copolymers, eqs. 9.2 and 9.3 give the true number average molecular weight of the copolymer (Figure 9.2). The method may be applied to a molecular weight range between 10,000 and 300,000 due to membrane permeation by polymer at the lower molecular weight limits and low values of the osmotic pressure at larger molecular weights.

2.2.2. Vapor Pressure Osmometry. In this method two temperature-sensitive thermistors are placed in a chamber saturated by solvent vapor (Mays 1991). The solution of the copolymer is brought into contact with one of them, whereas, on the other, solvent is placed. The temperature difference between the two thermistors, as a result of solvent condensation on the thermistor in contact with the solution, is measured. Solvent condensation is driven by the requirement for solvent vapor pressure equilibration in the solution and the neat solvent. ΔT depends on solution concentration. Actually, the temperature difference, ΔT, is determined indirectly by measuring the difference in electrical resistance, ΔR, between the two thermistors by an appropriate electronic setup (Hill 1930). Because heat losses cannot be eliminated, complete saturation of the chamber cannot be achieved, and the instruments are calibrated using substances of known molecular weight. The equation that is used is analogous to the one derived for membrane osmometry, i.e.:

$$\Delta R/c = (\Delta R/c)_{c=0}(1 + \Gamma_2 c + \Gamma_3 c + \cdots) \tag{9.4}$$

where $(\Delta R/c)_{c=0} = K/M_n$, K is the calibration constant, and Γ_i coefficients that are related to the corresponding virial coefficients. The method is usually employed for the determination of M_n in the range of molecular weights lower than 10,000. As in the case of membrane osmometry this method leads also to the true M_n of the copolymers.

2.2.3. Static Light Scattering. Light scattering is routinely used for the determination of weight average molecular weight of polymers and, under certain conditions, it can give the absolute M_w of block copolymers (Huglin 1972).

The absolute scattering intensity from a polymer solution at a certain scattering angle is often expressed in terms of the Reyleigh ratio defined as:

$$R(\theta, c) = I(\theta, c)r^2/I_o V$$

where $I(\theta, c)$ is the light-scattering intensity at a distance r, V is the scattering volume, and I_o is the intensity of the incident beam. The scattering angle θ is formed between the directions of the incident and the scattered light. The excess Reyleigh ratio:

$$\Delta R_\theta = R_{\theta,\text{solution}} - R_{\theta,\text{solvent}}$$

is related to the weight average molecular weight of the dissolved polymer via the following equation:

$$Kc/\Delta R_\theta = 1/M_w P(\theta) + 2A_2 c + 3A_3 c^2 + \cdots \tag{9.5}$$

where c is the polymer concentration, K is an optical constant ($K = 4\pi^2 n^2 (dn/dc)^2/\lambda^4 N_A$), n is the refractive index of the solvent at the wavelength used, (dn/dc) is the specific refractive index increment, λ is the wavelength of light used and N_A is the Avogadro number, and $P(\theta)$ is a function called the particle form factor, which describes the angular dependence of the scattered light. By extrapolation of $Kc/\Delta R_\theta$, both at $c = 0$ and $\theta = 0$, with a graphical procedure known as the Zimm plot (Zimm 1948), the M_w of the solute can be obtained.

In the case of block copolymers, the M_w obtained by light scattering is an apparent value, $M_{w,app}$, due to the different specific refractive increments of each part of the molecule in the same solvent. The specific refractive index increment of an AB block copolymer is related to the dn/dc of each block through the relationship:

$$(dn/dc)_{cop} = w_A(dn/dc)_A + (1 - w_A)(dn/dc)_B \tag{9.6}$$

where w_A, w_B are the weight fractions of blocks A and B, respectively. Benoît and Bushuk (Bushuk 1958) and Stockmayer (1950) proposed the following relation:

$$M_{w,app} = 1/v_C^2[w_A M_w^A v_A^2 + (1 - w_A)M_w^B v_B^2 + 2v_A v_B M_w^{AB}] \tag{9.7}$$

where M_w^A, M_w^B are the molecular weights of each block, v_C, v_A, v_B are the specific refractive index increments of the copolymer, and A and B parts. The quantity M_w^{AB} has no direct physical significance, but it can be expressed as a function of M_w^C, M_w^A, and M_w^B

$$M_w^{AB} = 1/2[M_w^C - w_A M_w^A + (1 - w_A)M_w^B] \tag{9.8}$$

with M_w^C being the true molecular weight of the copolymer. Eq 9.7 can be written in the form:

$$M_{w,app} = M_w^C + 2[(v_A - v_B)/v_C]P + [(v_A - v_B)/v_C]^2 Q \tag{9.9}$$

where P, Q are parameters that are related to the molecular weight and chemical heterogeneity of the copolymer. For an AB diblock copolymer, P, Q are given by:

$$P = w_B(1 - w_B)[(M_w^A - M_n^A) - (M_w^B - M_n^B)] \tag{9.10}$$

and

$$Q = w_B(1 - w_B)[(1 - w_B)(M_w^A - M_n^A) + w_B(M_w^B - M_n^B)] \tag{9.11}$$

From the above equations it is obvious that, as the heterogeneity in molecular weight and composition of a block copolymer decreases, the values for P and Q decrease, and the measured M_w of the copolymer tends towards the true molecular weight. On the other hand, by measuring $M_{w,app}$ of a block copolymer in three different solvents, it is possible to determine the true molecular weight of the copolymer and obtain information about its chemical and molecular weight heterogeneity. For block copolymer samples synthesized by a living polymerization mechanism, e.g., anionic polymerization, if the following conditions are held:

i) $|v_A|$, $|v_B|$ very large
ii) sign of v_A is the same as the sign of v_B

in a certain solvent, then the M_w determined for the AB copolymer in this solvent is very close to the true one (Tuzar 1970) (Figure 9.3).

2.2.4. Size Exclusion Chromatography (SEC). SEC has emerged as a powerful analytical technique for the determination of molecular weight as well as molecular weight and composition distributions of block copolymers (Balke 1991, Meira 1991, Pasch 1997). In SEC a dilute polymer solution is injected into

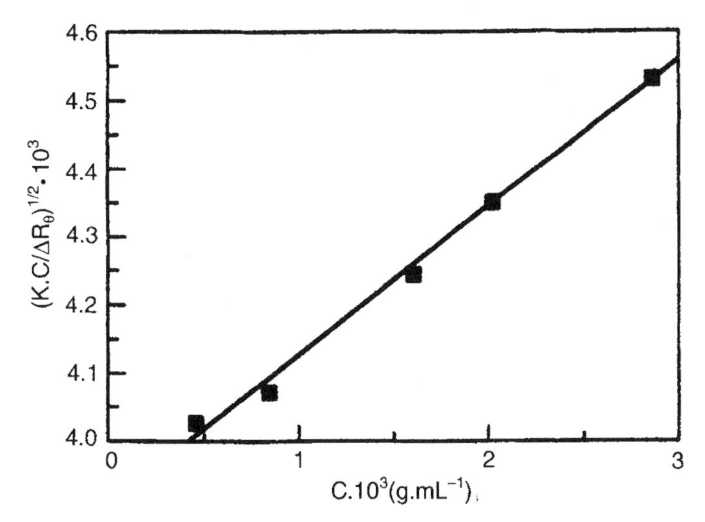

Figure 9.3. $Kc/\Delta R_\theta$ versus concentration plot for the same 4-miktoarm star of Figure 9.2 in THF, a solvent having a high light-scattering contrast (high dn/dc values) for all monomeric units, resulting from low-angle laser light measurements. Extrapolation to zero concentrations results in the determination of the apparent weight average molecular weight, M_w, of the copolymer. The second virial coefficient, A_2, can be also deduced from the slope of such a plot. The good agreement between M_w (65,500) and M_n (62,500) confirms the low degree of compositional and molecular weight heterogeneity of the sample (Iatrou 1993).

the solvent stream, which then flows through a series of columns packed with an inert porous material having a controlled pore size (usually polystyrene/divinylbenzene or silica gel). The size of the pores is comparable to the size of the macromolecules that are being analyzed. The smaller polymer molecules are able to pass through most of the pores, and they follow a relatively long flow path through the columns. The larger molecules are excluded from most of the pores, and, therefore, they follow a short path, and they elute first. During sample passage through the column, only size exclusion can take place; other interactions, like sample adsorption on the packing material, are inactive. As the polymer molecules exit the columns, they are detected by aid of one or more suitable detectors producing an elution curve (usually concentration versus elution time or elution volume). The elution curve is an ideal fingerprint of the polymer. The molecular weight of each eluted fraction can be determined by calibration of the instrument with standards of the same chemical composition with the sample under investigation. In this way the distribution of the sample can be obtained. Because the method separates species by their hydrodynamic volume, it is difficult to obtain true molecular weights in the case of block copolymers and nonlinear homopolymers. In these cases the so-called universal calibration curve can be used (Grubisic 1967). It is well known that the hydrodynamic volume (v_h) of a polymer molecule is proportional to the product of the intrinsic viscosity and the molecular weight.

$$[\eta]M = 0.025 N_A v_h \tag{9.12}$$

If the $[\eta]$ of the standards and the polymer under investigation have been determined independently, the molecular weight of the unknown sample can be obtained by use of a calibration curve of $[\eta]M$ versus elution volume (v_e). The universal calibration curve has been shown to be valid for a variety of homopolymers and copolymers with different chemical compositions and architectures (Figure 9.4).

A number of different types of detectors have been developed so far for use in SEC instruments. They can be divided into mass-concentration detectors and molecular-mass detectors. Mass-concentration detectors include the differential refractometer (DRI), the ultraviolet adsorption detector, and the IR detector.

The DRI detector is considered a universal detector because it is sensitive to differences in the refractive index of the solute and the carrier solvent. Although this detector can be used at high temperatures and gives a linear response over a wide range of concentrations and minimum band broadening, it is very sensitive to ambient temperature changes.

The UV detector is used when a part of the analyzed polymer contains chromophores that absorb in the UV region. It also offers a wide range of concentration, and it is sensitive at lower concentrations compared with the DRI detector. It can be operated at moderate temperatures and with solvents that present no absorption over a relatively wide range of wavelengths. Typical variable wavelength UV detectors operate in the range of 190 nm to 400 nm. They are valuable in monitoring the composition change as a function of elution volume in block copolymers with one UV-absorbing block, in conjunction with a DRI detector. Recent developments in photodiode array UV detectors allow the acquisition of the full absorption spectrum (from 200 nm to 700 nm) practically at each elution volume, making possible the observation of chemical changes (both qualitative and quantitative) in complex copolymer systems.

The infrared detector is also a concentration detector that is relatively insensitive to fluctuations in the set temperature and sometimes is the detector of choice for analyses carried out at high temperatures. Due to its ability to detect absorption of specific groups, it is useful in measuring polymer composition as a function of molar mass by monitoring specific functional groups in a block copolymer.

The molecular mass detectors include the light-scattering detector and the on-line viscometer.

Light-scattering detectors include small and multiangle detectors. Their application is useful in the determination of absolute molecular weight distributions and absolute molecular weight at every elution volume, always taking into account the limitation of light-scattering theory when applied to block copolymers. They are extremely sensitive to high molecular weight components in the sample but relatively insensitive to low molecular weight components. Multiangle detection allows continuous determination of the size of the analysate at every elution volume.

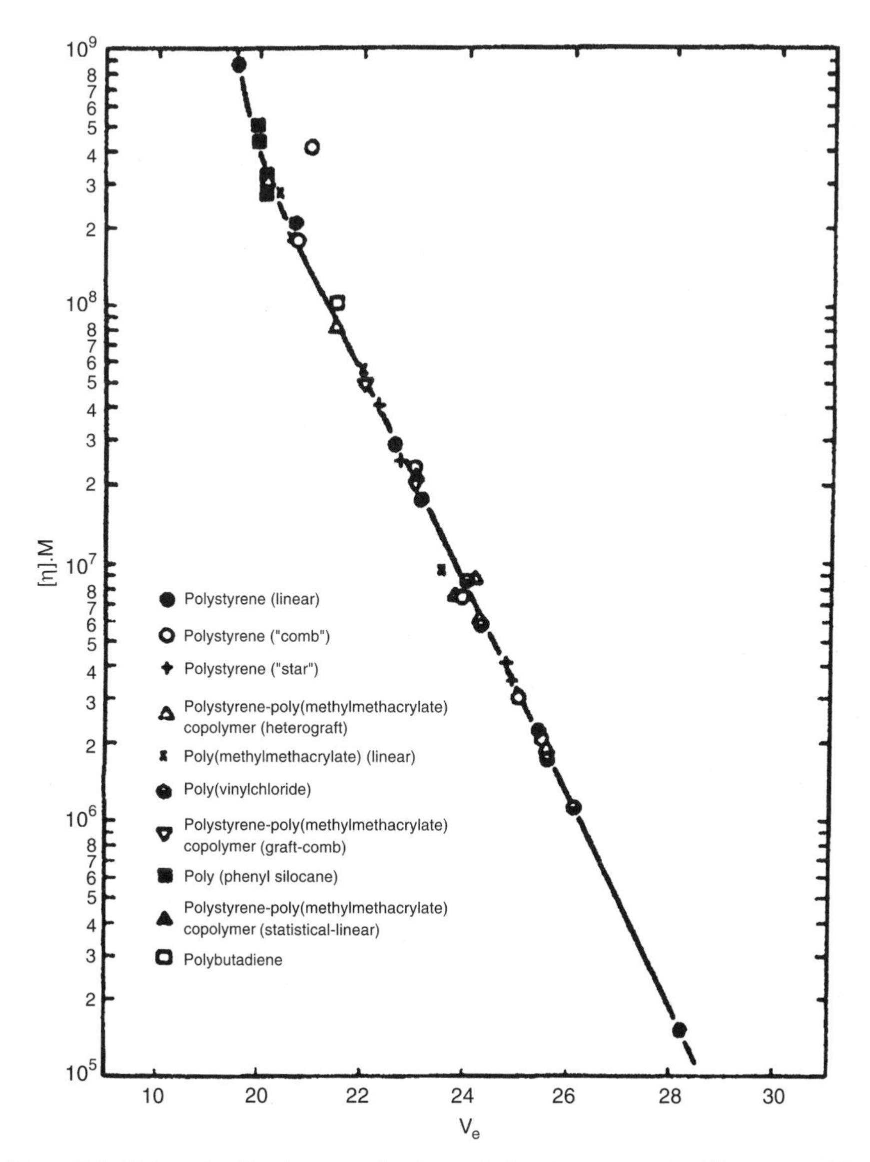

Figure 9.4. Universal calibration curve for size exclusion chromatography. The superposition of data corresponding to copolymers of different chemical nature and composition as well as macromolecular architecture confirms the applicability of the method to various kinds of copolymers (Grubisic 1967).

The on-line viscometer detectors, when used in series with a mass-concentration detector, can yield information on absolute molar mass averages, intrinsic viscosity (size), and long chain branching. In these detectors the solution viscosity is measured by the pressure drop across a flow-through capillary and is monitored

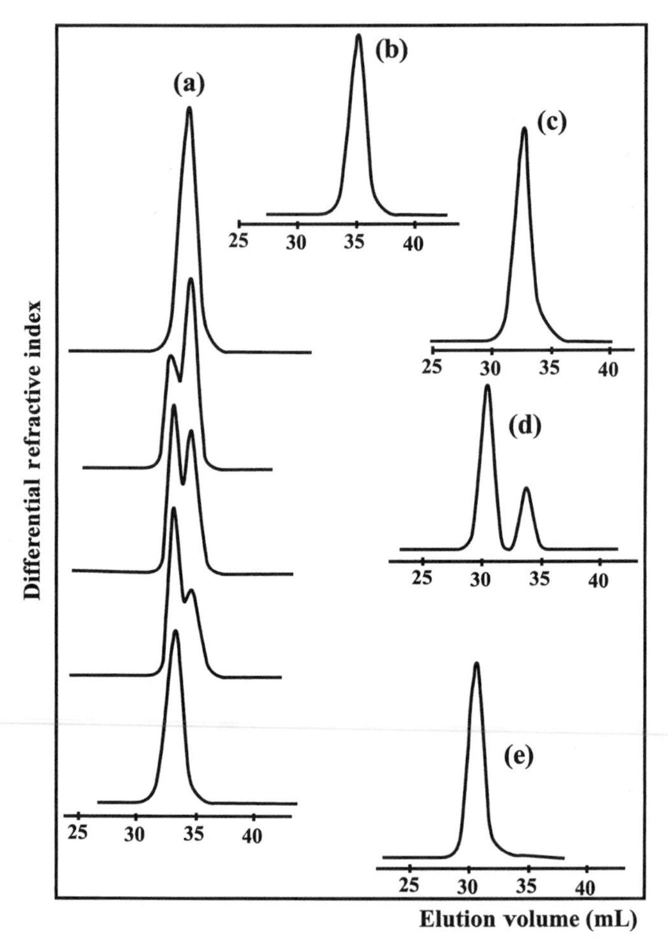

Figure 9.5. Monitoring the synthesis of a (PS)(PI)(PBd) 3-miktoarm star terpolymer by size exclusion chromatography with RI detection. (a) Titration of the macromolecular linking agent $PISiCl_2$ with the living PSLi arm. Samples have been withdrawn from the reaction mixture after every addition of the PSLi solution and analyzed. The low elution volume peak corresponds to the (PS)(PI)Si(CH_3)Cl block copolymer, and its intensity is increasing as the linking reaction proceeds. Meanwhile, the second high elution volume peak corresponding to the $PISiCl_2$ macromolecule is decreasing because this block is incorporated in the diblock. (b) PS arm; (c) PB arm; (d) unfractionated crude reaction product; (e) fractionated ABC miktoarm star (Iatrou 1992) (see Chapter 8).

by a differential pressure transducer. The sensitivity in the low molecular weight range is better than the light-scattering detector. One of the most useful features of an on-line viscometer is the ease of obtaining molecular weight data by the use of a universal calibration curve from chemically heterogeneous polymers. A representative example of the use of SEC in the monitoring of the synthesis and the analysis of a complex block copolymer is given in Figure 9.5. The successful application of

SEC with multiple detection in the case of block copolymers has been demonstrated (Jordon 1984, Cotts 1983, Huang 2000). Recently developed chromatographic techniques will also aid in the problem of block copolymer characterization (Falkenhagen 2000, Sato 1996, Pasch 1997, Lee 2001).

2.2.5. Matrix-assisted Laser Desorption Ionization-time of Flight Mass Spectroscopy (MALDI-TOF MS). Matrix-assisted laser desorption ionization-time of flight mass spectroscopy (MALDI-TOF MS) is a recently developed polymer analysis technique. Information about absolute molecular weight, molecular weight distributions, compositional homogeneity, and chemical identity of end groups can be gained in one run, especially if low molecular weight samples of narrow molecular weight distribution are analyzed. The method is considered to be complementary to SEC and absolute methods for molecular weight determination (like LS and MO). Problems related to the correct choice of the matrix in the case of block copolymers are more difficult. MALDI-TOF MS has been successfully employed from block copolymer sample analysis, and information related to the polymerization mechanism was extracted (Wilczek-Vera 1996 and 1999).

Hong et al. (Hong 2001) used MALDI-TOF MS to determine molecular weights and polydispersities of block copolymers of PS and poly(1,3-cyclohexadiene) (PCHD). Because both PS and polydienes are ionized by addition of silver salts, MALDI-TOF MS, in general, works well on low molecular weight PS-polydiene block copolymers.

Yang et al. (Yang 2001) recently applied MALDI-TOF MS to the characterization of poly(styrene sulfonate-b-tert-butyl styrene) copolymers (NaPSS-PtBuS). These materials are difficult to characterize because they form micelles at very low concentrations in solution (ppm level) due to their amphiphilic nature, precluding the use of light scattering, SEC, or osmometry for determining single-chain characteristics. However, MALDI-TOF MS revealed the molecular weights and polydispersities of the single chains. By comparing the MALDI-TOF MS results for the NaPSS-PtBuS samples and their precursor PS-PtBuS diblocks, the extent of sulfonation was also calculated (Yang 2001).

2.3. Molecular Size Determination

In many cases the molecular size of the block copolymer in solution must be known. Several methods exist for determination of the size of macromolecules in solution. Some of them are reported in the following paragraphs, and their relevance to the overall dimensions of block copolymer chains in solution is discussed.

2.3.1. Static Light Scattering. The general equation for static light scattering (eq. 9.5) can be modified if the particle form factor is expressed as (Huglin 1971):

$$1/P(q) = 1 + q^2\langle R_g^2\rangle/3 + \cdots \qquad (9.13)$$

where R_g is the radius of gyration of the macromolecule, i.e., the root mean square distance of every point of the macromolecular chain from its center of mass, q is the scattering vector $(q = (4\pi n/\lambda)\sin(\theta/2)$. This approximation for $P(\theta)$ is used in the case where R_g is becoming comparable with 1/q (or in other words when the dimensions of the particle become comparable or larger than $\lambda/20$). For polydisperse polymers, R_g is the z-average radius of gyration $(\langle R_g^2 \rangle = M_i c_i R_{g,i}^2 / M_i c_i)$. Thus, taking into account eq. 9.13, eq. 9.5 can be written as:

$$Kc/\Delta R_\theta = 1/M_w[1 + 16\pi^2 n^2 \langle R_g^2 \rangle \sin^2(\theta/2)/3\lambda^2] + 2A_2 c \qquad (9.14)$$

This equation constitutes the basis for a Zimm plot (Zimm 1948), where $Kc/\Delta R_\theta$ is plotted as function of $\sin^2(\theta/2) + kc$, with k being an arbitrary constant of the order $\sim c^{-1}$, chosen in such a way that experimental points are well-distributed along the x axis. The initial slope of the curve produced by points extrapolated to $c = 0$ is proportional to $\langle R_g^2 \rangle$, whereas the slope of the curve of points extrapolated to $\theta = 0$ is proportional to A_2. Double extrapolation at $c = 0$ and $\theta = 0$ yields an intercept inversely proportional to M_w, as was mentioned before. As can be understood, $\langle R_g^2 \rangle$ is a geometric radius of the macromolecule. As was also explained before, static light-scattering measurements on block copolymers are complicated by the fact that, in general, copolymers are polydisperse in composition and molecular weight (and, therefore, in size). The $\langle R_g^2 \rangle$, like M_w, is an apparent quantity in this case. The problem can be eliminated if all blocks have the same refractive index or the sample is completely homogeneous in composition and molecular weight—something unachievable in true block copolymer systems.

In any case the apparent radius of gyration is given by the eq.:

$$\langle R_g^2 \rangle_{app} = 1/v_C^2[w_A^2 v_A^2 \langle R_g^2 \rangle_A + (1 - w_A)^2 v_B^2 \langle R_g^2 \rangle_B + 2w_A(1 - w_A)v_A v_B \langle R_g^2 \rangle_{AB}] \qquad (9.15)$$

where $\langle R_g^2 \rangle_A$, $\langle R_g^2 \rangle_B$ are the radii of gyration of the A and B blocks and

$$\langle R_g^2 \rangle_{AB} = 1/2(\langle R_g^2 \rangle_A + \langle R_g^2 \rangle_B + l^2) \qquad (9.16)$$

with l being the distance between the centers of mass of the A and B parts of the molecule. l is a characteristic parameter of the molecular architecture (Huglin 1971). According to eq. 9.15, measurements of $\langle R_g^2 \rangle_{app}$ in three different solvents can give $\langle R_g^2 \rangle_A$, $\langle R_g^2 \rangle_B$, $\langle R_g^2 \rangle_{AB}$ and l. Additionally, it is possible to determine the $\langle R_g^2 \rangle$ of one part of the molecule if the measurements are contacted in a solvent that is isorefractive (has the same refractive index) with the other part of the molecule (which, in this case, becomes "invisible") (Hadjichristidis 1978). $\langle R_g^2 \rangle_{app}$ becomes similar to the true $\langle R_g^2 \rangle$ of the copolymer when $|v_A|$, $|v_B|$ are large and close to each other having the same sign. In spite of the difficulties arising from sample heterogeneity, static light scattering has been employed in several cases for the determination of the size of block copolymers in solution (Prud'homme 1972, Cirolamo 1972, Nguyen 1986, Hadjichristidis 1978).

2.3.2. Dynamic Light Scattering. The technique of dynamic light scattering (or quasilelastic light scattering or photon correlation spectroscopy) is based on the fact that the intensity of the light scattered from a collection of scatterers (macromolecules) is a result of the interactions of the scattered radiation by each one of them with the radiation scattered by the others. The phase and polarization of the scattered rays from each scattering center depends on its position and direction in space. Because the position and direction of the scatterers change with time due to the Brownian motion, the intensity, the phase, and the polarization of the scattered radiation change with time. As a result, the fluctuations of scattered light contain information about the dynamics of the scatterers (Berne 1976).

The variation of light-scattering intensity with time can be described by a time-correlation function. In the simple case of a group of identical, spherical, and noninteracting particles in solution, the correlation function is given by:

$$G^{(2)}(t) = \langle I(0)I(t)\rangle = A + B\exp(-\Gamma t) = A + B\exp(Dq^2t) \tag{9.17}$$

where A and B are constants related to the experimental setup, q the scattering vector, Γ the decay rate, and D the translational diffusion coefficient, which can be related to the size of the particle. Because Γ can be determined by an appropriate analysis of the experimental data (Koppel 1972, Phillies 1988, Provencher 1982), the diffusion coefficient can be evaluated directly from the correlation function. In most cases when polymer solutions are examined, there exists a heterogeneity in size within the group of molecules, each one of them having its own diffusion coefficient. In this case $G^{(2)}(t)$ becomes a sum of exponentials

$$G^{(2)}(t) = A + B[N_i a_i^2 P_i \exp(-q^2 D_i t)]^2 \tag{9.18}$$

where N_i is the number of particles, a_i the polarizability, and P_i the form factor of each particle. The measured diffusion coefficient is a z-average value. Many different methods have been proposed for the analysis of dynamic light-scattering data from polydisperse systems (Koppel 1972, Phillies 1988, Provencher 1982).

In real polymer solutions, D depends on concentration according to:

$$D = D_o(1 + k_D c + \cdots) \tag{9.19}$$

where D_o is the diffusion coefficient at infinite dilution and k_D a parameter that depends on the polymer-solvent-temperature system and contains thermodynamic and hydrodynamic terms of the system, i.e.:

$$k_D = 2A_2M - k_f - (N_AV/M) = 2A_2M - k_f - v \tag{9.20}$$

where A_2 is the second virial coefficient, M the molecular weight, N_A Avogadro's number, V the volume of the polymer molecule, v the partial specific volume of the polymer, and k_f the proportionality constant on the concentration-dependence of the friction coefficient ($f = f_o(1 + k_f c)$). D_o can be related to the hydrodynamic radius of the equivalent sphere through the Stokes-Einstein relation:

$$R_h = kT/6\pi\eta D_o \tag{9.21}$$

where k is the Boltzmann constant, T the absolute temperature, and η the solvent viscosity. R_h is a measure of the hydrodynamic size of the molecule in solution as is determined by polymer-solvent interactions and temperature of the system. D_o and, therefore, R_h are apparent quantities in the case of block copolymer in solution due to their inherent compositional and molecular weight heterogeneity. This case has been dealt with theoretically and experimentally in a way analogous to the static light-scattering case, and analogous conclusions have been drawn (Burchard 1984). The quantities D_o and R_h are close to the true ones if the polydispersity of the copolymer is low and both parts of the copolymer have a high dn/dc in the respective solvent. Dynamic light scattering has been shown to be useful in the determination of the size of block copolymers under these conditions (Burchard 1984, Pispas 1994, Pispas 1996).

2.3.3. Small Angle Neutron Scattering. Small angle neutron scattering has been applied in the study of polymer systems for the last 30 years with considerable success (Pynn 1990, Higgins 1994). Its employment for the study of block copolymer solutions is growing. Although the method is able to provide the molecular weight of the solute, it is almost exclusively used in the determination of chain dimensions. The total intensity of neutrons scattered per second by a polymer solution is given by:

$$I = I_o N\sigma \tag{9.22}$$

where I_o is the incident neutron flux in neutrons per second per square centimeters, N is the number of nuclei in the scattering volume, and σ is the total scattering cross-section per nucleus (a quantity analogous to the refractive index in light-scattering experiments). In a usual neutron scattering experiment, the scattering pattern is observed as a function of the scattering vector (angle), and the scattered neutrons are collected over a solid angle $d\Omega$. For this experimental arrangement, eq. 9.22 is written as:

$$I = I_o N(d\sigma/d\Omega)\Delta\Omega \tag{9.23}$$

where $d\sigma/d\Omega$ is the differential scattering cross-section. This is the parameter that determines the scattering pattern in elastic neutron scattering and can be related to the features of the scattering material. It can be proven that, in small angle neutron scattering:

$$d\sigma/d\Omega = 1/N_p(\rho_p - \rho_m)2\left|\int_V \exp(iqr)d^3r\right|^2 \tag{9.24}$$

where ρ_p and ρ_m are the scattering length densities of the particles and the matrix (solvent) and r the position inside the system. The integral in eq. 9.24 is the scattering law S(q), for the system that contains all the information on correlations within one particle (particle form factor) and correlations between particles (structure factor) depending on the q range used.

The theoretical equations derived for light scattering from block copolymers in solutions can be readily applied to small angle neutron scattering taking into account the different contrast factors. Thus, the scattering intensity from a block copolymer solution can be written:

$$I(q) = KcM_{app}(C - s)S(q) \tag{9.25}$$

where M_{app} is the apparent molecular weight of the copolymer with scattering length density C dissolved in a solvent having a scattering length density s,

$$C = w_A A + (1 - w_A)B \tag{9.26}$$

where A and B are the scattering length densities of parts A and B of the copolymer. Setting

$$K_A = A - s,\ K_B = B - s,\ \text{and}\ K_C = C - s \tag{9.27}$$

for the different contrast factors, then the apparent molecular weight is given by:

$$M_{app} = (K_A K_B/K_C^2)M_C + (K_A(K_A - K_B)/K_C^2)w_A M_A + (K_B(K_B - K_A))/K_C^2 w_B M_B \tag{9.28}$$

and the scattering law by:

$$S(q) = 1/K_C^2[w_A^2 K_A^2 S_A(q) + (1 - w_A)^2 K_B^2 S_B(q) + 2w_A(1 - w_A)K_A K_B S_{AB}(q)] \tag{9.29}$$

The apparent radius of gyration of the copolymer is given by:

$$\langle R_{g,app}\rangle = Y^2\langle R_{g,A}^2\rangle + (1 - Y)^2\langle R_{g,B}^2\rangle + Y(1 - Y)\langle R_{g,AB}^2\rangle \tag{9.30}$$

where $Y = w_A K_A/K_C$ and $\langle R_{g,AB}^2\rangle = \langle R_{g,A}^2\rangle + \langle R_{g,B}^2\rangle + l^2$, and the symbols have their usual meaning.

The technique of contrast variation is very frequently used in SANS. Thus, the solvent scattering length density can be adjusted by mixing hydrogenous and deuterated analogues of the desired solvent without changing appreciably its thermodynamic quality. In this way either K_A or K_B become zero, and the equations are simplified considerably. However, it should be noticed that, in order to determine accurately the true M or R_g of the copolymer, measurements should be made under at least three different contrasts (different K_C, K_A, and K_B values) (Ionescu 1981, Han 1977, Matsusita 1992).

2.3.4. Dilute Solution Viscometry. The viscosity, η, of a liquid is a measure of its resistance to flow. If spherical particles are introduced in the flowing liquid, having a size much larger than the size of the molecules of the liquid, the resistance of the solution to flow increases over that of the pure liquid, i.e., viscosity of the medium, η, increases. Einstein connected the specific viscosity $\eta_{sp} = (\eta_{solution} - \eta_{solvent})/\eta_{solvent}$ of a dilute solution of noninteracting spheres with the volume fraction of the spheres by (Flory 1953):

$$\eta_{sp} = (5/2)\phi \tag{9.31}$$

For more concentrated solutions and independent of the specific shape of the particles, η_{sp} can be calculated by:

$$\eta_{sp} = a_1\phi + a_2\phi^2 + \cdots \tag{9.32}$$

The volume fraction of particles is equal to $(v_h/M)/c$, where v_h is the effective hydrodynamic volume of the particle, M its molecular weight, and c the mass concentration of the particles. Then eq. 9.32 is written as:

$$\eta_{sp} = a_1(v_h/M)/c + a^2[(v_h/M)/c]^2 + \cdots = [\eta]c + k'[\eta]^2c^2 + \cdots \tag{9.33}$$

where $[\eta]$ is the intrinsic viscosity, a measure of the effective hydrodynamic volume of the particle (macromolecule) in solution divided by the molecular weight. Eq. 9.33 can be written as:

$$\eta_{sp}/c = [\eta] + k_H[\eta]^2c + \cdots \tag{9.34}$$

which is the Huggins equation, with k_H the Huggins coefficient. k_H is a measure of the hydrodynamic interactions experienced by polymer chains in solution and, therefore, a measure of solvent quality (Hadjichristidis 1991). It is obvious that a hydrodynamic radius of an equivalent sphere can be calculated from intrinsic viscosity assuming that the polymer chains (with solvent molecules entrapped within polymer coils) behave like spheres in solution. That is:

$$R_v = (3/10N_A)^{1/3}([\eta]M)^{1/3} \tag{9.35}$$

Eq. 9.35 can be used in the determination of a hydrodynamic radius for isolated block copolymer molecules in solution as was done in several cases (Prud'homme 1972, Pispas 1994, Iatrou 1995, Pispas 1996).

REFERENCES

Balke S.T. (1991) in Modern Methods of Polymer Characterization, Barth H. G., Mays J. W. (Eds), Wiley Interscience, , Chapter 1.

Barth H. G., Mays J. W. (1991) Modern Methods of Polymer Characterization, Wiley Interscience, .

Berne B. J., Pecora R. (1976) Dynamic Light Scattering, Wiley Interscience, New York.

Bovey F. A. (1972) High Resolution NMR of Macromolecules, Academic Press, New York.

Bovey F. A. (1982) Pure Appl. Chem. 54, 559.

Bu L., Xu Z., Wan Y., Mays J. W. (1995) Polym. Bull. 39, 510.

Burchard W., Kajiwara K., Nerger D., Stockmayer H. W. (1984) Macromolecules 17, 222.

Burge D. E. (1979) J. Appl. Polym. Sci. 24, 293.

Bushuk W., Benoît H. (1958) Can. J. Chem. 36, 1616.

Cheng H. N. (1991) in Modern Methods of Polymer Characterization, Barth H. G., Mays J. W. (Eds), Wiley Interscience, , Chapter 11.

Cirolamo M., Urwin J. R. (1972) Eur. Polym. J. 8, 1159.

Cotts D. B. (1983) Org. Coat. Appl. Polym. Sci. Proc. 48, 750.

Entelis S. G., Evreinov V. V., Gorshkov A. V. (1986) Adv. Polym. Sci. 76, 129.

Falkenhagen J., Much H., Stauf W., Muller A. H. E. (2000) Macromolecules 33, 3687.

Flory P. J. (1953) Principles of Polymer Chemistry, Cornell Univ. Press, Ithaca.

Grubisic Z., Rempp P., Benoit H. (1967) J. Polym. Sci. Polym. Lett. Ed. B5, 753.

Hadjichristidis N., Mays J. W. (1991) in Modern Methods of Polymer Characterization, Barth H. G., Mays J. W. (Eds), Wiley Interscience., Chapter 7.

Hadjichristidis N., Roovers J. (1978) J. Polym. Sci. Polym. Phys. Ed. 16, 851.

Han C. C., Mozer B. (1977) Macromolecules 10, 44.

Higgins J. S., Benoit H. C. (1994) Polymers and Neutron Scattering, Clarendon Press, Oxford.

Hill A. V. (1930) Proc. Roy. Soc. London Ser. A 127, 9.

Holland-Moritz K., Siesler H. W. (1976) Appl. Spectroscopy Rev. 11, 1.

Hong K., Wan Y., Mays J. W. (2001) Macromolecules 34, 2482.

Huang Y., Bu L., Zhang D., Su C., Xu Z., Bu L., Mays J. W. (2000) Polym. Bull. 44, 301.

Huglin M. B. (1972) Light Scattering From Polymer Solutions, Academic Press, New York.

Iatrou H., Hadjichristidis N. (1992) Macromolecules 25, 4649.

Iatrou H., Hadjichristidis N. (1993) Macromolecules, 26, 2479.

Iatrou H., Siakali-Kioulafa E., Hadjichristidis N., Roovers J., Mays J. W. (1995) J. Polym. Sci. Part B: Polym. Phys. 33, 1925.

Ionescu L., Picot C., Duplessix R., Duval M., Benoit H., Lingelser J. P., Gallot Y. (1981) J. Polym. Sci. Polym. Phys. Ed. 19, 1033.

Ivin K. J. (1983) Pure Appl. Chem. 55, 1529.

Jordon R. C., Silver S. F., Sehon R. D., Rivard R. J. (1984) Size Exclusion Chromatography, ACS Symp. Ser. 245, 295.

Koppel D. E. (1972) J. Chem. Phys. 57, 4814.

Lee W., Cho D., Chang T., Hanley K. J., Lodge T. P. (2001) Macromolecules 34, 2353.

Matsusita Y., Shimizu K., Noda I., Chang T., Han C. C. (1992) Polymer 33, 2412.

Mays J. W., Hadjichristidis N. (1991) in Modern Methods of Polymer Characterization, Barth H. G., Mays J. W. (Eds), Wiley Interscience., Chapter 6.

Meehan J. (1946) J. Polym. Sci. 1, 175.

Meira G. R. (1991) in Modern Methods of Polymer Characterization, Barth H. G., Mays J. W. (Eds), Wiley Interscience., Chapter 2.

Nguyen A. B., Hadjichristidis N., Fetters L. J. (1986) Macromolecules 19, 768.

Nuwaysir C. L., Wilkins C. L., Simonsick W. J. (1990) J. Am. Soc. Mass. Spectrom. 1, 66.

Pasch H., Trathnigs B. (1997) HPLC of Polymers, Springer, Heidelberg, Germany.

Phillies G. D. J. (1988) J. Chem. Phys. 89, 91.

Pispas S., Hadjichristidis N., Mays J. W. (1994) Macromolecules 27, 6307.

Pispas S., Hadjichristidis N., Mays J. W. (1996) Macromolecules 29, 7378.

Provencher S. W. (1982) Comput. Phys. Commun. 23, 649.

Prud'homme J., Roovers J. E. L., Bywater S. (1972) Eur. Polym. J. 8, 901.

Prud'homme J., Bywater S. (1972) Macromolecules 4, 543.

Pynn R. (1990) Mat. Res. Soc. Symp. Proc. 166, 15.

Riess G, Callot P. (1977) in Fractionation of Synthetic Polymers: Principles and Practices, Marcel Dekker, New York, p. 445.

Sato S., Ogino K., Darwint T., Kiyokawa I. (1996) Macromol. Symp. 110, 177.

Senn W. L. (1963) Analytica Chimica Acta 29, 505.

Stockmayer W. H. (1950) J. Chem. Phys. 18, 58.

Stockmayer W. H., Casassa E. F. (1952) J. Chem. Phys. 20, 1560.

Tung L. H. (1977) Fractionation of Synthetic Polymers: Principles and Practices, Marcel Dekker, New York.

Tung L. H. (1987) in Encyclopedia of Polymer Science and Engineering, Vol. 7, p. 298.

Tuzar Z., Kratochvil P., Strakova D. (1970) Eur. Polym. J. 6, 1113.

Wilczek-Vera G., Danis P. O., Eisenberg A. (1996) Macromolecules 29, 4036.

Wilczek-Vera G., Yu Y., Waddell K., Danis P. O., Eisenberg A. (1999) Macromolecules 32, 2180.

Yang J., Nonidez W. K., Mays J. W. (2001) Int. J. Polym. Anal. Charac. 6, 547.

Zimm B. H. (1948) J. Chem. Phys. 16, 1099.

PART III

SOLUTION PROPERTIES OF BLOCK COPOLYMERS

CHAPTER 10

DILUTE SOLUTIONS OF BLOCK COPOLYMERS IN NONSELECTIVE SOLVENTS

Most of the experimental studies of block copolymer dilute solutions in nonselective solvents, i.e., those that are thermodynamically good for all parts of the macromolecule, have been mainly concerned with the conformation of the copolymers (Matsusita 1992, Pitsikalis 1998, and references therein). A number of techniques have been employed including viscometry, light scattering, and neutron scattering. Generally, two models have been proposed for molecularly dissolved block copolymers (Figure 10.1).

The first one assumes a segregated conformation where the chemically different blocks occupy distinct regions in space due to the unfavorable thermodynamic interactions (heterocontacts) between the dissimilar segments. Thus, the configuration of each respective homopolymer is retained in the block copolymer. The second one describes a diblock copolymer as consisting of two mutually interpenetrating random coils. This interpenetration leads to more expanded conformations.

Monte Carlo calculations on a self-avoiding lattice walk were used in order to predict the preferred conformation of a diblock AB copolymer and a BAB triblock copolymer (Tanaka 1980 and 1977). It was concluded that, in contrast to the diblock case, the A block in a BAB triblock is more expanded compared with the identical homopolymer chain in the same solvent. The overall dimensions of a multiblock copolymer were found to increase with the number of blocks at the same composition and molecular weight. These theoretical calculations were complemented with light scattering and viscosity measurements on PS-PMMA diblocks, PMMA-PS-PMMA triblocks, and random copolymers.

Some investigators arrived at the conclusion that block copolymers have segregated conformations, based on light scattering and viscosity measurements

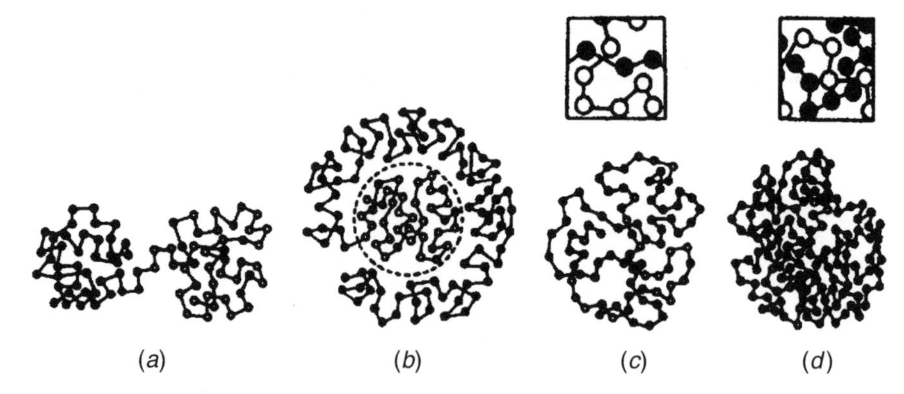

Figure 10.1. Possible conformations of a diblock copolymer chain in solution. Cases (a) and (b) correspond to segregated conformations [dumbbell (a) or core-shell (b)]. Cases (c) and (d) correspond to conformations, where mixing of the dissimilar segments occurs to various degrees resulting in smaller (c) or larger (d) contact between the two types of segments (and more or less expanded conformations) as shown in the insets (Urwin 1969).

(Ho-Duc 1976), whereas others, using the same techniques, arrived at the opposite conclusion (Utracki 1968). Bywater and collaborators have pointed out that a near-Gaussian distribution of block chains will tend to keep the number of hetero-contacts, responsible for a more segregated structure, small (Prud'homme 1972a). The effects of heterocontacts become measurable in poor solvents because their number increases in these solvents. Additionally, the effects of incompatible interactions on the conformation of a block copolymer would be relatively more pronounced in the case of a copolymer with highly incompatible blocks and especially in the case of a solution in a common theta solvent.

Prud'homme and Bywater (Prud'homme 1972b) conducted light scattering measurements in PS-PI diblock copolymers in solvents with different refractive indexes with respect to the two blocks. They estimated the distance between the two centers of mass, and this was twice the total radius of gyration in most cases. From their results they concluded that some separation of the two types of the chain segments must be present in solution, especially in poor solvents for the blocks.

Using light scattering, Tanaka et al. (Tanaka 1979) investigated a few PS-PMMA block copolymers in 2-butanone. They compared the values for the distance between the centers of mass and the sum of the individual radii of gyration of the two blocks. They found a considerable effect of the sample heterogeneity on the final results. Making the appropriate corrections, they concluded that segregated conformations are not consistent with their experimental results.

In a SANS investigation of near-monodisperse and polydisperse diblock copolymers of styrene and isoprene by Ionescu et al. (Ionescu 1981), the overall and individual dimensions of each block were determined by using the contrast variation technique. The solvents used were toluene, a common good solvent for both blocks, and cyclohexane, a theta solvent for PS and a good solvent for PI. The authors found no evidence for intermolecular segregation upon analyzing the experimental results. Using the same technique, Han and Mozer (Han 1977) arrived at the opposite conclusion. However, their experimental protocol received considerable criticism. Additional SANS measurements on PS-P2VP diblock copolymers showed that there is no change in the conformation of one block due to the presence of the other (Matsusita 1992).

Both theoretical and experimental conformations of nonlinear block copolymers have been investigated recently. These complex architectures allow for a larger number of heterocontacts between blocks, and, therefore, their effect on block copolymer dimensions is expected to be more pronounced.

Vlahos et al. (Vlahos 1995a) studied the case of miktoarm star copolymers of the type $A_xB_{f\text{-}x}$, with f being the total functionality of the star, using renormalization group theory and Monte Carlo calculations. They were able to estimate the dimensions of each arm and those of the whole molecule as a function of molecular weight and number of the dissimilar arms. They predicted a monotonic expansion of the arms of one kind limited only by the increase in the size of the other kind of arms. The dimensions of the whole star increase as the molecular weight and the functionality of the star increase.

The same group presented theoretical calculations on the behavior of ring diblocks in common good, theta, and selective solvents (Vlahos 1995b). They found that, in contrast with the linear diblock case, the dimensions of each part of a ring diblock are affected by both solvent quality and length of the other part. The differences in behavior vanished at the limit of infinite molecular weight. Cross-interactions were found to be more important in determining the dimensions of the ring diblock, with the effect being greater in the symmetric case.

Hadjichristidis and Roovers (Hadjichristidis 1978) conducted light-scattering measurements on poly(isoprene-g-styrene) graft copolymer in three thermodynamically good solvents for both parts. One of them was isorefractive with PS (bromoform), and the other was isorefractive with PI (chlorobenzene). In this way the overall dimensions of the copolymer and of each part of the macromolecule could be determined independently. By comparing the different radii of gyration, the authors arrived at the conclusion that these graft copolymers adopt a more or less core-shell like conformation in dilute solution having the PI backbone in the core and the PS branches in the shell (Table 10.1).

The specific graft copolymer was prepared by anionic polymerization (see Chapter 8), and it is comprised of 27.5 PS branches on average of Mn = 24,500 and a PI backbone of Mw = 546,000. Notice that, due to its low molecular weight and compositional heterogeneity, weight average molecular weights are similar in all solvents (see also Chapter 9). The dimensions of the copolymer measured in chlorobenzene (isorefractive for PI) are close to the ones measured in THF (high dn/dc values for both components), whereas those in bromoform (isorefractive for PS) are smaller. These results suggest that the molecule adopts a conformation where the PI backbone resides in the inner part of the molecular volume and PS branches reside at the periphery of the structure. A more elongated structure was

TABLE 10.1. Light Scattering Results on the Apparent Weight Average Molecular Weight and Radius of Gyration for a Poly(isoprene-g-styrene) Copolymer in Three Different Solvents (Hadjichristidis 1978).

Solvent	$M_{w,\ app} \times 10^{-6}$	$\langle S^2 \rangle_{z,\ app} \times 10^{-4}$ ($\AA^2$)
THF	1.18	19.6
Chlorobenzene	1.12	18.4
Bromoform	1.11	14.6

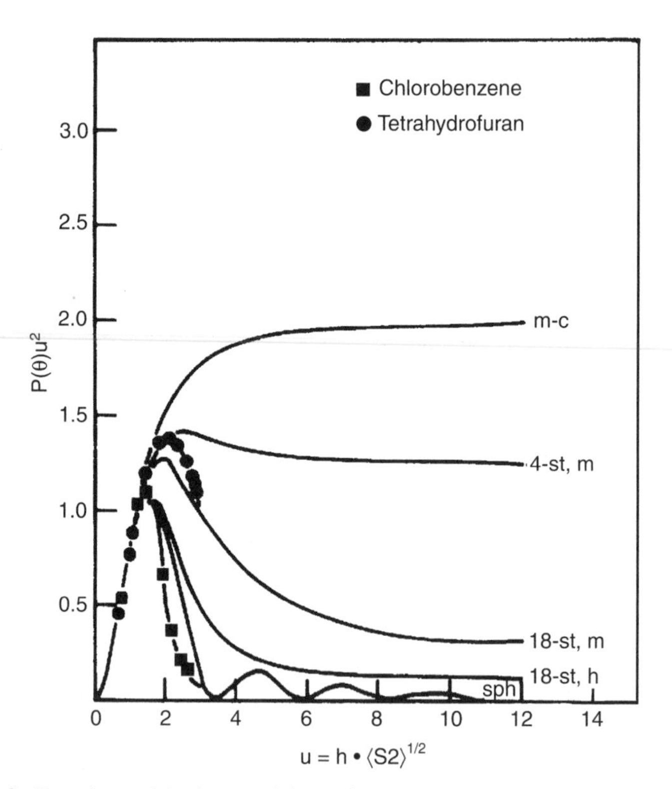

Figure 10.2. Experimental (points) and theoretical (curves) Kratky plots of 18-arm styrene-isoprene star block copolymers in two solvents with different light-scattering contrasts in relation to the star segments. Theoretical curves are calculated for polydisperse (p), monodieperse (m), and hollow (h) star (st) polymers. Tetrahydrofuran shows high scattering contrast for both segments, and the good agreement of the experimental results with the theoretical curve for a monodisperse star polymer (m) is expected. Chlorobenzene is almost isorefractive to the polyisoprene part of the molecule, and the good agreement of the experimental point with the curve calculated for hollow stars confirms the core-shell morphology of this kind of block copolymer in dilute solutions in good solvents (Nguyen 1986).

also observed for H and π graft copolymers of styrene and isoprene in the common good solvent THF (Pispas 1996).

Fetters et al. investigated the conformation of a star block copolymer having 18 arms (Nguyen 1986). They determined the dimensions of each part of the molecule by using static light scattering and isorefractive solvents. The results led to the conclusion that star block copolymers with many arms have a segregated, vesicle-like structure (Figure 10.2), in agreement with the theoretical and experimental results of Burchard et al. (Burchard 1984).

Hadjichristidis and coworkers have conducted a number of investigations on miktoarm star and model graft copolymers in solution using static and dynamic light scattering, as well as viscometry, in order to verify the segregated structure of these complex star architectures (Iatrou 1995, Vlahos 1996, Pispas 1996, Pispas 1999). Experimental results show that, for this kind of molecule where arms of a different chemical nature are forced to be close in space due to a common branching point, segregation of dissimilar segments can occur, the phenomenon being more pronounced as the branching density increases. The shape of the graft macromolecules becomes more elongated (Pispas 1996). These findings are in qualitative agreement with theoretical calculations (Vlahos 1995a,b, Vlahos 1996).

The dilute solution behavior of polymacromonomers, i.e., graft copolymers where each monomeric unit on the backbone carries a grafted chain, has been studied (Ishizu 1997, Kawaguchi 1998, Wintermantel 1994 and 1996, Tsukahara 1995, Nemoto 1995). These macromolecules have interesting solution properties because the crowding of the grafted chains induces stiffness to the backbone and the molecules adopt a bottle-brush conformation, which may be useful in applications involving polymer solutions.

REFERENCES

Burchard W., Kajiwara K., Nerger D., Stockmayer H. W. (1984) Macromolecules 17, 222.

Hadjichristidis N., Roovers J. (1978) J. Polym. Sci. Polym. Phys. Ed. 16, 851.

Han C. C., Mozer B. (1977) Macromolecules 10, 44.

Ho-Duc N., Prud'homme J. (1976) Intern. J. Polym. Mater. 4, 303.

Iatrou H., Siakali-Kioulafa E., Hadjichristidis N., Roovers J., Mays J. W. (1995) J. Polym. Sci. Part B: Polym. Phys. 33, 1925.

Ionescu L., Picot C., Duplessix R., Duval M., Benoit H., Lingelser J. P., Gallot Y. (1981) J. Polym. Sci. Polym. Phys. Ed. 19, 1033.

Ishizu K., Tsubaki K., Ono T. (1997) Polymer 39, 2935.

Kawaguchi S., Akaike K., Zhang Z-M., Matsumoto H., Ito K. (1998) Polymer J. 30, 1004.

Matsusita Y., Shimizu K., Noda I., Chang T., Han C. C. (1992) Polymer 33, 2412.

Nemoto N., Nagai M., Koike A., Okada S. (1995) Macromolecules 28, 3854.

Nguyen A. B., Hadjichristidis N., Fetters L. J. (1986) Macromolecules 19, 768.

Pispas S., Avgeropoulos A., Hadjichristidis N., Roovers J. (1999) J. Polym. Sci. Part B: Polym. Phys. 37, 1329.

Pispas S., Hadjichristidis N., Mays J. W. (1996) Macromolecules 29, 7378.

Pitsikalis M., Pispas S., Mays J. W., Hadjichristisis N. (1998) Adv. Polym. Sci. 135, 1.

Prud'homme J., Roovers J. E. L., Bywater S. (1972a) Eur. Polym. J. 8, 901.

Prud'homme J., Bywater S. (1972b) Macromolecules 4, 543.

Tanaka T., Kotaka T., Ban K., Hattori M., Inagaki H. (1977) Macromolecules 10, 960.

Tanaka T., Kotaka T., Inagaki H. (1976) Macromolecules 9, 561.

Tanaka T., Omoto M., Inagaki H. (1979) Macromolecules 12, 146.

Tanaka T., Omoto M., Inagaki H. (1980) J. Macromol. Sci. Phys. B17(2), 229.

Tsukahara Y., Ohta Y., Senoo K. (1995) Polymer 36, 3413.

Urwin J. R. (1969) Aust. J. Chem. 22, 1649.

Utracki L. A., Simha R., Fetters L. J. (1968) J. Polym. Sci. Part A-2, 6, 2051.

Vlahos C., Hadjichristidis N., Kosmas M., Rubio A., Freire J. J. (1995b) Macromolecules 28, 6854.

Vlahos C., Horta A., Hadjichristidis N., Freire J. J. (1995a) Macromolecules 28, 1500.

Vlahos C., Tselikas Y., Hadjichristidis N., Roovers J., Rey A., Freire J. J. (1996) Macromolecules 29, 5599.

Wintermantel M., Gerle M., Fisher K., Schmidt M., Wataoka I., Urakawa H., Kajiwara K., Tsukahara Y. (1996) Macromolecules 29, 978.

Wintermantel M., Schmidt M., Tsukahara Y., Kajiwara K., Kohjiya S. (1994) Macromol. Rapid Commun. 15, 279.

CHAPTER 11

DILUTE SOLUTIONS OF BLOCK COPOLYMERS IN SELECTIVE SOLVENTS

Block copolymers, like low molecular weight surfactants, form micelles in solvents selective for one of the blocks. The insoluble block forms the core of such structures, whereas the soluble block forms the corona (Tuzar 1996a). The phenomenon of micellization has been studied for many decades (Hamley 1998), both from theoretical as well as from the experimental points of view, due to the potential applications of micellar systems in areas like viscosity modification of fluids, cosmetics, pharmacy, lubrication of surfaces, emulcification, solubilization of insoluble substances, drug delivery and environmental purification methodologies (Tuzar 1996a, Haulbrook 1993, Harada 1995). Several reviews and chapters in books have focused on this important phenomenon (Tuzar 1993, Tuzar 1976, Hamley 1998, Chu 1995, Cast 1996, Booth 2000). Some general considerations of this field connected to block copolymer systems will be given in this chapter.

1. THERMODYNAMICS OF MICELLIZATION

1.1. Association Equilibria

According to the classification of Elias (Elias 1972 and 1973), two extreme models of self-association can be distinguished. In the case of the open association model, several equilibria exist described by the following equations:

$$\begin{aligned} M_1 + M_1 &\longrightarrow M_2 \\ M_2 + M_1 &\longrightarrow M_3 \\ M_3 + M_1 &\longrightarrow M_4 \\ &\cdots\cdots\cdots \\ M_p + M_1 &\longrightarrow M_{p+1} \end{aligned} \tag{11.1}$$

Depending on the relative values of the individual equilibrium constants, supramolecular species (M_2, $M_3, \ldots M_p$) with different number, p, of unimers (individual molecules or polymeric chains, M_1) can be present simultaneously in the solution. Association of soap surfactants approach this kind of behavior.

In the other extreme case, the model of closed association predicts only one equilibrium between unimers (isolated molecules) and aggregates (micelles) of a well-defined aggregation number p(molecules per aggregate), i.e.:

$$pM_1 \longrightarrow (M_1)_p \tag{11.2}$$

In this case micelles have a very low polydispersity in molecular weight and size, and the position of equilibrium in a certain selective solvent depends on concentration and temperature. Experiments so far have shown that the association of block copolymers in a selective solvent can be described, in the majority of the cases, by the closed association mechanism (Tuzar 1993).

The dependence of the inverse apparent average molecular weight of the species in solution on solute concentration in each case can be described by Figure 11.1. In the closed association model, three regimes exist (Fig. 11.1, a). In region I only unimers exist. In region II there is coexistence of unimers and micelles. The

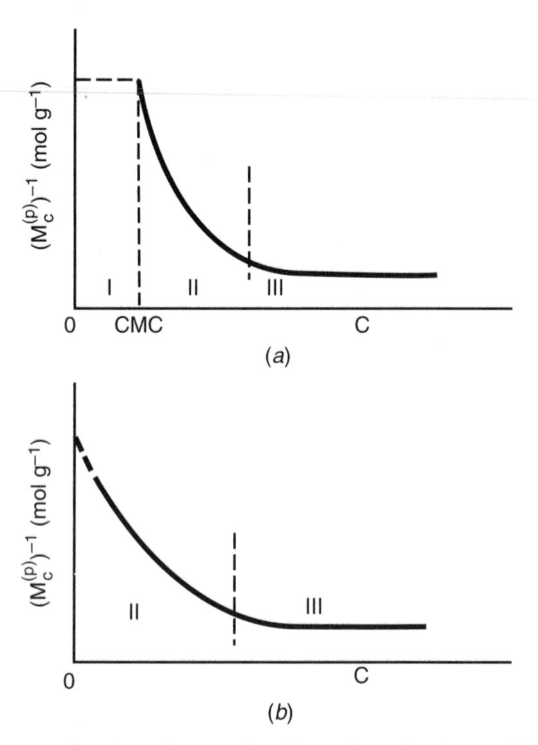

Figure 11.1. Concentration-dependence of reciprocal molecular weight in associating systems obeying the models of closed (a) and open (b) association (Elias 1972).

concentration at which micelles first appear is defined as the critical micelle concentration (cmc). Micellar concentration increases as concentration increases until region III is reached where all unimers have been incorporated in the micelles, except for a small amount equal to the amount of unimers at cmc. In this region the properties of the solution are dictated almost entirely from the properties of the micelles. In the case of open association (Fig. 11.1, b), no cmc exists, and there is a continuous change of properties with concentration.

For the closed association model and for a relatively large aggregation number, p, it has been shown that, in an ideal solution, the equilibrium association constant is given by:

$$K_m = [M_p]/[M_1]^p = \exp(-p\Delta G^0 RT) \tag{11.3}$$

where ΔG^0 is the free energy change per chain because of the formation of micelles from association of unimers. The standard free energy of micellization at any given concentration, c, is given by:

$$\begin{aligned}\Delta G^0 = {} & (p-1)p^{-1}RT\ln c - RTp^{-1}\ln(M_w M_1^{-1} - 1) + RT\ln(p - M_w M_1^{-1}) \\ & + p^{-1}RT\ln p - (p-1)p^{-1}RT\ln(p-1)\end{aligned} \tag{11.4}$$

where M_w is the average molecular weight of the total species in solution and M_1 the molecular weight of the unimer. In the limit of large association number and at c very close to cmc, eq. 11.4 can be written as:

$$\Delta G^0 = RT\ln \mathrm{cmc} \tag{11.5}$$

If p is independent of temperature, an assumption that is not always true for real micellar systems,

$$\Delta H^0 \sim RT^2(\mathrm{d}\ln \mathrm{cmc}/\mathrm{d}T) \tag{11.6}$$

where ΔH^0 is the standard enthalpy of micellization. Additionally, if ΔH^0 is independent of temperature, integration of 11.6 gives:

$$\ln \mathrm{cmc} = \Delta H^0/RT + \mathrm{const} \tag{11.7}$$

The above equation provides a means of determining ΔG^0 and ΔH^0 for micellar systems by determining the dependence of cmc on temperature. Alternatively, the critical micellization temperature (cmt) can be defined as the temperature at constant concentration at which micelles first appear. Experimental results on several block copolymer/organic solvent systems provide evidence that the enthalpic contribution to the Gibbs energy for micelle formation dominates over the entropic contribution, and, therefore, micellization in this case is an enthalpy-driven process (Price 1985, Prochazka 1988, Tuzar 1976 and 1993). In contrast, micellization of block copolymers in water has been shown to be an entropy-driven process (Tuzar 1996a,b).

A more detailed theoretical treatment has shown that the main contribution to the standard free energy of micellization comes from the formation of the micellar core from the insoluble blocks.

2. PHENOMENOLOGY OF BLOCK COPOLYMER MICELLAR STRUCTURE

Two of the main questions that predominate in the study of micelle formation by block copolymers are the question of the structure of an isolated micelle and the question of the configuration of a block copolymer chain incorporated in a micelle. It has to be kept in mind that most of the techniques used in the study of micellar solutions probe the solution as a whole and not just the micelles. Complications arising from the presence of unimers (unassociated copolymer chains), possible long-range ordering of the micelles, intermicellar interactions, etc. should be considered.

The generally accepted idea is that most block copolymer micelles are spherical with a relatively compact core consisting of the insoluble blocks and a corona consisting of the soluble ones (Fig. 11.2). The chains in the core can be considered to be in a state resembling the homopolymer melt if the solvent is very poor for them (i.e., highly selective). Some swelling of the core by solvent molecules cannot be ruled out, especially in cases of low selectivity (Tuzar 1993).

The corona is formed by the soluble blocks, which are in a well-solvated state but considerably stretched because of geometrical constrains, especially in the region near the core.

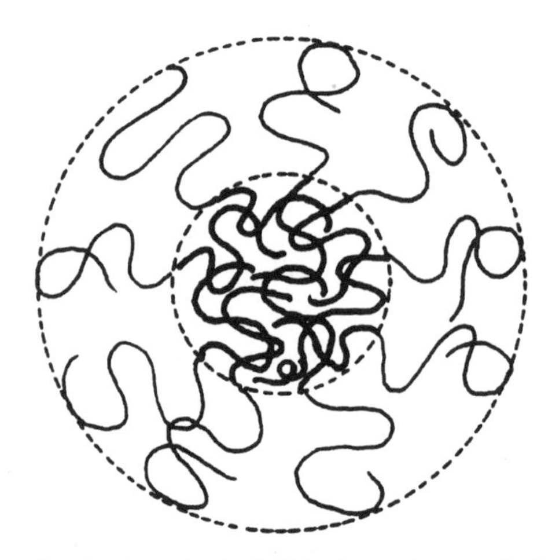

Figure 11.2. Multimolecular spherical diblock copolymer micelle (Tuzar 1996a).

Figure 11.3. Unimolecular diblock copolymer micelle (Tuzar 1996a).

Cylindrical micelles have also been observed in several cases (Nakano 1999, Mortensen 1996, Schillen 1994, Zhao 1995, Ding 1997). This change in the overall structure may be due to the high content of the block copolymer in the insoluble part, to the block copolymer architecture, and also to the selectivity of the solvent. In some studies clusters of micelles have been identified (Moffit 1996, Zhang 1995).

The structure of unimolecular micelles, i.e., the structure of an isolated block copolymer chain in a selective solvent, has also attracted the attention of the investigators, mainly because this situation has many similarities with protein structure in aqueous solutions in terms of solvophobic-solvophilic interactions (Tsunashima 1990, Goldmints 1997). In this case the insoluble block has a collapsed conformation, and it is protected from the hostile environment of the selective solvent by the soluble block. In this configuration insoluble block-solvent interactions are minimized (Fig. 11.3). The transition region close to cmc is also of considerable interest due to the changes in conformation of individual block copolymer chains (Tsunashima 1990, Tuzar 1993).

3. EXPERIMENTAL TECHNIQUES FOR STUDYING MICELLE FORMATION

A number of experimental techniques have been used so far for the study of micellar systems in an effort to shed some light on different aspects of the micellization process as well as on the structures of the micelles as a whole and those of their cores and coronas individually (Tuzar 1993). It has to be emphasized that different methods give different kinds of information for the system under investigation and on the various parameters responsible for the observed behavior. Additionally, each one of them is influenced in a different way by changes in micelle-unimer equilibria and has different sensitivity in different concentration ranges. Furthermore, because sample preparation for employment of a certain technique may involve different steps, the influence of sample preparation protocol must always be taken into account (Munk 1996a). Sometimes even the process of measurement itself may induce changes in the thermodynamics or the structural parameters of the system complicating the results (Antonietti 1994). In the following text, the techniques available for the study of micelles from block copolymers will be briefly described, and the information gained from each technique regarding micellar properties will be emphasized.

3.1. Static Light Scattering

Static light scattering (SLS) has been widely employed in the study of block copolymer micelles (Munk 1996b, Price 1974, Tuzar 1990, Raspud 1994, Prochazka 1988, Khougaz 1994). SLS provides the average molecular weight of the micelles if measurements can be performed far away from the cmc, where the micellization equilibrium is shifted towards micelles and the amount of unimers in solution is very small, and, therefore, their contribution to the light-scattering intensity is negligible (Fig. 11.4). Under these conditions the second virial coefficient associated with the micelles can also be estimated from the concentra-

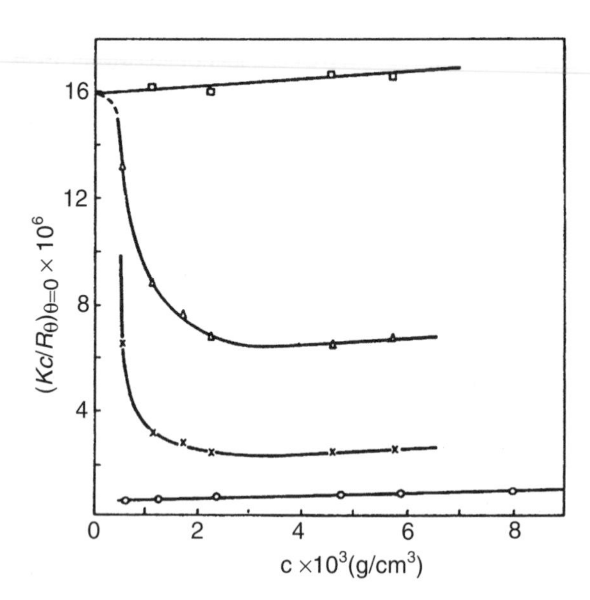

Figure 11.4. Light-scattering plots for a poly(styrene-b-isoprene) copolymer in n-decane, a selective solvent for the polyisoprene block, at different temperatures [25°C (○), 45°C (×), 55°C (Δ), and 65°C (□)]. In the concentration range studied, only unimers are present in solution at 65°C as evidenced by the calculated M_w by extrapolation to $c = 0$. Micelles are the dominant species at 25°C. At intermediate temperatures, a coexistence of unimers and micelles is evidenced by the curved nature of the plot at low concentrations (Price 1974).

tion-dependence of the reduced light-scattering intensity, in a manner similar to the case of molecularly dissolved chains in a good solvent. At the same concentration range, the radius of gyration, R_g, of the micelles can also be determined in principle if scattering from both micellar core and corona are not enormously different. It should be mentioned that, as in the case of the determination of M_w, A_2, and R_g for block copolymers in a good solvent by SLS, these quantities should be considered as apparent ones because they are greatly influenced by the copolymer inhomogeneity and also the differences in scattering contrast between solvent and the individual components of the macromolecule. Combining different copolymer molecules into micelles tends to reduce these differences, but the demand of equal scattering contrast from both core and corona is very difficult to achieve in practice.

Due to its greater sensitivity in large particles, SLS has been used for the determination of cmc, cmt, and thermodynamic parameters of several block copolymer/selective solvents systems. Apart from the difficulties associated with the determination of a true value for R_g, the measurement of the dissymetry ratio, Z, defined as the ratio of the scattered intensity from the solution at 45° to that at 135° ($Z = I_{45°}/I_{135°}$), as a function of concentration and temperature, has been used by experimentalists to deduce qualitative information about size and structural changes of block copolymer micelles (Mandema 1979).

3.2. Quasielastic Light Scattering

The method of quasielastic or dynamic light scattering (DLS), especially in the form of photon correlation spectroscopy, is now routinely employed in the study of block copolymer micellar systems (Balsara 1991, Tuzar 1983, Zhou 1996a,b, Desjardins 1992).

Having a sensitivity and flexibilty comparable to that of SLS, the technique can give an estimate of the hydrodynamic radius, R_h, of the micelles in solution through measurements of their diffusion coefficient. R_h is usually considered as giving the actual physical size of the micelles (due to their spherical nature), and its dependence on concentration and temperature can give valuable information on changes in micellization equilibria as well as changes of the size and shape of the associated species. Measurement of the depolarized light-scattering spectrum can produce results on the rotational diffusion coefficient of nonspherical micelles.

Information on frictional and thermodynamic interactions in micellar systems can also be obtained through the concentration-dependence of the diffusion coefficient (Tuzar 1996). Thanks to developments in the analysis of dynamic light-scattering data, information on the size distribution of the micelles and the relative amounts of individual species in aggregating systems can be obtained (Booth 1997, Yang 1996, Zhou 1996a,b) (Fig. 11.5).

3.3. Small Angle X-ray Scattering

Small angle X-ray scattering has also been used in the study of micellar systems (Plestil 1990, Bluhm 1984, Plestil 1975). Parameters like the molecular weight,

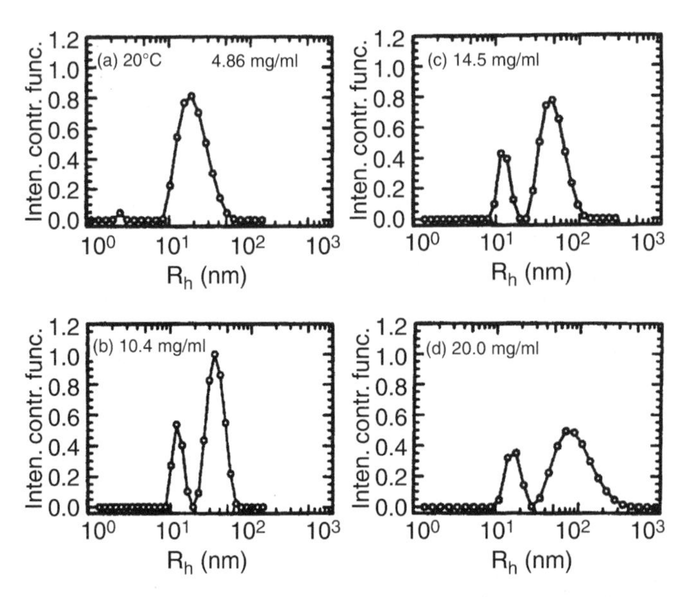

Figure 11.5. Distribution of apparent hydrodynamic radii measured by dynamic light scattering for a poly(butylene oxide-b-ethylene oxide-b-butylene oxide) triblock copolymer in water at 20°C and different concentrations. The peaks at high R_h values correspond to micelles, whereas those at low R_h values to unimers. Coexistence of unimers and micelles is evident at higher concentrations (Zhou 1996b).

overall size, and internal structure of individual micelles can be determined as a function of concentration and temperature. SAXS can be employed in micellar solutions of low and high concentrations (Fig. 11.6). Because the principle of the technique is based on the electron density differences between solvent and the different parts of the soluble species, individual dimensions of the micellar core and those of the corona can be determined, usually through the assumption of a model. Through model fitting, additional information on core/corona interface sharpness and corona/solvent fuzziness can be obtained. The method has been successfully employed in the study of intermicellar ordering in semidilute and concentrated solutions as well as in the study of intermicellar interactions.

3.4. Small Angle Neutron Scattering

Small angle neutron scattering has evolved into a powerful technique for investigating micelle internal structure (Higgins 1994 and 1986, Goldmints 1999, Liu 1998, Moffit 1998) (Fig. 11.7). In most cases block copolymer samples having one of the blocks selectively labeled with deuterium are examined. Using deuterated or hydrogenous solvents and solvent mixtures, the scattering cross-section of the core or the corona can be made invisible, making possible the determination of the dimensions of the visual part of the micelles (Goldmints 1997, Nakano 1999). This situation is almost impossible to achieve in SLS experiments because refractive

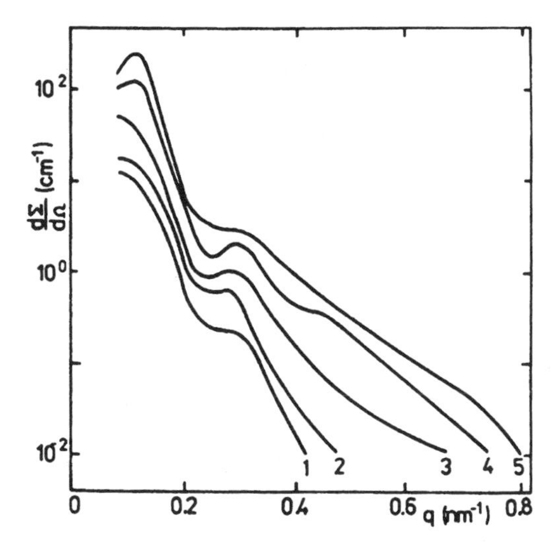

Figure 11.6. SAXS scattering curves for a poly(styrene-b-hydrogenated butadiene-b-styrene) copolymer in a 75% 1,4-dioxane/25% heptane mixture at various concentrations (concentration increases from 1 to 5) (Plestil 1990).

index matching in this case with different solvents would alter the micellar properties of the system. However, deuterium labeling can modify, to some extent, the thermodynamic parameters of the system. Unfortunately, deuterium-labeled samples with the desired chemical constitution are not always available. On the other hand, if carefully synthesized block copolymers can be examined, a variety of information, including overall micelle size, individual core and corona dimensions,

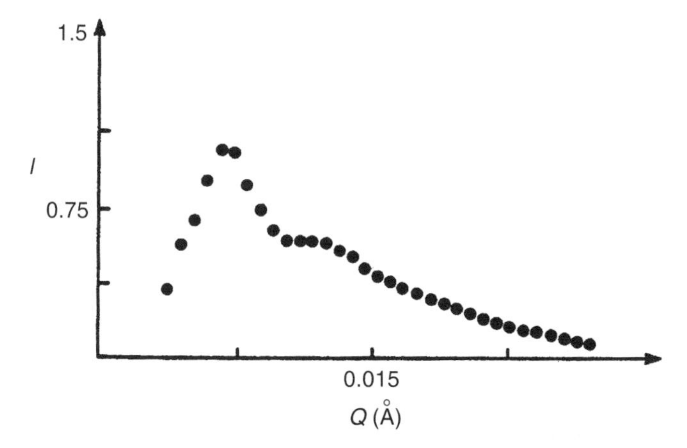

Figure 11.7. SANS scattering curve for a poly(d_8-styrene-b-hydrogenated isoprene) copolymer in dodecane. The pattern, obtained under core contrast, can be used for the determination of the core dimensions (Higgins 1986).

core/corona and corona/solvent interface sharpness, and core swelling by the selective solvent, can be obtained by model fitting of the experimental scattering curves. In this respect the method is analogous to SAXS but less universally applicable, mostly due to technical difficulties regarding availability of instrumentation and of isotopically labeled samples.

3.5. Dilute Solution Capillary Viscometry

Dilute solution viscometry has been routinely used for many years in the study of micellar systems (Munk 1996b, Antonietti 1994, Price 1974). Solutions of spherical micelles are usually characterized by very low intrinsic viscosity values as a result of their compact structures. Some semiquantitative information about micelle compactness can be deduced from the values of the Huggins constants, k_H, derived from the concentration-dependence of the specific viscosity of micellar solutions. Usually k_H values are higher than the ones obtained for polymers in good and theta solvents, due to increased hydrodynamic interactions between chains incorporated in the micelles. In conjuction with molecular weight measurements by SLS, values for the hydrodynamic (viscometric) radius, R_v, of the micelles can be obtained and compared with R_h, determined by DLS and other radii determined by SLS, SAXS, and SANS. Comparisons of this kind can be useful in extracting information on the anisotropy of micellar shape and on micellar stability in a shear flow, a point that is sometimes ignored in investigations of micellar systems (Antonietti 1994). Because the method is sensitive to the size of the micelles through the modification of solvent's viscosity, size and shape changes with concentration and temperature can be followed, and viscosity modification of fluids can be evaluated.

3.6. Membrane Osmometry

Membrane osmometry has been sparcely used for the study of micelles from block copolymers, mainly due to its low upper molecular weight limit (Tuzar 1993). However, when applicable, it is possible to determine the absolute molecular weight distribution of micelles in solution, in conjuction with SLS, and cmc values. Molecular weight polydispersity of micelles is a central parameter in all theories developed for micellar systems. Other limitations of the method, like the narrow range of solvents compatible with a certain membrane and high equilibration times at higher molecular weights, together with its lower sensitivity in the same molecular weight range, make the use of this method in micellar studies difficult.

3.7. Transmission Electron Microscopy (TEM)

TEM has been utilized for the characterization of the size, shape, size distribution, and internal structure of block copolymer micelles (Price 1979, Yu 1998, Zheng 1999, Gao 1994, Esselink 1998). The technique relies on the production of phase contrast between the micellar core and corona by selectively staining one of the micelle's parts. Two methodologies are used for sample preparation. In the first

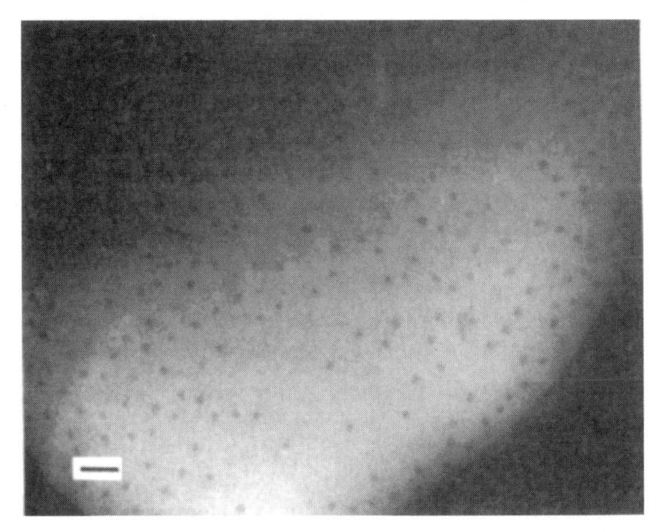

Figure 11.8. Cryo-TEM picture of a solution of a PS-P2VP diblock copolymer in toluene. The core of the micelles, comprised of P2VP blocks, has been stained with iodine. The scale bar corresponds to 200 nm (Esselink 1998).

method, a drop of a dilute solution of micelles is spread on a carbon film, and the solvent is allowed to evaporate. Then the dry isolated micelles can be stained and observed under the microscope. In a second, more recent, method called cryo-TEM, the solution is rapidly frozen, by liquid nitrogen, stained, and observed under appropriate low-temperature conditions. In both cases the final specimen for examination is a representative collection of micelles similar to their state in solution. However, in the first case, the dry micelles are in a collapsed state, i.e., an almost two-dimensional replica of the three-dimensional micelles. This sample preparation protocol has some obvious limitations in the correct determination of the micelllar size in solution, but valuable semiquantitative information about the size distribution, shape, and internal structure of the micelles can be obtained and compared with data on micellar solutions.

Ultracryomicroscopy allows freezing of the solution maintaining the micelles in their unperturbed solution state, and the information obtained reflects better their true size, structure, and shape in solution (Fig. 11.8). However, the technique requires elaborate instrumentation.

3.8. Ultracentrifugation

Sedimentation velocity analysis of micellar solutions can give valuable information on the size and size distribution of block copolymer micelles (Munk 1996b, Selb 1981). In this method the velocity at which each solute species is displaced, due to the influence of a strong centrifugal force, is measured. The sedimentation velocity depends on the molecular weight, size, and density of the species itself and the friction forces developed on the solute by the solvent medium.

The method has been employed for the study of micelle-unimer equilibria in a variety of block copolymer/selective single or mixed organic solvent systems (Tian 1993). Micellization kinetics and formation of mixed block copolymer micelles have also been studied by this method.

3.9. Size Exclusion Chromatography

Due to its ability for detecting hydrodynamic size distributions, this separation technique has been used in several cases for the study of micellar systems (Booth 1978, Price 1982, Prochazka 1989). Experiments were focused on the determination of the relative amounts of micelles and unimers as well as the size distribution of the micelled formed. Because, in the course of the analysis, large species (micelles) are continuously separated from smaller ones (unimers), micellization equilibria can be disturbed, and erroneous results about micelle-unimer equilibrium can be drawn. The problem is diminished if one works at concentrations far away from the cmc. However, in the case of systems like block copolymer micelles with ionic cores in organic media, where micelle-unimer equilibrium is essentially frozen (due to the very low mobility of the core blocks and their extreme solvent incompatibility exchange of block copolymer chains between micelles and free unimers does not take place), the method is more successful and gives more realistic information (Desjardins 1991).

3.10. Spectroscopic Methods

A number of spectroscopic methods, especially NMR (Speracek 1982, Kriz 2000) and fluoresence spectroscopy (Watanabe 1988, Major 1990, Wang 1995, Calderara 1994, Prochazka 1992), have been employed in the study of micellar solutions. These methods are suitable for investigation of local phenomena within micelles. Application of NMR spectroscopy is based on the fact that block copolymer segment mobility is related to the intensity of respective NMR spectrum peaks. Mobility of insoluble segments is very much reduced when the core of the micelle is formed, resulting in reduced intensity of the corresponding NMR peaks. In fluorescence spectroscopy, probe molecules are used that are attached with a covalent bond at specific sites on the copolymer chain or can be easily distributed within the core of a micelle. Using energy transfer, fluorescence quenching, and time-dependent fluorescence depolarization techniques, the microenvironment of the micelles can be explored. Due to their sensitivity on the local scale, in contrast with global scale sensitivity of scattering methods, the use of spectroscopic techniques involves determination of very small cmc values, inaccessible by scattering techniques, probing internal viscosity of the micellar core and observation of solubilization of low molecular weight substances into micelles. Micelle-unimer equilibria or chain exchange between micelles using specially labeled block copolymers with fluorophore groups can also be studied. Additionally, information on the modification of chain conformations or core/corona interface due to the formation of the micelles can also be investigated. Self-diffusion of micelles can be investigated by pulsed field gradient NMR (Fleischer 1999).

4. EQUILIBRIUM STRUCTURE OF BLOCK COPOLYMER MICELLES

4.1. Theory and Experiment

A large number of theories have been developed for the description of the micellization process and the dependence of fundamental properties of block copolymer micelles at equilibrium, like aggregation number, overall size, and core and corona dimensions, on the molecular characteristics of the block copolymers and solvent quality. They can be classified into two major categories: scaling and self-consistent mean field theories.

One of the first scaling theories was developed by deGennes (deGennes 1978) based on the Alexander-deGennes theory for polymer brushes (Alexander 1977, deGennes 1976). For uniform stretching of the core chains and short coronal blocks, the core radius, R_B, scales according to deGennes as:

$$R_B \sim aN_B^{2/3}(\gamma a^2/T)^{1/3} \tag{11.8}$$

where N_B is the number of segments of the insoluble block, γ is the interfacial tension between A and B segments, and a is the segment length. The scaling for the aggregation number, p, was found:

$$p \sim N_B \gamma a^2/T \tag{11.9}$$

The model for star-like polymers, derived by Daoud and Cotton (Daoud 1982), has also been applied to the description of star-like micelles, i.e., micelles with small cores and large coronas. This model predicts that the total radius of the micelle, R, depends on the number of monomeric units of the soluble block and the aggregation number according to:

$$R \sim N_A^{3/5} p^{1/5} \tag{11.10}$$

In their scaling model, Zhulina and Birnstein (Zhulina 1985) identified four different types of polymeric micelle structures depending on the composition of the diblock copolymer (Fig. 11.9). Their predictions are summarized in Table 11.1. They also found that the core chains are stretched compared with the unperturbed state.

Halperin's model (Halperin 1987) was also developed for the case of star-like micelles. His predictions for the scaling of the total micellar radius can be described by eq. 11.11:

$$R/a \sim N_B^{4/5} N_A^{3/5} \tag{11.11}$$

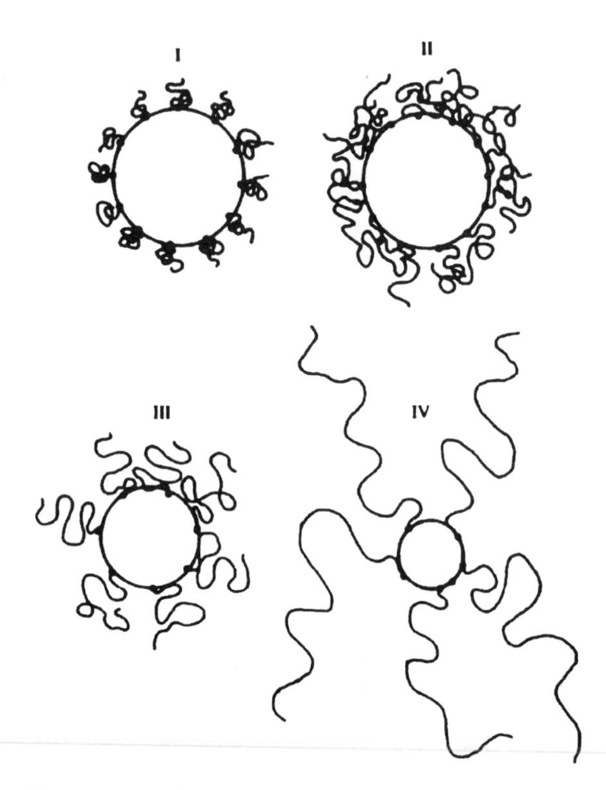

Figure 11.9. Different types of micelles of diblock copolymers according to Zhulina and Berstein (Zhulina 1985).

TABLE 11.1. Predictions of the Zhulina-Birstein Theory on the Scaling of the Micellar Characteristics With the Degrees of Polymerization of the Blocks

Region	Composition of the Diblock Copolymer	R_B	R_A	p
I	$N_A < N_B^{\nu/6}$	$N_B^{2/3}$	N_A^{ν}	N_B
II	$N_B < N_A < N_B^{(1+2\nu)/6\nu}$		$N_A N_B^{(\nu-1)/6\nu}$	
III	$N_B^{(1+2\nu)/6\nu} < N_A < N_B^{(1+2\nu)/5\nu}$	$N_B N_A^{-2\nu/(1+2\nu)}$	$N_A^{3\nu/(3\nu+1)}$	$N_B^2 N_A^{-6\nu/(1+2\nu)}$
IV	$N_A > N_B^{(1+2\nu)/5\nu}$	$N_B^{3/5}$	$N_A^{\nu} N_B^{2(1-\nu)/5}$	$N_B^{4/5}$

AB block copolymer micelle in a solvent selective for the A block. R_A = the thickness of the corona; R_B = the radius of the core; p = the aggregation number; ν = the Flory exponent in the relation between the radius of dyration and the molecular weight. $\nu = 0.5$ for theta solvents and $\nu = 0.588$ for good solvents.

denoting a stronger dependence of the micellar dimensions on the length of the soluble block. This theory predicts, like the rest of the theories examined so far, that, for star-like micelles, the aggregation number and core dimensions are dependent only on the length of the core-forming block.

Noolandi and Hong (Noolandi 1982) developed a self-consistent mean field theory for spherical micelles in solution, based on their earlier developed model for block copolymer/homopolymer blends. Taking into account the known molecular characteristics of the copolymer, its concentration in the solution, and making a simple approximation for the core/corona interfacial tension, they were able to calculate the size of the micelles in each case. The core radius was found to depend only on the length of the core block, i.e.:

$$R_B \sim N_B^{0.64} \tag{11.12}$$

The aggregation number was also found to depend only on the length of the insoluble block as:

$$p \sim N_B^{0.9} \tag{11.13}$$

In a later development of this theory, Whitmore and Noolandi (Whitmore 1985) found R_B and R_A to scale as:

$$R_B \sim N_B^{\beta} N_A^{\mu}, \qquad R_A \sim N_A^{\omega} \tag{11.14}$$

with $0.67 \leq \beta \leq 0.76$, $-0.1 \leq \mu \leq 0$ and $0.5 \leq \omega \leq 0.86$.

A number of self-consistent mean field theories, originally developed for polymer brushes, were modified in order to be applied in the case of spherical block copolymer micelles.

The theory developed by van Lent and Schentjens (van Lent 1989) predicts that the cmc for a diblock copolymer in a selective solvent depends mostly on the length of the core-forming block and the quality of the solvent. Spherical micelles were found to be the most thermodynamically stable in most cases, but a transition to a lamellar micellar structure was predicted in the case where the soluble block is much smaller than the insoluble one.

Leibler et al. (Leibler 1983) developed a mean field theory for spherical micelle formation by a diblock copolymer in a homopolymer matrix or in a low molecular weight solvent. They found the free energy of the micelle to be comprised of three terms, namely the free energy of the core, that of the corona, and that of the A/B interface, i.e.:

$$F_{mic} = F_{core} + F_{corona} + F_{int} \tag{11.15}$$

$$\text{with} \quad F_{core} = 3/2(pk_BT)[(R_B^2/N_B\alpha^2) + (N_B\alpha^2/R_B^2) - 2] \tag{11.16}$$

$$F_{corona} = 3/2(pk_BT)[(R_A^2/N_A\alpha^2) + (N_A\alpha^2/R_A^2) - 2] + [pk_BTN_A(1-\eta)\ln(1-\eta)]/\eta N_H] \tag{11.17}$$

$$F_{int} = 4\pi R_B^2[(k_BT/\alpha^2)(\chi_{AB}/6)^{1/2}] \tag{11.18}$$

In this way the size of a micelle, its aggregation number, and the fraction of copolymer chains in the micelles can be calculated (leading to the determination of the cmc) as a function of the molecular characteristics of the copolymer and A-B interactions.

The results of Leibler et al. were extended by Mayes and Olvera de la Cruz to the case of diblock copolymers/homopolymers mixtures in dilute solutions (Mayes 1988). They predicted a preference for the formation of cylindrical micelles by increasing the core block fraction and increasing homopolymer molecular weight.

Nagarajian and Ganesh (Nagarajian 1989a) developed a detailed mean field theory for micelle formation by diblock copolymers in dilute solutions of selective solvents. The solution was treated as a multicomponent system by taking into account any possible chemical species present (solvent molecules, unimer, and micelles of different aggregation numbers). They obtained an expression for the size distribution of equilibrium micelles, and they concluded that the length and chemical nature of the soluble block can strongly affect micellar characteristics, in contrast with the predictions of other theories. The dependence of micellar parameters on the characteristics of the soluble block was found to be stronger in the case where the solvent is a very good one for the soluble block. By compiling numerical results from true systems, they derived scaling expressions for the core radius, the aggregation number, and the ratio of the corona to core dimensions:

$$\mathrm{R_B} = [3\mathrm{N_B^2}(\gamma_{\mathrm{BS}}\alpha^2/\mathrm{k_B T}) + \mathrm{N_B^{3/2}} + \mathrm{N_B N_A^{1/2}}(\mathrm{R_B/R_A})^{1/3}]\alpha/[1 + \mathrm{N_{B}^{1/3}} + (\mathrm{N_B/N_A})(\mathrm{R_A/R_B})^2]^{1/3} \tag{11.19}$$

$$\mathrm{p} = [4\pi \mathrm{N_B}(\gamma_{\mathrm{BS}}\alpha^2/\mathrm{k_B T}) + (4\pi/3)\mathrm{N_B^{1/2}} + (4\pi/3)\mathrm{N_A^{1/2}}(\mathrm{R_B/R_A})]/[1 + \mathrm{N_B^{1/3}} + (\mathrm{N_B/N_A})(\mathrm{R_A/R_B})^2]^{1/3} \tag{11.20}$$

$$\mathrm{R_A/R_B} = 0.867[1/2 + (\mathrm{N_A N_B^2})/(\mathrm{N_A} + \mathrm{N_B})^3 - \chi_{\mathrm{AS}}]^{1/5}\mathrm{N_A^{6/7}}\mathrm{N_B^{-8/11}} \tag{11.21}$$

where γ_{BS} is the interfacial tension between solvent and the core-forming block.

Linse (Linse 1993) found cmc to decrease on going from a triblock to a diblock. The molecular mass polydispersity was shown to reduce the cmc and increase the overall size of the micelles. Chemical polydispersity was found to have similar effects (Linse 1994).

The results of Gao and Eisenberg (Gao 1993) on the effect of polydispersity of the block copolymer on micellar characteristics are in agreement with those of Linse. Additionally, they found that the dependence of cmc on each component of the system becomes less well-defined as polydispersity increases.

Recently, Pepin and Whitmore (Pepin 2000) studied the case of crew-cut micelles by mean field and Monte Carlo calculations. Mean field calculations gave scaling relations for equilibrium crew-cut micelles, and Monte Carlo calculations focused on the micellar size and its distribution in conjunction with micellar relaxation times in an effort to determine the variables controlling the freezing of structures in such systems. A stronger dependence of micellar characteristics on AB copolymer characteristics was obtained with mean field calculations as compared with the Monte Carlo ones. At fixed solvent quality, an exponent of 0.77 was found for the core radii-aggregation number relation. The system relaxation times were found to increase with molecular weight of the insoluble block and interaction energy. The weaker MC scaling can be attributed to this molecular weight dependence of relaxation times. The density profile of the corona was found to be nonuniform. Distributional analysis of the chain segments indicated that B ends were present in the corona region and A ends into the core region. The junction points were found to be located in a broad interfacial area. This, together with a large amount of unimers and of solvent in the core, was taken as evidence of weak segregation in the system under study. The calculated polydispersities of the micellar size were narrow, with a long tail attributed to clusters of micelles due to the short A block.

A number of studies, aiming to the investigation of the applicability of theoretical predictions on micellar systems, have been undertaken.

Tao et al. (Tao 1997) studied polystyrene-polycinnamoethyl methacrylate block copolymer micelles. Due to the composition of their samples, star-like micelles were formed in most cases. They found that the aggregation number scaled as $p \sim N_B^{0.92}$, which is in good agreement with the predictions of Zhulina and Birstein for this type of micelle. The core radius was found to scale as $R_B \sim N_B^{0.63}$ as the theories of deGennes, Zhulina-Birstein, Halperin, Noolandi, and Nagarajian predict. The scaling of the corona thickness was found to be in agreement with the predictions of Daoud and Cotton.

Xu et al. (Xu 1992) studied a series of PS-PEO diblock and PEO-PS-PEO triblock copolymer micelles in water, having a PS core. They found good agreement between their experimental results and the predictions of Halperin on the dependence of micellar radius on the lengths of the soluble and insoluble blocks.

Special mention has to be made in the experimental work of Forster et al. (Förster 1995) on the PS-P4VP system in toluene, a selective solvent for PS. They found a scaling relation for the aggregation number $p \sim N_{P4VP}^{1.93}\ N_{PS}^{-0.8}$ characteristic for strongly segregating block copolymers. The exponent for N_B (the solvent insoluble block) was very close to the value of 2 predicted by theory. The coronal dimension was found to scale as $D \sim p^{0.21} N_{PS}^{0.63}$ in agreement with the theoretical predictions of $D \sim p^{1/5} N_B^{3/5}$ for star-like micelles. The major contribution of this work, however, is the development of a simple micellization model that unifies and predicts the aggregation behavior of low molecular weight surfactant (nonionic, anionic, and cationic), diblock, triblock, and graft copolymers in selective solvents.

5. EFFECT OF ARCHITECTURE

So far the focus of the theoretical developments has concerned the micellization of linear diblock copolymers in selective solvents. Another parameter that can strongly influence the micellization of block copolymers in solution is the architecture of the block copolymer. Unfortunately, not much theoretical work has been done in this area.

The case of triblock BAB or ABA copolymers in selective solvents has been considered by some investigators. In the case of a triblock copolymer BAB in a solvent selective for the middle block, there is a possibility for loop formation if both end blocks are incorporated into the same micelle (Fig. 11.10). An extra energy term must be added to the total free energy of the micelle.

ten Brinke and Hadziioannou (ten Brinke 1987) calculated the free energy associated with the formation of loops as:

$$F_{loop} = 3/2\beta kT \ln N_A \tag{11.22}$$

where β is a correction factor, and concluded that this term is so large that it may lead to absence of micelles in these systems.

Balsara et al. (Balsara 1991) estimate this F_{loop} as:

$$F_{loop} = -pkT \ln q \tag{11.23}$$

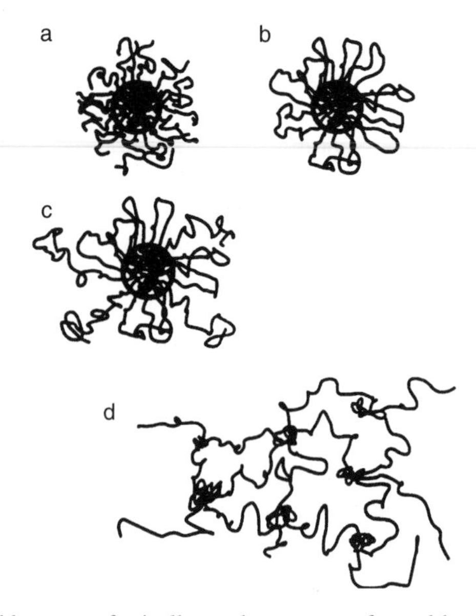

Figure 11.10. Possible types of micelles and aggregates formed by AB diblock and ABA triblock copolymers in selective solvents: a) the solvent is selective for the A blocks, and normal micelles are formed by AB and ABA block copolymers. Cores are formed of the B blocks and coronas from the A blocks; b) the solvent is selective for the B block. Flower-like micelles are formed if all A blocks of the ABA triblock are incorporated to the core of the same micelle; c) the solvent is selective for the B block, but some of the A blocks of the triblock can be found free in the solution; d) the solvent is selective for the B block of the ABA, and the A blocks of the same ABA chain are incorporated in the cores of different micelles. A network is formed in this case (Balsara 1991).

where q is the fraction of chains that form loops, and their ends are connected to the core/corona interface. This estimation is considerably lower than that of Hadziioannou et al., allowing the formation of micelles from BAB triblocks in solvents selective for the A block.

The existence of micelles in systems like the ones mentioned above has been observed experimentally, indicating that the entropy penalty for loop formation cannot inhibit micelle formation (Tuzar 1993). However, in these systems gels are formed at higher concentrations (Tuzar 1990), leading to the conclusion that some fraction of the chains must have a nonlooped conformation (Booth 2000). At intermediate concentrations entangled micelles have been observed.

Kratochvil et al. (Prochazka 1991) modified the theory of Nagarajan and Ganesh for the case of ABA triblock copolymers in solvents selective for the outer blocks. In an analogous way, they considered several factors affecting the micellization process, i.e., the change in the dilution and deformation of the A and B block due to the incorporation of the chain into a micelle, the localization of the two junction points at the core/corona interface, and the formation of this interface. The calculated aggregation numbers for the triblock were found to be lower than the ones for the diblock copolymer of the same overall molecular weight and composition. Additionally, the aggregation number for the ABA block copolymer was found to be between half and one third of the value for a diblock having half the molecular weight and the same composition of the triblock (in the same selective solvent). The theoretical predictions are in agreement with experimental results on triblock copolymers.

More complex block copolymer architectures have been studied in solution in selective solvents. Graft copolymers with many branches usually form unimolecular micelles in solvents selective for the branches, whereas, in solvents selective for the backbone, multimolecular micelles having aggregation numbers lower than those formed by diblocks of the same molecular weight and composition are present (Tuzar 1976 and 1993, Price 1974, Candau 1979a,b, Selb 1981a,b and 1985, Tuzar 1989, Kikuchi 1996, Ma 1998). This is due to the enhanced steric hindrance in the corona of the micelles due to the crowding of the branches along the backbone in the latter case and the increased protection of the backbone from the large number of branches, which lowers the free energy of a unimolecular micelle, in the former.

Other studies have been concerned with the micellization behavior of well-defined nonlinear block copolymer architectures including miktoarm star copolymers (Pispas 1998 and 2000, Tsitsillianis 1995a,b, Voulgaris 1999), H-, super-H, π, and pom-pom-shaped copolymers (Pispas 1996, Iatrou 1996, Bayer 1994). However, a concise theory that deals in detail with the effect of nonlinear block copolymer architecture has not yet been developed, although some attempts to explain theoretically the results obtained in individual cases have been made (Iatrou 1996, Pispas 2000).

An elaborate description of experimental results on PEO-containing block copolymers of different architectures is given in a recent review (Booth 2000).

6. KINETICS OF MICELLIZATION

The kinetics of micelle formation and decomposition have attracted much attention recently both theoretically and experimentally. These studies are closely related to the questions of how micelles are formed under certain conditions, what are the equilibrium micellar characteristics as a function of time after some perturbation of a micellar system, and what are the properties and parameters characterizing mixed micelles derived from different copolymers. Some aspects of micelle formation may also have tremendous impact on applications of micellar systems.

In analogy to low molecular weight surfactants, block copolymer micelle formation (or decomposition) can take place by two different mechanisms (Dormidontova 1999). The first one is the micelle-unimer exchange process, where unimers are associated to form a micelle (or unimers are expelled from a micelle during its decomposition). This process is expected to be completed at relatively short times. The second process involves micelle fusion or fission. In this mechanism micelles of the same or different aggregation numbers can be fused in order to produce a new micelle (when micelle formation is considered) as the system is driven closer to equilibrium, or a large micelle can decompose to smaller ones (when micelle decomposition is considered). Due to the molecular weight of the species involved in each process, the second one is expected to be completed at longer times. Of course, differences in behavior are expected between low molecular weight surfactant and block copolymer micelles, because, in micelles of block copolymers, long entangled chains are present in the core in contrast to the short chains of surfactants.

Halperin and Alexander (Halperin 1989) developed a scaling model for analyzing micellar kinetics for A-B neutral diblock copolymer micelles. They assumed that micellar size distribution can be affected only by single chain insertion/expulsion. They found that the faster relaxation time scales differently for star-like and crew-cut micelles for concentrations above cmc. The different scalings were attributed to the fact that passage of a single chain through the corona is the rate-determining step for unimer insertion/expulsion. For star-like micelles, the faster relaxation time was found to scale as:

$$\tau_1 \sim \exp(N_A^{2/3}\gamma a^2/kT) \tag{11.24}$$

where N_A is the number of segments in the insoluble block. Different preexponential factors were also calculated for the different type of micelles:

$$A \sim N_A^{7/3} \tag{11.25}$$

for crew-cut micelles and

$$A \sim N_B^{9/5} N_A^{22/5} \tag{11.26}$$

for star-like micelles.

Computer simulation by Mattice et al. (Haliloglu 1996) have indicated that, even for block copolymer micelles, both unimer insertion/expulsion and micelle fusion/fission are operative during micelle evolution.

Recently, a more detailed analytical theory for micelle formation/decomposition kinetics was presented by Dormidontova (Dormidontova 1999), which takes into account both mechanisms. This theory is based on Kraemer's theory for the calculation of association/dissociation rates. According to this theory, after coupling of unimers at the early stages of micelle formation, fusion of micelles becomes the dominant mechanism for system equilibration. At this point unimer exchange is effectively frozen due to the high activation energy required for unimer release. However, at the latter stages of micellization, both mechanisms are operative, with the micelle fusion being considerably slowed down as the average size of the micelles increases. Micelle fission plays an important role when the micellar system is driven away from equilibrium (i.e., by an increase in temperature). It was shown that micelle fusion is strongly concentration dependent.

Hybridization of micelles, i.e., micelle formation by mixtures of block copolymers in dilute solutions, has also been studied theoretically by Borovinski and Khokhlov (Borovinskii 1998). They examined the case of a mixture containing AB and A′B diblocks, with B being a long soluble block and A, A′ blocks of the same chemical nature but of different lengths. By calculating the free energy of the core of the mixed micelles and taking into account the association-dissociation reactions describing the micellization process, they were able to identify the conditions under which pure micelles and comicelles can be formed. It was shown that the block copolymers having the longer insoluble block are always the first to form micelles. In these micelles small diblock chains are incorporated at concentrations lower than the cmc of the shorter chains. Over a wide range of concentration and temperature, coexistence of mixed micelles and micelles made exclusively of the short chains has been predicted.

Extensive experimental results on the micellization kinetics of the system poly(α-methylstyrene-b-vinylphenethyl alchohol) (PαMS-b-PVA) in benzyl alchohol, a solvent selective for PVA, have been presented by Nose et al. (Honda 1994). Using time-resolved light-scattering experiments, they have found that micellization proceeds in two stages. In the first stage of the process, association of unimers takes place rapidly, and quasi-equilibrium micelles are formed. The increase in the number of micelles dominates the process of micellar size growth. In the second slower step, the system proceeds to equilibrium by increase in micellar size, accompanied by the fission of the quasi-equilibrium micelles, resulting in a gradual decrease in the number of micelles in solution. The time constants for the two stages could be evaluated. The characteristic time for the first stage, τ_1, was found to depend on polymer concentration, while τ_2 was almost independent of polymer concentration. Both characteristic times decreased by increasing temperature.

In their micelle dissociation experiments, Nose et al. performed temperature jumps within the micellar region (Honda 1996). They found that, when the system was brought from a temperature close to the cmt to a temperature well above the cmt, the unimers tended to form new micelles (I-type jump). In temperature jumps from a temperature well beyond the cmt to a temperature close to cmt, they found that micelles first decompose (a decrease in micellar size was observed), and then the micellar size increases to the equilibrium one (II-type jump). In the I-type jumps, the time constant associated with $M_{w, app}$ changes was equal to τ_1 of the fast process and decreased with concentration. In the II-type jumps, the time constant of micelle decomposition, τ_D, and micelle reformation, τ_R, were estimated, and τ_D was found to be much smaller than τ_R and independent of concentration, whereas τ_R was concentration-dependent and equal to τ_1.

In a related study, Tuzar et al. (Bednar 1988) compared the kinetics of micelle formation in solutions of a PS-PBd diblock and a PS-PBd-PS triblock in heptane/dioxane mixtures. A stopped flow technique was used where a micellar solution was mixed with a good solvent (heptane) for the core-forming block (PBd), and micelle dissociation was monitored by measuring the light-scattering intensity as a function of time. In another type of experiment, a unimer solution was mixed with the selective solvent (dioxane), and micelle formation was monitored. The authors concluded that the relaxation time for micelle formation was shorter for the diblock case. The micellar dissociation relaxation time was also smaller for the diblock compared with the triblock. Additionally, they found that micelle dissociation was much faster than micelle formation in the diblock case, but it was two times slower than micelle formation in the case of the triblock. The observed behavior was explained on the basis that, in triblock copolymer micelles, two junction points are located in the core/corona interface, and disentanglement of the middle block from the core is more difficult compared with the diblock.

7. SOLUBILIZATION OF LOW MOLECULAR WEIGHT SUBSTANCES IN BLOCK COPOLYMER MICELLES

Considerable attention has recently been given in the solubilization of low molecular weight substances in the cores of block copolymer micelles because this aspect of micellar properties is related to many technologically important applications (Chapter 21) like controlled drug delivery (Harada 1995) and environmental purification methodologies (Haulbrook 1993).

Nagarajian and Ganesh (Nagarajan 1989b) developed a theory dealing with the solubilization of low molecular weight compounds (LMWC) in the cores of spherical block copolymer micelles. They calculated the cmc, aggregation number, and size of the micelles containing the LMWC as well as the extent of micellization, the core radius, and the corona thickness. They concluded that micelles remain monodisperse in size, and the LMWC distributes equally to all micellar cores. Parameters like thermodynamic interactions between LMWC and the core-forming block, as well as LMWC/solvent interfacial tension, were found to play a major role in the solubilization process and the geometrical characteristics of the loaded micelles.

A large number of experimental studies dealing with solublization of LMWC in block copolymer micelles have appeared in the literature (Nagarajan 1996, Teng

1998, Arca 1995, Cao 1991, Stepanek 1998). Emphasis has been given to solubilization of hydrophobic LMWC in aqueous micellar solutions. A variety of methods were employed with scattering and spectroscopic methods, giving the most valuable information about the systems under investigation.

8. IONIC BLOCK COPOLYMER MICELLES

Micellization of ionic block copolymers have attracted the interest of the scientific community in recent years (Selb 1985, Moffit 1996). Due to the presence of a hydrophilic ionic and a hydrophobic nonionic block, these copolymers can form micelles both in organic and aqueous media. In particular, the formation of micelles in water opens the way to many possible applications of these materials. The micellar properties of nonionic block copolymers containing blocks of PEO, PPO, and other polymers of this family, in water and organic solvents, have been reviewed recently (Booth 2000).

In addition to the theories of micellization discussed so far, theoretical investigations dealing with ionic block copolymer micelles have also appeared due to the complex behavior of polyelectrolyte chains in solution.

Marko and Rabin (Marko 1992) proposed a model to describe monodisperse micelles of AB copolymers having a fully charged A block, forming the corona, in dilute salt-free solutions. They assumed that the micelles are neutral, and the condensed counterions can move freely in the corona. They predict that the aggregation number, p, at cmc is given by:

$$p \sim (1-f)^2/f^3 N \tag{11.27}$$

where f is the fraction of A-charged segments and N the total number of monomers within the chain. According to their calculations, only short chains with $f \ll (1-f)$ are able to form micelles.

Dan and Tirrell (Dan 1993) addressed the case of micelles with ionic corona in the limit of high-salt concentration. Under these conditions the charged chains behave similar to nonionic ones. The predicted behavior is similar to the behavior of noncharged diblock copolymer micelles in nonpolar solvents.

Guenoun et al. (Huang 1997) used a thermodynamic approach to study theoretically dilute salt-free aqueous solutions of AB copolymer chains having long charged A blocks and short uncharged B blocks. They considered the possibility of the polydispersity in the aggregation number as well as the extent of counterion condensation. They concluded that the cmc and the degree of polydispersity of ionic micelles are large. They also found that micelles with large p are nearly neutral and that, as the overall concentration of the copolymer increases above cmc, the concentration of free chains decreases in contrast to nonionic micelle systems. Compared with uncharged AB diblock, under the same thermodynamic conditions for the corona blocks, the cmc for the ionic block copolymers was found to be larger.

Khokhlov et al. studied AB ionic-nonionic block copolymer micelles in dilute solutions by using a lattice mean field theory (Shusharina 2000). They focused on the case of highly asymmetric (in composition) diblock copolymers where the ionic block is long and weakly charged and the hydrophobic block is short. The thermodynamic stability and size of the micelles were examined as a function of the length and charge of the ionic block as well as the solvent quality and salt concentration. The concentration of unimers and the aggegation were found to be essentially constant above cmc. Cmc was found to increase and aggregation number to decrease as the molecular weight and charge of the solvophilic block were increased and as the salt concentration decreased. On the other hand, under the same conditions, the size of the micelle increased, mainly due to an increase in the corona thickness. Aggregation number was also increased as the solvent became poorer for the corona block. These authors concluded that electrostatic repulsion between solvated blocks in the corona is important in controlling the aggregation number and the cmc of the system. The scaling dependencies of aggregation number, core radius, and corona thickness on the various parameters were also calculated. A scaling theory of ionic block copolymer micelles has also been presented by the same group (Shusharina 1996).

Many experimental studies have also been conducted on the micellization behavior of ionic-nonionic block copolymers.

Selb and Gallot (Selb 1985) were the first to investigate extensively the behavior of block polyelectrolytes with linear diblock and graft architectures. In their systems PS was the nonionic block and quaternized poly(4-vinyl pyridine), a cationic polyelectrolyte, was the ionic block. A strong dependence of the micellar properties on the length of the PS block, temperature, and salt concentration was observed.

Webber and coworkers have used a number of techniques, including static and dynamic light scattering, NMR, and fluoresence spectroscopy, to investigate the behavior of PS-poly(methacrylic acid) diblock and PMA-PS-PMA triblock copolymers in aqueous media as a function of ionic strength and p^H. Mixtures of dioxane and water have been used as solvents in several cases (Qin 1994, Arca 1995, Cao 1991, Prochazka 1992, Teng 1998, Tian 1993).

The group of Eisenberg has investigated in detail several ionic-nonionic block copolymer systems in selective organic solvents and in aqueous solutions (Moffit 1996, Astafieva 1993, Zhang 1995 and 1996). The systems they have studied include PS-quaternized P4VP and PS-PAA, PS-PMA diblocks and their salts with various metals. They were primarily concerned with the dependence of aggregation number, cmc, and size of the micelles on the molecular characteristics of the copolymers. Star-like and crew-cut micelles formed in neat or mixed solvents were characterized extensively with a variety of experimental techniques such as SEC, light, X-ray and neutron scattering, NMR, and TEM. A variety of different aggregate morphologies were identified, and their potential applications were discussed.

Guenoun and coworkers (Guenoun 1996a) have also studied ionic/neutral block copolymer micelles in aqueous media. They have focused their attention in block copolymers containing polystyrene sulfonate blocks. SLS, DLS, as well as electron microscopy, were used for determining the aggregation number and size of the poly(sodium styrene sulfonate)/poly(tert-butyl styrene) block copolymer micelles in water (Guenoun 2000), whereas fluorescence spectroscopy was applied in order to estimate their cmc (Guenoun 1996b). They have found that individual polyelectrolyte chains in the micelle corona adopt a rodlike conformation (Guenoun 1998).

Due to the great variability of ionic block copolymer micellar systems, further studies are needed for complete understanding their behavior. The study of such systems is still under development.

Recently, the micellization of model ionomers, i.e., nonpolar block copolymers with a small number of polar groups positioned at specific sites along the block copolymer chain, have been investigated (Pispas 1996, Mendes 1997, Schadler 2000). The presence of the polar groups have been found to have a profound effect in micellar characteristics (aggregation number, size, structure, and cmc). These novel block copolymer structures bridge the gap between nonpolar and ionic block copolymers and will help in the understanding of the basic structure-properties relationships.

REFERENCES

Alexander S. (1977) J. Phys. 38, 977.

Antonietti M., Heinz S., Schmidt M., Rosenauer C. (1994) Macromolecules 27, 3276.

Arca E., Tian M., Webber S. E., Munk P. (1995) Int. J. Polym. Anal. Charact. 2, 31.

Astafieva I., Zhong Y. F., Eisenberg A. (1993) Macromolecules 26, 7339.

Bahadur P., Sastry N. V., Marti S., Riess G. (1985) Colloids Surf. 16, 337.

Balsara N. P., Tirrell M., Lodge T. P. (1991) Macromolecules 24, 1975.

Bayer U., Stadler R. (1994) Macromol. Chem. Phys. 195, 2709.

Bednar B, Edwards K., Almgren M., Tormod S., Tuzar Z. (1988) Makromol. Chem. Rapid Comm. 9, 785.

Bluhm T. L., Whitmore M. D. (1985) Can. J. Chem. 63, 249.

Booth C., Attwood D. (2000) Macromol. Rapid Commun. 21, 501.

Booth C., Naylor T. D., Price C., Rajab N. S., Stubberfield R. B. (1978) J. Chem. Soc. Far. Trans. I 74, 2352.

Booth C., Yu G. E., Nace V. M. (1997) in Amphiphilic Block Copolymers: Self-Assembly and Applications, Alexandritis P., Lindman B. (Eds), Elsevier, Amsterdam.

Borovinskii A. L., Khokhlov A. R. (1998) Macromolecules 31, 7636.

Calderara F., Hruska Z., Hurtrez G., Lerch J. P., Nugay T., Ries G. (1994) Macromolecules 27, 1210.

Candau S., Boutillier J., Candau F. (1979a) Polymer 20, 1237.

Candau S., Guenet J-M., Boutillier J., Picot C. (1979b) Polymer 20, 1227.

Cao T., Munk P., Ramireddy C., Tuzar Z., Webber S. E. (1991) Macromolecules 24, 6300.

Chu B. (1995) Langmuir 11, 414.

Dan N., Tirrell M. (1993) Macromolecules 26, 4310.

Daoud M., Cotton J. P. (1982) J. Phys. 43, 531.

deGennes P. G. (1976) J. Phys. 37, 1443.

deGennes P. G. (1978) in Solid State Physics, Liebert L. (Ed), Academic, New York, Vol 14.

Desjardins A., Eisenberg A. (1991) Macromolecules 24, 5779.

Desjardins A., van de Ven T. G. M., Eisenberg A. (1992) Macromolecules 25, 2412.

Ding J., Liu G., Yang M. (1997) Polymer, 38, 5497.

Dormidontova E. E. (1999) Macromolecules 32, 7630.

Elias H.-G. (1972) in Light Scattering from Polymer Solutions, Huglin M. B. (Ed), Academic, London, p. 397.

Elias H.-G. (1973) J. Macromol. Sci. Chem. 7, 601.

Esselink F. J., Dormidontova E., Hadziioannou G. (1998) Macromolecules 31, 2925.

Fleischer G., Puhlmann A., Ritting F., Konak C. (1999) Colloid Polym. Sci. 277, 986.

Förster S., Zisenis M, Wenz E., Antonietti M. (1996) J. Chem. Phys. 104, 9956.

Gao Z., Eisenberg A. (1993) Macromolecules 26, 7353.

Gao Z., Varshney S. K., Wong S., Eisenberg A. (1994) Macromolecules 27, 7923.

Gast A. P. (1996) Langmuir 12, 4060.

Goldmints I., von Gottberg F. K., Smith K. A., Hatton T. A. (1997) Langmuir 13, 3659.

Goldmints I., Yu G., Booth C., Smith K.A., Hatton T. A. (1999) Langmuir 15, 1651.

Guenoun P., Davis H. T., Tirrell M., Mays J. W. (1996a) Macromolecules 29, 3965.

Guenoun P., Davis H. J., Lipsky T., Tirrell M., Mays J. W. (1996b) Langmuir 12, 1425.

Guenoun P., Muller F., Tirrell M. (1998) Phys. Rev. Lett. 81, 3872.

Guenoun P., Davis H. T., Zheng T. (2000) Langmuir 16, 4436.

Haliloglu T., Bahar I., Erman B., Mattice W. L. (1996) Macromolecules 29, 4764.

Halperin A. (1987) Macromolecules 20, 2943.

Halperin A., Alexander S. (1989) Macromolecules 22, 2403.

Hamley I. W. (1998) The Physics of Block Copolymers, Oxford Univ. Press, Oxford, Chapter 3.

Harada A., Kataoka K. (1995) Adv. Drug Delivery Rev. 16, 295.

Haulbrook W. R., Feerer J. L., Hatton T. A., Tester J. W. (1993) Environ. Sci. Technol. 27, 2783.

Higgins J. S., Benoit H. C. (1994) Polymers and Neutron Scattering, Clarendon Press, Oxford.

Higgins J. S., Dawkins J. V., Maghami G. G., Shakir S. A. (1986) Polymer 27, 931.

Holtzer A., Holtzer M. F. (1974) J. Phys. Chem. 78, 1443.

Honda C., Abe Y., Nose T. (1996) Macromolecules 29, 6778.

Honda C., Hasegawa Y., Hirunuma R., Nose T. (1994) Macromolecules 27, 7660.

Huang C., Olvera de la Cruz M., Delsanti M., Guenoun P. (1997) Macromolecules 30, 8019.

Iatrou H., Willner L., Hadjichristidis N., Halperin A., Richter D. (1996) Macromolecules 29, 581.

Khougaz K., Gao Z., Eisenberg A. (1994) Macromolecules 27, 6341.

Kikuchi A., Nose T. (1996) Polymer 37, 5889.

Kriz J., Brus J., Plestil J., Kurkova D., Masar B., Dybal J., Zune C., Jerome R. (2000) Macromolecules 33, 4108.

Leibler L., Orland H., Wheeler J. C. (1983) J. Chem. Phys. 79, 3550.

Linse P. (1993) Macromolecules 26, 4437.

Linse P. (1994) Macromolecules 27, 6404.

Liu Y., Chen S. H., Huang J. S. (1998) Macromolecules 31, 2236.

Ma Y., Cao T., Webber S. E. (1998) Macromolecules 31, 1773.

Major M. D., Torkelson J. M., Brearly A. M. (1990) Macromolecules 23, 1700.

Mandema W., Emeis C. A., Zeldenrust H. (1979) Makromol. Chem. 180, 2163.

Marko J. F., Rabin Y. (1992) Macromolecules 25, 1503.

Mayes A. M., Olvera de la Cruz M. (1988) Macromolecules 21, 2543.

Mendes E., Schadler V., Marques C. M., Lindner P., Wiesner U. (1997) Europhys. Lett. 40, 521.

Moffit M., Khougaz K., Eisenberg A. (1996) Acc. Chem. Res. 29, 95.

Moffit M., Yu Y., Nguyen D., Graziano V., Schneider D. K., Eisenberg A. (1998) Macromolecules 31, 2190.

Mortensen K. (1996) J. Phys.: Condens. Matter. 8, A103.

Munk P. (1996a) in Solvents and Self-Organization of Polymers, Webber S. E., Tuzar Z., Munk P. (Eds) Kluwer Academic Publishers, The Netherlands, p. 19.

Munk P. (1996b) in Solvents and Self-Organization of Polymers, Webber S. E., Tuzar Z., Munk P. (Eds) Kluwer Academic Publishers, The Netherlands, p. 367.

Nagarajan R. (1996) in Solvents and Self-Organization of Polymers, Webber S. E., Tuzar Z., Munk P. (Eds) Kluwer Academic Publishers, The Netherlands, p. 121.

Nagarajan R., Ganesh K. (1989a) J. Chem. Phys. 90, 5843.

Nagarajan R., Ganesh K. (1989b) Macromolecules 22, 4312.

Nakano M., Matsuoka H., Yamaoka H., Poppe A., Richter D. (1999) Macromolecules 32, 697.

Noolandi J., Hong K. M. (1982) Macromolecules 15, 482.

Pepin M. P., Whitmore M. D. (2000a) Macromolecules 33, 8644.

Pispas S., Hadjichristidis N., Mays J. W. (1996a) Macromolecules 29, 7378.

Pispas S., Allorio S., Hadjichristidis N., Mays J. W. (1996b) Macromolecules 29, 2903.

Pispas S., Hadjichristidis N., Potemkin I., Khokhlov A. (2000) Macromolecules 33, 1741.

Pispas S., Poulos Y., Hadjichristidis N. (1998) Macromolecules 31, 4177.

Plestil J., Baldrian J. (1975) Die Makromol. Chem. 176, 1009.

Plestil J., Hlavata D., Hrouz J., Tuzar Z. (1990) Polymer 31, 2112.

Price C., Canham P. A., Duggleby M. C., Naylor T., Rajab N. S., Stubberfield R. B. (1979) Polymer 20, 615.

Price C., Chan E. K. M., Mobbs R. H., Stubberfield R. B. (1985) Eur. Polym. J. 21, 355.

Price C., Hudd A. L., Booth C., Wright B. (1982) Polymer 23, 650.

Price C., McAdam J. D. G., Lally T. P., Woods D. (1974) Polymer 15, 228.

Price C., Woods D. (1974) Polymer 15, 389.

Prochazka K., Bednar B., Tuzar Z., Kocirik M. (1989) J. Liq. Chromatogr. 12, 1023.

Prochazka K., Delcros H., Delmas G. (1988) Can. J. Chem. 66, 915.

Prochazka K., Tuzar Z., Kratochvil P. (1991) Polymer 32, 3038.

Prochazka K., Kiserow D., Ramireddy C., Webber S. E., Munk P., Tuzar Z. (1992) Makromol. Chem. Macromol. Symp. 58, 201.

Qin A., Tian M., Ramireddy C., Webber S. E., Munk P., Tuzar Z. (1994) Macromolecules 27, 120.

Ramzi A., Prager M., Richter D., Efstratiadis V., Hadjichristidis N., Young R. N., Allgaier J. B. (1997) Macromolecules 30, 7171.

Raspaud E., Lairez D., Adam M., Carton J-P. (1994) Macromolecules 27, 2956.

Schadler V., Nardin C., Wiesner U., Mendes E. (2000) J. Phys. Chem. B 104, 5049.

Schillen K., Brown W., Johnsen R. M. (1994) Macromolecules 27, 4825.

Selb J., Gallot Y. (1981a) Makromol. Chem. 182, 1513.

Selb J., Gallot Y. (1981b) Makromol. Chem. 182, 1775.

Selb J., Gallot Y. (1985) in Developments in Block Copolymers, Goodman I. (Ed), Elsevier, London, Vol. 2, p. 27.

Shusharina N. P., Linse P., Khokhlov A. R. (2000) Macromolecules 33, 3892.

Shusharina N. P., Nyrkova I. A., Khokhlov A. R. (1996) Macromolecules 29, 3167.

Speracek J. (1982) Makromol. Chem. Rapid Commun. 3, 697.

Stepanek M., Krijtora K., Prochazka K., Teng Y., Webber S. E., Munk P. (1998) Acta Polym. 49, 96.

Tao J., Stewart S., Liu G., Yang M. (1997) Macromolecules 30, 2738.

ten Brinke G., Hadziioannou G. (1987) Macromolecules 20, 486.

Teng Y., Morrisson M. E., Munk P., Webber S. E., Prochazka K. (1998) Macromolecules 31, 3578.

Tian M., Qin A., Ramireddy C., Webber S. E., Munk P., Tuzar Z. (1993) Langmuir 9, 1741.

Tsitsillianis C., Kouli O. (1995a) Macromol. Rapid Commun. 16, 591.

Tsitsiliannis C., Papanagopoulos D., Lutz P. (1995b) Polymer 36, 3745.

Tsunashima Y., Hirata M., Kawamata Y. (1990) Macromolecules 23, 1089.

Tuzar Z. (1996a) in Solvents and Self-Organization of Polymers, Webber S. E., Tuzar Z., Munk P. (Eds) Kluwer Academic Publishers, The Netherlands, p. 1.

Tuzar Z. (1996b) in Solvents and Self-Organization of Polymers, Webber S. E., Tuzar Z., Munk P. (Eds) Kluwer Academic Publishers, The Netherlands, p. 309.

Tuzar Z., Konak C., Stepanek P., Plestil J., Kratochvil P., Prochazka K. (1990) Polymer 31, 2118.

Tuzar Z., Kratochvil P. (1976) Adv. Colloid Interface Sci. 6, 201.

Tuzar Z., Kratochvil P. (1993) Surface & Colloid Sci. 15, 1.

Tuzar Z., Kratochvil P., Prochazka K., Contractor K., Hadjichristidis N. (1989) Makromol. Chem. 190, 2967.

Tuzar Z., Plestil J., Konak C., Hlavata D., Sikora A. (1983) Makromol. Chem. 184, 2111.

van Lent B., Scheutjens J. H. M. (1989) Macromolecules 22, 1931.

Voulgaris D., Tsitsiliannis C., Grayer V., Esselink F. J., Hadziioannou G. (1999) Polymer 40, 5879.

Wang Y., Kausch C.M., Chun M., Quirk R. P., Mattice W. L. (1995) 28, 904.

Watanabe A., Matsuda M. (1986) Macromolecules 19, 2253.

Whitmore M. D., Noolandi J. (1985) Macromolecules 18, 657.

Xu R., Winnik M. A., Riess G., Chu B., Croucher M. D. (1992) Macromolecules 25, 644.

Yang Y-W., Yang Z., Zhou Z-K., Attwood D., Booth C. (1996) Macromolecules 29, 670.

Yu G., Eisenberg A. (1998) Macromolecules 31, 5546.

Zhang L., Eisenberg A. (1995) Science 268, 1728.

Zhang L., Eisenberg A. (1996) J. Am. Chem. Soc. 118, 3168.

Zhao J. Q., Pearce E. M., Kwei T. K., Jeon H. S., Keseni P. K., Balsara N. P. (1995) Macromolecules 28, 1972.

Zheng Y., Won Y-Y., Bates F. S., Davis H. T., Scriven L. E., Talmon (1999) J. Phys. Chem. B 103, 10331.

Zhou Z., Yang Y-W., Booth C., Chu B. (1996a) Macromolecules 29, 8357.

Zhou Z-K., Chu B., Nace V. M., Yang Y-W., Booth C. (1996b) Macromolecules 29, 3663.

Zhulina E. B., Birnstein O. V. (1985) Vysokomolekulyarnye Soedineniya 27, 511 [English translation: Polym. Sci. USSR (1987) 27, 570].

CHAPTER 12

ADSORPTION OF BLOCK COPOLYMERS AT SOLID-LIQUID INTERFACES

Block copolymer adsorption is a very interesting and important phenomenon from both an academic as well as a technological point of view (Fleer 1993). It can be described as the accumulation (or depletion) of polymer chains at an interface, like a polymer solution-solid substrate interface. Adsorption of block copolymers at solid-liquid interfaces is strongly related to a variety of technological important processes like steric stabilization of colloids, flocculation, adhesion, and lubrication between surfaces, as well as surface coating. The behavior of polymeric chains at solid-liquid (as well as liquid-liquid and air-liquid) interfaces, and more specific, parameters like chain chemical nature and configuration, adsorbed amount of polymer, and structure of the adsorbed layer, play a crucial role in these processes. Due to its importance, the study of block copolymer adsorption began many years ago, and many theoretical and experimental investigations have appeared in the literature in an effort to better understand this phenomenon (Fleer 1983 and 1993, Cohen Stuart 1986, deGennes 1987, Patel 1989, Kawaguchi 1990 and 1992).

1. PHENOMENOLOGY OF BLOCK COPOLYMER ADSORPTION

Block copolymers (in the most simple case, diblock copolymers) are adsorbed onto a surface from solution if one of the blocks has a high affinity for the surface (anchor block) while the other is standing in the solution (buoy block) (Fleer 1993). The situation is similar, in some respects, to the case of terminally attached polymer chains (end-tethered chains) (Alexander 1977, deGennes 1980), the main difference being that because the anchoring block has a finite size (much greater than an anchoring point), it can influence the structure of the adsorbed layer by changing, for example, the distance between buoy blocks depending on its size. Thus, the adsorbed polymer layer can be described as consisting of a more or less continuous collapsed layer of the anchoring block and a layer of well-solvated buoy block in a more or less extended conformation depending on surface coverage, solvent quality, and interchain interactions between the buoy blocks (Fleer 1993, Halperin 1992).

Furthermore, the adsorption of a block copolymer can be complicated by the possibility of micelle formation in the bulk solution (van Lent 1989, Johner 1990), depending on the solvent quality with respect to the anchoring block and the overall concentration of the block copolymer in solution compared with its critical micelle concentration (Tirrell 1996). Investigations so far tend to conclude that, irrespective of the block copolymer concentration in the bulk solution, adsorption of polymer onto the surface is accomplished by anchoring of individual chains because, even at a concentration above cmc, some free unimers exist in solution, and the micelle-unimer equilibrium can always be shifted in a way to accommodate unimer depletion. Additionally, direct micelle adsorption onto the surface seems to be improbable because the well-solvated corona chains, encapsulating the core formed by the anchoring blocks, tend to be repelled from the surface, at least in most cases.

The kinetics of block copolymer adsorption in the case where micelles are not present in solution can be described as a two-stage process (Fig.12.1). In the first stage, at early times, the transport of polymer molecules to the surface is due to diffusion towards the surface (Fleer 1993, Motschmann 1991). Thus, the adsorption of polymer chains onto the bare surface occurs instantaneously when they are in close proximity to the surface, and the adsorption rate is kinetically controlled (transport-limited regime). The adsorbed amount of polymer as a function of time, A(t), can be given as the integral of the flux of the material to the surface, j_s, in respect to time, i.e.:

$$A(t) = \int_{0,t} j_s \, dt = (2/\pi^{1/2}) c_o (Dt)^{1/2} \tag{12.1}$$

where D is the diffusion coefficient of the adsorbing chains and c_o their bulk concentration. Therefore, a $t^{1/2}$ dependence of the adsorbed amount is predicted for this regime. At this low coverage, the block copolymer chains are widely separated, so there is no overlap between them, and the conformation of the buoy blocks is not significantly different from their conformation in solution.

At later times the growth of the adsorbed layer by diffusion of the material to the surface will continue until all empty area on the surface has been occupied by polymer chains. At this stage adsorption of additional chains on the surface will require the penetration of chains through the barrier created by chains already attached to the surface. As a consequence the adsorption rate will be decreased (brush-limited regime). At this regime a deviation of A(t) versus $t^{1/2}$ plot from linearity should be observed. The adsorbed amount, A_{trans}, at this point can be used to calculate the average spacing between chains, σ, at the point of the transition. In

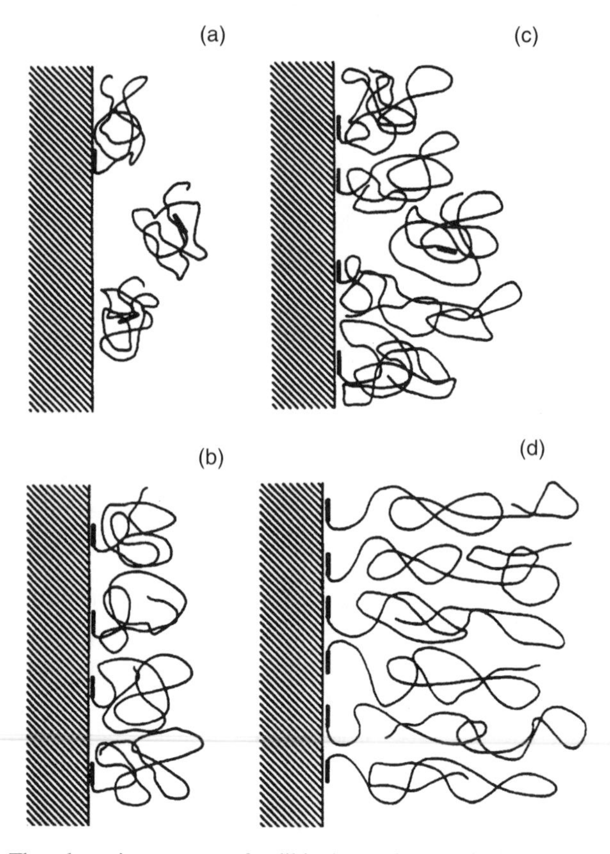

Figure 12.1 The adsorption process of a diblock copolymer with one adsorbing block on a planar solid surface: (a) diffusion-controlled regime with small interaction between chains, (b) formation of a complete monolayer, (c) further adsorption of chains requires penetration of the layer and conformational rearrangement of the adsorbed chains, (d) final state at high coverage where the nonadsorbing blocks are in a brushlike conformation (Motschmann 1991).

general, the average intermolecular spacing at a given adsorbed amount, A, is given by:

$$\pi(\sigma/2)^2 = M/AN_A \tag{12.2}$$

where M is the molecular weight of the block copolymer, and N_A is the Avogadro number. When the average distance between anchored chains becomes smaller than the size of the buoy block in solution (its diameter is a good measure of this size), chains begin to overlap. The osmotic pressure on the solvated layer increases and causes the buoy blocks to stretch, creating a brush-like structure. In order for the adsorbed amount to increase in this regime, the energy cost associated with buoy chain stretching must be balanced by the energy gain because of adsorption on the surface (i.e., the favorable interactions between the anchoring block and the surface). The chains in solution have to penetrate the barrier of the brush formed by the already adsorbed chains in order to reach the surface. This becomes progressively more difficult, as increased coverage increases chain stretching. As a result the adsorption rate decreases, and the adsorbed amount reaches a saturation value. The increase of the adsorbed amount in the brush-limited regime should follow an exponential time dependence:

$$A(t) = A(\infty)(1 - e^{-kt}) + A_{exp} \quad (12.3)$$

where k is the rate constant for the adsorption process in the brush-limited regime and A_{exp} is the value of the adsorbed amount beyond which the exponential law is followed. It should be noted that the adsorption process in both regimes depends on the molecular weight and composition of the block copolymer, the structure of the existing adsorbed layer, the surface coverage, and the interactions between the polymer molecules, the solvent molecules, and the surface (Fleer 1993, Tirrell 1996).

2. EXPERIMENTAL TECHNIQUES FOR STUDYING BLOCK COPOLYMER ADSORPTION

A variety of experimental techniques are currently available for the investigation of behavior of block copolymers adsorbed at solid-liquid interfaces, including spectroscopic techniques (IR, NMR, EPR, and labeling techniques), ellipsometry, X-ray and neutron reflectivity, photon correlation spectroscopy, neutron scattering, surface force apparatus, and atomic force microscopy (Kawaguchi 1992). These techniques provide valuable information on the adsorbed amount of polymer, the thickness of the adsorbed layer, the segment density profile within the adsorbed layer, the interaction energy between adsorbed chains and surfaces, and the forces acting between surfaces bearing adsorbed block copolymers chains. The employment of different techniques has helped in the understanding of structure and interactions of adsorbed chains and the parameters that govern block copolymer adsorption. The working principles of some of them are described briefly in the following text.

2.1. Ellipsometry

Ellipsometry is one of the oldest characterization techniques involved in the study of polymer adsorption (Azzam 1977). It is based on the fact that a beam of well-defined polarized light (usually a laser source) changes its state of polarization after being reflected from a surface or an interface. Usually the variation of the technique used to determine the state of light polarization is that of null ellipsometry (Dorgan 1993a). In this case, the experimentally determined parameters are two ellipsometric angles associated with the phase difference between incoming and reflected beam, Δ, and their amplitude ratio, $\tan \Psi$. The basic equation for ellipsometry in the case of a thin polymer film deposited on a substrate can be written as:

$$\tan \Psi^{e-i\Delta} = r_p/r_s = f(n_0, n_1, n_2, d_2, \varphi_0, \lambda) \tag{12.4}$$

where r_p and r_s are the complex reflectances for the substrate parallel and perpendicular to the plane of incidence, n_0, n_1, n_2 are the refractive indices of the surrounding solution, the substrate, and the polymer film, respectively, d_2 is the thickness of the polymer film, φ_0 is the angle of incidence, and λ is the wavelength of the light used. The refractive index of the solution can be calculated, taking into account the contributions of all components (e.g., solvent and copolymer):

$$n_0 = n_{solvent} + (dn/dc)_{polymer}\, c_{polymer} \tag{12.5}$$

where dn/dc is the specific refractive index of the polymer and c its concentration in the solution. Equation 12.4 is based on the assumption of a perfectly planar substrate and a homogeneous polymer film of uniform thickness, a situation far from real polymer adsorbed films. In this respect n_2 and d_2 can be regarded as effective numbers.

The adsorbed amount of polymer A (in mg/m^2) can be calculated by the following equation:

$$A = d_2(n_2 - n_0)/(dn/dc)_{polymer} \tag{12.6}$$

Usually the differences in refractive indices between the polymer and solution are small, and an independent determination of n_2, d_2 is difficult, but they can be evaluated simultaneously. The product n_2d_2 does not depend on the model adopted for the adsorbed polymer layer. Also, the adsorbed amount does not depend on the type of the concentration profile near the surface of the substrate (step, parabolic, or exponential). Thus, ellipsometry is a simple and useful technique for the determination of layer thickness and the adsorbed amount. The kinetics of polymer adsorption can also be studied using this technique by monitoring the change in the ellipsometric angles as a function of time.

2.2. Reflectometric Techniques

The adsorption of polymer molecules from solution and their depletion near a solid substrate can be studied by using evanescent wave ellipsometric techniques (Kim 1989). In this case the phase difference as a function of the angle of incidence is monitored around the critical angle for total internal reflection, $\theta_c =$

$\sin^{-1}(n_2/n_1)$. It has been proven that the dielectric constant $\varepsilon(z)$ is related to the polymer segment profile $\phi(z)$ at a distance z from the surface with:

$$\phi(z) = \varepsilon_{solvent} + [\phi(z)/\phi_b](\varepsilon_{solution} - \varepsilon_{solvent}) \tag{12.7}$$

assuming that $\varepsilon_{solution}$ is proportional to polymer concentration in solution.

Scanning angle reflectometry is based on the measurement of the reflection intensity of a light wave polarized parallel to the incidence plane, as a function of the angle of incidence around the Brewster angle (Leermakers 1991). Transparent surfaces must be used in these cases. The data are fitted with a Fresnel function, and information analogous to that obtained from ellipsometry is gained. Time resolved reflection intensities can be used for the study of block copolymer adsorption kinetics.

Internal reflection interferometry has been used for the study of block copolymer adsorption on dielectric surfaces (Munch 1990). In this technique phase changes in the two polarizations are measured after reflection on the adsorbed polymer layer-substrate and polymer layer-bulk solution interfaces, at a fixed angle of incidence. The measured intensities can be analyzed using Fresnel reflection coefficients for each interface, and information on the average thickness of the adsorbed polymer layer and its refractive index can be obtained.

The surface plasmon technique (Tassin 1989) measures changes in the reflectivity due to the progressive adsorption of the polymer on a metal surface. It can give information on the adsorbed amount and the thickness of the adsorbed polymer layer.

2.3. Neutron Reflectivity

In general, radiation incident on an interface will undergo refraction and reflection, provided the refractive indices of the media on the two sides of the interface are different. Neutrons can also be reflected from a surface in a way similar to light and, thus, neutron reflectivity can be analyzed by the methods used for optical reflectivity (Russell 1990). The refractive index (scattering contrast), n_n, of a medium for neutrons is related to the scattering length density (b/V) as:

$$n_n^2 = 1 - (\lambda^2/2\pi)(b/V) \tag{12.8}$$

where λ is the wavelength of neutrons. When a neutron beam is incident on a surface at an angle θ, it will be reflected at the corresponding angle θ, and the wave vector normal to the surface is given by:

$$k_{z,o} = (2\pi/\lambda)\sin\theta \tag{12.9}$$

The reflectance or the reflection coefficient at an infinitely sharp interface separating two media, i and i + 1 is given by:

$$r_{i,i+1} = (k_{z,i} - k_{z,i+1})/(k_{z,i} + k_{z,i+1}) \tag{12.10}$$

and in the case of a free polymer layer surface (in contact with the bulk solution)

$$r_{0,1} = (k_{z,0} - k_{z,1}/(k_{z,0} + k_{z,1}) \tag{12.11}$$

The reflectivity, R, is given by

$$R = r_{0,1}\, r^*_{0,1} \tag{12.12}$$

where the asterisk denotes the complex conjugate. It is obvious that R is a function of k_z. In the case of a polymer layer adsorbed on a substrate, the reflectance takes the form:

$$R = [r_{0,1} + r_{1,2}\exp(2ik_{z,1}d)]/[1 + r_{0,1}r_{1,2}\exp(2ik_{z,1}d)] \tag{12.13}$$

where d is the thickness of the polymer layer and $r_{o,1}$ and $r_{1,2}$ are the reflection coefficients between the solution and the polymer layer and the polymer layer and the substrate, respectively. In this case:

$$R(k_{z,0}) = rr^* = [r^2_{0,1} + r^2_{1,2} + 2r_{0,1}r_{1,2}\cos(2k_{z,1}d)]/[1 + r^2_{0,1}r^2_{1,2} + 2r_{0,1}r_{1,2}\cos(2k_{z,1}d)] \tag{12.14}$$

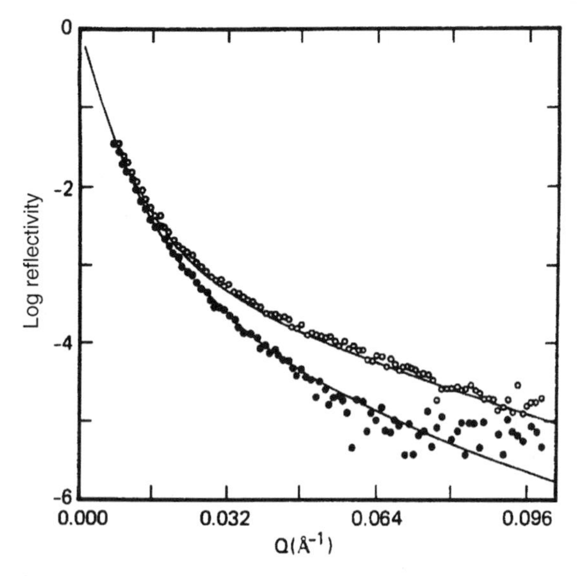

Figure 12.2 Neutron reflectivity curves for poly(styrene-b-deuterated methyl methacrylate) (○) and poly(styrene-b-methyl methacrylate) (•) having the same molecular characteristics and at the same concentration, adsorbed on quartz from carbon tetrachloride. The lines are the fits to the data using the parabolic concentration profile (Satija 1990).

The reflectivity profiles contain a series of maxima and minima, which characterize the thickness of the film and are given by:

$$D = \pi/\Delta k_{z,1} \propto \pi/\Delta k_{z,0} \tag{12.15}$$

Thus, if the reflectivity is obtained as a function of the scattering vector, it can be fitted with any model of scattering length density profile perpendicular to the reflection plane (substrate surface) within the specimen under investigation (Fig. 12.2).

2.4. Techniques for Studying Polymer Layers Adsorbed on Particles in Solution

Dynamic light scattering (or photon correlation spectroscopy) is a well-established technique for determining the hydrodynamic radius of polymers and particles suspended in solution through measurements of the time fluctuations of light-scattering intensity. First, the diffusion coefficient of the particles is determined by measuring the decay rate of the autocorrelation function, and it is converted to hydrodynamic radius through the Stokes-Einstein relationship. If a polymer layer is adsorbed on a particle, its hydrodynamic radius increases. The hydrodynamic thickness of the adsorbed polymer layer is determined as the difference between the hydrodynamic radius of the bare particle and that of the particle coated with the adsorbed layer (Garvey 1976). The method has been frequently used for the determination of the hydrodynamic thickness, d_h, of polymer chains adsorbed onto colloidal particles, latexes, and inorganic particles (d'Oliveira 1993, Killmann 1988).

The hydrodynamic thickness of the polymer layer adsorbed onto spherical particles in solution can be also determined by dilute solution viscometry (Rowland 1966). In a dilute solution of these particles, it is assumed that no interparticle interactions exist, and each particle is covered with an identical polymer layer. The difference in the hydrodynamic thickness of the layer can be determined by the difference in the hydrodynamic radius of the bare and the coated particle because this is correlated with the intrinsic viscosity of the solution and the respective molecular weights of the two types of the particle. In a more straightforward way, the coated and uncoated particles can be modeled as rigid spheres through Einstein relation for viscosity:

$$\eta = \eta_o(1 + 2.5\phi_0) \tag{12.16}$$

where η is the viscosity of the suspension of the bare particles, η_o the viscosity of the solvent, and ϕ_0 is their volume fraction. After adsorption of the polymer chains, the effective volume fraction of the suspended particles would change to ϕ'. If the coefficient 2.5 is assumed to be held, then:

$$\eta' - \eta_o/\eta - \eta_o = \phi'/\phi_o = (R_h + d_h)^3/R_h^3 \tag{12.17}$$

where R_h is the hydrodynamic radius of the bare particles.

The viscosity technique can also be used to study the properties of polymer layers adsorbed onto pores of porous media or on the inner surface of capillaries (Webber 1990, Cohen 1982). By measuring the permeability of the material to solvent (estimation of flow rates) before and after adsorption of the polymer chains and having prior knowledge about the size of the pores or the capillary, one can gain information on the hydrodynamic layer thickness.

Small angle neutron scattering has also been used in an analogous manner for investigating the properties of polymer layers adsorbed onto particles. By using the contrast matching technique, the adsorbed layer can be made visible to neutrons, and, by the appropriate analysis of the scattering profiles, information about the thickness of the layer as well as the segment density profile within this layer can be obtained.

2.5. Direct Surface Force Measurements

The forces acting between two surfaces, bearing adsorbed macromolecules, as they approach each other will be greatly affected by the nature of the adsorbed polymer layer (Patel 1989). The measurement of how force changes with distance can be interpreted with respect to the polymer segment distribution within the layers. Depending on the nature of the solvent-polymer interaction and the distance between surfaces, attractive or repulsive forces can result. The main experimental requirement for the appropriate operation of a surface force apparatus is the smoothness of the substrate, allowing, essentially, point contact between the surfaces at zero distance. For this purpose mica sheets are used as substrates that provide atomically smooth surfaces. The mica sheets are mounted in a cross cylinder configuration on cylindrical quartz lenses. The distance between the mica surfaces is controlled by a positioning mechanism consisting of micrometers and a piezoelectric crystal, and it can be measured with high accuracy by interferometry (Israelachvili 1987). White light is passed between the mica cylinders, which is multiply reflected, and destructive interference occurs for all wavelengths different from those that correspond to the separating distance between the surfaces producing the well known pattern of fringes (lines). The difference in wavelength between fringes is a very precise measurement of the distance between the separated surfaces, whereas the difference between the odd and even number of half wavelengths is a measure of the mean refractive index of the surrounding medium, i.e., the mean concentration of polymer. When one of the cylinders is rigidly mounted, whereas the other is held on a calibrated spring, a precise measurement of the force acting between the surfaces can be made. The technique

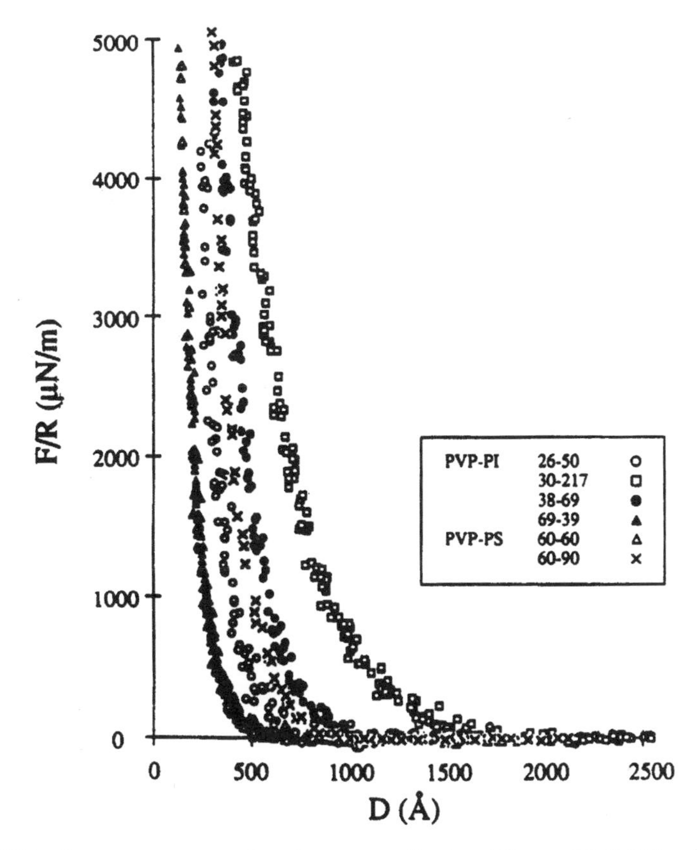

Figure 12.3 Force versus separation distance profiles for different PS-P2VP and PI-P2VP block copolymers adsorbed onto mica from toluene. The number of segments of each block are given in the inset (Watanabe 1993).

allows the determination of the force-distance curve or the mean concentration-distance curve in the case of irreversible adsorption (Hadziioannou 1986, Watanabe 1993) (Fig.12.3). Force measurements can be made in the presence or the absence of polymer dissolved in the surrounding solvent. Apart from the measurement of forces at the equilibrium of the adsorption process, the forces acting between the surfaces when the two surfaces are moved laterally past each other can also be monitored, and dynamic effects can be studied.

2.6. Spectroscopic Techniques

Spectroscopic techniques are very useful in investigating the structure and characteristics of adsorbed block copolymer chains because they are, in general, sensitive to changes in the molecular or submolecular level. Binding of functional groups on a surface can create changes in the IR (or the NMR) spectra of the molecule adsorbed through these functional groups on the surface. Thus, the number of adsorbed polymer segments can be directly measured by IR by monitoring the changes in the spectrum of the polymer chain or the number of occupied and unoccupied binding sites on the surface by monitoring the spectrum associated with the surface (Kawaguchi 1988).

NMR can probe directly the mobility of block copolymer segments through magnetic relaxation times and bandwidth values. Because this mobility changes when a polymer segment is attached to a surface, the method can provide direct information about the bound and unbound parts of a polymer chain and indirectly about chain conformation near the surface. EPR functions in a similar way (Barnett 1981, Robb 1974).

Labeling techniques that involve the employment of specially labeled, with radioactive or fluorescent groups, polymer chains together with the appropriate spectroscopic technique have also been used (Huguenard 1991). Due to their high sensitivity, they can accurately determine very low concentrations of adsorbed chains and are useful in the study of polymer dynamics at the molecular level.

3. THEORIES OF BLOCK COPOLYMER ADSORPTION

A number of theoretical studies dealing with the adsorption of amphiphilic block copolymers at solid-liquid interfaces have appeared in the literature in an effort to theoretically predict and understand the structure of the adsorbed polymeric layer as well as aspects related to the kinetics of block copolymer adsorption.

A mean field model for block copolymer adsorption on an excess adsorbing surface was developed by Munch and Gast (Munch 1988). They examined different regimes of surface-polymer attraction energy, solvent-polymer interactions, and ranges of A and B block lengths. They found, in analogy to the critical micelle concentration, that a critical adsorption concentration exists beyond which adsorption of chains takes place. The critical adsorption concentration decreases as the surface attraction increases and as the solvent becomes poorer for the adsorbing block, B, of the copolymer. When the solvent quality for the B block decreases in a level where micelles are formed, micelle formation precludes adsorption. The critical adsorption concentration also decreases as the length of the soluble A block decreases and the solvent molecular size increases. The polymer layer thickness, L_A was found to increase as the length of block A increased (following the relationship, $L_A \sim N_A^{0.7}$), as the surface attraction for block B increased and as the solvent size decreased. The surface polymer density was found to depend on both the size of the A and B block according to the equations $\sigma \sim N_A^{-0.3}$ and $\sigma \sim N_B^{-0.5}$, where N_A and N_B are the number of segments for blocks A and B, respectively.

van Lent and Scheutjens (van Lent 1989) presented a modification of Scheutjens-Fleer self-consistent field theory for homopolymer adsorption for the case of block copolymer adsorption from selective solvents. The association state of the block copolymers was found to greatly influence their adsorption properties. A strong increase in the adsorbed amount was observed just below the cmc. Beyond the cmc the rate of adsorption on the surface was found to be constant. Thick adsorption layers could be formed. They calculated that micellization and adsorption are competitive phenomena in the case of weakly adsorbing B blocks in very poor solvents.

Evers et al. (Evers 1990) generalized the theory of Scheutjens-Fleer to the case of block copolymer adsorption from a multicomponent system. The cases of AB diblock and ABA triblock copolymers with a strongly A adsorbing block were discussed. In the case of diblocks, the adsorbing A block adopts a rather flat conformation on the surface, whereas the B block dangles free in the solution far from the surface. The density profile of the A block was found to be very similar to that of an A absorbed homopolymer, while that of the B block is similar to the segment density profile of terminally anchored chains. The stretching of the B block depends strongly on the solvent quality in the case of high-adsorbed amounts. It was determined that the ABA triblocks tend to adopt a nonlooped conformation, having dangling tails protruding into the bulk solution with sticky ends.

The same approach was used in order to predict the interactions between two adsorbed block copolymer layers (diblocks and triblocks) at equilibrium (Evers 1991). At full equilibrium, when the diblock copolymer chains are free to diffuse out of the gap between the surfaces, repulsive interactions were found. Attractive interactions were observed only in the case where both blocks have some affinity for the surfaces. In the case of restricted equilibrium, where the chains cannot move out of the gap, repulsive interactions were also observed in good solvents, originating from osmotic effects. In poor solvents for the nonadsorbing block, attractive interactions were calculated at large separations. At short distances interactions are repulsive because of incompressibility of the segments. The onset of interaction was found to be close to distances twice the hydrodynamic thickness of the layers. This thickness increases as the solvent quality for the B block increases and, in good solvents, depends on the length of the B block (increases as the length of the B block increases) and also increases as the adsorbed amount increases (i.e., as the length of the A block increases). In the case of ABA block copolymers in good solvents for the B nonadsorbing block, attractive interactions were found at large distances due to bridging.

Marques et al. (Marques 1988) presented a scaling theory for AB diblock copolymer adsorption from a selective solvent for the B block. They studied the structure of the adsorbed polymer layer as a function of block copolymer asymmetry and the structure of the solution. They considered four different adsorption regimes depending on the value of the chemical potential of the bulk solution. In very dilute solutions, there is a threshold for adsorption that depends on the van der Waals interaction of the adsorbing block A and its spreading power. Above this threshold the thickness d of the A block layer scales as $|\log \phi_b|^{-1/3}$, where ϕ_b is the volume fraction of the polymer in the bulk solution, and the van der Waals interactions, between the A block and the wall, dominate the adsorption process (Rollin regime). At much higher concentrations, micelles can be formed in the bulk solution. When the asymmetry of the block copolymer is low ($\beta \sim 1$), adsorption is governed by the elastic energy of the B block (buoy-dominated regime), and the thickness of the A layer is almost equal to its radius of gyration, whereas the thickness of the B layer is analogous to its unperturbed radius. When the asymmetry of the diblock is high ($\beta > 1$), the van der Waals interaction of block A and the wall competes with the elastic energy of the B block, and the thickness of the grafted layer is analogous to the molecular weight of the diblock. At even higher concentration, where the polymer chains in solution are organized in lamellae microdomains, theory predicts that, when the van der Waals forces are small, the structure of the adsorbed layer is that of the lamellae.

In the case of adsorption from a nonselective solvent, Marques and Joanny (Marques 1989) predicted that the structure of the adsorbed layer is primarily governed by the asymmetry of the copolymer. For low asymmetry ($\beta < N_A^{1/2}$, where N_A is the number of segments in the A block), the adsorbing A block forms a fluffy layer considerably swollen by the solvent, and the segment profile in the A layer is similar to that of an adsorbed homopolymer. In this regime the surface density scales as $\sigma \sim 1/N_A$, the thickness of the A layer as $d \sim N_A^{1/2}/\beta$, and the thickness of the B layer as $L \sim N_B^{2/3}\beta^{1/3}$. For short adsorbing A blocks, relative to the B blocks ($\beta > N_A^{1/2}$), the layer brakes into isolated copolymer chains. The anchor layer has the thickness of a monomer size, α, and the two-dimensional solution can be either dilute or semidilute. The scaling relations become $\sigma \sim \beta^{-2}$, $d \sim \alpha$, and $L \sim N_B^{3/5} N_A^{2/5} \alpha$.

Whitmore and Noolandi (Whitemore 1990) presented a mean field theory for block polymer adsorption. They studied the cases of adsorption on a single surface and that of adsorption on two parallel surfaces at finite distance apart. They calculated the shape of the segment density profiles within the polymer layer and its thickness as a function of solvent quality, relative block lengths, and the average surface per chain. In the case of the two surfaces, the forces acting between the layers as a function of the separation distance, as well as the interpenetration of the layers, were calculated (Fig.12.4).

The adsorption kinetics of diblock copolymers from a solution in a selective solvent onto a wall have been studied theoretically by Johner and Joanny (Johner 1990). They considered the case where micelles are present in the solution. They have found that only free copolymer chains are able to adsorb on the wall. This adsorption creates a depletion zone near the wall where the number of unimers is decreased, and, thus, the micelle-unimer equilibrium is distorted. As a result micelles start to expel chains in order to reach a new equilibrium state. Taking into account this scenario of single chain diffusion to the wall and micelle relaxation, four different time regimes were distinguished. At early times the adsorption process is dominated by the diffusion of the free chains to the wall (first regime), whereas, at longer times, micelle relaxation through unimer expulsion is the limiting step (second regime). Surface coverage increases linearly with time. When a dense brush is formed, the single chains have to overcome this barrier in order to adsorb on the wall. This then becomes the limiting step for adsorption (third regime). At even longer times, the adsorbed layer has to relax in order to acquire its equilibrium structure (fourth regime).

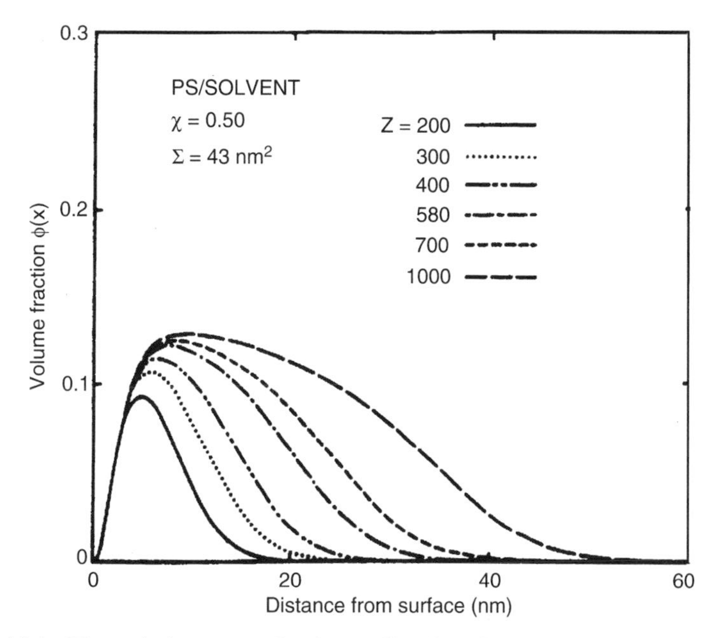

Figure 12.4 Theoretical segment density profiles for PS brushes of different degree of polymerization at fixed surface density in a theta solvent. The segment density rises from zero to a maximum at a finite distance from the surface. Then it decreases, in a parabolic fashion, at low degrees of polymerization, but, at a high number of segments, a smooth tail is evident (Whitmore 1990).

Dan and Tirrell (Dan 1993) have presented a numerical self-consistent field model for the calculation of the equilibrium properties of bimodal polymer brushes in order to investigate the effect of molecular weight distribution of the buoy block on the properties of the adsorbed layer. They found that the longer chains are more stretched than the shorter ones in the inner region near the surface, independent of the difference in the molecular weights and composition of the layer. Thus, the degree of polydispersity can affect significantly the inner structure of the polymer brush. In the case where the difference in molecular weight is small, the ends of the longer chains were found to be located at the edges of the brush. They presented calculations on the segment density profiles of two bimodal brushes under compression.

The adsorption of ionic AB block copolymers has been investigated theoretically by Israels et al. (Israels 1993). In the case examined, the A anchoring block is uncharged, and the buoy B block has a fixed charge, whereas the surface is uncharged. The molecules in the layer were assumed to be in equilibrium with those in the bulk solution (nonfixed density of adsorbed chains). Different regimes were identified with respect to the anchor density (high and low anchor density regimes) and to the charge of the B block. Adsorbed amounts were found to be lower (by a factor of 10) compared with the case of uncharged block copolymers.

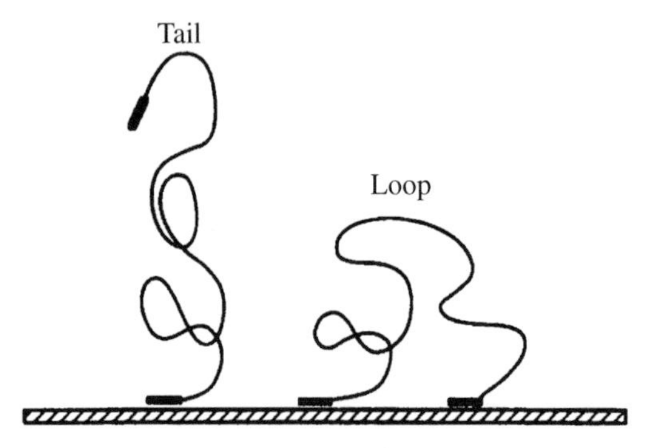

Figure 12.5 Possible conformations of an ABA triblock copolymer adsorbed to a surface through the A blocks (Dorgan 1993).

The thickness of the brush is, however, large even at low surface coverage due to the electrostatic interactions between the B blocks.

ten Brinke and Hadziioannou (ten Brinke 1987) presented Monte Carlo calculations on the conformations of adsorbed triblock copolymers of the ABA type, where A is a short strongly adsorbing block. They assumed irreversible adsorption and absence of interaction of the main chain segments with the surface and tried to determine the conformation of chains in the dilute and concentrated regime. Differences were observed in the percentage of knotted conformations between triblock copolymers and rings (Fig.12.5).

4. EXPERIMENTS ON BLOCK COPOLYMER ADSORPTION

Many experimental investigations have been concerned with the adsorption of block copolymers onto surfaces from solution. A variety of techniques have been used for the determination of adsorption kinetics, adsorbed polymer amounts, structure of the adsorbed layer (layer thickness, segment density profiles), as well the interaction forces between layers. Some representative examples will be discussed in the following text.

Parsonage et al. (Parsonage 1991) investigated the adsorption behavior of poly(2-vinylpyridine)-polystyrene diblock copolymers with a strongly adsorbing P2VP block from toluene solutions onto silicon oxide and mica surfaces. They used scintillation counting techniques and X-ray photoemission spectroscopy to determine the adsorbed amounts for a series of diblocks with various compositions. For relatively low asymmetry (in composition) copolymers, the surface density of the adsorbed polymer chains was found to depend more strongly on the molecular weight of the P2VP block. For copolymers with high asymmetry, the surface density was depended primarily by the molecular weight of the nonadsorbing PS block. The results were, in general, in agreement with the theory of Marques et al.

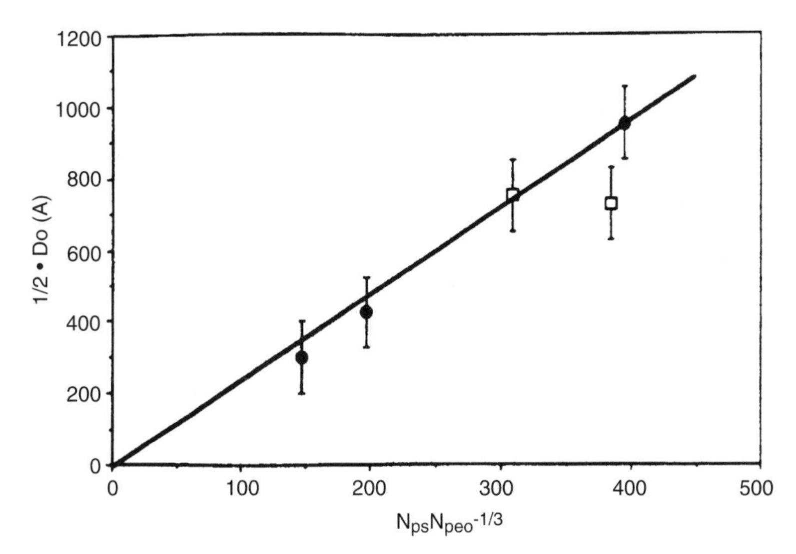

Figure 12.6 Scaling plot for the dependence of the brush height, L, taken as half of the distance at which the repulsive force is first measureable, by a force apparatus, on the molecular characteristics of PS-PEO diblock copolymers adsorbed on mica from toluene solutions (Hair 1991).

Guzonas et al. (Guzonas 1992, Hair 1991) used the surface force apparatus and IR spectroscopy to study the adsorption of PS-PEO diblocks from toluene onto mica. For moderately asymmetric copolymers, the adsorbed layer thickness was found to scale as $L \sim N_{PS} N_{PEO}^{1/3}$. The surface density of the copolymers with moderate asymmetry, determined by the IR technique, was found to be described rather well by the scaling relation of Marques and Joanny. Deviations were observed for copolymers with high asymmetry (Fig.12.6).

Cosgrove et al. (Cosgrove 1991) have used neutron reflectivity in order to study the conformation of PS-P2VP block copolymers adsorbed on mica from toluene solutions. Information about the segment density profile within the adsorbed layer was obtained by fitting the reflectivity curves using various models. For diblock copolymers, the best fits were obtained using a parabolic segment density profile or a profile with a maximum. In all cases the profiles decayed monotonically.

Satija et al. (Satija 1990) used the same technique to investigate symmetric deuterium-labeled PS-PMMA block copolymers adsorbed on a quartz surface from carbon tetrachloride. The anchoring block in this case was the PMMA block because CCl_4 is a poor solvent for this block. The thickness of the PMMA layer was found to be comparable to the radius of gyration of the respective block.

Ansarifar and Luckman (Ansarifar 1988) measured the interaction forces between mica surfaces bearing poly(2-vinylpyridine)-poly(tert-butyl styrene) block copolymer. Repulsive interactions were observed, which increased exponentially by decreasing the separation distance. Comparison with scaling theory showed good agreement as far as the dependence of the adsorbed layer thickness on molecular weight was concerned. However, scaling predictions could not describe the force-distance profile well.

Hadziioannou et al. (Hadziioannou 1986) also used the surface force apparatus for investigating the interactions between mica surfaces covered with PS-P2VP diblock layers. The shape of the force-distance profile as well as the separation range where interaction forces were observed depended on the molecular weight of the PS block and the thermodynamic quality of the solvent with respect to this block (buoy block). The repulsive forces were observed at larger separations as the molecular weight of the PS block was increased. Similarly in toluene, a good solvent for PS, the acting forces were more long-ranged than in the theta solvent cyclohexane. This was attributed to the larger extension of the PS in toluene, whereas, in cyclohexane, the PS block is contracted, and the binary interactions between segments must be diminished.

Taunton et al. (Taunton 1988 and 1990) have used the same technique to study adsorbed layers of highly asymmetric PS-PEO blocks onto mica from toluene and xylene solutions. PEO was the anchoring block in this case. They concentrated on the scaling behavior of the adsorbed layer thickness with molecular weight of the buoy block, and they found that their data could be well fit both by scaling and mean field predictions, i.e., linear dependence on molecular weight was found.

The kinetics of adsorption of PS-PEO from cyclopentane solutions onto dielectric surfaces were investigated by Munch and Gast (Munch 1990) as a function of concentration. It was found that, above cmc, the adsorbed amount increases rapidly, indicating direct adsorption of micelles. This can be explained only if some rearrangement within the micelle takes place that permits interaction of the PEO core with the surface. However, adsorption of chains below the cmc resulted in a more homogeneous layer. The evolution of surface concentration, layer thickness, and refractive index of the layer with time indicated that the chains first adsorb in a mushroom conformation. The space occupied by the buoy chains is decreased as the surface coverage increases, and, finally, a homogeneous layer is formed. It was observed that, as the molecular weight of the tail increases, the surface density decreases.

Motschmann et al. (Motschmann 1991) used ellipsometry in order to study the kinetics of PS-PEO diblock adsorption on silicon wafers. The adsorbed amount profiles showed two processes. At early times a diffusion-controlled process takes place that results in a layer where copolymer chains interact very little with each other. At longer times chains penetrate the already formed layer, and subsequent rearrangement of the chains on the surface leads to a more brush-like conformation. It was observed that the interactions between the nonadsorbing PS tails determine the maximum value for the adsorbed amount and that the block copolymer chains can be readily displaced completely by a PEO homopolymer having a molecular weight comparable with the molecular weight of the PEO block in the copolymer (Fig.12.7).

Huguenard et al. (Huguenard 1991) used radiolabeled P2VP/PS diblocks to investigate their adsorption kinetics from toluene on silica below the cmc. Initially,

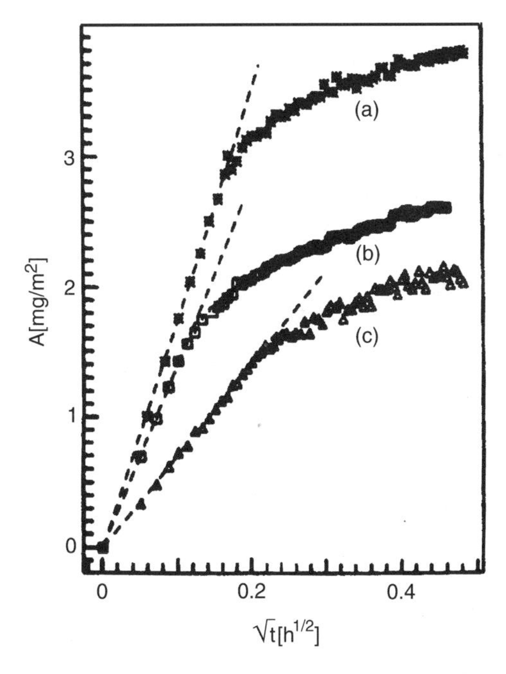

Figure 12.7 Adsorbed amount as a function of time for a series of PS-PEO diblock copolymers at the initial stages of adsorption (concentrations of the bulk solutions are similar) on SiO_2, measured by ellipsometry. The curves correspond to different lengths of the two blocks described by the degrees of polymerization (numbers in parentheses) (a) PS (730)-PEO (90), (b) PS (1700)-PEO (167), (c) PS (4788)-PEO (91). The dashed lines indicate the calculated $t^{1/2}$ increase of the adsorbed amount, assuming that each molecule reaching the surface is adsorbed. Notice the dependence of the adsorbed amount on the molecular characteristics of the block copolymers (Motschmann 1991).

fast adsorption was observed until the surface was fully covered with P2VP blocks. At later times (over time periods of 2 months), a slow rearrangement of the chains in the layer was observed. This rearrangement was faster for the copolymers having the larger P2VP blocks. It was concluded that adsorption of block copolymers below the cmc need long periods of time in order for an equilibrium state to be attained. Slowly, the layer thickens to the larger dimensions that is permitted by the surface and internal cohesive forces of the adsorbed block and that is allowed by the dimensions of the solvated block.

Leermakers and Gast (Leermakers 1991) also observed slow relaxation of the initially produced layers of PS-PEO block copolymers at low concentrations using scanning angle reflectometry. By increasing concentration of the bulk solution in a step procedure, stable fast relaxing layers were obtained. Their characteristics were found to be in agreement with the predictions of the theory of Scheutjens and Fleer.

Amiel et al. (Amiel 1995) studied the adsorption of highly asymmetric hydrophobic-hydrophilic poly(tert-butylstyrene)-poly(sodium styrene sulfonate) from aqueous solutions on silica surfaces. They used ellipsometry and atomic force microscopy to determine the kinetics and the structure of the adsorbed layers. Adsorption was observed only after the addition of salt (NaCl) at relatively high concentrations (1 M). This was taken as a proof of the effect that electrostatic interactions can have on the adsorption process of block polyelectrolytes. These interactions prohibit the formation of an adsorbed layer in the absence of charge screening. The kinetic investigation showed a two-stage process: a diffusion-controlled process at early stages and a slower increase in the surface coverage at later stages. The higher molecular weight species showed the lower surface density as expected from scaling theories. This is associated with the fact that longer solvated chains pay a higher elastic penalty for stretching at a given surface density at the surface. Zhang et al. (Zhang 1996) studied the same polymers using a variety of different salts.

Recently, Abraham et al. (Abraham 2000) presented results on the adsorption kinetics of hydrophilic-hydrophobic block copolymers of poly(tert-butyl methacrylate) and poly(glycidyl methacrylate sodium sulfonate) on hydrophobic surfaces from aqueous solutions. They used ellipsometry to study the effects of salt concentration and monovalent counterion size on the adsorption process. The experimental results indicated a three-stage adsorption process: an incubation period, a second period characterized by fast growth of the adsorbed polymer layer, and a plateau region (equilibrium state). The formation of a bound ionic layer on the substrate at short times was proposed as the reason for the observation of the incubation period. The three stages were found to be dependent on the salt concentration and the size of the counterion. The equilibrium-adsorbed amount increased as the salt concentration was increased due to the increase in the electrostatic screening effect. An Avrami type growth process was used to model the adsorption process. Analysis of the kinetic data within this framework indicated that the diffusion of chains to the surface is not the rate-controlling step of the adsorption process, but, rather, a slow nucleation and fast growth of the layer determine the whole process. The Avrami analysis also showed that the formation of the block polyelectrolyte layer structure depends strongly on the added salt concentration.

Dorgan et al. (Dorgan 1993b) used ellipsometry to study the adsorption kinetics of triblock copolymers PEO-PS-PEO on silicon oxide surfaces from toluene. A diffusion-controlled first stage was observed. At low bulk solution concentrations, the saturation of the surface was approached monotonically, while, at larger concentrations, an overshoot in the absorbed amount was observed. The adsorbed amount at equilibrium for triblock copolymers was found to be lower compared with that of diblocks with the same length of anchoring block. This was attributed to the formation of looped conformations in the case of the triblock copolymers that require more surface area.

The adsorption of P2VP-PS-P2VP triblock copolymers was also investigated by Dai et al. (Dai 1995) using the surface force apparatus. During compression-decompression cycles, rapid conformational transitions between looped and tailed conformations were observed for these triblock copolymers. Furthermore, it was

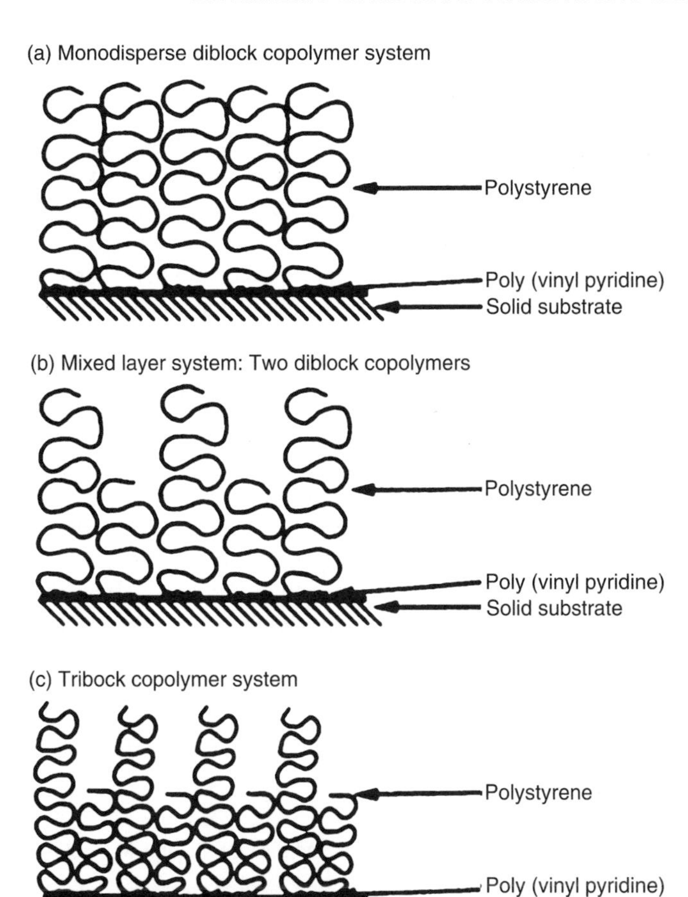

Figure 12.8 Types of PS brushes formed by the adsorption of styrene-2-vinylpyridine block copolymers from toluene solutions: a) adsorbed PS-P2VP block copolymer, b) adsorbed mixture of two PS-P2VP diblocks having different PS blocks, c) adsorbed PS-P2VP-PS' asymmetric triblock copolymer (Dhoot 1994).

shown that the tailed conformations could, under certain conditions, be stretched beyond the equilibrium dimensions of the chains. The kinetics of both loop-tail and collapsed-stretched conformational transitions were found to be fast. In the same study, adsorbed PS-PEO layers were also investigated. The PS-PEO chains adsorbed onto mica from toluene were found to collapse upon replacement of the good solvent for PS, toluene with the theta solvent cyclohexane.

The competitive adsorption of binary mixtures of block copolymers onto surfaces was investigated by Dhoot and Tirrell (Dhoot 1995). They used two types of binary mixtures. In the first type, PS-P2VP specimens were used where the lengths of the anchoring block (P2VP) was kept almost constant and that of the buoy (PS) was varied or vise versa. In the second type, PS-P2VP and PI-P2VP diblocks were used where the P2VP blocks were similar in molecular weight. Toluene, a good solvent for both PI and PS and poor for P2VP, was used as the solvent. The concentrations of the bulk solutions were always lower than the cmc of all block copolymers. It was found that when the P2VP blocks were of similar lengths, the adsorption of the diblocks having the shorter PS block was favored. In the case where the PS blocks were similar in length, the adsorption of the diblock with the longer P2VP block was favored. In the case of PS-P2VP and PI-P2VP mixtures, the authors suggested that the appropriate parameter for comparing the preference for adsorption was the contour length of the nonadsorbing block. They also concluded that the composition of the adsorbed layer can be controlled by carefully selecting the composition of the bulk solution and the molecular characteristics of the block copolymers.

Dhoot et al. (Dhoot 1994) studied the interactions between bimodal layers of PS and P2VP containing block copolymers adsorbed onto mica, using the surface force apparatus. These bimodal brushes were formed either by competitive adsorption of PS-P2VP diblocks from toluene or by adsorption of PS-P2VP-PS triblocks containing PS blocks of different lengths (Fig. 12.8). The force-distance profiles for the bimodal brushes were compared with those of pure (monomodal) brushes using two different scaling procedures. The analysis, based on the Alexander-de Gennes analysis of bimodal brushes, brought the data for monomodal and bimodal brushes onto the same curve at low and high compressions (Dhoot 1994).

REFERENCES

Abraham T., Giasson S., Gohy J. F., Jerome R., Muller B., Stamm M. (2000) Macromolecules 33, 6051.

Alexander S. (1977) J. Phys. (Paris) 38, 977.

Amiel C., Sikka M., Schneider J. W., Tsao Y-H., Tirrell M., Mays J. W. (1995) Macromolecules 28, 3125.

Ansarifar M. A., Luckman P. F. (1988) Polymer 29, 329.

Azzam R. M. A., Bashara N. N. (1977) Elipsometry and Polarized Light, North Holland, Amsterdam.

Barnett K. G., Cosgrove T., Vincent B., Sissons D. S., Cohen Stuart M. A. (1981) Macromolecules 14, 1018.

Cohen Stuart M. A., Cosgrove T., Vincent B. (1986) Adv. Colloid Interface Sci. 24, 143.

Cohen Y., Metzner A. B. (1982) Macromolecules 15, 1425.

Cosgrove T., Heath T. G., Phipps J. S., Richardson R. M. (1991) Macromolecules 24, 94.

Dai L., Toprakcioglu C., Hadziioannou G. (1995) Macromolecules 28, 5512.

Dan N., Tirrell M. (1993) Macromolecules 26, 6467.

deGennes P. G. (1980) Macromolecules 13, 1069.

deGennes P. G. (1987) Adv. Coll. Interface Sci. 27, 189.

Dhoot S., Tirrell M. (1995) Macromolecules 28, 3692.

Dhoot S., Watanabe H., Tirrell M. (1994) Colloids & Surfaces A: Physicochem. Eng. Aspects 86, 47.

d'Oliveira J. M. R., Xu R., Jensma T., Winnik M. A., Hruska Z., Hurtrez G., Riess G., Martinho J. M. G., Croucher M. D. (1993) Langmuir 9, 1092.

Dorgan J. R., Stamm M., Toprakcioglou C., Jerome R., Fetters L. J. (1993a) Macromolecules 26, 5321.

Dorgan J. R., Stamm M., Toprokcioglu C. (1993b) Polymer 34, 1554.

Evers O. A., Scheutjens J. M. H. M., Fleer G. J. (1990) Macromolecules 23, 5221.

Evers O. A., Scheutjens J. M. H. M., Fleer G. J. (1991) Macromolecules 24, 5558.

Fleer G. J., Cohen Stuart M. A., Scheutjens J. M. H. M., Cosgrove T., Vincent B. (1993) Polymers at Interfaces, Chapman & Hall, London.

Fleer G. J., Lyklema J. (1983) in Adsorption from Solution at the Solid/Liquid Interface, Parfitt G. D., Rochester C. H. (Eds) Academic Press, New York, p. 153.

Garvey M. J., Tadros T. T., Vincent B. (1976) J. Colloid Interface Sci. 55, 440.

Guzonas D. A., Boils D., Tripp C. P., Hair M. L. (1992) Macromolecules 25, 2434.

Hadziioannou G., Patel S., Granick S., Tirrell M. (1986) J. Am. Chem. Soc. 108, 2869.

Hair M. Z., Guzonas D., Boils D. (1991) Macromolecules 24, 341.

Halperin A., Tirrell M., Lodge T. P. (1992) Adv. Polym. Sci. 100, 31.

Huguenard C., Varoqui R., Pefferkorn E. (1991) Macromolecules 24, 2226.

Israelachvili J. (1987) Proc. Natl. Acad. Sci. USA 84, 4722.

Israels R., Scheutjens J. M. H. M., Fleer G. J. (1993) Macromolecules 26, 5405.

Johner A., Joanny J. F. (1990) Macromolecules 23, 5299.

Kawaguchi M. (1990) Adv. Coll. Interface Sci. 32,1.

Kawaguchi M., Kawarabayashi M., Nagata N., Kato T., Yoshioka A., Takahashi A. (1988) Macromolecules 21, 1059.

Kawaguchi M., Takahashi A. (1992) Adv. Coll. Interface Sci. 37, 219.

Killmann E., Maier H., Baker J. A. (1988) Colloids & Surfaces 31, 51.

Kim M. W., Peiffer D. G., Chen W., Hsiung H., Rasing T., Shen Y. R. (1989) Macromolecules 22, 2682.

Leermakers F. A. M., Gast A. P. (1991) Macromolecules 22, 718.

Marques C. M., Joanny J. F. (1989) Macromolecules 22, 1454.

Marques C., Joanny J. F., Leibler L. (1988) Macromolecules 21, 1051.

Motschmann H., Stamm M., Toprakcioglu C. (1991) Macromolecules 24, 3681.

Munch M. R., Gast A. P. (1988) Macromolecules 21, 1366.

Munch M. R., Gast A. P. (1990) Macromolecules 23, 2313.

Parsonage E., Tirrell M., Watanabe H., Nuzzo R. G. (1991) Macromolecules 24, 1987.

Patel S. S., Tirrell M. (1989) Annu. Rev. Phys. Chem. 40, 597.

Robb I. D., Smith R. (1974) Eur. Polym. J. 10, 1005.

Rowland R. W., Eirich F. R. (1966) J. Polym. Sci. A-1, 4, 2033.

Russell T. P. (1990) Mater. Sci. Reports 5, 171.

Satija S. K., Majkrzak C. F., Russell T. P., Sinha S. K., Sirota E. B., Hughes G. J. (1990) Macromolecules 23, 3860.

Tasin J. F., Siemens R. L., Tang W. T., Hadziioannou G., Swalen J. D., Smith B. A. (1989) J. Phys. Chem. 93, 2106.

Taunton H. J., Toprakcioglu C., Fetters L. J., Klein J. (1988) Nature 332, 712.

Taunton H. J., Toprakcioglu C., Fetters L. J., Klein J. (1990) Macromolecules 23, 571.

ten Brinke G., Hadziioannou G. (1987) Macromolecules 20, 489.

Tirrell M. (1996) in Solvents and Self-Organization of Polymers, Webber S. E., Tuzar Z., Munk P. (Eds), Kluwer Academic Publishers, The Netherlands.

van Lent B., Scheutjens J. M. H. M. (1989) Macromolecules 22, 1931.

Watanabe H., Tirrell M. (1993) Macromolecules 26, 6455.

Webber R. M., Anderson J. L., Shon M. S. (1990) Macromolecules 23, 1026.

Whitmoore M. P., Noolandi J. (1990) Macromolecules 23, 3321.

Zhang Y., Tirrell M., Mays J. W. (1996) Macromolecules 29, 7299.

PART IV

PHYSICAL PROPERTIES OF BLOCK COPOLYMERS

CHAPTER 13

THEORY

An essential parameter in all theories of block copolymer segregation is the interaction parameter χ_{AB} that provides the driving force for the phase separation. The segment–segment interaction parameter (Flory–Huggins) described the free energy cost per monomer of conducts between the A and B monomeric units and is given by: $\chi_{AB} = (Z/k_B T)[\varepsilon_{AB} - (\varepsilon_{AA} + \varepsilon_{BB})/2]$, where ε_{AB} is the interaction energy per monomer units between A and B monomers and Z is the number of nearest neighbor monomers to a copolymer configuration cell. Thus, positive χ_{AB} (which is the vast majority of cases) shows repulsion between the A and B monomers, whereas a negative value signifies mixing of unlike monomers. Moreover, χ_{AB} usually varies inversely with temperature. Another parameter that strongly influences the block copolymer behavior is the total degree of polymerization N. For large N the loss of translational and configurational entropy leads to a reduction of the A-B monomer contacts and thus to local ordering. Since the entropic and enthalpic contributions to the free energy scale as N^{-1} and χ, respectively, it is the product χN that is of interest in the block copolymer phase state.

The phase state of block copolymers can be discussed with respect to three regimes according to the value of the product χN: strong segregation limit (for $\chi N > 100$), weak segregation limit ($\chi N \sim 10$) with an intermediate range called intermediate segregation limit. We discuss below each regime separately.

1. STRONG SEGREGATION LIMIT (SSL)

This corresponds to the situation $\chi N \gg 10$. In this regime, narrow interfaces are formed with width depending on the χ-parameter as $\alpha\chi^{-1/2}$. Central to the theory of strong segregation in block copolymers is the concept of an extended block conformation imposed by the combined effects of localization of the block junction points at a narrow interface and an overall uniform density. The effect of the extended chain conformation can be observed in the molecular weight dependence of the microdomain period, which scales as $d \sim N^{\delta}$, with $\delta \sim 2/3$ in the SSL, as opposed to $\delta = ½$ for the unperturbed (Gaussian) chains.

We describe below two SSL theories. The first was developed by Helfand and Wasserman (Helfand 1975, 1976, 1978) who introduced a self-consistent field theory with three principal contributions to the free energy: (i) contact enthalpy between the pure A and B microdomains at the interface, (ii) entropy loss due to the stretching of chains, and (iii) confinement entropy due to the localization of the junction points at the interface. Their result for the interfacial thickness was:

$$\Delta_{\infty} = \frac{2\alpha}{\sqrt{6\chi}} \tag{13.1}$$

and for the microdomain period (for $N \rightarrow \infty$)

$$d \sim \alpha N^{\delta} \chi^{\nu} \tag{13.2}$$

with $\delta \sim 9/14$ and $\nu = 1/7$.

An analytical method for estimating the free energy in the asymptotic limit $\chi N \rightarrow \infty$ was developed by Semenov (Semenov 1985). Semenov argued that the copolymers are strongly stretched that their chain-ends are distributed in excess in the domain interiors. The situation resembles grafted polymer brushes and surfactant interfaces, which brought about a high simplification to the problem. Based on Semenov's SSL theory, the domain contribution to the free energy per chain was found to have (surprisingly) the same scaling as for a Gaussian chain, i.e.:

$$\frac{F_{domain}}{kT} \sim \frac{d^2}{\alpha^2 N} \tag{13.3}$$

and the observed nonuniform stretching comes only as a prefactor. The domain free energy is balanced by the interfacial energy per chain,which is given by:

$$\frac{F_{interface}}{kT} \sim \gamma\sigma \sim \frac{N\alpha\chi^{1/2}}{d} \tag{13.4}$$

In the above equation, the results for the interfacial tension, $\gamma \sim \chi^{1/2}\alpha^{-2}$, and for the area per chain, $\sigma \sim N\alpha^3/d$, have been inserted. Balancing the two free energy contributions results in Semenov's prediction for the domain period in the asymptotic limit ($\chi N \rightarrow \infty$):

$$d \sim \alpha N^{2/3} \chi^{1/6} \tag{13.5}$$

Equation 13.5 predicts a weak dependence of the microdomain spacing on χ and a strong dependence on N. Because of the weak $d(\chi)$ dependence, the experimental verification of eq. 13.5 is not an easy task, but the $N^{2/3}$ dependence can be more easily checked. However, experiments in symmetric diblock copolymers (Papadakis 1997) have shown a weaker dependence as $d \sim N^{0.61}$, for $\chi N > 29$, with a peculiar N-dependence ($d \sim N^{0.8}$) at weaker segregation. These experiments suggest that the SSL predictions are only correct for $\chi N > 100$.

2. WEAK SEGREGATION LIMIT (WSL)

This is the regime where most of the experiments are made with $\chi N \sim 10$. Leibler (Leibler 1980), in his seminal work, considered a monodisperse AB diblock copolymer melt with interaction parameter χ, degree of polymerization N, and equal monomer volumes and statistical segment lengths ($\alpha_A = \alpha_B$). Leibler constructed a Landau expansion of the free energy to fourth order in a composition order parameter field:

$$\psi(r) = \langle \phi_A(r) - f \rangle \qquad (13.6)$$

where $\delta\varphi_A(r) = \varphi_A(r) - f$, is the fluctuation in the local volume fraction of the type A monomers at position r. Leibler provided microscopic expressions for the Landau expansion coefficients as functions of the copolymer composition f and the incompatibility χN. The coefficients were calculated using the random phase approximation (RPA) introduced earlier by de Gennes (de Gennes 1970).

The Landau approximation for the thermodynamic potential (the free energy) has the form (Leibler 1980, Fredrickson 1989):

$$f(A) = \tau A^2 + \frac{u}{4} A^4 \qquad (13.7)$$

where $\tau = 2(\chi_S N - \chi N)/c^2$ is a reduced temperature variable that is a measure of the distance from the spinodal, and u depends on f and N. Figure 13.1 gives the free energy density in the Landau approximation of Leibler. At temperatures above the mean field spinodal, i.e., $\chi N < \chi_S N$, the disordered phase with zero amplitude ($A = 0$) is stable. At the spinodal, the curvature of the $A = 0$ well vanishes, and a second order transition to the lamellar phase takes place. Below the spinodal, the free energy density develops two symmetric minima, at $A = \pm(2\tau/u)^{1/2}$ and free energy density $f(A) = -\tau^2/u < 0$, that describe the stable lamellar phase.

Leibler used only the leading harmonics in the Fourier representation of the various ordered phases in the Landau free energy expansion. He was able to construct the phase diagram by comparing the free energies of the three classical ordered phases (lamellar, spheres in a body centered cubic lattice, cylinders packed in a hexagonal lattice) with respect to the disordered phase. The phase diagram so obtained is shown in Figure 13.2, in the parameter space χN and f.

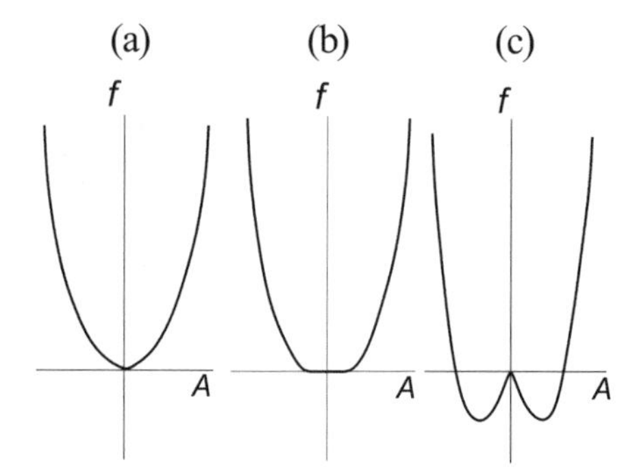

Figure 13.1. The free energy density in the Landau approximation. (a) When $\chi N < \chi_S N$, the disordered phase is stable. (b) At the spinodal a second-order transition to the lamellar phase occurs. (c) When $\chi N > \chi_S N$, the stable lamellar phase is indicated by the two symmetric minima.

The Landau theory predicts a critical point at $(\chi N)_c = 10.5$ and $f_c = 0.5$, where a compositionally symmetric diblock melt is expected to undergo a second-order phase transition from the disordered to the lamellar phase. The period of the lamellar phase is predicted to scale as $D \sim N^{1/2}$ at the order-to-disorder transition, which is not surprising, given the initial assumption that the copolymers are only weakly perturbed by the inhomogeneous composition field. For $f \neq 0.5$, the Landau

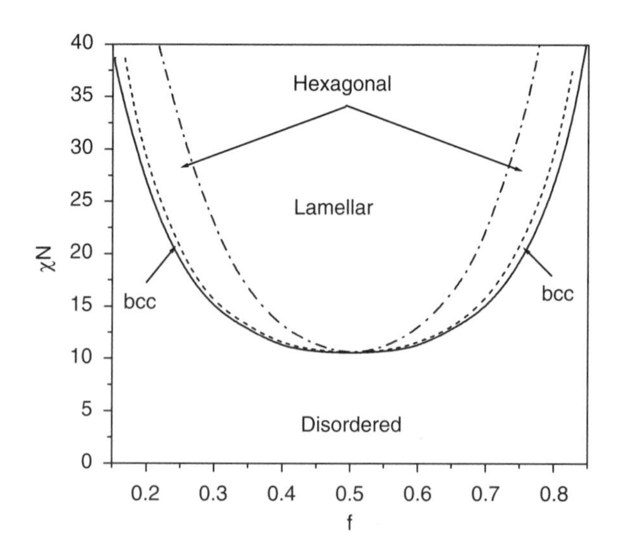

Figure 13.2. Theoretical phase diagram for diblock copolymers calculated by Leibler within the mean field theory. The phase diagram assumes equal monomer volumes and statistical segment lengths for the two blocks.

theory predicts a weak first-order transition from the disordered to the bcc phase. Furthermore, the mean field theory (MFT) allows for order-to-order transitions (OOT) between the classical ordered phases by changing χN.

The order-to-disorder transition can best be illustrated by showing small-angle X-ray scattering (SAXS) images from macroscopically oriented block copolymers. In the ordered phase, the SAXS patterns appear highly anisotropic, the anisotropy reflecting the symmetry of the microstructure. Upon heating, the anisotropy weakens, and, at the order-to-disorder transition (T_{ODT}), the scattering becomes completely isotropic.

3. STRUCTURE FACTOR

Leibler, in addition to the phase diagram (Fig. 13.2), provided an expression for the disordered phase structure factor $S(q) = \langle \delta\varphi_A(q)\, \delta\varphi_A(-q) \rangle$ (Leibler 1980). Within the RPA this function is given by:

$$\frac{1}{S(q)} = \frac{F(x, f)}{N} - 2\chi \tag{13.8}$$

where

$$F(x, f) = \frac{D(1, x)}{D(f, x)D(1-f, x) - \frac{1}{4}[D(1, x) - D(f, x) - D(1-f, x)]} \tag{13.9}$$

and D(f,x) is the Debye function, defined as:

$$D(f, x) = \frac{2}{x^2}\left[fx - 1 + e^{-fx}\right] \tag{13.10}$$

and

$$x = \frac{q^2 N l^2}{6} = q^2 R_g^2 \tag{13.11}$$

At $q = 0$, S(q) is zero because the system is assumed to be incompressible. For large q ($qR \gg 1$), S(q) is independent of χ and tends to zero like $1/q^2$, i.e.:

$$S(q) \sim \frac{2N}{q^2 R^2} f(1-f) \tag{13.12}$$

For small q, ($qR \ll 1$), S(q) is again independent of χ but now tends to zero like q^2:

$$S(q) \sim 2Nf^2(1-f)^2 \frac{q^2 R^2}{3} \tag{13.13}$$

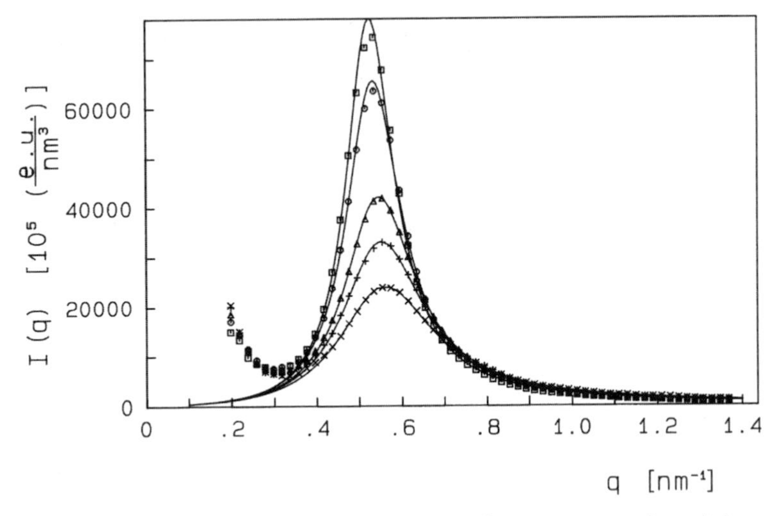

Figure 13.3. Small-angle X-ray scattering profiles from a symmetric poly(styrene-b-isoprene) diblock copolymer with an order-to-disorder transition temperature of 350 K. The different spectra shown are taken at temperatures: (x): 410, (+): 390, (Δ): 380, (O): 361, and (□): 355 K. The solid lines give the result of the fit of Leibler's mean field theory to the scattering profiles.

The $S(q) \sim q^2$ dependence originates from the "correlation hole" effect because each monomer is surrounded by a correlation hole, inside of which the concentration of monomers from other chains is slightly reduced. This correlation hole has a size comparable with the size of one chain. At intermediate q values, S(q) develops a maximum which, at high temperatures, results solely from a fixed length scale reflecting the average distance between the centers of the A and B blocks.

Although the presence of the peak in S(q) is independent of segregation effects (especially at high T), the shape of the curve strongly depends on the product χN. An example is given in Figure 13.3, where the SAXS scattering intensity I(q) is plotted as a function of q, for a symmetric poly(styrene-b-isoprene) diblock copolymer, at different temperatures corresponding to the disordered regime. Decreasing temperature produces a narrower peak up to a χN value, $(\chi N)_S$ where, according to the MFT, the S(q) diverges at $q = q^*$; this is the *spinodal* point.

The S(q) being sensitive to the interaction parameter can be used to extract the $\chi(T)$. In the fitting procedure to the MFT, both $R_g(T)$ and $\chi(T)$ are used as adjustable parameters. Thus, the good agreement between theory and experiment should not be taken as a proof of the accuracy of Leibler's theory. Such a proof requires an independent determination of $R_g(T)$ and $\chi(T)$.

An accurate check of the validity of the MFT can be provided by plotting 1/S(q) versus $(qR_g)^2$, which, for a true MFT behavior, should result in a series of parallel curves with a relative distance depending on the value of χN. Thus, a vertical displacement results in a "master curve." Again, the construction of such a master curve is a check of the validity of the MFT.

4. INTERMEDIATE SEGREGATION LIMIT (ISL) AND SELF-CONSISTENT FIELD THEORY (SCFT)

The complete MFT for block copolymers was developed by Helfand and coworkers (Helfand 1975, 1976, 1978). Important advances were made in the limits of weak and strong segregation by Leibler and Semenov, respectively. The above works established the basic physics controlling block copolymer phase behavior, which involves the competition between interfacial tension and the entropic penalty for stretching of polymer chains so as to fill space uniformly. These approaches were very successful in predicting the three classical phases but failed to account for the more complex phases, such as the double gyroid structure. Later, Matsen (Matsen 1994) has shown that a full SCFT results at intermediate segregation in more complex phases. Figure 13.4 gives the calculated phase diagram at the ISL.

The notable aspect of this phase diagram, as opposed to the MFT work by Leibler, is that, in addition to the usual lamellar (L), hexagonal (C), and spherical (S) phases, there exists a new phase, the double gyroid phase. Unlike the other phases, which all extend to the critical point, the double gyroid phases end at triple points: $\chi N = 11.14$ with $f = 0.452$ and 0.548. The two triple points are shown in Figure 13.5.

For intermediate segregation, the SCFT calculations predict the sequence: Lam-gyroid-hexagonal-spheres-disordered as f progresses from 0.5 to 0 or 1. Notice that the perforated layer (PL) phase is absent from the phase diagram because it is marginally unstable in the Lam/gyroid phase boundary. On the other hand, the ordered bicontinuous double diamond structure (OBDD) (i.e., a two fourfold coordinated lattice) was found to be totally unstable.

Earlier theories have shown that phase transitions in block copolymers are driven by the tendency to curve the interface as the copolymer becomes asymmetric in

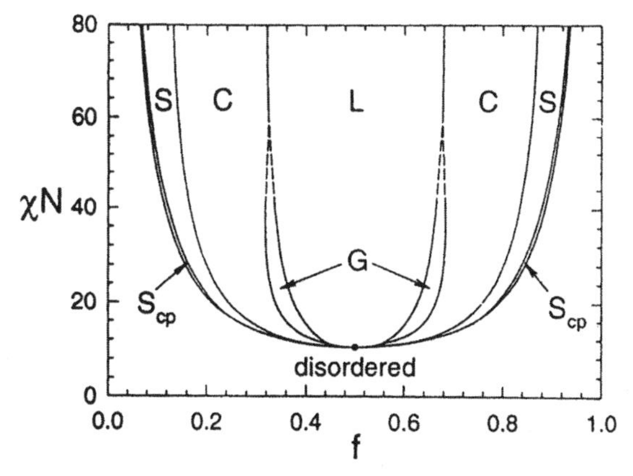

Figure 13.4. Mean field phase diagram within the SCFT approximation for conformationally symmetric diblock copolymers constructed by Matsen. Notice the differences with Leibler's MFT phase diagram (Fig. 13.2). The different phases are: L: lamellar, C: hexagonally packed cylinders, S: spheres packed in a bcc lattice, G: bicontinuous $Ia\bar{3}d$ cubic (double gyroid), S_{cp}: closed packed spheres (Matsen 1996b).

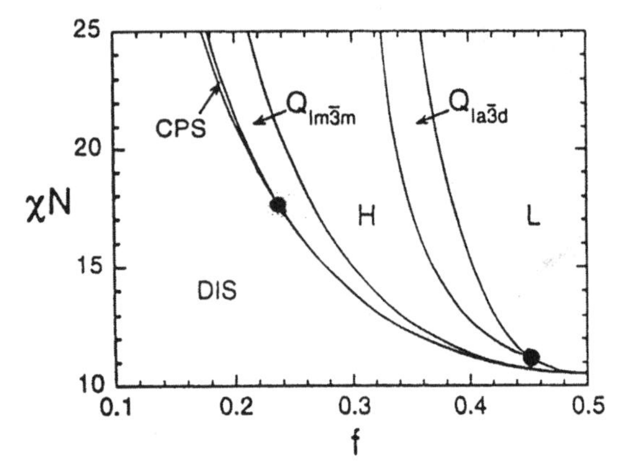

Figure 13.5. Expanded view of the phase diagram shown in the Figure 13.5. The two triple points are denoted with dots (Matsen 1996b).

composition. The curvature allows the molecules to balance the degree of stretching between the A and B blocks. It is, therefore, thought that constant mean curvature (CMC) interfaces are good models for block copolymer morphologies. To illustrate this Figure 13.6 depicts the area-average $\langle H \rangle$ of the mean curvature $H = 1/2(C_1 + C_2)$ for each structure, where C_1 and C_2 are the principal curvatures at a

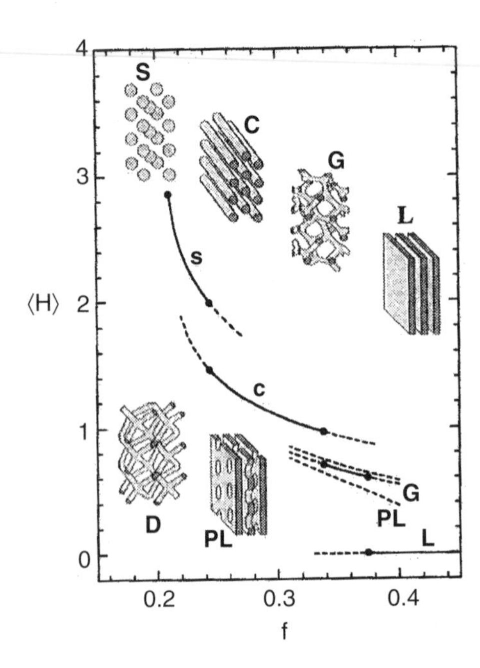

Figure 13.6. Area-averaged mean curvature $\langle H \rangle$ as a function of volume fraction for each of the structures shown calculated using SCFT. The stable and metastable phases are shown with solid and dashed lines, respectively, and the transitions are denoted with dots (Matsen 1996).

given point on the surface of the phase. Figure 13.6 shows that, as f deviates from ½, transitions take place to structures possessing more interfacial curvature. However, based on the $\langle H \rangle$ variations alone, one would expect the following sequence of phases: Lam-PL-gyroid-diamond-hexagonal-spheres-disordered, in contrast with the SCFT result and the experimentally found sequence where the PL and double diamond (OBDD) structures are absent.

This inconsistency was resolved by Matsen (Matsen 1996 and 1997) who argued that while the average $\langle H \rangle$ of the mean curvature controls the sequence of phases, it is the standard deviation σ_H of the mean curvature that governs the phase selection. The second requirement is imposed by the packing frustration, i.e., the tendency to form domains of uniform thickness so that none of the chains are excessively stretched, a mechanism that causes the σ_H to deviate locally from zero. This situation is best illustrated in Figure 13.7, where the interfacial area is shown for a cylindrical structure. The shape of the interface can be parameterized as $r(\theta) = r_o(1 + \delta\cos(6\theta))$, where δ measures the deviation from CMC. While the interfacial tension and packing considerations in the minority domain favor $\delta = 0$, the majority domain prefers $\delta > 0$ so as to produce a more uniform thickness.

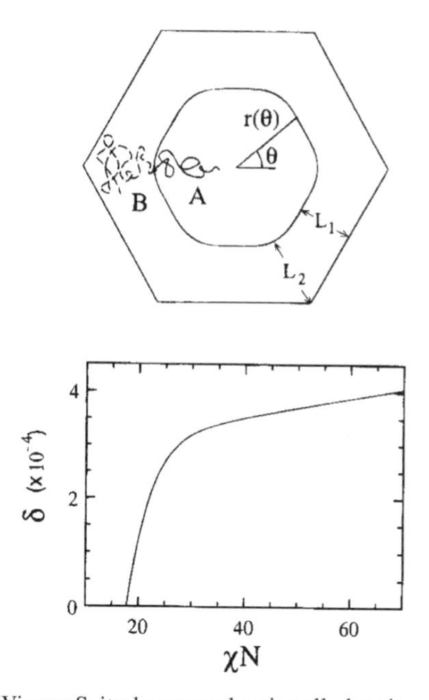

Figure 13.7. Top: The Wigner-Seitz hexagonal unit cell showing the interface between A and B domains of a cylindrical phase. The interfacial shape is described by $r(\theta) = r_o(1 + \delta \cos(6\theta))$. The stretching energy of the B domain prefers $\delta > 0$ in order to produce a uniform thickness (i.e., $L_1 \sim L_2$), but the interfacial energy favors $\delta = 0$ (i.e., a constant mean curvature surface). Bottom: The values of δ that minimize interfacial tension and packing frustration for different segregations at $f = 0.3378$. Notice the small values of δ, which reflect that the packing frustration in the hexagonal phase is minimal (Matsen 1996).

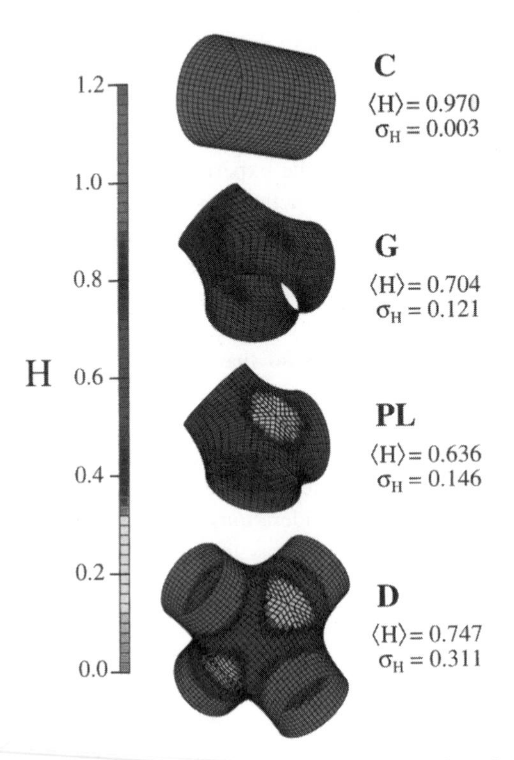

Figure 13.8. Interfacial surfaces associated with elementary units of the hexagonal (C), double gyroid (G), perforated layer (PL), and double diamond (D) structures calculated at the C/G phase boundary ($\chi N = 20$, $f = 0.3378$). See color plates. For each structure, the distribution of mean curvature H over the surface is indicated using the color scale, and the area-average $\langle H \rangle$ and standard deviation σ_H of H is provided. Notice the large values of σ_H in the PL and G phases, which imply a large degree of packing frustration perturbing the interface away from CMC (Matsen 1996).

Moreover, the shape of the interface becomes less CMC-like at strong segregations as shown in the figure. The observation that increasing incompatibility amplifies the packing frustration may also explain the absence of complex phases such as the double gyroid and the perforated layer structures in the strong-segregation regime.

Figure 13.8 shows the interfacial curvature over elementary units of the cylindrical, the double gyroid, the perforated layer, and the double diamond structures calculated at the cylinder/gyroid boundary. The distribution of H is shown with the color scale, and the average $\langle H \rangle$ and the standard deviation σ_H are calculated for each structure. Notice that the hexagonal phase is nearly CMC (with $\sigma_H \sim 0$), whereas all complex phases exhibit significant variation in H. It is thought that the value of σ_H reflects the inability of a structure to simultaneously minimize surface area and packing frustration and, as such, is coupled to the stability of phases. Based on the SCFT calculations, it was postulated that the packing frustration prevents the stability of the double diamond and perforated layer structures in the strong-segregation limit, which agrees with the experimental observations.

REFERENCES

de Gennes P.G. (1970) J. Phys. (Paris) 31, 235.

Fredrickson G.H., Binder K. (1989) J. Chem. Phys. 91, 7265.

Helfand E. (1975) Macromolecules 8, 522.

Helfand E., Wasserman Z.R. (1976) Macromolecules 9, 879.

Helfand E., Wasserman Z.R. (1978) Macromolecules 11, 961.

Leibler L. (1980) Macromolecules 13, 1602.

Matsen M.W., Schick M. (1994) Phys. Rev. Lett. 72, 2660.

Matsen M.W., Bates F.S. (1996) Macromolecules 29, 7641.

Matsen M.W., Bates F.S. (1996b) Macromolecules 29, 1091.

Matsen M.W., Bates F.S. (1997) J. Chem. Phys. 106, 2436.

Papadakis C.M., Almdal K., Mortensen K., Posselt D. (1997) J. Phys. II 7, 1829.

Semenov A.N. (1985) Soviet Phys. JETP 61, 733.

CHAPTER 14

STRUCTURE FACTOR AND CHAIN ARCHITECTURE

Block copolymers of diverse molecular architectures have been synthesized such as diblocks with tapered interfaces, triblocks, graft and star copolymers, which show enhanced mechanical and viscoelastic properties and can be used as compatibilizers in polymer blends. This chapter examines how molecular architecture modifies the phase behavior of block copolymer melts.

Chemically joining two homopolymers to form a diblock copolymer increases the compatibility, and this is reflected in the reduction of the critical temperature for phase separation as compared to a homopolymer blend. For example, a binary mixture of A and B homopolymers phase separates at a critical value of the interaction parameter χ such as $(\chi N)_c = 4$, each of the homopolymers having N/2 monomer units. As we have seen in the previous chapter, Leibler demonstrated that a symmetric AB diblock copolymer with N monomer units phase separate at a critical value of $(\chi N)_c = 10.5$. Thus, the formation of a diblock copolymer results in the reduction of the critical temperature for phase separation by a factor of 2.625. Because, in the simplest approximation, χ is invertly proportional to temperature, the higher critical value in the copolymers implies that phase separation is more difficult, and, therefore, diblock copolymers are more compatible than polymer blends. Compatibility can be further increased by changing the molecular architecture. For this purpose, triblock, star, and, recently, graft and miktoarm copolymers have been synthesized and theoretical (de la Cruz 1986, Dobrynin 1993, Mayes 1989, Milner 1994, Matsen 1994, Read 1998 and 1999) and experimental efforts (Thomas 1986, Hashimoto 1988, Hadjichristidis 1993, Floudas 1994, 1996, 1997, 1998, Hajduk 1995, Pochan 1996, Tselikas 1996, Buzza 1999 and 2000) revealed that, in general:

$$(\chi N_t)_{c,\ blend} < (\chi N_t)_{ODT,\ diblock} < (\chi N_t)_{ODT,\ graft} < (\chi N_t)_{ODT,\ triblock} < (\chi N_t)_{ODT,\ star} \quad (14.1)$$

where $(\chi N_t)_{c \text{ or ODT}}$ is the critical value of χN_t for the stability of the disordered state (N_t is the total degree of polymerization). This again means that a melt of triblock or graft copolymers will remain in the homogeneous (disordered) phase for temperatures where a melt of linear diblocks is already microphase separated. In the following text we will review some of the pertinent theoretical and experimental works on this subject, which demonstrate the validity of eq. 14.1 for the different block copolymer architectures.

1. GRAFT COPOLYMERS

An example of a nonlinear block copolymer undergoing an order-to-disorder transition is shown in Figure 14.1 (Floudas 1994). The copolymer is a 3-miktoarm star (graft) copolymer of the SI_2 type, where S is polystyrene and I is isoprene. The patterns shown in the figure are two-dimensional SAXS images with X-rays parallel (tangential view) and perpendicular (normal view) to the shear direction from a

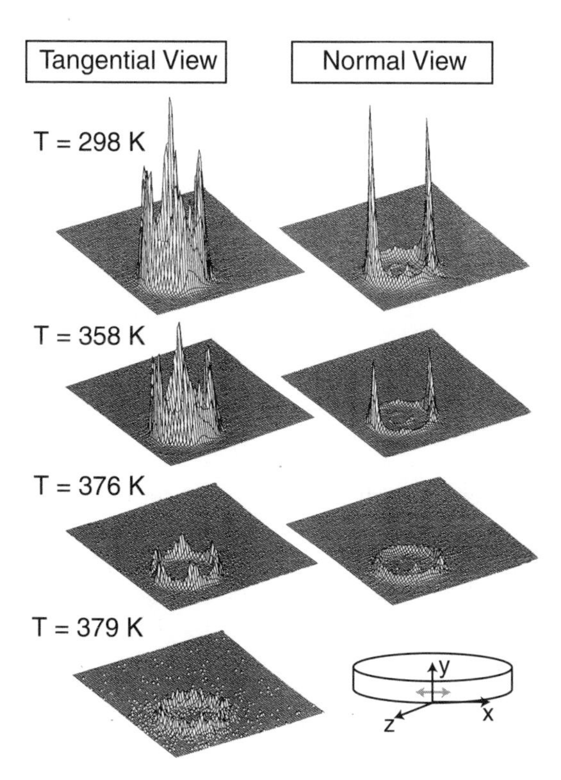

Figure 14.1 Representative SAXS patterns from an oriented SI_2 (S: polystyrene, I: polyisoprene) miktoarm (graft) copolymer [$M_n = 23700$; $S = 32\%$ (w/w)]. The tangential and normal views are shown. The data indicate the formation of a hexagonal structure with cylinders aligned along the direction of shear. The order-to-disorder transition is at 379 K (Floudas 1994).

shear-oriented sample at different temperatures. The patterns display a hexagonal symmetry with cylinders aligned along the direction of shear. A pertinent feature is that the cylindrical morphology persists up to $T_{ODT}-T \sim 3K$. At the ODT ($T_{ODT} = 379$ K), the scattering pattern turns to isotropic. When heating an asymmetric but linear diblock copolymer, MFT would predict the following sequence of phases: hexagonal-bcc-disordered. The absence of the intermediate bcc structure from the scattering patterns of weakly segregated asymmetric diblocks has been discussed in terms of fluctuations effects (see Chapter 15). Thus, the microphase separation in the miktoarm star copolymer shown in Figure 14.1 is reminiscent of asymmetric, but linear, diblock copolymers.

Despite these similarities, graft copolymers have some distinct features; first they undergo an order-to-disorder transition at lower temperatures as compared with the linear diblock copolymers, and, second, their structure factor in the disordered phase depends on *three* variables: χN, f and τ, where the latter is the fractional position along the A chain backbone at which the B graft is chemically linked. The structure factor now has the form (de la Cruz 1986):

$$\frac{N}{S(q)} = \frac{D_f(x) + D_{1-f}(x) + 2A_{\tau f}(x)A_{1-f}(x)}{D_f(x)D_{1-f}(x) - [A_{\tau f}(x)A_{1-f}(x)]^2} - 2\chi N \tag{14.2}$$

where $D_f(x) = f^2 D(f, x)$, $D(f, x)$ is the Debye function

$$D(f, x) = \frac{2}{(fx)^2}[fx + e^{-fx} - 1] \tag{14.3}$$

and

$$A_{\tau f}(x) = \frac{2 - e^{-f\tau x} - e^{-f(1-\tau)x}}{x} \tag{14.4}$$

$$A_{1-f}(x) = \frac{1 - e^{-(1-f)x}}{x} \tag{14.5}$$

When $\tau = 0$ or 1, the graft copolymer degenerates to a linear diblock, and S(q) becomes Leibler's structure factor for linear diblocks. An example of the use of this structure factor is given in Figure 14.2, where the SAXS profiles of the same SI_2 copolymer, now in the disordered phase, are fitted according to the MFT.

The structure factor given by eq. 14.2 can be approximated for the limiting values of $x = qR_g$ as:

$$S(x \ll 1) \simeq \frac{2}{3}Nf^2(1-f)^2[1 - 3f\tau(1-\tau)]x + \cdots \tag{14.6}$$

and

$$S(x \gg 1) \simeq \frac{2Nf(1-f)}{x} + \cdots \tag{14.7}$$

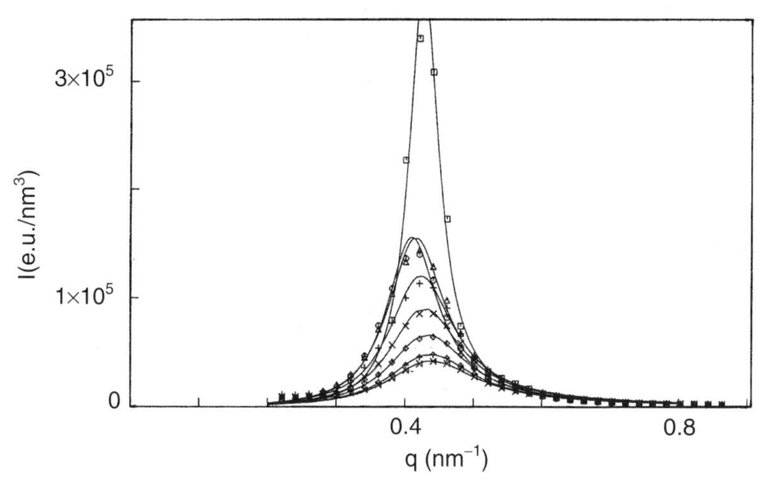

Figure 14.2 SAXS profiles of the graft copolymer SI_2 at different temperatures in the disordered phase fitted to the MFT predictions. The most intense spectrum is taken at 378 K, i.e., 1 K below the ODT. Notice the failure of the calculated profiles to describe the profiles as the temperature is lowered near the transition (Floudas 1994).

Hence, the S(q) increases as q^2 for low q, whereas, for large q, drops like q^{-2}. At intermediate q it goes through a maximum at $q = q^*$, which can be estimated by equating the above expressions:

$$q^* R_g \simeq \left\{ \frac{3}{f(1-f)[1 - 3f\tau(1-\tau)]} \right\}^{1/2} \tag{14.8}$$

The exact value of q* can be obtained from $\partial S(q)/\partial q = 0$ and is only 5% higher than the one given in eq. 14.8. One of the main predictions of the MFT is that q*(graft) > q*(diblock), which means that the spacing of the mesophase will be smaller in a graft copolymer than in the corresponding diblock copolymer. The increase in q* results from the reduction in R_g, which, for a graft copolymer, is reduced by a factor

$$g = 1 - 6f^2\tau(1-\tau)(1-f) \leq 1 \tag{14.9}$$

as $R_g^2 = gR_0^2(R_0^2 = N_0 l^2/6)$. Another important MFT prediction is that $(\chi N_t)_s$ $(graft) > (\chi N_t)_S(diblock)$ for all $0 < \tau < 1$ (N_t is the total degree of polymerization of the block copolymer). This is depicted in Figure 14.3 in the following text.

For any given τ, $(\chi N)_S$ reaches a minimum at f = 1/2, and, for τ = f = 1/2, the $(\chi N)_S$ value is 13.5 as compared with 10.5 for the diblock copolymer case. This implies that it is more difficult to phase separate any graft copolymer as compared with a linear diblock copolymer. This effect is largely of entropic origin; although the entropies of the graft and diblock melts are comparable, the entropy of a phase separated graft copolymer is lower than of the corresponding diblock copolymer. Thus, the entropy change, ΔS, associated with the disorder-to-order transition, is

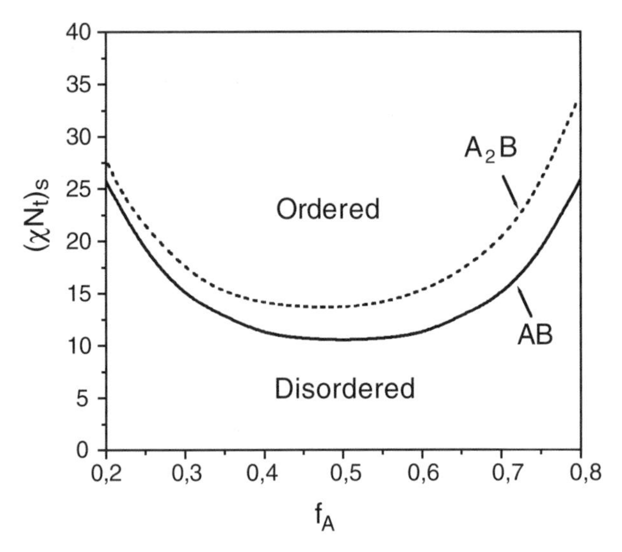

Figure 14.3 Comparison of the variation of $(\chi N_t)_S$ with composition of the A block in graft and diblock copolymers. Notice the higher values and the asymmetry in the A_2B case (de la Cruz 1986).

greater in a graft copolymer. Assuming that the heats of the transition are comparable, we are led to the conclusion that the transition temperature is reduced, which implies that $(\chi N_t)_S$ is larger in the graft copolymer in qualitative agreement with the result shown in Figure 14.3.

The MFT was subsequently generalized (Floudas 1997) to treat the most general case of miktoarm stars, i.e., A_mB_n with $m \neq n$. The calculated spinodal curves for the case $m = 1$, i.e., in AB_n miktoarm stars with an increasing number of B blocks, results in the dependence shown in Figure 14.4. Notice that all curves are asymmetric due to the inherent asymmetry of the miktoarm stars, and they become more asymmetric with increasing n. There are some interesting trends found within the class of the miktoarm copolymers with increasing n. For n up to 3, there is a considerable increase in the critical value of (χN_t), suggesting an increasing compatibility; however, for $n > 3$, there is a reversal in the $(\chi N_t)_c$, which starts to decrease towards the diblock case. This suggests that the AB_3 has the highest intrinsic compatibility among all AB_n miktoarm stars.

Chain packing and elasticity considerations for miktoarm stars have also been proposed to explain their phase state (Milner 1994). Consider, for example, an AB_2 star; at $f = 1/2$ the two B arms are half as long as the single A arm. If such a copolymer formed a lamellar phase at strong segregation, the B arms would have to stretch a lot in order to fulfill the requirement of a constant mass density, which forces the thickness of the A and B layers to be equal. To compensate for the stretching of the crowded B chains, there is a tendency for the interface to curve away from the B blocks. This convexity provides the required volume to the B chains near the A-B interface, which now do not have to stretch so much. Milner

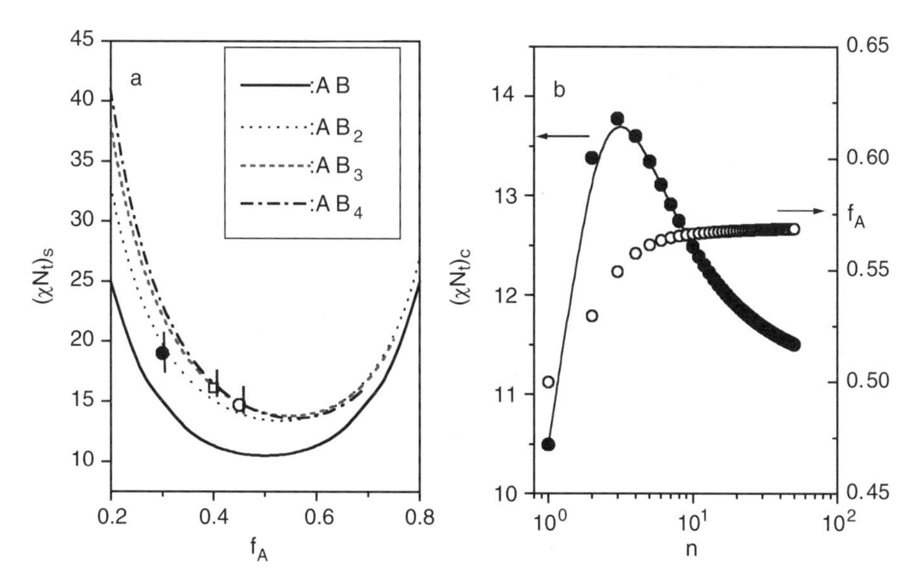

Figure 14.4 (a) Comparison of the spinodal lines $(\chi N_t)_S$ for a diblock and different AB_n miktoarm copolymers with n = 2, 3, and 4. The theoretical results are compared with experimental values. (b) Critical values of (χN_t) plotted as a function of the number of arms of the B-blocks. The dependence of the optimal composition corresponding to the minima of the spinodal curves is also shown (Floudas 1997).

calculated the volume fraction for which the layer has no tendency to bend by treating the miktoarm star layers as brushes having twice as many chains per unit area in the B brush as in the A brush. The "phase diagram" thus obtained is shown in Figure 14.5.

The free energies of four phases were considered: lamellar, hexagonal, bicontinuous double diamond, and spherical, and the crossings of these free energy curves determined the phase diagram as a function of volume fraction φ and the asymmetry parameter ε. The latter is defined as $\varepsilon = (n_A/n_B)(l_A/l_B)^{1/2}$ (where n_A and n_B are the number of A and B arms, respectively, and $l_i = V_i/R_i^2$, where V_i and R_i are the molecular volume and radius of gyration of the respective blocks) and is the single parameter used to describe the different architectures and the elastic asymmetry in the strong-segregation regime. Notice that, if the constituent blocks are conformationally symmetric, then ε reduces to the ratio of the number of A and B arms in the star.

The phase diagram reveals a strong dependence of the phase boundaries on the number of arms; for example, at a constant volume fraction of $\varphi = 0.4$, and assuming elastically symmetric blocks, an AB diblock ($n_A = n_B = 1$) would form lamellae, an A_2B miktoarm star would form cylinders, and an A_3B miktoarm star polymer would form spheres. Experiments (Pochan 1996, Hadjichristidis 1993) have shown a qualitative agreement with the predicted phases; however, systems at intermediate and weak segregation are not expected to be described by the theory.

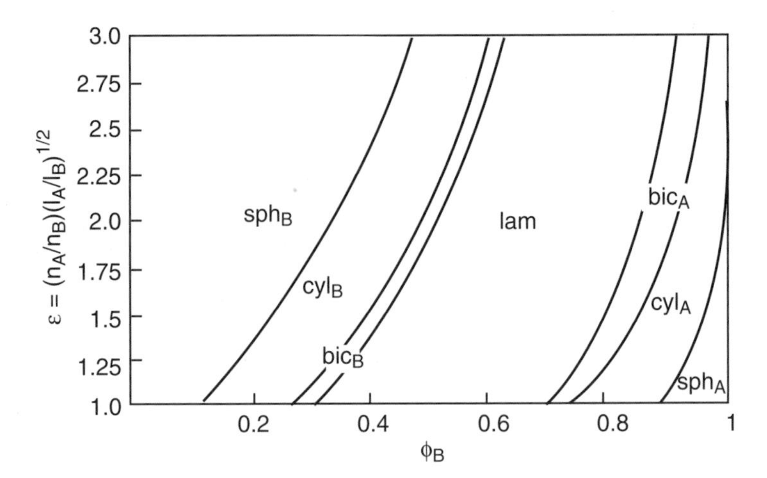

Figure 14.5 Phase diagram in the strong segregation limit for miktoarm star copolymers with n_A number of A arms and n_B number of B arms plotted as a function of the volume fraction of B (Milner 1994).

2. A_nB_n STAR BLOCK COPOLYMERS

The MFT was also formulated for this case (de la Cruz 1986), and the calculated structure factor in the disordered state is given in Table 14.1. The interesting prediction within this class of polymers was that the $(\chi N_0)_c$ should be independent of n and equal to 10.5 (N_0 is the number of monomers in the A_1B_1 diblock). The variation of the $(\chi N_0)_S$ with composition is given in Figure 14.6 in the following text. Opposing entropic effects are thought to be responsible for this dependence; the entropy of the star melt is smaller than that of the corresponding diblock melt because of the additional constraints on the A-B junction point. However, the A_nB_n mesophase also has a lower entropy than the corresponding diblock mesophase. At the critical point, the two effects should result in a constant $(\chi N_0)_c$, but, for noncritical compositions, there will be a lowering of the transition entropy caused by the junction constraint. Assuming that $(\chi N_0)_S$ is directly proportional to the entropy of the transition results in the prediction that: $(\chi N_0)_S(A_nB_n) < (\chi N_0)_S$ $(A_{n-1}B_{n-1})$. Similarly, the MFT predicts that q* should be identical to the diblock case for a symmetric composition.

These theoretical predictions have been examined experimentally (Buzza 1999 and 2000), and serious discrepancies were found; a pair of symmetric AB and A_2B_2 block copolymers with matched arm molecular weights was found to have significantly different transition temperatures and different q* values (with q*(AB) > q*(A_2B_2). The difference was too large to be accounted for by polydispersity, segmental asymmetry, or fluctuation effects (see Chapter 15). However, by incorporating the different degree of stretching in the two systems at the

TABLE 14.1. Structure Factors of Different Block Copolymer Architectures

(I) (AB)

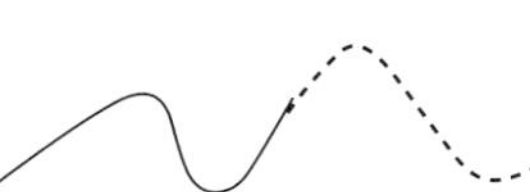

$$S(q)^{-1} = \frac{F(x,f)}{N} - 2\chi$$

$$F(x,f) = \frac{D_1}{D_f D_{1-f} - \frac{1}{4}[D_1 - D_f - D_{1-f}]}$$

$$D_f(x) = D(f,x) = \frac{2}{x^2}[fx - 1 + e^{-fx}], \quad x = \frac{q^2 N l^2}{6}$$

(II) (A_2B)

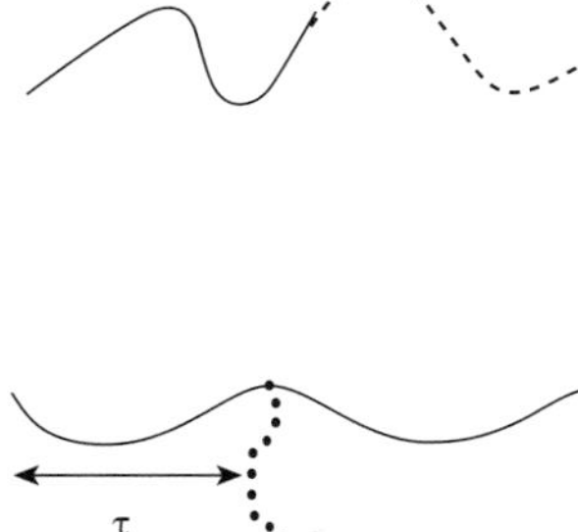
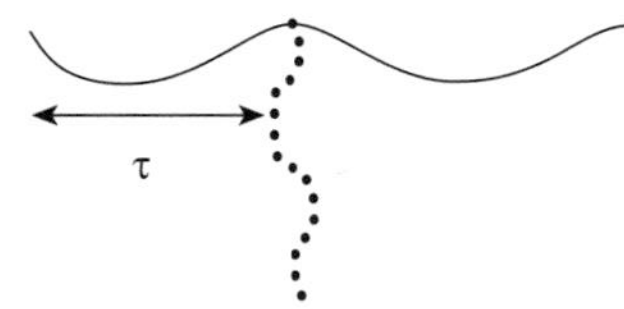

$$S(q)^{-1} = \frac{F(x,f)}{N} - 2\chi$$

$$F(x,f) = \frac{D_f + D_{1-f} + 2A_{\tau f}A_{1-f}}{D_f D_{1-f} - (A_{\tau f}A_{1-f})^2}$$

$$D_f(x) = f^2 D(f,x)$$

$$A_{\tau f}(x) = \frac{2 - e^{-f\tau x} - e^{-f(1-\tau)x}}{x}$$

$$A_{1-f}(x) = \frac{1 - e^{-(1-f)x}}{x}$$

(III) A_nB_n

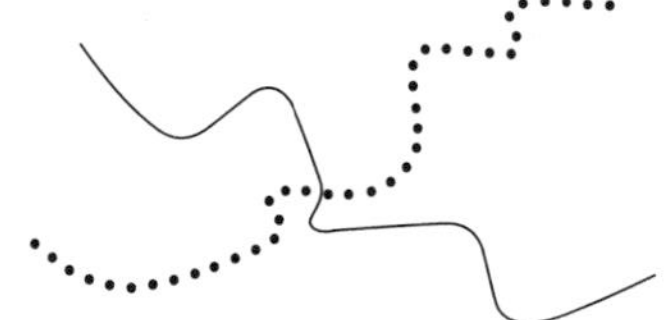
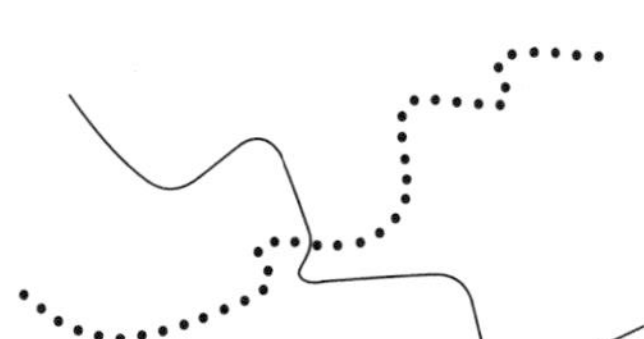

$$S(q)^{-1} = \frac{F(x,f)}{N_0} - 2\chi$$

$$F(x,f) = \frac{D_f + D_{1-f} + 2nA_fA_{1-f} + (n-1)(A_f^2 + A_{1-f}^2)}{D_f D_{1-f} + (n-1)(D_f A_{1-f}^2 + D_{1-f}A_f^2) - (2n-1)(A_fA_{1-f})^2}$$

$$D_f(x_0) = f^2 D(f, x_0)$$

$$A_f(x_0) = \frac{1 - e^{-fx_0}}{x_0}$$

$$D(f, x_0) = \frac{2}{x_0^2}[fx_0 - 1 + e^{-fx_0}], \quad x_0 = \frac{q^2 N_0 l^2}{6}$$

TABLE 14.1. (*Continued*)

(IV) $(AB)_n$

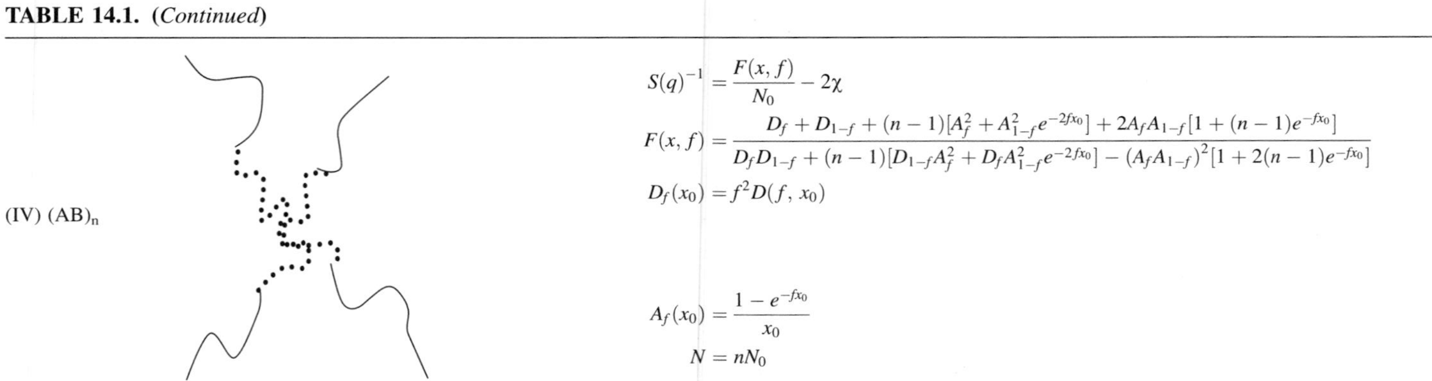

$$S(q)^{-1} = \frac{F(x, f)}{N_0} - 2\chi$$

$$F(x, f) = \frac{D_f + D_{1-f} + (n-1)[A_f^2 + A_{1-f}^2 e^{-2fx_0}] + 2A_f A_{1-f}[1 + (n-1)e^{-fx_0}]}{D_f D_{1-f} + (n-1)[D_{1-f}A_f^2 + D_f A_{1-f}^2 e^{-2fx_0}] - (A_f A_{1-f})^2[1 + 2(n-1)e^{-fx_0}]}$$

$$D_f(x_0) = f^2 D(f, x_0)$$

$$A_f(x_0) = \frac{1 - e^{-fx_0}}{x_0}$$

$$N = nN_0$$

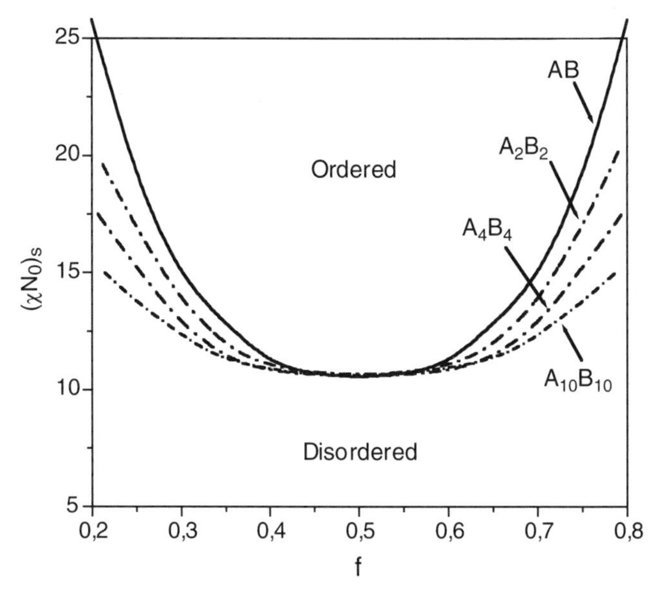

Figure 14.6 Variation of $(\chi N_0)_S$ with composition and arm number for A_nB_n star copolymers (de la Cruz 1986).

transition into the fluctuation theory, one could account for this effect (Buzza 2000).

3. $(AB)_n$ STAR COPOLYMERS

Star block copolymers of the $(AB)_n$ type have attracted many experimental investigations (Thomas 1986, Alward 1986, Hashimoto 1988, Ijichi 1989, Archer 1994, Hajduk 1995, Floudas 1996) and were the first class of copolymers where the ordered bicontinuous double diamond (OBDD) structure was suggested (Thomas 1986). Later (Hajduk 1995), a careful reevaluation of the scattering from this cubic phase revealed that the two interwoven minority component networks were threefold coordinated rather than fourfold coordinated, suggesting that the structure actually has the $Ia\bar{3}d$ space group symmetry (double gyroid).

According to the MFT, a lowering of the transition entropy is expected in the case of star copolymers because of the junction constraints. The net transition entropy change per $(AB)_n$ star molecule is proportional to $\Delta S_0 - [(n-1)/n]\ln(1/f)$, where ΔS_0 refers to the transition entropy change in the case of a diblock. Because (χN_0) can be, again, assumed to be proportional to the transition entropy, the expectation is that the spinodal will be lower than for a diblock copolymer. Thus, even at $f = 1/2$, the $(\chi N_0)_S$ for $(AB)_n$ stars will be lower than for an AB diblock. For example, for $f = 0.5$, (χN_0) equals 10.5, 8.86, and 7.07 for $n = 1$, 2, and 4, respectively. Moreover, the spinodal curves are expected to be more asymmetric than in the corresponding diblocks, and q^* is expected to pass through a minimum as the block composition changes. The $(\chi N_0)_S$ variation with the core

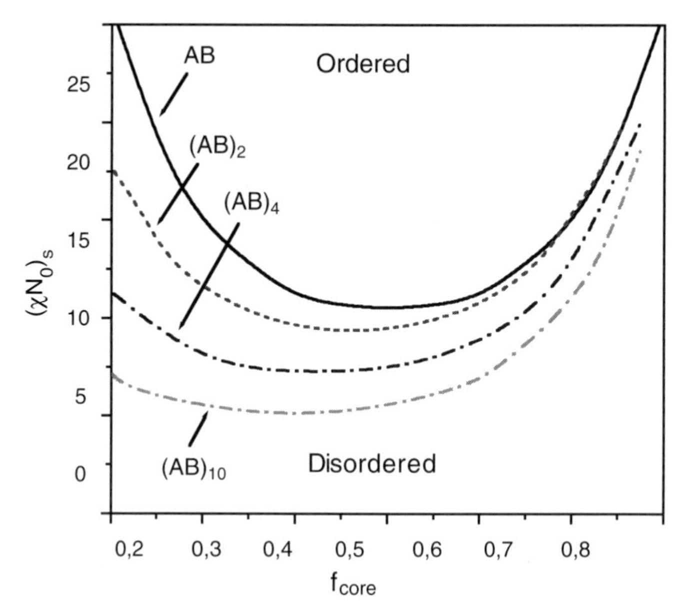

Figure 14.7 Variation of $(\chi N_0)_S$ with composition and arm number for $(AB)_n$ star copolymers. The volume fraction corresponds to the block forming the core of the star (de la Cruz 1986).

composition and arm number for $(AB)_n$ star copolymers is shown in Figure 14.7. These theoretical predictions motivated experimental investigations by scattering and other techniques. In particular, in the scattering studies (Hashimoto 1988, Ijichi 1989, Floudas 1996), the profiles were analyzed using the MFT structure factor (Table 14.1.), and a quantitative agreement was found. A surprising result was the strong n dependence of the interaction parameter.

Experiments using scattering and rheology confirmed the basic MFT predictions. An example is given in Figure 14.8 where the viscoelastic properties of a linear poly(styrene-b-isoperene) copolymer are compared with a four-arm star copolymer with the same arm molecular weight and comparable composition (Floudas 1996). As seen in the figure, the order-to-disorder transition temperature in the star block copolymer has increased by about 50 K as compared with the corresponding diblock, in qualitative agreement with the MFT prediction.

The complete phase diagram of star block copolymers was also examined first by using a Landau-Ginzburg expansion with a truncated set of wave vectors (Dobrynin 1988, Floudas 1996). More recently, the phase diagram of star block copolymers was calculated within the self-consistent field theory (SCFT) (Matsen 1994) for copolymers with one, three, five, and nine arms. The topology of phases was found by comparing the free energies of the different phases. The obtained phase diagram was similar with the corresponding in diblock copolymers, in that the lamellar, gyroid, hexagonal, and cubic phases were found to be stable. The calculated phase diagram for a five-arm star block copolymer melt is depicted in

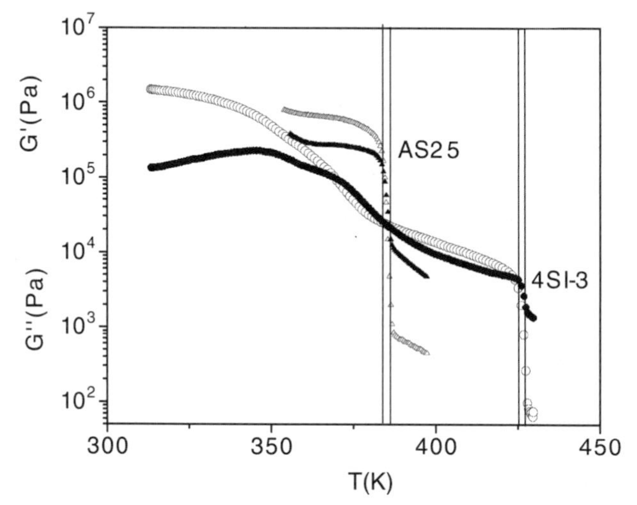

Figure 14.8 Isochronal measurements of the storage (open symbols) and loss (filled symbols) moduli of an asymmetric linear poly(styrene-b-isoprene) diblock copolymer denoted as AS25 ($M_n = 23800\,g/mol,\ f_{PS} = 0.22$) with a four-arm star block copolymer denoted as 4SI-3 ($M_n^{arm} = 24000\,g/mol, M_n^{star} = 98000\,g/mol, f_{PS} = 0.25$). Vertical lines indicate the respective T_{ODT} (Floudas 1996).

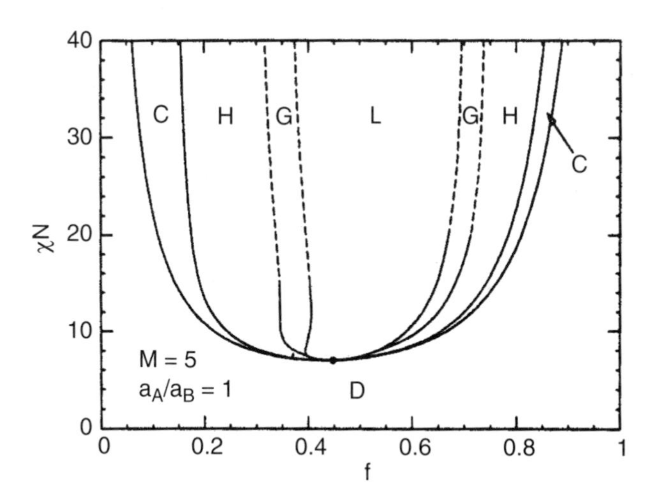

Figure 14.9 Phase diagram for a five-arm star block copolymer melt calculated assuming equal segment lengths within the SCFT plotted as a function of the volume fraction of the inner block forming the core. All transitions are first order except of the critical point, which is marked by a dot (Matsen 1994).

Figure 14.8. All transitions, except of the critical points, were found to be first order. One important observation was that all phase diagrams were asymmetric showing a preference for the inner block of the arms to locate inside of the cylindrical and spherical structures. This preference was found to increase with the number of arms. This can clearly be seen in Figure 14.9; whereas for $f < 1.2$ the region of stability of the hexagonal and bcc phases has increased, the opposite happens for $f > 1/2$. Finally, regions in which the double-diamond and catenoid-lamellar phases are nearly stable were also located.

4. ABA TRIBLOCK COPOLYMERS

The MFT approach proposed by Leibler was also used in the calculation of the triblock copolymer phase diagram (Mayes 1989). A first-order transition to bcc spheres was found for all compositions except for $f = f_c$. The calculated phase diagram is shown in Figure 14.10 for two ABA triblocks with $\tau = 0.25$ and $\tau = 0.5$, where τ is a symmetry parameter ($\tau = f_1/f$, and the cases of $\tau = 0$ or 1 reduce to a pure diblock case). The triblock copolymer phase diagrams below are highly asymmetric. The reason for this asymmetry is the high deformation of the central B blocks in order to accommodate the outer A blocks into A domains. Increasing τ gives rise to important differences in the window of stability for each morphology at a given composition. For example, at $f = 0.5$, a diblock copolymer goes from the disordered phase directly to the lamellar, whereas, for a triblock copolymer, with

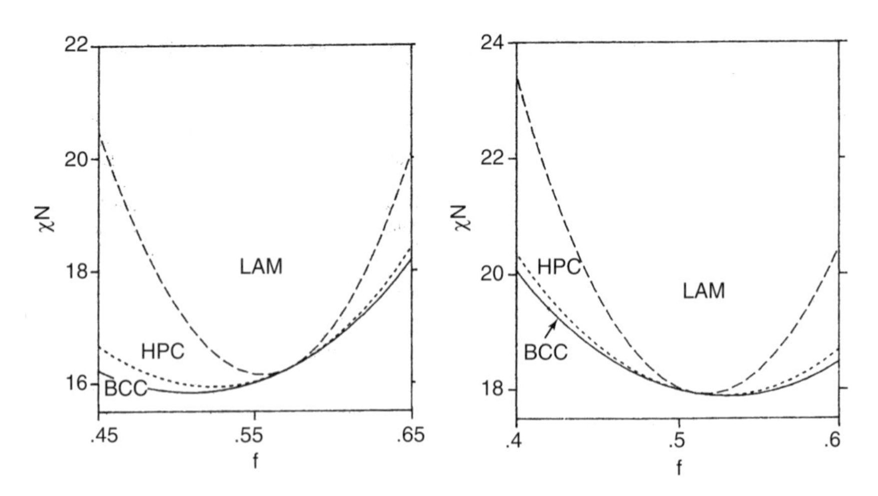

Figure 14.10 Phase diagrams for ABA triblock copolymer melts with $\tau = 0.25$ (left) and $\tau = 0.5$ (right). Solid lines give the disorder-to-order transition as (χN_t) (f). Dotted lines give the transitions between bcc and hex, and dashed lines the transition from hex to lam (Mayes 1989).

$\tau = 0.25$ at the same composition, there are large zones of bcc and hex phases. These regions become narrower for the more symmetric triblock copolymers. These differences suggest that transitions between different morphologies with decreasing temperature are more likely to be seen experimentally in asymmetric triblock copolymers with $\tau = 0.25$ and $f < 0.5$.

5. TAPERED BLOCK COPOLYMERS

Typical thermoplastic elastomers (such as Kraton, an SBS triblock copolymer) are composed of a rubber matrix containing inclusions of the hard phase (as spheres or cylinders) and exhibit low modulus, high elongation, and excellent recovery. There has been a challenge to develop more stiff resins with a lower rubber content while preserving the ability of plastic deformation and impact strength. Reduction of the rubber content to about 25% in triblock copolymers leads to higher moduli, but the material becomes brittle instead. A step forward in this direction was made with the synthesis of a linear SBS copolymer with a B/S tapered interface. Subsequently, experimental studies (Hashimoto 1983, Hodrokoukes 2001) and computer simulations (Pakula 1996) have shown a strong influence of the presence of the tapered interface on the phase state, the mechanical properties, and, therefore, the processability of block copolymers. A study of the influence of the interface on the structure and viscoelastic properties of tapered block copolymers can best be made by: (i) systematically varying the amount of interfacial material and (ii) varying the sequence of appearance of monomers. Figure 14.11 gives, in a schematic way, the structure of a four triblock copolymer with a tapered middle block, three of which (denoted as T1, T2, and T3) have *inverse* tapered middle blocks, while the fourth (denoted as T4) has a regular tapered midblock (Hodrokoukes 2001). Such block copolymers based on styrene and isoprene were actually synthesized via anionic polymerization with a nearly symmetric composition. With such systems one can study both the effect of increasing the amount of the interfacial material (in going from T1 to T2 to T3) and the effect of block sequencing (by comparing T2 and T4 with the same overall molecular weight). It is noteworthy that increasing the amount of the tapered middle block in the inverse tapered copolymers systematically increases the compatibility. This is depicted in the TEM images of Figure 14.11, which reveal a disordered phase in the inverse taper copolymer with the broader interface (T3) and a lamellar structure in the T1, T2, and T4.

Block sequencing in the interface with respect to the outer blocks is found to be an important factor controlling compatibility. Inverse tapered block copolymers are much more compatible than the corresponding normal tapered block copolymers. The order-to-disorder transition was investigated, and all copolymers were found to exhibit a first-order transition at the T_{ODT}, as evidenced by the existence of latent heat in calorimetry and from the discontinuous changes of the structure factor and viscoelastic properties. An interesting finding was that an inverse tapered copolymer with about 20% of the segments within the tapered interface possesses the same

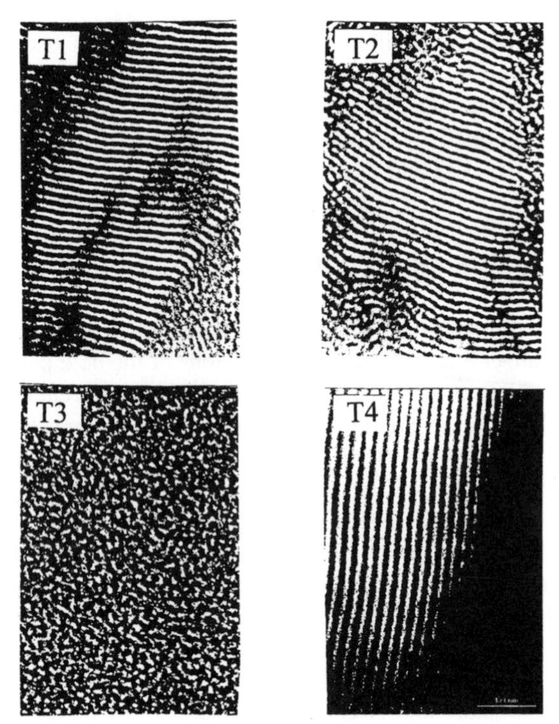

Figure 14.11 Schematic structure of the inverse taper (T1, T2, and T3) and the normal taper (T4) block copolymers. TEM images on sections stained by OsO_4 of the microdomain structure are shown. A lamellar structure is formed in T1 and T2 and in the T4, whereas the broader interface in T3 gives rise to a disordered structure (Hodrokoukes 2001).

transition temperature with a normal diblock copolymer of the same overall molecular weight, suggesting substantial interfacial mixing in block copolymers.

These results on the phase state are summarized in Figure 14.12, where the spinodals of the four copolymers are shown. In conclusion, the increase of the amount of material within the tapered interface enhances the compatibility in a systematic way. On the other hand, inverse tapered block copolymers are much more compatible than the corresponding normal tapered copolymers.

These results provide new means of designing new block copolymers with a highly controlled compatibility at the synthesis level, which can be triggered by two independent parameters: the block sequencing and the amount of tapered midblock.

6. MULTIBLOCK COPOLYMERS

A theory of microdomain structures in random copolymers and in highly polydisperse multiblock copolymers has been developed (Fredrickson 1991,

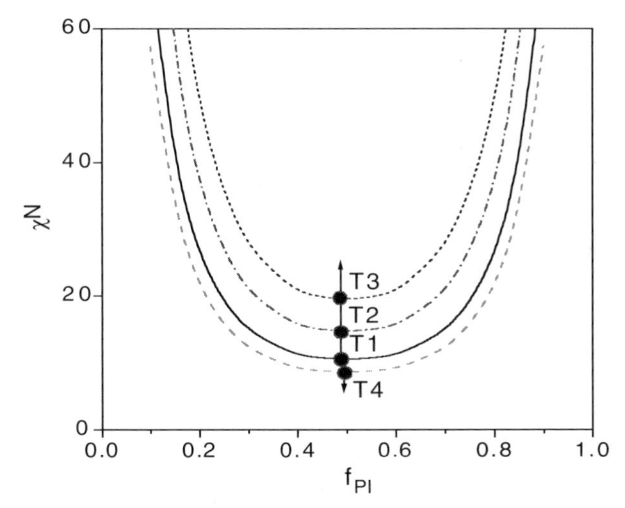

Figure 14.12 Schematic phase diagram showing the location (spinodals) of the inverse and normal tapered copolymers. Increasing the molecular weight of the midblock in the inverse tapered copolymers (T1, T2, and T3) increases the compatibility, whereas the opposite is happening in the normal tapered copolymer T4 (Hodrokoukes 2001).

Dobrynin 1997, Sfatos 1995). In the case of random copolymers, it has been suggested that the lamellar and liquid-like disordered phases are separated by a two-phase region (Dobrynin 1997). The existence of such two-phase regions can be understood by an increasing length of critical fluctuations at lower temperatures, which becomes comparable with the size of the largest blocks that start to form domains. However, this takes place in the presence of the smaller blocks, which stay in the liquid-like phase and create the biphasic region.

Recently (Semenov 1998), it was shown that the effect of polydispersity in multiblock copolymers gives rise to the formation of unusual secondary periodic structures that form on top of the primary domain structures characteristic of diblock copolymers in some windows located in the vicinity of morphological transitions between different primary structures. These windows essentially replace biphasic regions between different domain structures predicted for polydisperse diblock copolymers. The driving force for the formation of such secondary structures comes from the competition between the fractionation effect of a first-order morphological transition and the chemical connectivity of chain fragments with different composition and/or different primary structure. Thus, secondary domain structures have been predicted for weakly polydisperse multiblock copolymers and for completely random block copolymers.

Some of the secondary morphologies resulting from the competition between cylindrical and spherical primary structures are shown in Figure 14.13. Experimental efforts in the near future are likely to focus on this problem.

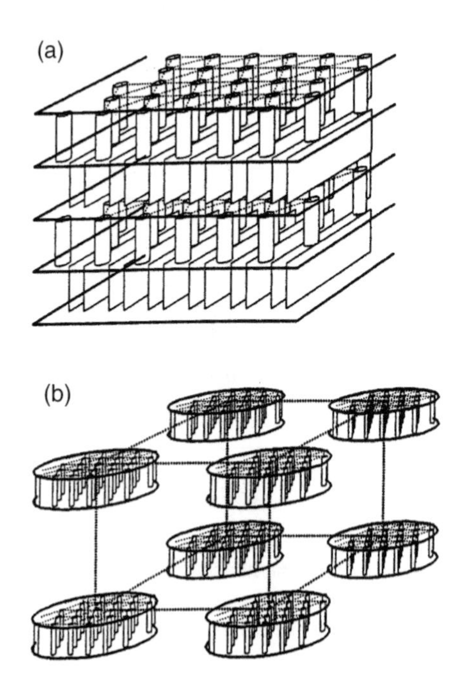

Figure 14.13 Secondary structures observed near the hexagonal/lamellar line: (a) lamellar sheets, (b) asymmetric disks arranged in an fcc superlattice (Semenov 1998).

In another case, multiple homogeneous phases, and periodic mesophases as well as an isotropic Lifshitz point, have been identified in random copolymers (Fredrickson 1991).

REFERENCES

Alward D.B., Kinning D.J., Thomas E.L., Fetters L.J. (1986) Macromolecules 19, 215.

Archer L.A., Fuller G.G. (1994) Macromolecules 27, 4804.

Buzza D.M.A., Hamley I.W., Fzea A.H., Moniruzzaman M., Allgaier J.B., Young R.N., Olmsted P.D., McLeish T.C.B. (1999) Macromolecules 32, 7483.

Buzza D.M.A., Fzea A.H., Allgaier J.B., Young R.N., Hawkins R.J., Hamley I.W., McLeish T.C.B., Lodge T.P. (2000) Macromolecules 33, 8399.

de la Cruz O.M., Sanchez I.C. (1986) Macromolecules 19, 2501.

Dobrynin A.V., Erukhimovich I.Ya. (1993) Macromolecules 26, 276.

Dobrynin A.V. (1997) J. Chem. Phys. 107, 9234.

Floudas G., Hadjichristidis N., Iatrou H., Pakula T., Fischer E.W. (1994) Macromolecules 27, 7735.

Floudas G., Pispas S., Hadjichristidis N., Pakula T., Erukhimovich (1996) Macromolecules 29, 4142.

Floudas G., Hadjichristidis N., Tselikas Y., Erukhimovich I. (1997) Macromolecules 30, 3090.

Floudas G., Hadjichristidis N., Iatrou H., Avgeropoulos A., Pakula T. (1998) Macromolecules 31, 6943.

Fredrickson G.H., Milner S.T. (1991) Phys. Rev. Lett. 67, 835.

Hadjichristidis N., Iatrou H., Behal S.K., Chludzinski J.J., Disko M.M., Garner R.T., Liang K.S., Lohse D.J., Milner S.T. (1993) Macromolecules 26, 5812.

Hajduk D.A., Harper P.E., Gruner S.M., Honeker C.C., Thomas E.L., Fetters L.J. (1995) Macromolecules 28, 2570.

Hashimoto T., Tsukahara Y., Tachi K., Kawai H. (1983) Macromolecules 16, 648.

Hashimoto T., Ijichi Y., Fetters L.J. (1988) J. Chem. Phys. 89, 2463.

Hodrokoukes P., Floudas G., Pispas S., Hadjichristidis N. (2001) Macromolecules 34, 650.

Ijichi Y., Hashimoto T. (1989) Macromolecules 22, 2817.

Matsen M.W., Schick M. (1994) Macromolecules 27, 6761.

Mayes A.M., de la Cruz M.O. (1989) J. Chem. Phys. 91, 7228.

Milner S.T. (1994) Macromolecules 27, 2333.

Pakula T., Matyjaszewski K. (1996) Macromol. Theory Simul. 5, 987.

Pochan D.J., Gido S.P., Pispas S., Mayes J.W. (1996) Macromolecules 29, 5099.

Read D.J. (1998) Macromolecules 31, 899.

Read D.J. (1999) Eur. Phys. J. B 12, 431.

Sfatos C.D., Gutin A.M., Skakhnovich E.I. (1995) Phys. Rev. E 51, 4727.

Semenov A.N., Likhtman A.E. (1998) Macromolecules 31, 9058.

Thomas E.L., Alward D.B., Kinning D.J., Martin D.C., Handlin D.L., Fetters L.J. (1986) Macromolecules 19, 2197.

Tselikas Y., Iatrou H., Hadjichrisridis N., Liang K.S., Mohanty K., Lohse D.J. (1996) J. Chem. Phys. 105, 2456.

CHAPTER 15

BLOCK COPOLYMER PHASE STATE

According to Leibler's MFT, the two variables discussed earlier, f and χN, completely determine the equilibrium thermodynamic state of a diblock copolymer melt. For a symmetric diblock copolymer, the theory predicts a second-order phase transition from the disordered to the lamellar phase by lowering temperature at the critical point ($\chi N = 10.495$, $f = 1/2$). For asymmetric diblock copolymers, the theory predicts a first-order phase transition to a bcc microphase. In addition, Leibler's mean-field theory provided an expression for the disordered phase structure factor S(q) (eq. 13.8), which was employed in the description of the scattering profiles from disordered copolymers.

Despite the success of the theory, there have been several cases where the predicted phase diagram was inadequate in explaining the rich experimental features and the suggested structure factor was insufficient to describe the actual experimental data near the ODT. For example, according to Leibler's structure factor, 1/S(q) should be proportional to 1/T in the disordered phase (given that, in the simplest approximation, χ is inversely proportional to T). Contrast this with the experimentally obtained $S(q^*)$ from two nearly symmetric poly(styrene-b-isoprene) diblock copolymers depicted in Figure 15.1.

At $T > T_{ODT}$ there exists a pronounced curvature, which cannot be accounted for by the theory. Furthermore, the peak intensity at the transition remains finite, and the $S(q^*)$ is discontinuous at the transition. Careful thermal measurements (Stühn 1992) were able to identify a heat of fusion at the T_{ODT} associated with the order-to-disorder transition, suggesting a first-order transition even for symmetric block copolymers. Lastly, experimental phase diagrams have shown the possibility of a direct disordered to lamellar transformation (i.e., without the intermediate bcc and hexagonal phases) even for nonsymmetric compositions. The above experimental

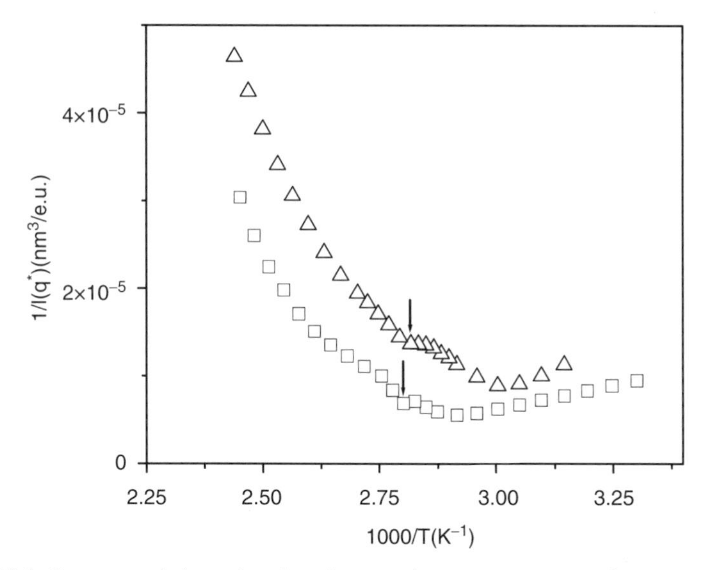

Figure 15.1. Inverse peak intensity plotted versus inverse temperature for two symmetric poly(styrene-b-isoprene) diblock copolymers. The arrows indicate the order-to-disorder transition temperatures. Notice the pronounced curvature of $I(q^*)^{-1}$ at temperatures above the transition that contradict the MFT predictions.

results motivated Fredrickson and Helfand (Fredrickson 1987) to introduce fluctuation corrections to Leibler's MFT for weakly segregated diblock copolymers.

1. FLUCTUATION EFFECTS

Fredrickson and Helfand (Fredrickson 1987) extended Brazovskii's Hartree approximation to demonstrate that systems in the new universality class exhibit a fluctuation-induced first-order transition in place of the continuous second-order transition predicted by the MFT. The theory assumes identical statistical segment lengths for the two components and incompressible system. The free energy density in the Hartree approximation has the form (Fredrickson 1989):

$$f_H(A) = \tau_R A^2 + \frac{u_R}{4} A^4 + \frac{w_R}{36} A^6 \tag{15.1}$$

where τ_R, u_R, and w_R are temperature-dependent renormalized parameters. For example, τ_R is of the form: $\tau_R = \tau + du\ \tau_R^{-1/2}$, where $d = 3x^*/2\pi$ and gives rise to the nonlinear dependence between τ_R and $1/T$ (τ is defined as previously $\tau = 2(\chi_S N - \chi N)/c^2$). The qualitative $f_H(A)$ dependence is plotted in Figure 15.2. For $\chi N < \chi_0 N$, where χ_0 is the stability limit (the spinodal) of the lamellar phase in the disordered phase, the lamellar phase is unstable, and f_H has a single minimum at $A = 0$. As χN approaches the value of $\chi_0 N$, f_H develops two additional minima that

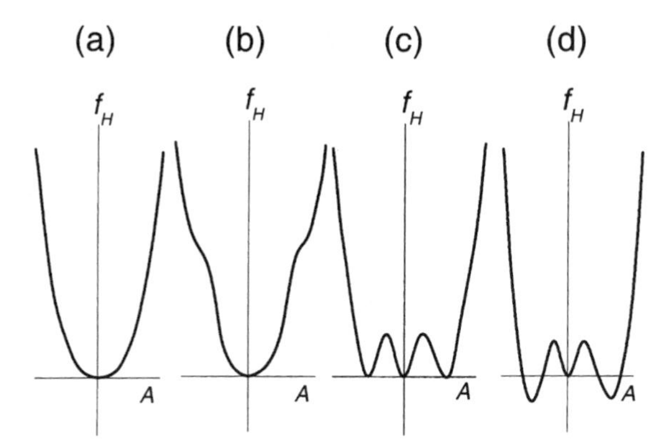

Figure 15.2. The free energy density in the Hartree approximation: (a) when $\chi N < \chi_0 N$, the free energy has a minimum at $A = 0$, indicating the stability of the disordered phase. (b) At $\chi_0 N$ (the stability limit of the lamellar phase in the disordered phase), two additional minima develop in f, which remain metastable. (c) At $\chi_t N$, the fluctuation-induced first-order transition takes place with three minima having $f = 0$. (d) Below the transition the two symmetric minima with $A \neq 0$ describing the lamellar phase are stable, and the minimum with $A = 0$ is metastable. The predicted metastability is of key importance for the observed ordering kinetics (Fredrickson 1989).

are unstable with respect to the disordered phase minimum. At $\chi N = \chi_t N$, a first-order transition takes place, and all three minima have $f_H = 0$. As χN is raised above $\chi_t N$, the free energy density of the two symmetric minima change signs, and the lamellar phase becomes thermodynamically stable, while the disordered phase with $f_H = 0$ becomes metastable. The predicted metastability has been confirmed through the experimental results on the ordering kinetics.

According to Fredrickson and Helfand (Fredrickson 1987), fluctuation corrections apply to both the disordered and ordered phases in the vicinity of the transition. In the disordered phase, the structure factor is modified as:

$$\frac{N}{S(q)} = \varepsilon + F(x, f) - F(x^*, f) \tag{15.2}$$

where

$$\varepsilon = F(x^*, f) - 2\chi N + \frac{c^3 d\lambda}{(\varepsilon N)^{1/2}} \tag{15.3}$$

$d = 3x^*/2\pi$ and c, λ are composition-dependent coefficients. Combining the above equations results in the expression:

$$\frac{N}{S(q)} = F(x, f) - 2\chi N + \frac{c^3 d\lambda}{(\varepsilon \mathrm{N})^{1/2}} \tag{15.4}$$

or equivalently:

$$\frac{N}{S(q)} = F(x, f) - 2\chi N + \frac{c^3 d\lambda}{\bar{N}^{1/2}} \frac{\sqrt{S(q^*)}}{\sqrt{N}} \tag{15.5}$$

where $\bar{N} = N\alpha^6/u^2$ and α, u are the statistical segment length and volume, respectively. Because the last term is independent of q, it is only the peak height that is affected. Thus, approaching the T_{ODT} from high temperatures, the predicted intensities are now lower than the ones expected from Leibler's theory. Furthermore, eq. 15.5 predicts a nonlinear dependence of 1/S(q) on 1/T, which is in qualitative agreement with the nonlinear dependence obtained experimentally (Fig. 15. 1).

For a symmetric composition (c = 1.10195, d = 1.8073, λ = 106.18), the location of the transition is predicted to be at:

$$(\chi N)_t = 10.495 + 41.022\bar{N}^{-1/3} \tag{15.6}$$

and the peak scattering intensity reaches a maximum at the first-order transition as $0.12328\ \bar{N}^{1/3}$. For the fluctuation correction to be really a correction to the leading term, high molecular weights are necessary. Actually, the Hartree approximation is only rigorously accurate for $\bar{N}$ of the order of 10^9. This is depicted in Figure 15.3, where the phase diagram for a diblock copolymer with $\bar{N}$ of the order of 10^4 is shown to be inaccurate, i.e., the Hartree approximation breaks down.

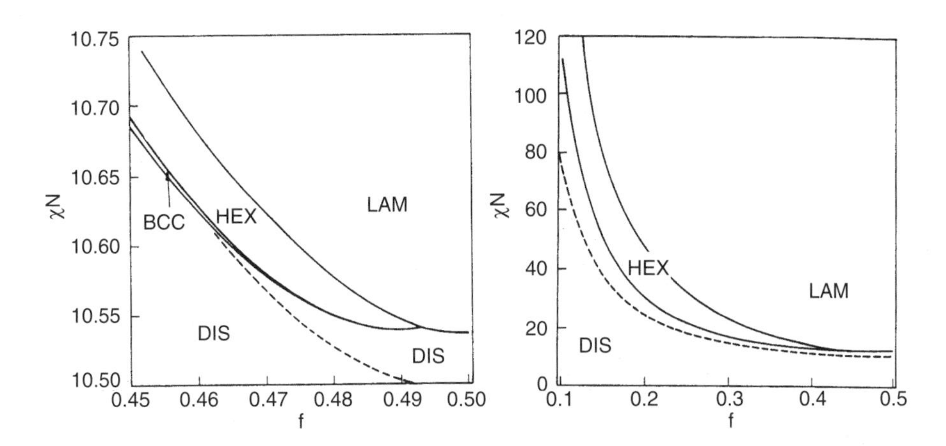

Figure 15.3. (Left): The phase diagram for a diblock copolymer with $\bar{N} = 10^9$ in the Hartree approximation. The dashed curve is the classical spinodal line calculated by Leibler. (Right): The breakdown of the Hartree approximation for a diblock copolymer with $\bar{N} = 10^4$. Notice that the low molecular weight in this case makes the phase diagram inaccurate, i.e., the bcc phase is missing from the phase diagram (Fredrickson 1987).

2. CONFORMATIONAL ASYMMETRY

Self-consistent field theory calculations (Matsen 1997) suggested that conformational asymmetry plays a dominant role on the order-disorder and order-to-order phase boundaries. Comparison the free energies of the classical and more complex structures found in block copolymers, produced the phase diagrams shown in Figure 15.4. These phase diagrams are calculated within the SCFT using $\alpha_A/\alpha_B = 1$ and 2, thus ranging from totally symmetric to very asymmetric statistical lengths. The main effect of conformational asymmetry is to shift the phase boundaries toward compositions richer in the segments with the higher asymmetry. This asymmetry has been attributed to differences in monomer volume and backbone flexibilities of the blocks, leading to an overall conformational asymmetry described by the ratio ζ as:

$$\zeta = \frac{\alpha_A^2/6u_A}{\alpha_B^2/6u_B} \tag{15.7}$$

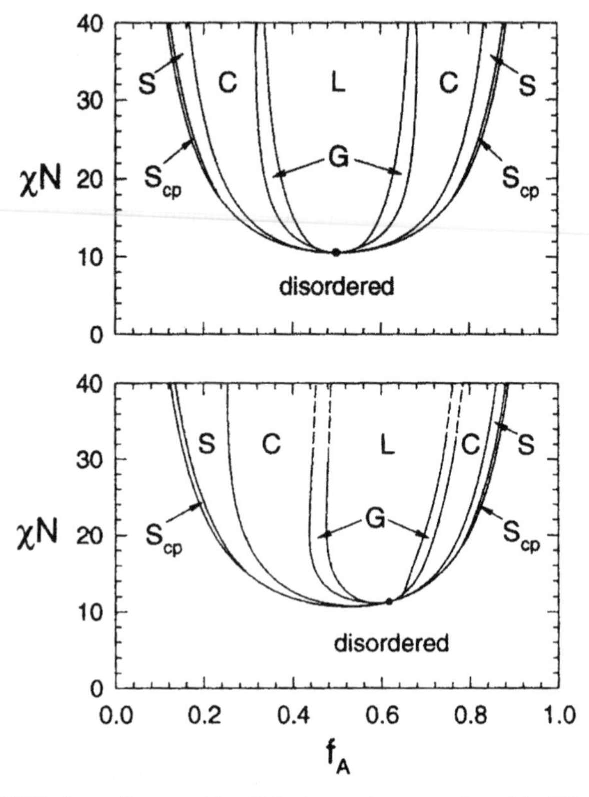

Figure 15.4. MFT phase diagrams for diblock copolymer melts with different conformational asymmetries. Top: $\alpha_A/\alpha_B = 1$, Bottom: $\alpha_A/\alpha_B = 2$. The ordered phases are L (lamellar), G (gyroid), C (hexagonally packed cylinders), S (spheres in bcc lattice), and S_{cp} (spheres in fcc lattice) (Matsen 1997).

As seen in Figure 15.4, the conformational asymmetry produces larger shifts in the order-to-order transitions (OOT) than in the order-to-disorder transition (ODT). The shift in the OOT is explained as follows: As α_A/α_B increases, the A blocks become easier to stretch, while the opposite is true for the B blocks. This creates a tendency to curve the interface towards the A blocks, which allows the B blocks to relax at the expense of the A blocks. At a fixed composition, f_A, the lamellar phase will tend to transform into the hexagonal phase where cylinders of A blocks are embedded in the matrix composed from B blocks. This tendency causes the OOTs to shift towards the larger A-block-volume fractions. The small effect of conformational asymmetry on the ODT can be explained as follows: Disordering from a spherical domain requires pulling the minority blocks forming the spheres until they disorder. The energy binding the A blocks to the A domain relative to the thermal energy is approximately $\chi N f_A$, thus independent of the statistical segment lengths. Therefore, the ODT is relatively unaffected by the conformational asymmetry. Apart from the differences in the phase state, conformational asymmetry strongly affects the relative domain spacing between structures along their boundaries. This is expected to affect the kinetics of the respective order-to-order transitions.

The result of the combined effects of fluctuation corrections and conformational asymmetry are discussed below with respect to the four known phase diagrams for diblock copolymers, and the corresponding parameters are summarized in Table 15.1.

TABLE 15.1. Interaction Parameters (χ), Fluctuation Correction $\bar{N}$ at the ODT, and Conformational Asymmetries (ζ) of Some Diblock Copolymer Systems Obtained for Nearly Symmetric Compositions

A-B Diblock	χ	$\bar{N}^a_{ODT}$	ζ	Gyroid Volume Fractions	Ref.
Poly(styrene-b-2-vinylpyridine) (PS-P2VP)	92/T-0.095		1	$0.35 < f_{PS} < 0.4$	(Schulz 1996)
Poly(ethyleneoxide-b-butyleneoxide)	48.5/T-0.0534				
Poly(isoprene-b-styrene) (PI-PS)	71/T-0.0857	1100	1.5	$0.35 < f_{PI} < 0.4$ $0.64 < f_{PI} < 0.69$	(Khandpur 1995)
Poly(ethylene-b-ethylene propylene) (PE-PEP)	7.9/T-0.0146	27000	1.5	-	(Bates 1994)
Poly(ethylene propylene-b-ethylethylene) (PEP-PEE)	$4.7/T+4.44\cdot10^{-4}$	3400	1.7	$0.38 < f_{PEP} < 0.42$	(Rosedale 1990)
Poly(ethylene-b-ethylethylene) (PE-PEE)	15/T-0.0055	3300	2.5	$0.39 < f_{PE} < 0.44$	(Zhao 1996)
Poly(ethyleneoxide-b-isoprene) (PEO-PI)	65/T+0.0125	960	2.7	$0.4 < f_{PEO} < 0.45$ $0.66 < f_{PEO} < 0.7$	(Floudas 2001)

a: Based on a segment volume of 118 Å^3, $\bar{N} = (\alpha^6_{AB}/u^2)N$, $\alpha^2_{AB} = f_A\alpha^2_A + (1 - f_A)\alpha^2_B$

3. THE KNOWN PHASE DIAGRAMS

The more extensively investigated diblock copolymers systems are the poly(isoprene-b-styrene) (PI-PS) (the phase diagram is based on 16 diblocks) (Khandpur 1995) and the poly(ethyleneoxide-b-isoprene) PEO-PI (based on 25 diblocks) (Floudas 2001). Other systems investigated to a smaller extent are the poly (ethylene-b-ethylene propylene) (PE-PEP) (Bates 1994), poly(styrene-b-2-vinylpyridine) (PS-P2VP) (Schulz 1996), poly(ethylene propylene-b-ethylethylene) (PEP-PEE) (Rosedale 1990), and poly(ethylene-b-ethylethylene) (PE-PEE) (Zhao 1996) systems. The four known phase diagrams are shown in Figures 15.5, 15.6, and 15.7 in the following text.

There are some intriguing features with respect to the appearance and origin of the complex phases in these phase diagrams: First, all phase diagrams are asymmetric around $f = 1/2$. There is a clear correlation of the phase asymmetry with the conformational asymmetry parameter ζ; the higher the value of ζ, the more

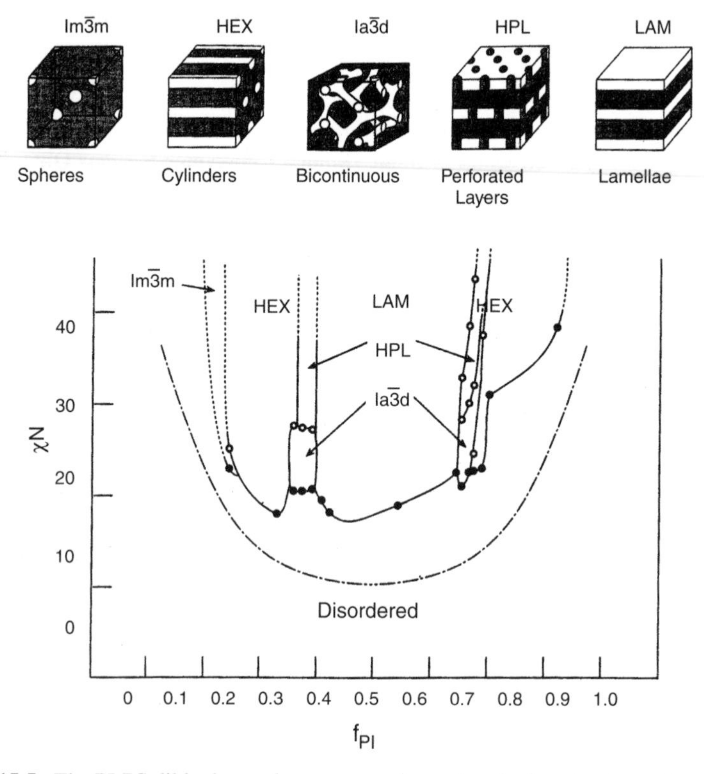

Figure 15.5. The PI-PS diblock copolymer respective ordered microstructures shown on top. The solid curves indicate the approximate boundaries between the ordered phases, and the dash-dot line is the MFT prediction for the ODT (Khandpur 1995).

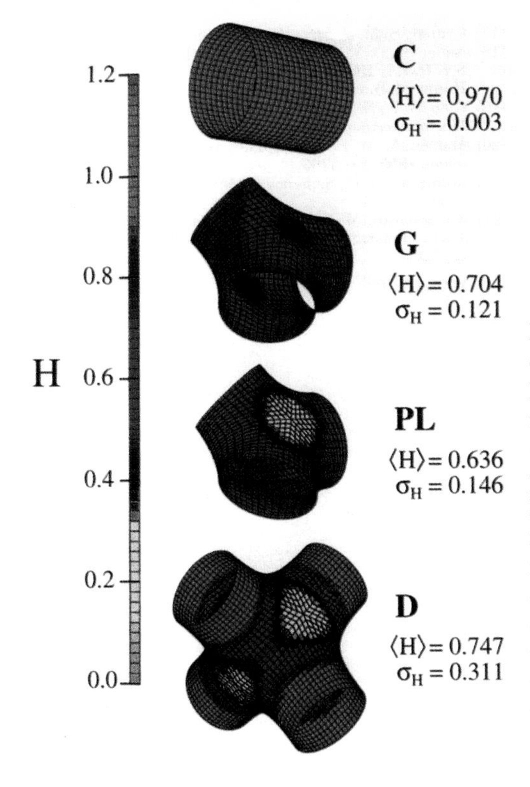

Figure 13.8. Interfacial surfaces associated with elementary units of the hexagonal (C), double gyroid (G), perforated layer (PL), and double diamond (D) structures calculated at the C/G phase boundary ($\chi N = 20$, $f = 0.3378$). For each structure, the distribution of mean curvature H over the surface is indicated using the color scale, and the area-average $\langle H \rangle$ and standard deviation σ_H of H is provided. Notice the large values of σ_H in the PL and G phases, which imply a large degree of packing frustration perturbing the interface away from CMC (Matsen 1996).

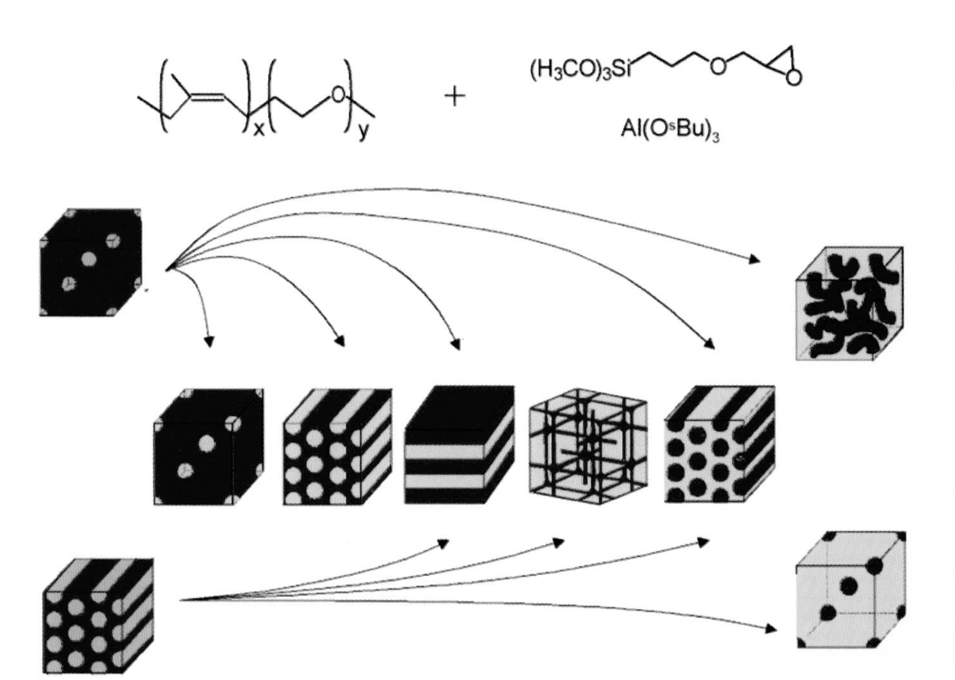

Figure 15.8. PEO-PI/inorganic phase diagram (Simon 2001).

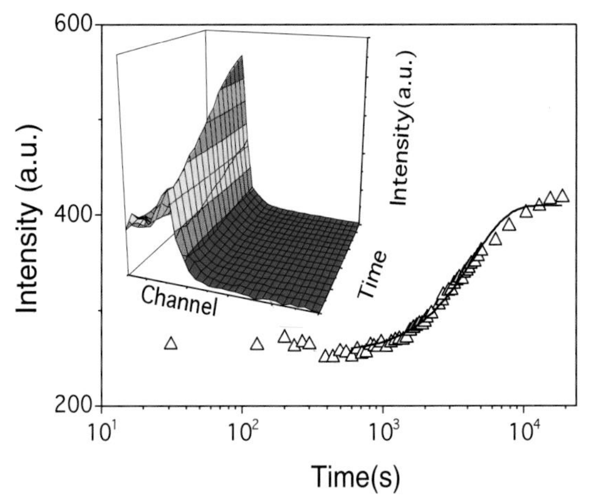

Figure 17.3. Evolution of the peak scattering intensity I(q*)(t) following a temperature jump from the disordered to the ordered state of a block copolymer. The line is a fit to the Avrami equation (eq. 17.3). The evolution of the scattering profiles is shown in the inset (Floudas 1994b).

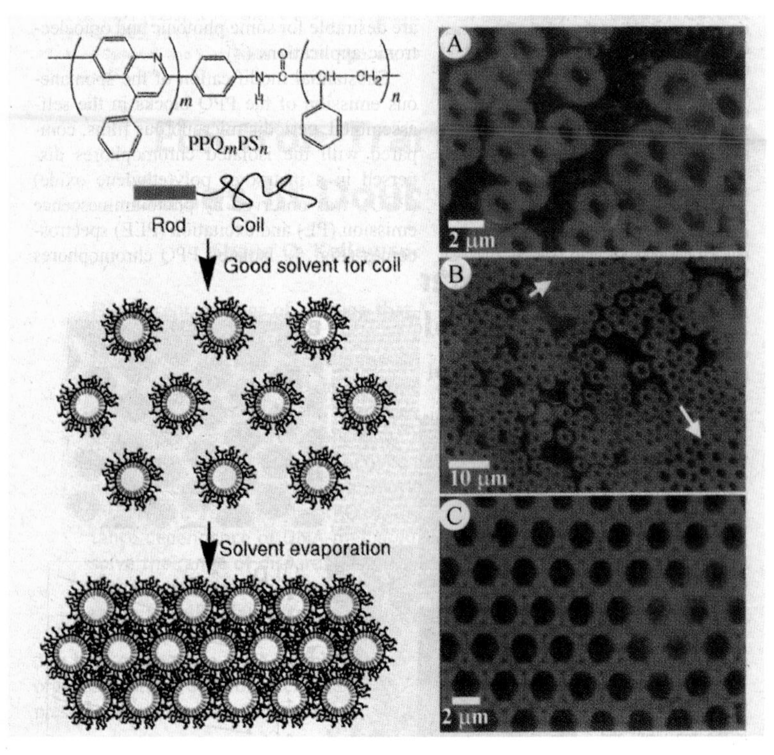

Figure 19.4. (Left) Molecular structure of rod-coil diblock copolymers and a highly schematic illustration of its hierarchical self-assembly into ordered microporous materials. (Right) Fluorescence photomicrographs of solution-cast micellar films revealing a two-dimensional hcp structure composed of air holes (Jenekhe 1999).

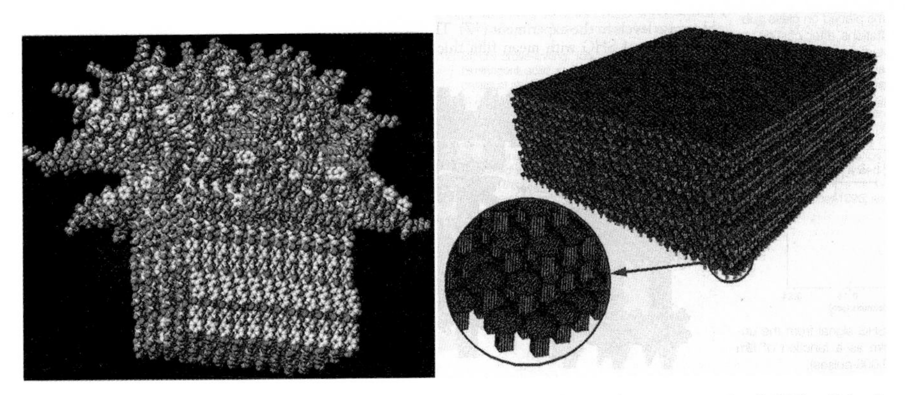

Figure 19.6. (Left) Molecular model of a supramolecular unit composed of 100 triblock copolymers with a rod-coil structure giving rise to mushroom-like nanostructures. (Right) Schematic representation of how these nanostructures of the triblock copolymers could organize to form a macroscopic film (Stupp 1997).

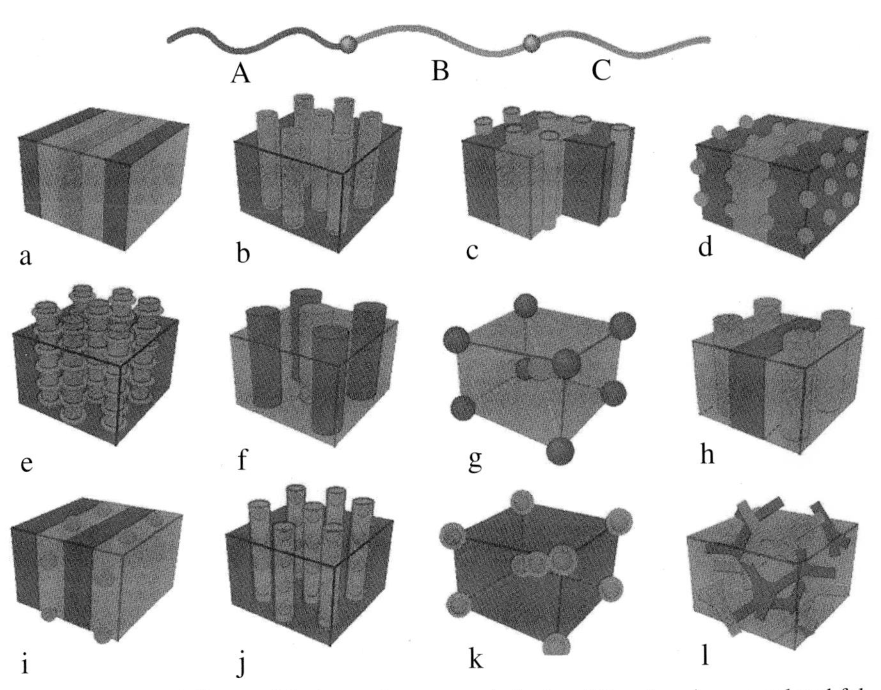

Figure 19.7. ABC linear triblock copolymer morphologies. Microdomains are colored following the code of the triblock molecule in the top (Stadler 1995, Zheng 1995, Bates 1999).

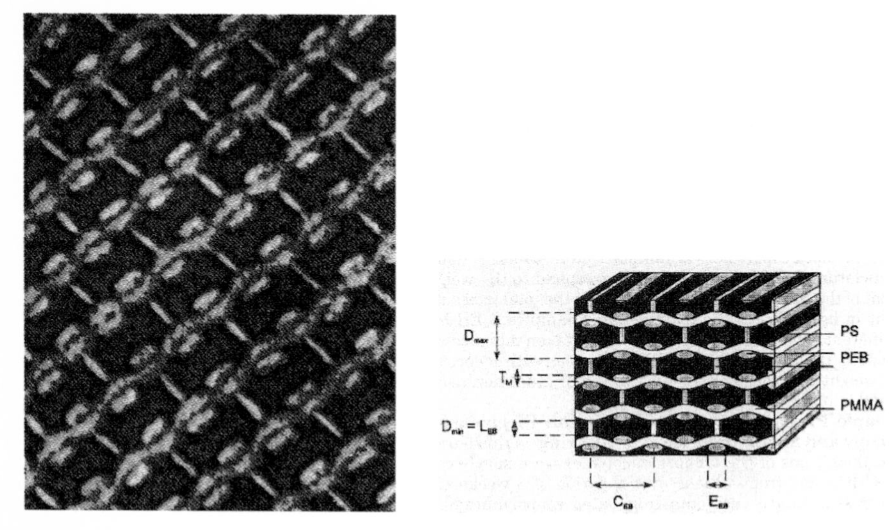

Figure 19.8. (Left) Bright field TEM of the SEM (35/27/38) triblock copolymer stained with RuO_4 showing the celebrated "knitting" pattern. (Right) Schematic description of the knitting pattern morphology (Breiner 1998).

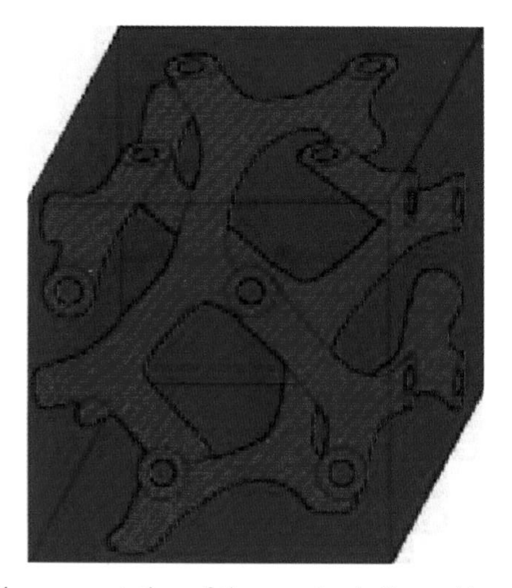

Figure 19.9. Schematic representation of the core-in-shell gyroid morphology found in the poly(isoprene-b-styrene-b-dimethylsiloxane) (ISD) triblock copolymer ($f_{PI} = 0.40$, $f_{PS} = 0.41$, $f_{PDMS} = 0.19$). Blue, red, and green regions correspond to I, S, and D domains, respectively (adapted by Shefelbine 1999).

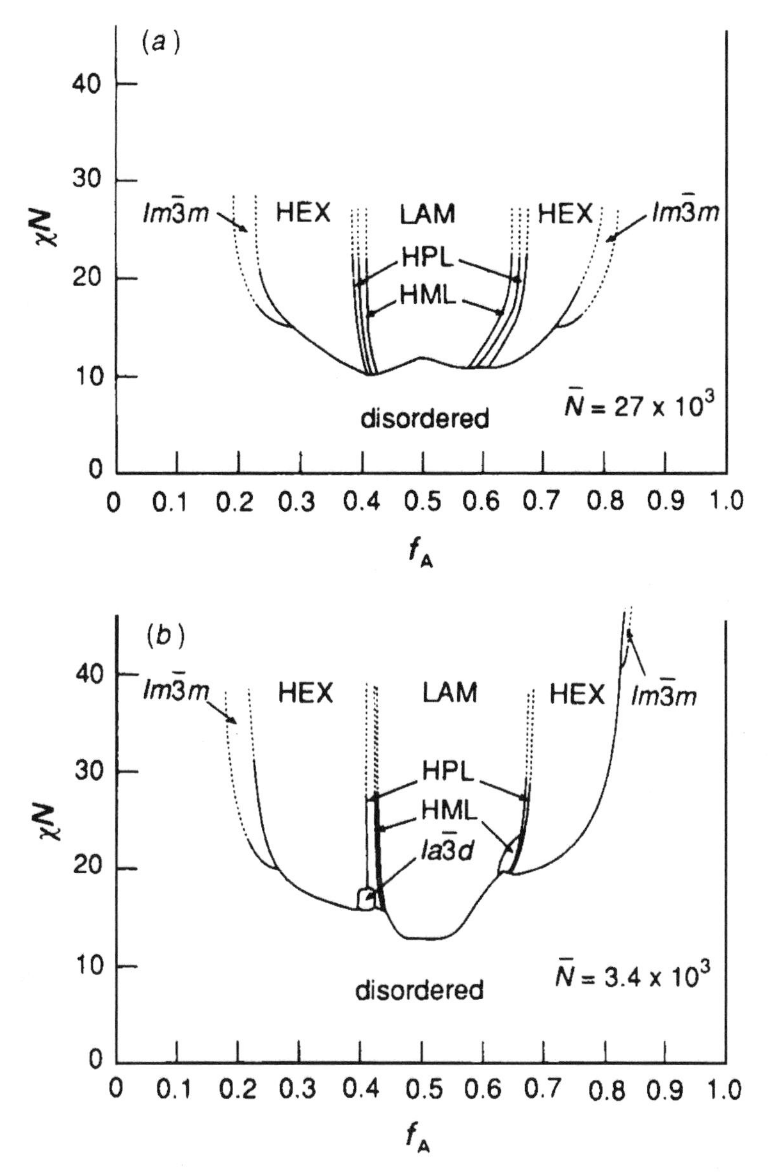

Figure 15.6. The PE-PEP (*a*) and combined PEP-PEE and PE-PEE (*b*) diblock copolymer phase diagrams. HML indicates the metastable hexagonally modulated lamellar structure and HPL the metastable hexagonally perforated lamellar structure (Bates 1994, Zhao 1996).

asymmetric the phase diagram (compare for example the PE-PEP and PI-PEO phase diagrams with corresponding ζ values of 1.5 and 2.7). The value of ζ, however, does not seem to affect the presence or absence of the gyroid phase from both sides of the phase diagram because, in the PEO-PI system, the double gyroid

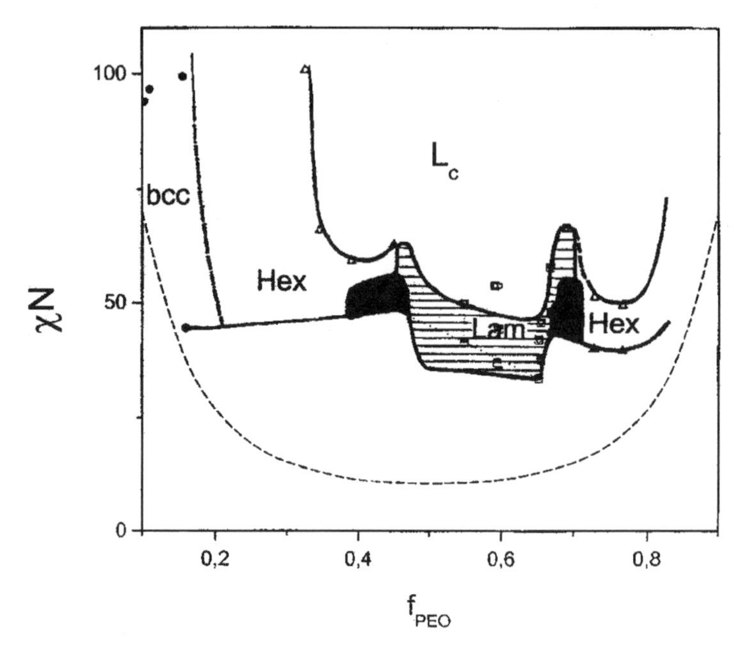

Figure 15.7. The PEO-PI diblock copolymer phase diagram. The phase notation is as follows: (L_c) crystalline lamellar, (Lam) amorphous lamellar, (Hex) hexagonally packed cylinders, (G) bicontinuous cubic structure with $Ia\bar{3}d$ symmetry (gyroid-shadowed areas). Only the equilibrium phases are shown, obtained on cooling from the disordered phase. The dashed line gives the spinodal line in Leibler's mean-field prediction (Floudas 2001).

phase was observed for both $f < 1/2$ and $f > 1/2$. Second, the gyroid phase is completely absent from the PE-PEP phase diagram possessing the highest $\bar{N}$ value ($\bar{N} = 27000$). Notice also that, in the PI-PS, and especially in the PEO-PI phase diagrams (with the lower $\bar{N}$ values), the gyroid structure is stabilized for a range of compositions and χN values. Thus, fluctuation corrections contribute in a systematic way to breaking the phase diagram symmetry and in the development of this cubic phase. Third, the experimental phase diagrams allow for direct transitions from the hexagonal and lamellar phases to the disordered phase.

4. THE PEO-PI PHASE DIAGRAM

The PEO-PI phase diagram is shown in Figure 15.7. The diagram is constructed based on 25 diblock copolymers spanning the composition range $0.05 < f_{PEO} < 0.8$. Five ordered phases have been identified based on their scattering patterns, TEM results, and viscoelasticity: crystalline lamellar (indicated as L_c), amorphous lamellar (denoted as Lam), cylinders packed in a hexagonal structure (denoted as Hex), spheres in a bcc lattice, and, finally, the gyroid phase (shaded areas in the figure). In addition, a perforated layered structure was found, which facilitates the lamellar-to-gyroid forward transformation. The PL phase, however,

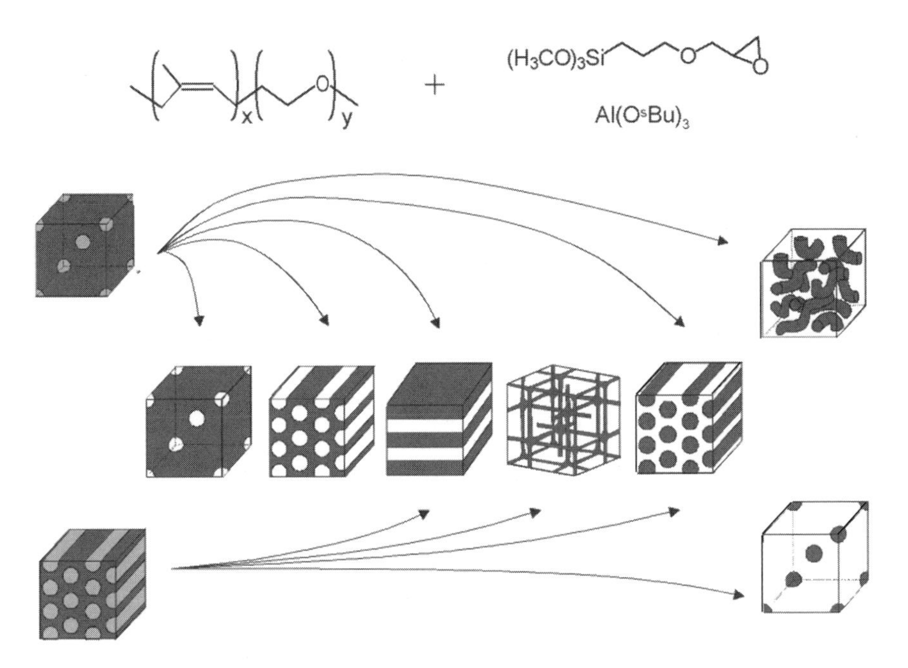

Figure 15.8. PEO-PI/inorganic phase diagram (Simon 2001). See color plates.

was absent on cooling and, therefore, is not an equilibrium phase, and, as such, was not included in the "equilibrium" PEO-PI phase diagram. The PEO-PI phase diagram allows for direct transitions between the Hex and gyroid phases for $0.38 < f_{PEO} < 0.46$ and between the lamellar and gyroid on cooling diblocks with compositions in the range $0.66 < f_{PEO} < 0.7$.

The PEO-PI system has a high interaction parameter (Table 15.1.), a low value of $\bar{N}_{ODT} = 960$, and very high conformational asymmetry ($\zeta = 2.72$). As a result of the high conformational asymmetry, the gyroid phase is observed for $0.4 < f_{PEO} < 0.45$, with the upper bound being the highest composition reported for a stable gyroid phase.

Recently, it was shown (Templin 1997) that the combination of the PEO-PI diblock copolymer system with organically modified ceramic precursors (ormocer) allows for a rational organic-inorganic hybrid morphology design. The basis for this morphological control is a unique polymer-ceramic interface. The hydrophilic blocks of the amphiphilic copolymers are completely integrated into the ceramic phase, giving rise to the rich phase diagram depicted in Figure 15.8 (Simon 2001).

5. THE PS-PI-PEO PHASE DIAGRAM

Adding one chemically distinct C block to a linear AB diblock copolymer entirely disrupts the AB diblock phase state. For example, experiments with ABC triblocks in the last decade revealed the formation of many new morphologies, unattainable

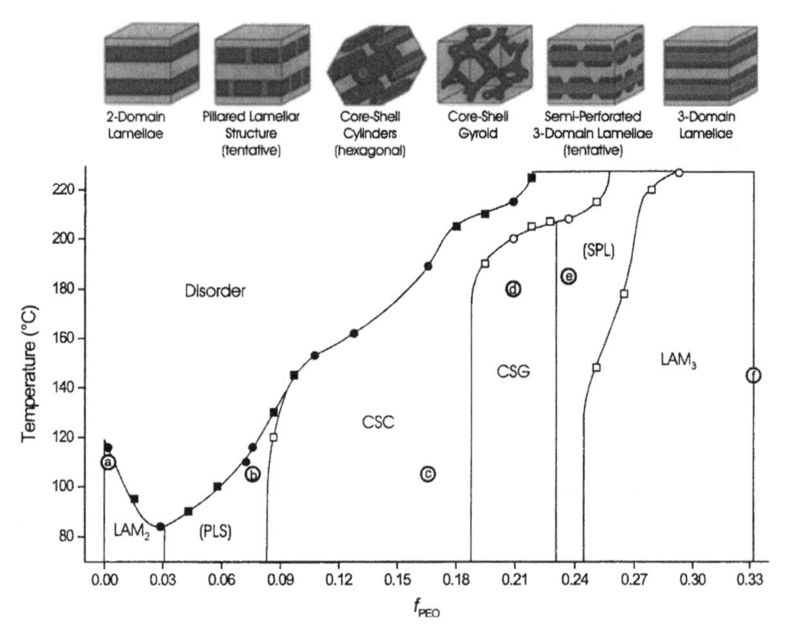

Figure 15.9. The morphology diagram of the PS-PI-PEO triblock copolymer system as a function of the PEO volume fraction. Filled symbols give the order-to-disorder transition temperatures and open symbols the respective order-to-order transitions (Bailey 2001).

with conventional ABA triblock copolymers (Stadler 1995, Zheng 1995). Linear ABC triblock phase behavior is controlled by three interaction parameters (χ_{AB}, χ_{BC}, and χ_{AC}), two composition variables, and three sequencing possibilities, leading to unparalleled possibilities for microstructural complexity. Although many such systems have been investigated in the past, in the SSL, there has recently been a triblock terpolymer case in the WSL. A sequence of 10 poly (styrene-b-isoprene-b-ethylene oxide) triblock terpolymers were synthesized (Bailey 2001) from a single symmetric poly(styrene-b-isoprene) diblock copolymer with a molecular weight of 18000 g/mol and final compositions in the range from 2.9 to 33.2 vol% PEO. The resulting morphologies are shown in Figure 15.9.

Within the investigated composition range, six regions of morphological structures were identified: two domain lamellae, hexagonally packed core-shell cylinders, pentacontinuous core-shell gyroid, and three domain lamellae, together with two nontypical structures containing PS/PEO interfaces that were not solely identified. The rich phase behavior in the PS-PI-PEO and the two morphologies containing PS/PEO contacts is thought to arise from the highly unfavorable PI/PEO interactions. Altering the block sequence is expected to relieve these frustrations. In accord with this expectation, the PI-PS-PEO system, where the PI and PEO interactions are screened by the presence of the mid-PS block, morphologies are formed with only two interfaces.

REFERENCES

Bailey T.S., Pham H.D., Bates F.S. (2001) Macromolecules 34, 6994.

Bates F.S., Schulz M.F., Khandpur A.K., Förster S., Rosedale J.H. (1994) Faraday Discuss. 98, 7.

Floudas G., Vazaiou B., Schipper F., Ulrich R., Wiesner U., Iatrou H., Hadjichristidis N. (2001) Macromolecules 34, 2947.

Fredrickson G.H., Helfand E. (1987) J. Chem. Phys. 87, 697.

Fredrickson G.H., Binder K. (1989) J. Chem. Phys. 91, 7265.

Hamley I.W., Podneks V.E. (1997) Macromolecules 30, 3701.

Khandpur A.K., Förster S., Bates F.S., Hamley I., Ryan A.J., Almdal K., Mortensen K. (1995) Macromolecules 28, 8796.

Matsen M.W., Bates F.S. (1997) J. Polym. Sci. Polym. Phys. 35, 945.

Simon P.F.W., Ulrich R., Spiess H.W., Wiesner U. (2001) Chem. Mater. 13, 3464.

Stadler R., Aushra C., Beckmann J., Krappe U., Voigt-Martin I., Leibler L. (1995) Macromolecules 28, 3080.

Stühn B. (1992) J. Polym. Sci. Polym. Phys. Ed. 30, 1013.

Rosedale J.H., Bates F.S. (1990) Macromolecules 23, 2329.

Schulz M.F., Khanpur A.K., Bates F.S., Almdal K., Mortensen K., Hajduk D.A., Gruner S.M. (1996) Macromolecules 29, 2857.

Templin M., Franck A., DuChesne A., Leist H., Zhang Y., Ulrich R., Schadler V., Wiesner U. (1997) Science 278, 1795.

Zhao J., Majumdar B., Schulz M.F., Bates F.S., Almdal K., Mortensen K., Hajduk D.A., Gruner S.M. (1996) Macromolecules 29, 1204.

Zheng W., Wang Z.-G. (1995) Macromolecules 28, 7215.

CHAPTER 16

VISCOELASTIC PROPERTIES OF BLOCK COPOLYMERS

Rheology is a very sensitive probe of: (i) the order-to-disorder transition, (ii) the different ordered phases, and (iii) the phase transformation kinetics between the disordered and ordered phases as well as of the transformation among the different ordered phases of block copolymers. This sensitivity originates from the large viscoelastic contrast that can be used as a dynamic probe of the different phases.

In addition, rheology serves to orient ordered phases on a macroscopic scale, which provides the means of a definite structure identification through scattering experiments. We will comment further below on the flow alignment of block copolymers. Here, we show an example of a lamellar-forming block copolymer in Figure 16.1. The multigrain structure shown in the figure is characteristic of unoriented block copolymers, the size of the grains being determined by the temperature and the type of microstructure (i.e., cubic phases tend to make larger grains). Application of a large amplitude oscillatory shear (LAOS) field (well in the nonlinear viscoelastic regime) results in the macroscopic orientation of grains through the elimination of interfaces perpendicular to the shear direction and reorientation of the domains along the shear direction. As we will see in the following text, however, this type of lamellar orientation is not the only possibility. During LAOS, the magnitude of the complex shear modulus $G^* = ((G')^2 + (G'')^2)^{1/2}$ is a decreasing function of time, and the magnitude of this decrease depends on temperature, frequency, strain amplitude, distance from the order-to-disorder transition, and the characteristic relaxation spectrum of the system.

The overall shape of $G^*(t)$ reflects the kinetics of the morphological reorientation and on complex relationships between the morphology and the mechanical properties of the system. In order to extract the information about the kinetics of

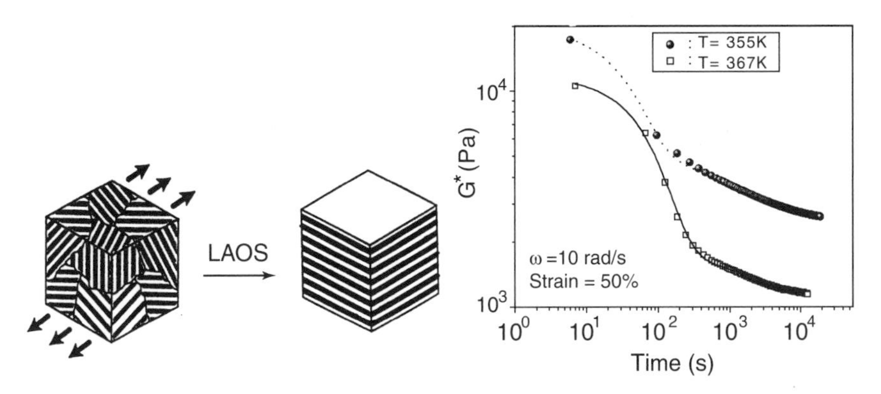

Figure 16.1. (Left): Effect of large amplitude oscillations on the morphology of a multigrain lamellar-forming block copolymer. (Right): Magnitude of the complex shear modulus during isochronal/isothermal experiments performed at two temperatures as indicated.

reorientation, mechanical models relating the morphological orientation parameters with the macroscopic mechanical properties are needed. There is a lack of such models, and, instead, the analysis of the G*(t) can be performed by means of a sum of two "stretched exponential" functions of the form:

$$G^*(t) = \sum_i \Delta G_i^* \exp\left[-(t/\tau)^{\beta_i}\right] \tag{16.1}$$

where ΔG*, τ, and β are the amplitudes of the modulus relaxation, the relaxation times, and the shape parameters of the "fast" and "slow" processes, respectively. The result of the fit to eq. 16.1 is depicted by the lines in the figure. The "fast" Debye-like process ($\beta = 1$) can be attributed to the domain reorientation towards the shear direction and the "slow" process to the perfection of the oriented morphology (Floudas 1994).

Subsequently, scattering experiments can be used for the identification of the morphology. An example is shown in Figure 16.2 where the macroscopically oriented lamellar-forming diblock copolymer is examined with small-angle X-ray scattering. In the figure the radial view is shown at different temperatures obtained on heating the copolymer. At 303 K, the scattering pattern exhibits a weak isotropic component (reflecting the residual miss-aligned grains), and on top there is a large anisotropic component reflecting the lamellar symmetry. Increasing temperature results in the increase of scattering contrast up to 323 K, reflecting the intervening glass temperature of the polystyrene block, whereas, at a higher temperature, the contrast is a decreasing function of temperature as the disordered phase is approached. At 363 K, the anisotropic component is completely lost, reflecting the dissolution of the lamellar structure at the (apparent) T_{ODT}. The scattering pattern still displays a weak isotropic peak, which can be described by Leibler's MFT.

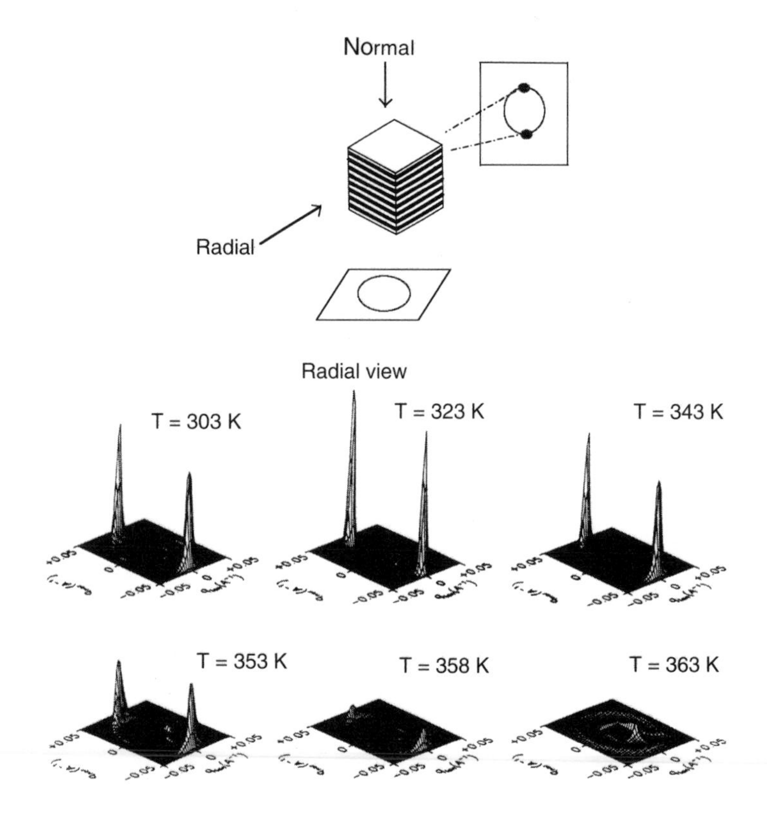

Figure 16.2. (Top): Scattering geometry showing the scattering patterns obtained with X-rays in the normal and radial directions from a well-oriented copolymer forming a lamellar structure. (Bottom): SAXS patterns obtained with the X-ray beam in the radial direction taken at different temperatures while heating the diblock copolymer from the low temperature ordered state to the disordered phase at higher temperatures. The diblock system is a nearly symmetric poly(styrene-b-isoprene) copolymer with an order-to-disorder transition temperature at 363 K.

1. LOCALIZATION OF THE (APPARENT) ORDER-TO-DISORDER TRANSITION

Because of the distinctly different viscoelastic contrast of the ordered and disordered states, low-frequency rheology is very sensitive for detecting the dissolution of ordered microstructures (Bates 1990, Rosedale 1990, Floudas 1994, Han 1995). Isochronal measurements of the storage modulus performed at low frequencies with low strain amplitudes by slowly heating the specimen provide a good way of locating the T_{ODT}. As we will see in the next chapter with respect to the kinetic studies, the temperature extracted in this way reflects some apparent temperature (T'_{ODT}), which is lower than the true equilibrium ODT (T^{o}_{ODT}).

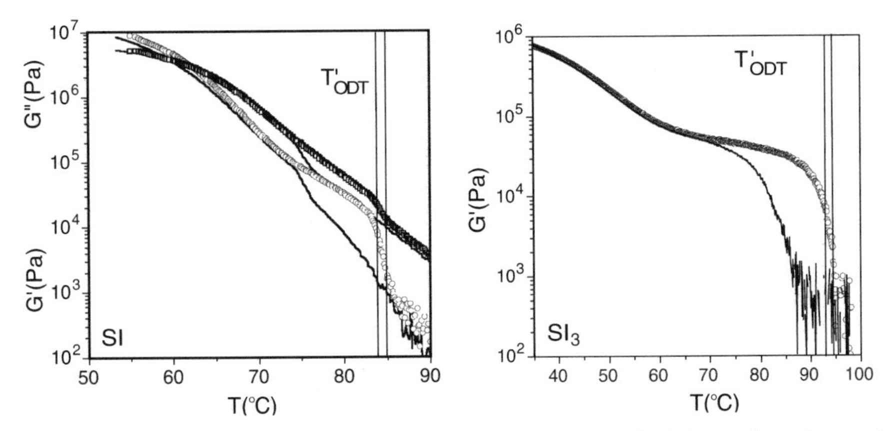

Figure 16.3. (Left): Isochronal measurement of the storage (circles) and loss (squares) moduli for the symmetric diblock copolymer SI ($M_n = 2200$, $f_{PS} = 0.51$) and (b) the isochronal temperature scan of the storage modulus for the miktoarm star SI3, ($M_n = 24000$, $f_{PS} = 0.3$). The dependencies are taken with $\omega = 1$rad/s and a strain amplitude of 2% on heating (symbols) and subsequent cooling (line). The heating and cooling rate was 0.2 K/min in all cases (Floudas 1998a).

Figure 16.3 shows the result of the isochronal measurements of the storage (G') and loss (G'') moduli at $\omega = 1$ rad/s obtained on heating (symbols) and subsequent cooling (lines) of a symmetric diblock copolymer [poly(styrene-b-isoprene), $M_n = 12200$ and $f_{PS} = 0.51$]. The change of the moduli within the indicated range signify an apparent ODT. There is a pronounced hysteresis on cooling, which extends to some 10°C below the transition and which is the signature of metastability.

In the same figure, the result of a similar experiment for an asymmetric and nonlinear miktoarm star block copolymer (SI_3, $M_n = 24000$, $f_{PS} = 0.3$) is shown. The isochronal ($\omega = 1$rad/s) temperature run reveals an apparent ODT and a pronounced hysteresis for about 15°C below the (apparent) transition. The well-defined loop calls for a pronounced supercooling effect. The hysteresis observed in all block copolymers around the ODT indicates a first-order transition (Floudas 1998a) and will be used in the next chapter as a tool to investigate the ordering kinetics and to localize the equilibrium ODT.

2. VISCOELASTIC SPECTRUM OF BLOCK COPOLYMERS

The viscoelastic response of disordered block copolymers is usually similar to that observed for homopolymer melts. At high and intermediate frequencies, two relaxation processes affect the viscoelastic response: the segmental and chain relaxations, respectively. However, in some cases there is a broadening or bifurcation in the range of segmental relaxation depending on the nature of monomers. At

$T > T_{ODT}$, the time-temperature superposition (tTs) works well, and the moduli exhibit typical terminal behavior ($G' \approx \omega^2$ and $G'' \approx \omega$). When examined over a broad temperature range, however, tTs is violated due to the order-to-disorder transition, which drives the system from the disordered state to a microphase-separated state. At $T < T_{ODT}$ and at low frequencies, the moduli exhibit weak frequency-dependencies of the order of $\omega^{1/2}$ (for symmetric block copolymers) to $\omega^{1/4}$. This results from the appearance of the new ultra slow relaxation process related to morphological rearrangements. Unfortunately, quite often a complete relaxation of the microdomains in block copolymers cannot be observed within the accessible frequency and temperature window.

An example of the influence of the order-to-disorder transition on the frequency-dependencies of G' and G' is shown in Figure 16.4. The "master curve" show the segmental relaxation of the block with the higher glass temperature (T_g) block, the chain relaxation corresponding to the same block, and, at lower frequencies, the structural relaxation. The breakdown of tTs is observed at low frequencies where the Newtonian behavior of the disordered state is replaced by a rubbery state related to unrelaxed morphology. Strictly speaking, the use of tTs is not permitted in systems with a T-dependent internal structure such as in block copolymers (Han 1993).

Furthermore, reporting a shift factor for spectra that do not shift is not of much use. Nevertheless, the result of the attempted tTs can be used as a guide of how the different ordered processes affect the viscoelastic response of the system. Another

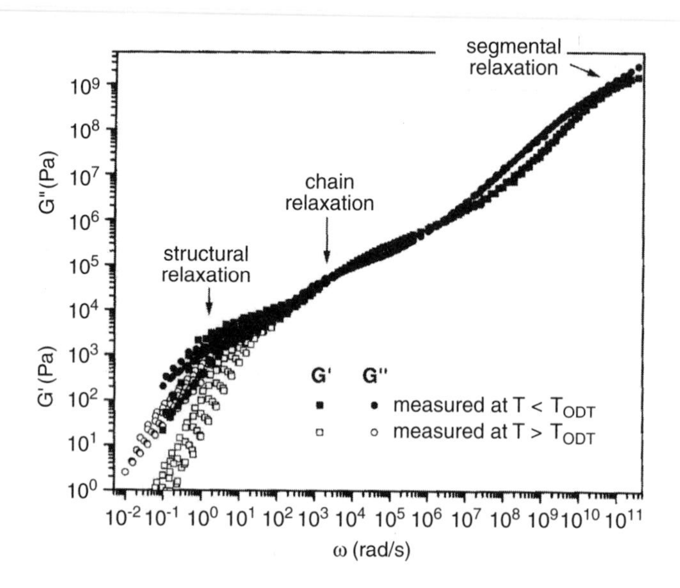

Figure 16.4. Frequency-dependencies of G' and G'' for a diblock copolymer (poly (methylphenylsiloxane-b-styrene)) measured at various temperatures and shifted horizontally to form "master curves" at the reference temperature $T_{ref} = 383$ K ($T_{ODT} = 403$ K, $f_{PS} = 0.36$). Below ODT the cylindrical morphology has been detected by SAXS (Gerharz 1991).

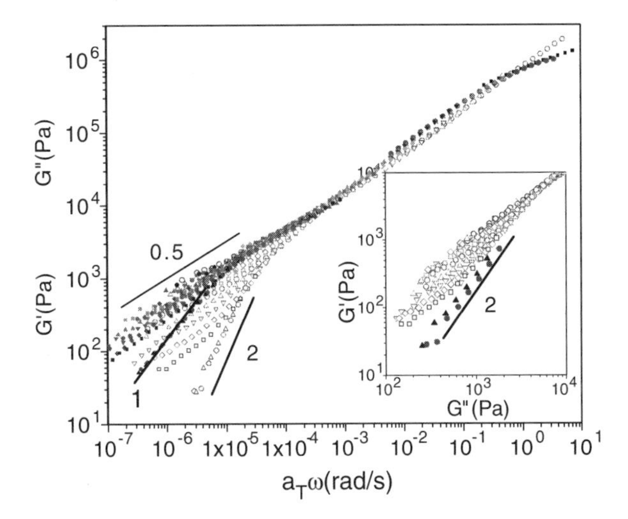

Figure 16.5. Reduced frequency plots for the storage (open symbols) and loss (filled symbols) moduli of a lamellar-forming block copolymer. Two lines with limiting slopes at low frequencies are shown (2 and 1 for the loss and storage moduli, respectively). The T_{ODT} is at 460 K. In the inset the same data are plotted in the G′ versus G″ representation and result in an identical transition temperature. The system is a PS-P(S/I)-PI copolymer with a tapered (S/I) interface ($M_n^{total} = 22800$, $f_{PS} = 0.53$, $T_{ODT} = 460$ K) (Hodrokoukes 2001).

way of determining the ODT, which does not require the use of tTs, is by plotting the logarithm of the storage moduli as a function of the logarithm of the loss moduli for the different temperatures, known as the Han representation (Han 1995, Sakamoto 1997). In this representation, for symmetric copolymers, the ODT corresponds to a temperature where the slope attains a value of 2. The application of this representation is shown for a diblock copolymer system in Figure 16.5. It is interesting to note that the two representations (shifted vs. unshifted data) provide identical results for the transition temperature. This type of representation is certainly advantageous to the use of tTs in the vicinity of the order-to-disorder transition.

3. VISCOELASTIC RESPONSE OF ORDERED PHASES

The linear viscoelastic properties of microphase-separated block copolymers have been the subject of theoretical studies. In the study by Rubinstein and Obukhov (Rubinstein 1993), both microscopic and mesoscopic mechanisms have been invoked, which were attributed, respectively, to the dispersion in the number of entanglements of a chain with the opposite brush (high-frequency response) and to the collective diffusion of copolymer chains along the interface. The latter mechanism is controlled by defects in lamellar orientation and contributes to the low-frequency side. For the disordered lamellar mesophase, they predict: $G'(\omega) \approx$

$G''(\omega) \approx \omega^{1/2}$, whereas, for the cylindrical mesophase: $G'(\omega) \approx G''(\omega) \approx \omega^{1/4}$. On the other hand, Kawasaki and Onuki (Kawasaki 1990) reached the same conclusion through a completely different approach; they proposed that overdamped second-sound modes in an orientationally disordered lamellar phase could result in a complex shear modulus proportional to $(i\omega)^{1/2}$.

As we discussed early in this chapter, rheology can be used not only in the identification of the ordered and disordered phases but also to "distinguish" between the different ordered phases. This, again, results from the different viscoelastic contrast of ordered phases. We mention, however, that the complete structure assignment can only be made with direct experiments that probe the actual morphologies [i.e., SAXS, SANS, transmission electron microscopy (TEM)]. To illustrate the distinct viscoelastic properties of some ordered phases, we show in Figure 16.6 a composite plot reflecting the structure, thermal, and viscoelastic

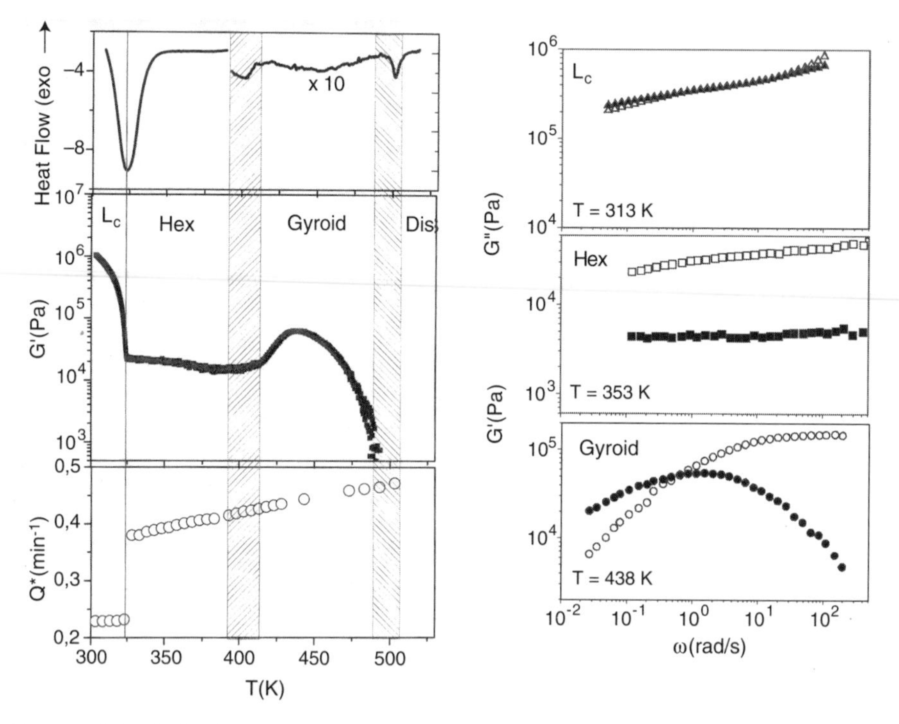

Figure 16.6. (Left): A composite plot showing the heat flow obtained during heating, the temperature dependence of the storage modulus, and the corresponding dependence of the most important wave vector obtained in SAXS for a poly(ethylene oxide-b-isoprene copolymer) ($M_n = 9800$, $f_{PEO} = 0.39$). The vertical lines separate the high temperature disordered phase (Dis), the bicontinuous cubic phase (Gyroid), the hexagonally packed cylinders (Hex), and crystalline lamellar (L_c) phases. (Right): Frequency-dependencies of the storage (open symbols) and loss (filled symbols) moduli at three temperatures: T = 313 K, T = 353 K, and T = 438 K corresponding to Lc, Hex, and the Gyroid phases, respectively (Floudas 1999 and 2000).

properties of the same diblock copolymer system. The system is a polyisoprene-poly(ethylene oxide) (PI-PEO) diblock copolymer ($M_n = 9800$, $f_{PI} = 0.61$) undergoing the following transitions up on heating: from a low temperature crystalline lamellar (L_c), to hexagonally packed cylinders (Hex), to the Gyroid phase, and, finally, to the disordered phase (Floudas 2000). Notice the strong endothermic peak reflecting the melting of the semicrystalline block (PEO). Two smaller in magnitude endothermic peaks around 400 K and 500 K with latent heats of 0.3 J/g and 4 J/g, respectively, can also be seen. The latter value, corresponding to the order-to-disorder transition, can be accounted for from (Stühn 1992, Hajduk 1996):

$$\Delta H \simeq \frac{RT_{ODT} f(1-f)(\chi N)_{ODT}}{M_n} \tag{16.2}$$

where R is the gas constant, N is the total degree of polymerization, and M_n is the total number-averaged molecular weight. Implicit in the DSC result is the notion of a first-order transition, not only for the order-to-disorder but also for the Hex-to-Gyroid transition. The structure investigation reveals that the Gyroid structure grows nearly epitaxially on the hexagonal structure as indicated by the continuity in the $q^*(T)$ dependence.

Returning to the viscoelastic properties of the ordered phases, we find that the L_c phase has the highest moduli, followed by the Gyroid, and, finally, the hexagonal phase. The L_c phase is formed by a semicrystalline block (PEO) and an elastomeric block (PI), both giving rise to a spherulitic superstructure. Therefore, the storage and loss moduli below the equilbrium melting temperature (T_m^o) of PEO correspond to the spherulitic superstructure, and the relatively low values of G' and G'', as compared with bulk semicrystalline PEO, are caused by the unentangled PI block. The frequency-dependence of the moduli (Floudas 1999) is also shown in the figure and displays a very weak dependence within the L_c phase.

The hexagonal phase, which exhibits the lower modulus, is characterized by an elastic response with a weak frequency-dependence. In contrast, the bicontinuous Gyroid phase, composed of two interpenetrating networks of the minority phase, exhibits a strong frequency-dependence with a crossover in $G'(\omega)$ and $G''(\omega)$ to a nonterminal regime. Near the crossover, G'' develops a maximum characteristic of all cubic structures. Recently, the moduli in bicontinuous ($I\alpha\bar{3}d$) and spherical ($Im3m$) morphologies in linear (Zhao 1996) and nonlinear (Floudas 1998b, Johnson 1995) block copolymers were found to exhibit a maximum, which signifies a new characteristic frequency for the relaxation of the ordered phase. The maximum in $G''(\omega)$ can be attributed to the grain relaxation.

4. FLOW-INDUCED ALIGNMENT OF BLOCK COPOLYMER MELTS

Figure 16.1 already shows that the local order in block copolymers can be extended to macroscopic scales using external mechanical fields. The resulting material exhibits highly directional properties such as optical, transport, and electrical properties. Furthermore, the direction of alignment can be switched between different orientations, which may be useful for designing switchable material properties. Flow is not the only way to induce macroscopic order; electric fields (Amundson 1993 and 1994) can also serve this purpose, but flow in its different forms (oscillatory and steady shear, extrusion, roll-casting, extensional flow) provides an efficient and versatile means of obtaining macroscopic alignment.

In using flow fields, there are many questions that need to be addressed, such as the type of flow and the parameters affecting the direction and degree of alignment with respect to the inherent viscoelastic properties of the different block copolymer systems. The effects of mechanical fields on block copolymers was pioneered by the work of Keller et al. (Keller 1970) using an extrusion flow field. Since then the process of flow alignment in block copolymers has attracted a lot of interest (Hadziioannou 1979, Pakula 1985 and 2000, Koppi 1992, Okamoto 1994, Riise 1995, Balsara 1994 and 1996, Wang 1997, Zhang 1995 and 1997, Chen 1997 and 1998). Herein, we will review recent work on the flow-induced alignment of block copolymer melts by means of oscillatory shear.

Until recently it was believed that, when subjecting a lamellar-forming block copolymer to reciprocating shear, the lamellae orient with their lamellar normal parallel to the shear gradient direction (parallel orientation). However, it was shown in two block copolymer systems (Koppi 1992, Zhang 1995 and 1997) that another orientation is also possible, namely the perpendicular orientation. Figure 16.7 depicts a sequence of SAXS patterns obtained from a poly(styrene-b-isoprene) diblock copolymer for the three orthogonal directions: radial, tangential, and normal, with respect to the sample geometry, which is also shown in the figure.

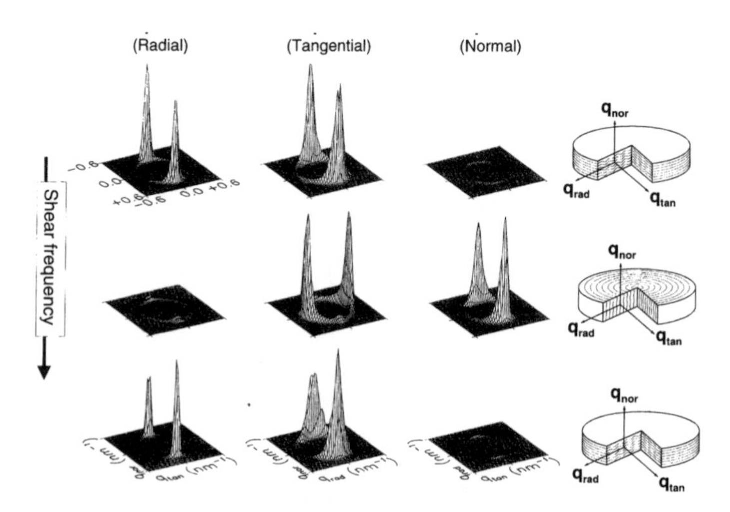

Figure 16.7. SAXS patterns taken at various directions showing the sequence of parallel-perpendicular-parallel orientations of lamellae with increasing deformation frequency. The sample is a poly(sryrene-b-isoprene) diblock copolymer ($M_n = 18400$, $f_{PS} = 0.43$) (Zhang 1995).

The figure shows that the parallel orientation is obtained at low and high frequencies, but, at intermediate frequencies, the perpendicular orientation is produced. These experiments demonstrated that the orientation of the lamellar microstructure of SI block copolymers under large amplitude oscillatory shear in the vicinity of the order-to-disorder transition temperature depends strongly on frequency. Annealing effects have shown to be important for the observation of the parallel orientation at low frequencies (Wiesner 1997). Based on these experiments, a "phase diagram" has been proposed for the different orientations in lamellar SI diblock copolymer in the parameter field strain amplitude and frequency, which is depicted in Figure 16.8. The analysis with respect to the order parameter within the different orientations indicated a higher orientational order within the parallel, as opposed to the perpendicular, orientation ($\langle P_2 \rangle$ of 0.58 and 0.78, respectively). Based on these experiments, it was proposed that a uniform orientation exhibiting high order parameters can result by annealing the sample at temperatures far below the T_{ODT} but above the high glass temperature of PS, for long times, before LAOS is applied.

Optical rheometry provides additional information on the flow-induced alignment of block copolymers (Chen 1997 and 1998). It was shown that these final orientations can be produced through different routes. An example is shown in Figure 16.9, where three families of alignment trajectories observed for a symmetric SI diblock copolymer are shown. The figure combines the optical trace together with ex situ SAXS images taken at different times during the alignment process. Notice that the parallel alignment (cases b and c) is obtained through two routes for different high frequencies. To explain the observed frequency-dependent alignment behavior, two characteristic frequencies have been proposed: the frequency ω_c, above which the distortion of chain conformation dominates the viscoelastic response, and the frequency ω_d, below which the relaxation of domains becomes important. The two frequencies are indicated in Figure 16.10 for entangled

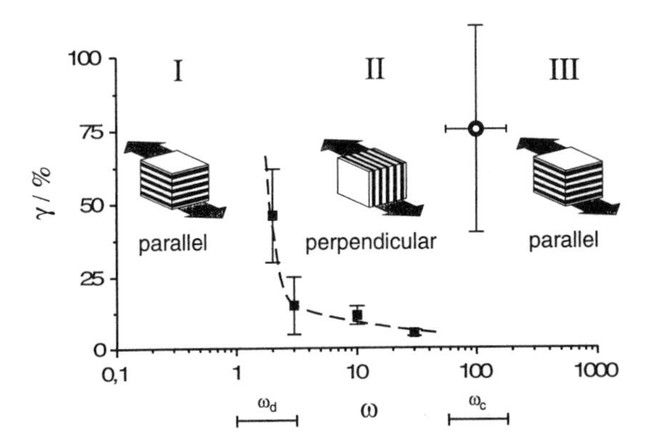

Figure 16.8. Proposed orientation diagram for the orientation of lamellar SI diblock copolymers under large amplitude oscillatory shear in the vicinity of the T_{ODT} (Wiesner 1997).

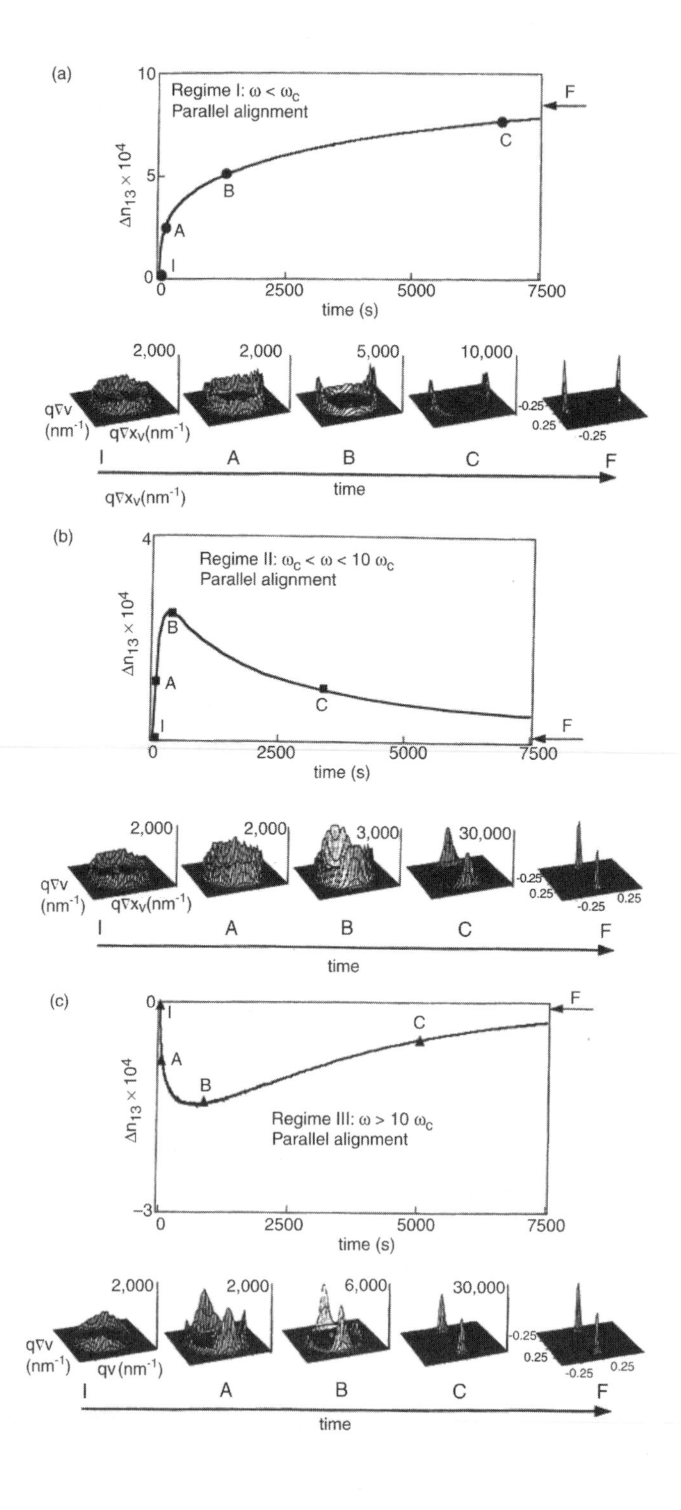
(a)
Regime I: ω < ω_c
Parallel alignment
Δn_13 × 10^4
time (s)
2,000
2,000
5,000
10,000
q∇v (nm⁻¹)
q∇x_v(nm⁻¹)
I A B C F
time
(b)
Regime II: ω_c < ω < 10 ω_c
Parallel alignment
Δn_13 × 10^4
time (s)
2,000
2,000
3,000
30,000
I A B C F
time
(c)
Regime III: ω > 10 ω_c
Parallel alignment
Δn_13 × 10^4
time (s)
2,000
2,000
6,000
30,000
qv (nm⁻¹)
I A B C F
time

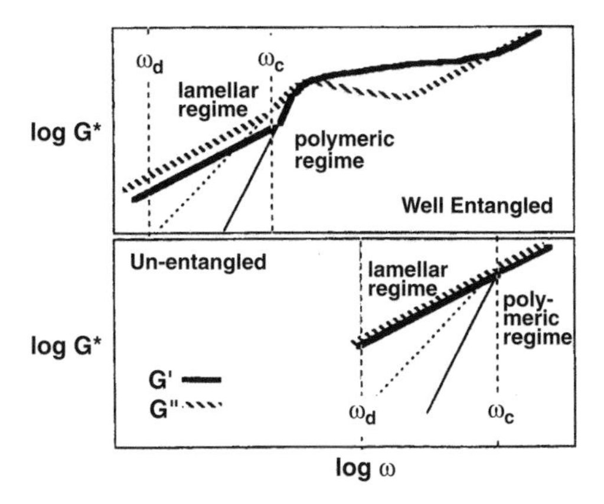

Figure 16.10 Identification of the frequencies ω_c and ω_d affecting the orientational behavior of entangled and unentangled block copolymers (Chen 1998).

and unentangled block copolymers. Although different methods for obtaining these frequencies have been proposed (i.e., from the reduced complex viscosity plots), we should mention that the actual value of ω_d and the dynamics that dominate at $\omega < \omega_c$ are still poorly understood.

The combined results from rheometry and SAXS in the PEP-PEE and PS-PI copolymers are shown in the "phase diagram," with parameters of the reduced temperature and frequency as well as the strain amplitude in Figure 16.11 in the following text. Four frequency regimes have been found. Perpendicular alignment is induced for frequencies in the range $\omega_d < \omega < \omega_c$ (Regime I), where coupling with the dynamics of the microstructure seem to dominate. Shearing at higher frequencies $\omega > \omega_c$ results in a parallel orientation but through two different routes: at frequencies slightly above ω_c, parallel orientation results through a transient orientation distribution consisting of parallel and perpendicular orientations as well as a range of orientation between them (Regime II). In contrast, for $\omega \gg \omega_c$, the parallel alignment results from a bimodal distribution rich in parallel and transverse orientations (Regime III). For low frequencies $\omega < \omega_d$, the parallel orientation is obtained (Regime IV).

Computer simulations have also been employed to address this point (Pakula 2000). It was shown that there are additional parameters that have been

Figure 16.9. Three alignment trajectories observed for a symmetric SI diblock copolymer. In the figure the real-time optical measurements are compared with the results from ex situ SAXS. (a) Regime I showing the route to perpendicular, (b) Regime II and route to parallel, (c) Regime III and route to parallel (Chen 1998).

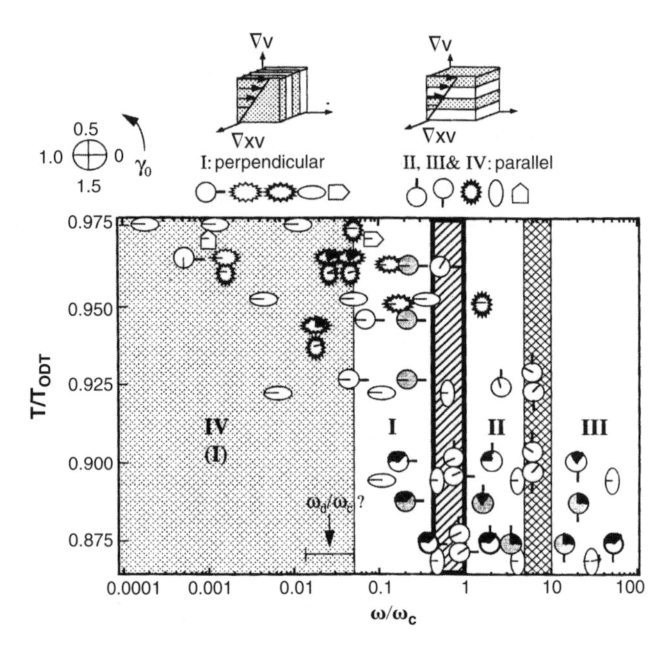

Figure 16.11 Flow-induced alignment regimes as a function of temperature, frequency, and strain amplitude. The strain amplitude is depicted within the symbols using the code in the upper left corner. Results on the PS-PI system are shown as circles, ellipses, and modulated ellipses. Results on the PEP-PEE system are indicated by rectangular arrowheads. The hatched region indicated the boundary between perpendicular alignment (Regime I) and parallel alignment (Regimes II and III). Strain-induced flipping occurs within this area: small strain values induce perpendicular alignment, whereas large strain values lead to parallel alignment. The cross-hatched area indicates the boundary between regimes, both resulting in a parallel orientation following different paths. Regime II: transient distribution rich in parallel, perpendicular, and mixed orientations, and Regime III: bimodal distributions rich in parallel and transverse orientations (Chen 1998).

neglected in the experimental phase diagrams and are of paramount importance in controlling the lamellar orientations: the affinity of the blocks to the surface and the layer thickness. A schematic phase diagram summarizing the results from the simulations is depicted in Figure 16.12.

Notice that only between neutral walls or at intermediate deformation rates is the perpendicular orientation observed. Overall, the experiments and the simulations on the flow-induced alignment of lamellar block copolymer melts have shown that the orientation depends on many parameters, giving rise to various sequences of structural rearrangements. Despite the different experimental efforts, it is likely that this issue will be explored further in the future.

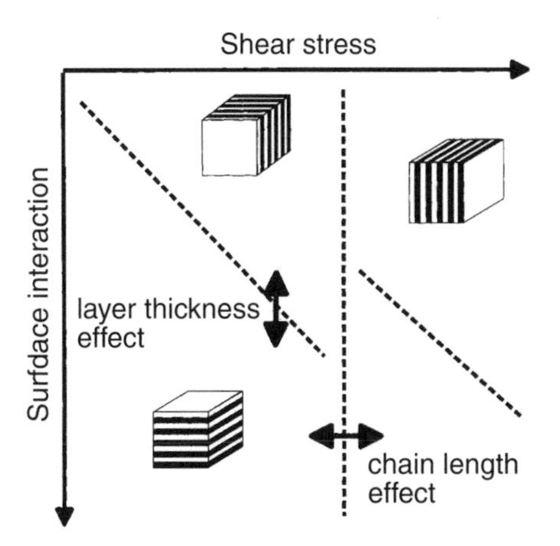

Figure 16.12 Schematic representation of the different lamellar orientations induced by shear within a parameter field of shear stress and the strength of selective interactions of one block to the layer surface. Dashed lines indicate the transitions between the different orientations. Positions of transitions between different structures represented by dashed lines can additionally be influenced by the layer thickness and by the chain length, as shown (Pakula 2000).

REFERENCES

Amundson K., Helfand E., Quan E., Hudson S.D., Smith S.D. (1994) Macromolecules 27, 6559.

Amundson K., Helfand E. (1993) Macromolecules 26, 1324.

Balsara N.P., Hammouda B., Kesani P.K., Jonnalagadda S.V., Straty G.C. (1994) Macromolecules 27, 2566.

Balsara N.P., Dai H.J., Watanabe H., Sato T., Osaki K. (1996) Macromolecules 29, 3507.

Bates F.S., Rosedale J.H., Fredrickson G.H. (1990) J. Chem. Phys. 92, 6255.

Chen Z-R., Kornfield J.A. (1998) Polymer 39, 4679.

Chen Z.-R., Issaian A., Kornfield J.A., Smith S.D., Grothaus J.T., Satkowski M.M. (1997) Science 277, 1248.

Chen Z.-R., Issaian A., Kornfield J.A., Smith S.D., Grothaus J.T., Satkowski M.M. (1997) Macromolecules 30, 7096.

Floudas G., Hadjichristidis N., Iatrou H., Pakula T., Fischer E.W. (1994) Macromolecules 27, 7735.

Floudas G., Pakula T., Velis G., Sioula S., Hadjichristidis N. (1998a) J. Chem. Phys. 108, 6498.

Floudas G., Hadjichristidis N., Iatrou H., Avgeropoulos A., Pakula T. (1998b) Macromolecules 31, 6943.

Floudas G., Ulrich R., Wiesner U. (1999) J. Chem. Phys. 110, 652.

Floudas G., Ulrich R., Wiesner U., Chu B. (2000) Europhys. Lett. 50, 182.

Gerharz B. (1991) PhD Thesis, Univ. Mainz.

Hadziioannou G., Mathis A., Skoulios A. (1979) Colloid Polym. Sci. 257, 136.

Hajduk D.A., Gruner S.M., Erramilli S., Register R.A., Fetters L.J. (1996) Macromolecules 29, 1473.

Han C.D., Baek D.M., Kim J.K., Ogawa T., Sakamoto N., Hashimoto T. (1995) Macromolecules 28, 5043.

Han C.D., Kim J.K. (1993) Polymer 34, 2533.

Hodrokoukes P., Floudas G., Pispas S., Hadjichristidis N. (2001) 34, 650.

Johnson J.M., Allgaier J.B., Wright S.J., Young R.N., Buzza M., McLeish T.C.B. (1995) J. Chem. Soc. Faraday Trans. 91, 2403.

Kawasaki K., Onuki A. (1990) Phys. Rev. A 42, 3664.

Keller A., Pedemonte E., Willmouth F.M. (1970) Nature 225, 538.

Koppi K.A., Tirrell M., Bates F.S., Almdal K., Colby R.H. (1992) J. Phys. II, 2, 1941.

Okamoto S., Saijo K., Hashimoto T. (1994) Macromolecules 27, 5547.

Pakula T., Saijo K., Kawai H., Hashimoto T. (1985) Macromolecules 18, 1294.

Pakula T., Floudas G. (2000) in Block Copolymers, Calleja F.J.B., Roslaniec Z. (Eds), Marcel Dekker Inc., New York.

Riise B.L., Fredrickson G.H., Larson R.G., Pearson D.S. (1995) Macromolecules 28, 7653.

Rosedale J.H., Bates F.S. (1990) Macromolecules 23, 2329.

Rubinstein M., Obukhov S.P. (1993) Macromolecules 26, 1740.

Sakamoto N., Hashimoto T., Han C.D., Kim D., Vaidya N.Y. (1997) 30, 1621.

Stühn B. (1992) J. Polym. Sci. Polym. Phys. 30, 1013.

Wang H., Kesani P.K., Balsara N.P., Hammouda B. (1997) Macromolecules 30, 982.

Wiesner U. (1997) Macromol. Chem. Phys. 198, 3319.

Zhang Y., Wiesner U. (1995) J. Chem. Phys. 103, 4784.

Zhang Y., Wiesner U. (1997) J. Chem. Phys. 106, 2961.

Zhao J., Majumdar B., Schulz M.F., Bates F.S., Almdal K., Mortensen K., Hajduk D.A., Gruner S.M. (1996) Macromolecules 29, 1204.

CHAPTER 17

PHASE TRANSFORMATION KINETICS

Kinetics of phase transitions is one of the outstanding problems in statistical physics. The most interesting example is that of the demixing process in alloys or binary liquids brought about by a sadden temperature change below the stability line known as spinodal decomposition. The softness of interactions and long relaxation times involved in block copolymers make the kinetics of phase transformation in this system richer and more interesting than spinodal decomposition. Questions of fundamental interest here include the mechanism of the disorder-to-order and order-to-order transformation, the existence and role of intermediate states, epitaxy, time scales, and activation barriers. The phase transformation kinetics in block copolymers have, in addition, engineering applications; understanding the physics behind the transformations may help in designing processing routes for obtaining suitable structures for nanotechnology applications.

The ordering process in block copolymers bears many similarities with the crystallization process of semicrystalline materials. A variety of different experiments can be used to monitor the crystallization process, i.e., dilatometry, X-ray diffraction, rheology, dielectric spectroscopy, etc., but, undoubtedly, the most direct evidence is provided by dilatometry by recording the discontinuous change of the specific volume (Kasten 1995) and of the isothermal compressibility (Floudas 1993) as the material passes through the melting point. Earlier work on this subject revealed, however, that this method is not a sensitive probe of the ordering process in diblock copolymers because the effect of the ODT on the density and the compressibility is very small as a result of the transition being *weakly* first-order. On the other hand, scattering techniques such as SAXS and rheology are very sensitive to the evolution of the structure and of the viscoelastic properties during the ordering process. Therefore, it is not surprising that most investigations of the order-to-disorder transition in block copolymers involved scattering and rheology experiments (Adams 1996, Floudas 1994a,b, 1996, and 1998, Hashimoto 1995, Stühn 1994). We should mention, however, that techniques such as electron microscopy (Hashimoto 1996) and birefringence (Balsara 1992 and 1998, Dai 1996, Floudas 1995) have shed light on the actual grain growth mechanism. In this chapter we will review the main theoretical and experimental results on the ordering mechanism as well as the current understanding of transformations involving ordered microphases.

1. DETECTION AND ANALYSIS OF THE ORDERING KINETICS

Rheology is a powerful method for the detection of the ordering kinetics in block copolymers. The success of the method is based on the large viscoelastic contrast between the disordered and ordered phases. In rheology the time-evolution of the storage (G′) and loss (G″) moduli can be monitored following temperature jumps from the disordered to the ordered phase, and its actual shape is used to analyze the ordering kinetics. Some typical G′(t) and G″(t) dependencies are shown in Figure 17.1 for a symmetric SI diblock copolymer ($M_n = 12200$, $f_{PS} = 0.51$) (Floudas 1994a). Both moduli show a sigmoidal shape with distinct plateaus at

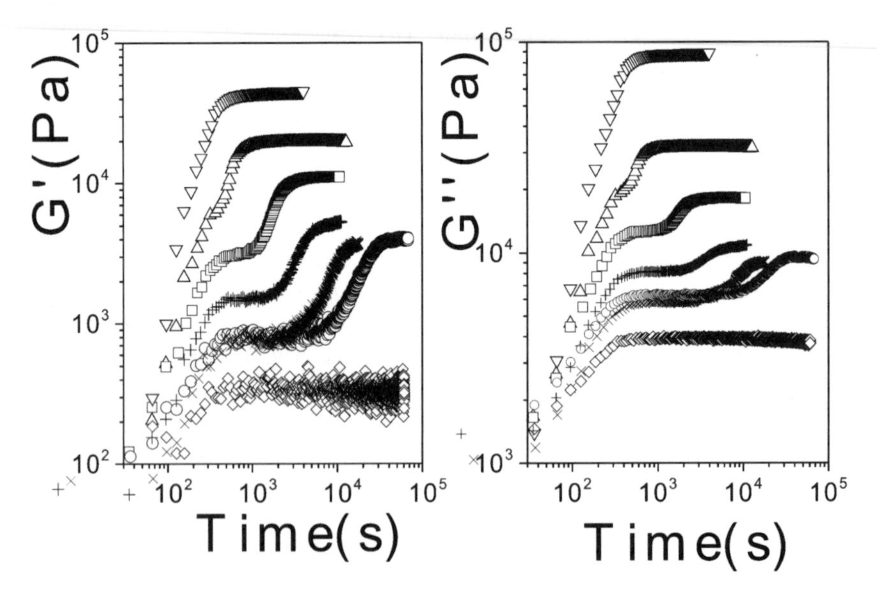

Figure 17.1. Time evolution of G′ and G″ for a symmetric poly(styrene-b-isoprene) (SI) diblock copolymer ($M_n = 12200$, $f_{PS} = 0.51$) following quenches from the disordered state (Ti = 363 K) to different final temperatures in the ordered phase: rhombus: 359, circles: 358, X: 356.5, +: 356, squares: 355, triangles: 353, and inverted triangles: 347.6 K (Floudas 1994a).

short and long times. The plateau of G′ and G″ at short times are assumed to describe the mechanical properties of the disordered phase at the quenched temperature. On the other hand, the plateaus at longer times describe the properties of the final microphase-separated state. In the intermediate time-range the system is regarded as a composite material made of two phases (ordered and disordered), the proportion of which changes with time, causing the observed changes in the rheological response. The long "incubation" time observed for the shallow quenches as well as the overall shape of the curves (see below) shown in Figure 17.1 point towards a *nucleation and growth* ordering mechanism.

In order to analyze the ordering kinetics in terms of a nucleation and growth mechanism, the time-dependence of the volume fractions of the constituent phases is needed. Several models have been developed to describe the properties of composite materials, but they do not provide any precise and unique solution to this problem. However, one can use the simplest "series" and "parallel" models, which provide the limits for the mechanical response of a two-phase system as a function of the properties of the constituent components and the composition. In the "series" model, the same stress is applied to both phases (ordered and disordered), and this results in different displacements. The modulus of the mixed phase is expressed as a linear combination of the compliances of the constituent phases:

$$\frac{1}{G(t)} = \frac{1-\phi(t)}{G_0} + \frac{\phi(t)}{G_\infty} \tag{17.1}$$

where $G(t)$ $(=(G'^2(t) + G''^2(t))^{1/2})$ is the absolute value of the complex modulus and G_0, G_∞ are the moduli of the initial ($t = 0$) disordered and final ($t = \infty$) ordered phases, respectively. In the "parallel" model, the two phases have the same displacement but different stresses, and the modulus of the mixed phase can be expressed as a linear combination of the moduli of the constituent phases:

$$G(t) = (1-\phi(t))G_0 + \phi(t)G_\infty \tag{17.2}$$

The time-dependence of ϕ obtained from both models at every temperature is analyzed by fitting the Avrami equation (Avrami 1939, 1940 and 1941):

$$\phi(t) = 1 - exp(-zt^n) \tag{17.3}$$

where z is the rate constant, and n is the Avrami exponent. The former gives a quantitative information on the course of "crystallization," and it is usually expressed in terms of the half-time (or completion time):

$$t_{1/2} = \left(\frac{ln\,2}{z}\right)^{1/n} \tag{17.4}$$

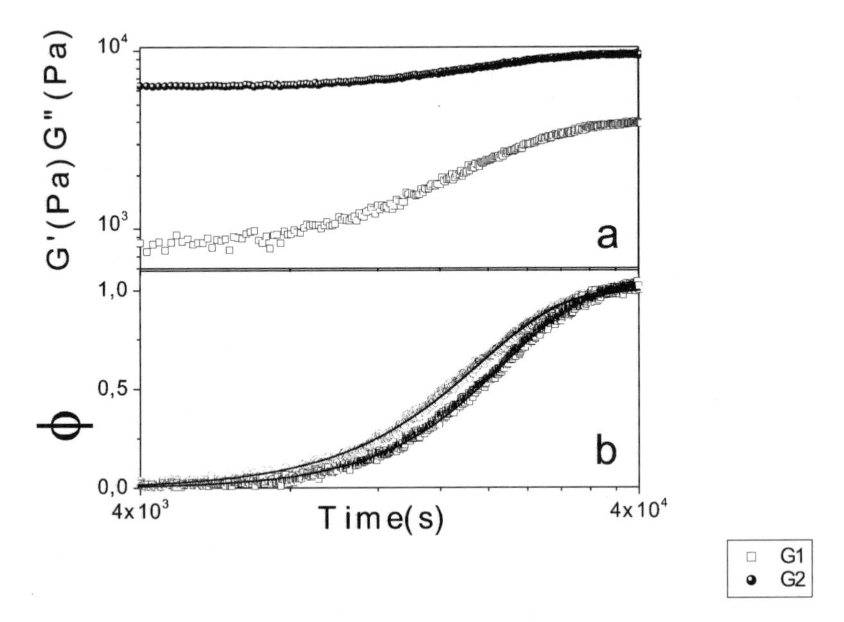

Figure 17.2. (a) Time evolution of G′ and G″ for the same diblock copolymer as in Figure 17.1 at T = 356.5 K following a quench from T = 363 K and (b) time-dependence of the volume fraction φ of the ordered phase calculated using the "paraller" (squares) and "series" (circles) models. Solid lines are fits to the Avrami equation (eq. 17.3).

The Avrami exponent, n, is a combined function of the growth dimensionality and of the time-dependence of the nucleation process and provides qualitative information on the nature of the nucleation and growth process. The Avrami parameters are usually extracted from a plot of $\log(-\log(1-\phi))$ versus $\log t$, from which it is possible to derive n and z from the slope and intercept, respectively. Figure 17.2 gives the evolution of the viscoelastic properties during ordering of a symmetric SI diblock copolymer (the same as in Fig. 17.1 at T = 356.5 K) and the corresponding volume fractions of the ordered phase (lamellar) using the two extreme models.

The analysis of the experimental data have shown that, for shallow quenches, the ordering process proceeds by heterogeneous nucleation and growth of three-dimensional objects with lamellar microstructure. At higher undercoolings the kinetic curves were steeper (higher n value), suggesting spinodal decomposition as a possible mechanism of structure formation (Floudas 1994a).

SAXS can also be used to follow the self-assembly process in block copolymers. With SAXS we can monitor the evolution of the structure factor at the relevant length scale (q^*), which is sensitive to the structural changes occurring during the local demixing process. An example is given in Figure 17.3, where the evolution of the SAXS profiles are shown for a block copolymer under isothermal conditions following a temperature jump from the disordered phase.

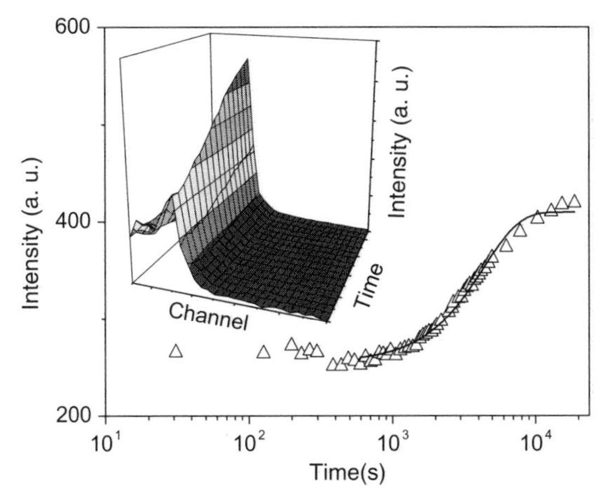

Figure 17.3. Evolution of the peak scattering intensity I(q*)(t) following a temperature jump from the disordered to the ordered state of a block copolymer. The line is a fit to the Avrami equation (eq. 17.3). The evolution of the scattering profiles is shown in the inset (Floudas 1994b). See color plates.

Upon ordering the scattering, profiles change from a rather broad liquid-like peak to a narrow solid-like peak, reflecting the characteristics of the ordered microdomain structure. The peak intensity, I(q*,t), changes by a factor of two in going from the disordered to the ordered state, and the shape of I(q*, t) can be fitted to the Avrami equation. A more exact comparison between the rheological and SAXS kinetic studies requires the extraction of the same quantity, namely, the volume fraction of the ordered phase. In SAXS this quantity can be extracted by measuring the total scattering power of the medium:

$$P = \frac{1}{2\pi^2}\int_o^\infty q^2 I(q)\, dq \tag{17.5}$$

which is evaluated from the scattering intensity integrated over all q. For an ideal two-phase system, the scattering power depends on the relative amounts of the volume fraction of the ordered (ϕ) and disordered (1−ϕ) phases:

$$P \sim \Delta n^2 \phi(1 - \phi) \tag{17.6}$$

where Δn is the electron density difference between the two phases. In that way the extracted volume fraction can be directly compared with the results obtained from rheology. A certain advantage in using rheology for the kinetic studies is that, by choosing a convenient (low) frequency, one can make the step in G*(t) to become large (usually one order of magnitude), whereas the observed change in I(q*,t) can be a factor of three at most.

2. THE EQUILIBRIUM ORDER-TO-DISORDER TRANSITION TEMPERATURE

One complication regarding the $T_{ODT,}$ which is usually obtained from a single heating experiment (in rheology or SAXS), is that the thus extracted temperature does not exactly correspond to the equilibrium transition temperature, $T_{ODT}{}^{0}$. The situation is very similar with the melting point in semicrystalline polymers. The way to obtain the equilibrium transition temperature is outlined below (Floudas 1998). Following the ordering kinetics, i.e., in rheology, isochronal temperature scans at low frequency aiming to disorder the system can be made. Should there be a single T_{ODT}, all heating curves would overlap. The result of such isochronal heating runs are shown in Figure 17.4 for a miktoarm star block copolymer of the SI_3 type. As shown in the figure, for every final ordered state, there is a different T_{ODT}; the lower the ordering temperature, the lower the apparent ODT. The pronounced dependence of the ODT on the ordering temperature is shown in the inset to Fig.17.4, where the apparent transition temperature (T_{ODT}') is plotted as a function of the corresponding ordering temperature (T_{ord}). The corresponding plot for semicrystalline polymers is known as the Hofmann-Weeks plot, where the apparent melting temperature (T_m') is plotted as a function of the crystallization

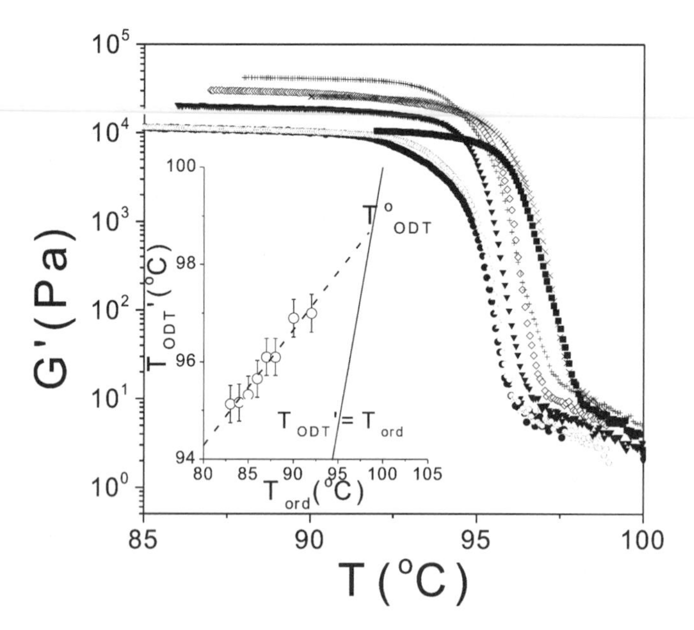

Figure 17.4. Isochronal (ω = 1rad/s) temperature scans (heating rate = 0.2 K/min) for the miktoarm star block copolymer SI_3 (M_n = 24000, f_{PS} = 0.61) obtained by heating the ordered phase formed at the kinetic experiments at different temperatures: filled circles: 83°C, open circles: 84°C, open triangles: 85°C, inverse triangles: 86°C, rhombus: 87°C, +: 88°C, X: 90°C, filled squares: 92°C. In the inset the corresponding Hofmann-Weeks plot is shown, and the $T_{ODT}{}^{o}$ is obtained by extrapolation (Floudas 1998).

temperature (T_c). The solid line in the figure signifies the $T_{ord} = T_{ODT}'$, and the equilibrium ODT can be obtained by extrapolation (dashed line). The extrapolation is based on the underlying Gibbs-Thomson equation which, for

$$T_m' = T_m^o\left(1 - \frac{2\sigma_e}{\Delta H_f}\frac{1}{d}\right) \tag{17.7}$$

semicrystalline polymers, gives the melting point depression due to the finite thickness (d) of the crystal. In eq. 17.7 σ_e is the surface free energy, and ΔH_f is the heat of fusion. Some typical values of these parameters for a semicrystalline polymer (i.e., PEO) are 20 erg/cm^2 and 200 J/g, respectively. Compare the latter value with the 1-2 J/g obtained during the order-to-disorder transition of a block copolymer.

These similarities between the order/disorder process and the crystallization/melting process suggest that we can simply rewrite eq. 17.7 for the former process, as:

$$L = \left(\frac{2\sigma}{\Delta H}\right)\left(\frac{T_{ODT}^o}{T_{ODT}^o - T_{ODT}'}\right) \tag{17.8}$$

where σ is now the interfacial tension, and ΔH is the change of enthalpy upon mixing. The above equation implies a strong effect of supercooling on sizes (L) of ordered regions.

Eq. 17.8 has been derived for lamellar crystallites where only the interactions at the lamellar surfaces play a role. In the case of grains, however, the whole intergrain surface should be considered and, thus, a generalization of eq. 17.8 is needed. The transition temperature is defined as $T_{ODT}' = \Delta H/\Delta S$, where ΔH (ΔS) is the enthalpy (entropy) difference between the ordered and disordered states. Under the assumption that the finite grain size causes the enthalpy reduction related to the disorder at the intergrain surfaces (constant entropy approximation), $\Delta H_o/T_{ODT}^o = \Delta H/T_{ODT}'$, where ΔH_o refers to the enthalpy of an infinitely large grain. The enthalpy reduction is related to a kind of disorder (defects) at the grain boundaries. This effect should be proportional to the surface-to-volume ratio of individual grains and the remaining enthalpy; limiting the stability of grains can be expressed as $\Delta H_o(1-(s/v^*)l)$, where s and v* are, respectively, the grain surface and volume, and l is the intergrain layer thickness. Based on these considerations, the grain dimensions, L, are:

$$L = \beta\frac{T_{ODT}^o}{T_{ODT}^o - T_{ODT}'} \tag{17.9}$$

where β is a function of the geometrical characteristics of the grain and of the intergrain layer thickness. Eq. 17.9 is a generalized Gibbs-Thomson equation for grain sizes, and knowledge of β allows the determination of the grain size. A pronounced grain size dependence on supercooling has been observed in experiments,

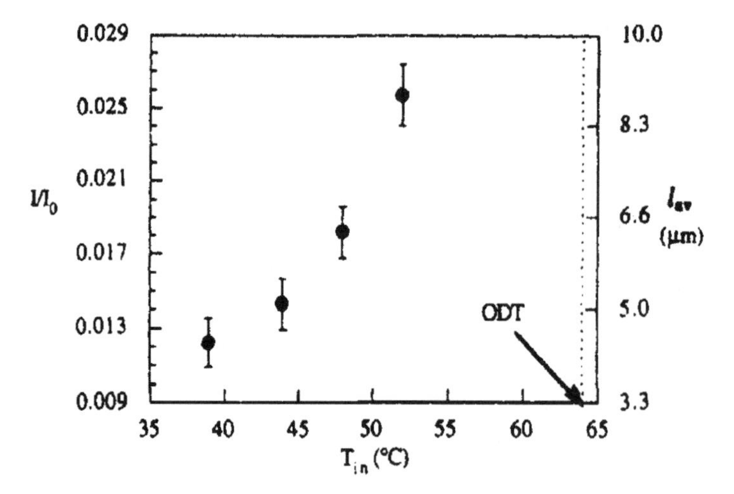

Figure 17.5. Dependence of birefringence and of the estimated grain size on the ordering temperature. Notice the strong grain-size dependence below the ODT (Balsara 1992).

which probe the development of birefringence (Balsara 1992) and optical anisotropy (Floudas 1995) following quenches from the disordered phase. The results of the birefringence experiment is depicted in Figure 17.5 and reveals a strong effect of the grain size on the ordering temperature as the one predicted by eq. 17.9. As a last remark, we note the similarities of the dependencies shown in Figures 17.4 and 17.5 with the crystallization process in semicrystalline polymers.

3. EFFECT OF FLUCTUATIONS

As we have seen in Chapter 15, in contrast to the Leibler's MFT, the FH theory allows for the existence of metastable states as observed in the experiments. The theoretical work of Fredrickson and Binder (Fredrickson 1989) predicted the existence of such metastable states and described the nucleation and growth of a lamellar phase from an undercooled disordered phase. The nucleation barrier for $f = 0.5$ was found to be unusually small when compared with polymer blends and equal to:

$$\left.\frac{\Delta F^*}{k_B T}\right|_s \approx \bar{N}^{-1/3}\delta^{-2} \tag{17.10}$$

where δ is the dimensionless undercooling parameter defined as $\delta = (\chi - \chi_{ODT})/\chi_{ODT}$, and χ_{ODT} and χ are values of the interaction parameter at the ODT and at the final temperatures. The characteristic ordering time is given by:

$$t_{1/2} \approx exp\left(\frac{\Delta F^*}{k_B T}\right) \tag{17.11}$$

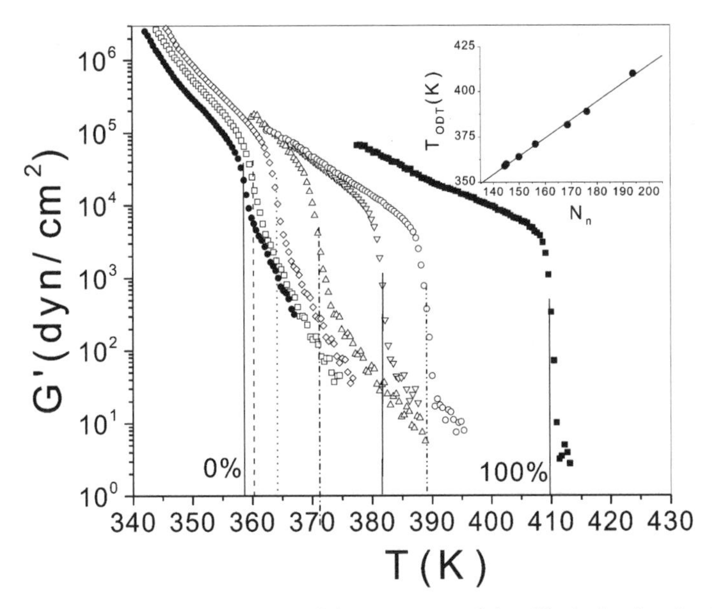

Figure 17.6. Temperature-dependence of the storage modulus G′ obtained at 1 rad/s while heating the pure diblock copolymers (with N_n of 144 and 194) and their blends with different compositions as indicated. Vertical lines indicate the different T_{ODT}'s. The inset gives the dependence of the order-to-disorder transition temperatures on the number average degree of polymerization N_n for all copolymers (Floudas 1996).

Binder has shown that asymmetry greatly affects the ordering kinetics as (Binder 1995):

$$\left.\frac{\Delta F^*}{k_B T}\right|_{AS} \approx \bar{N}^{1/2}|f - 0.5|^5\delta^{-2} \tag{17.12}$$

through the $|f - 0.5|^5$ term. A pertinent feature of the ordering kinetics as described by eq. 17.10 is that the characteristic ordering time scales with $\bar{N}^{-1/3}$, which is the usual fluctuation term in the FH theory. Testing this theoretical prediction requires a large number of exactly symmetric diblock copolymers with different molecular weights. Instead, one can employ just two symmetric diblock copolymers (i.e., two symmetric poly(styrene-b-isoprene) diblock copolymers) with ratio of molecular weights not more than 5 (to ensure the formation of a single lamellar). Mixtures of the copolymers can then be prepared at different compositions giving rise to "new" symmetric copolymers with intermediate molecular weights. The corresponding ODTs of these copolymers, as obtained from rheology, are shown in Figure 17.6 (Floudas 1996). These (apparent) ODTs are plotted in the inset to Figure 17.6 as a function of the number-averaged degree of polymerization and give rise to a linear dependence.

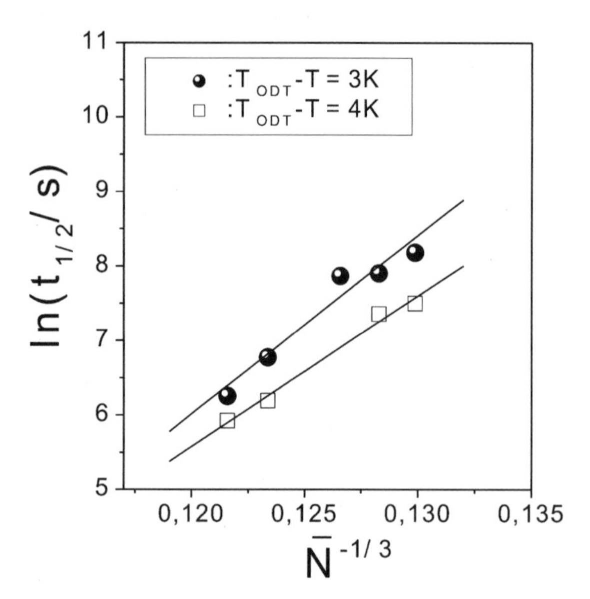

Figure 17.7. Molecular weight-dependence of the ordering times in the symmetric diblock copolymers. The characteristic time $t_{1/2}$ is plotted versus the fluctuation parameter at two temperatures, as indicated (Floudas 1996).

By performing kinetic studies for each one of the "new" copolymers, the characteristic ordering times can be obtained. These times are then plotted as a function of $\bar{N}^{-1/3}$ for two supercoolings in Figure 17.7. The linearity shown constitutes a proof of the theoretical predictions and shows that the ordering kinetics in symmetric diblocks are controlled by the composition fluctuations.

As we discussed earlier in the text, the kinetic studies provide a direct proof for the inadequacy of the MFT near the transition. In fact, the sole existence of kinetics during the disorder-to-order transformation contradict the MFT, which does not allow for metastable states. Furthermore, the molecular weight dependence of the characteristic times are in agreement with the fluctuation approach.

4. GRAIN GROWTH

The long incubation times and the overall shape of the kinetics obtained through SAXS and rheology revealed that, at least for shallow quenches, the ordering process proceeds by a heterogeneous nucleation and growth process. Recently, results from transmission electron microscopy (Hashimoto 1996, Sakamoto 1998) and birefringence (Balsara 1992 and 1998, Dai 1996) confirmed these findings and shed more light on the ordering process. Figure 17.8 gives a TEM micrograph from a nearly symmetric poly(styrene-b-isoprene) diblock copolymer ($M_n = 15000$, $f_{PS} = 0.45$) during the ordering process. The image clearly shows coexisting

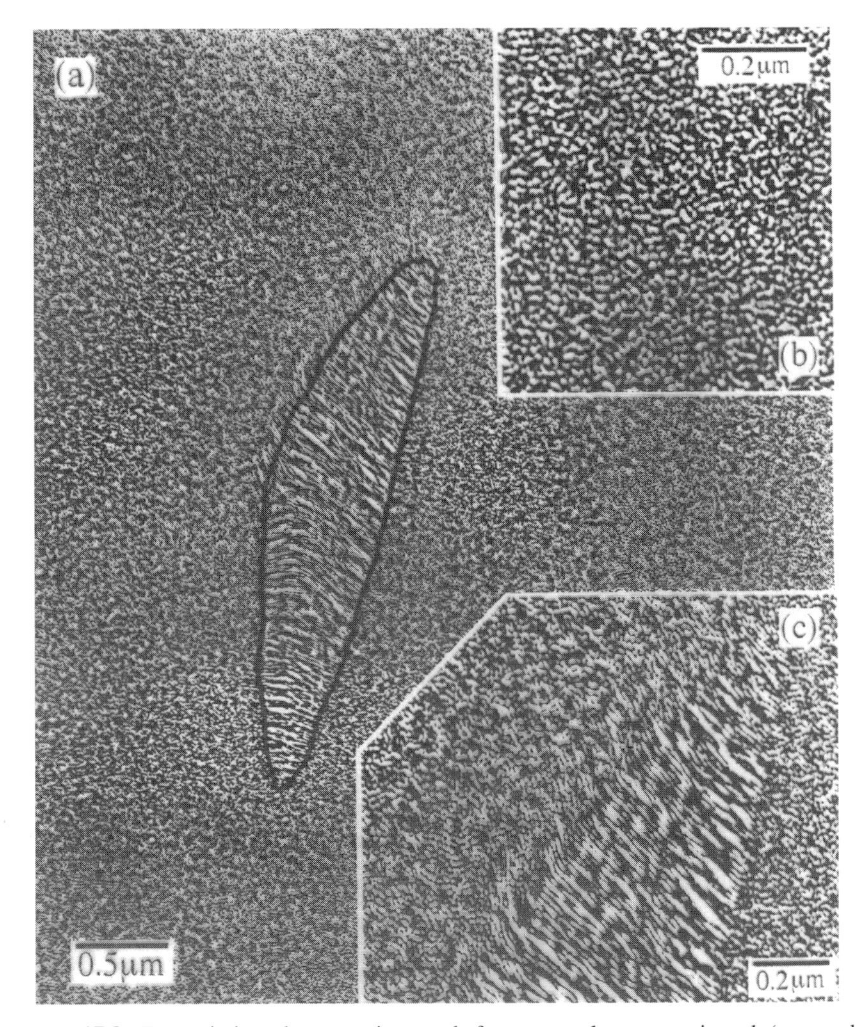

Figure 17.8. Transmission electron micrograph from a nearly symmetric poly(styrene-b-isoprene) diblock copolymer ($M_n = 15000$, $f_{PS} = 0.45$) following a quench from the disordered phase (Hashimoto 1996).

ordered lamellar grains with disordered regions, thus confirming the assignment as a nucleation and growth process. Furthermore, the shape of the lamellar grains is highly anisotropic with the size along the lamellar normal being much larger than that parallel to the lamellar interfaces. Finally, it was found that the lamellae at the beginning of the process are perforated, but, later on, they transform into the solid lamellae.

Depolarized light scattering was used to detect the size, shape, and concentration of ordered grains in a quenched block copolymer. The experiment revealed also anisotropic grains with grain size parameters changing with time. This is depicted

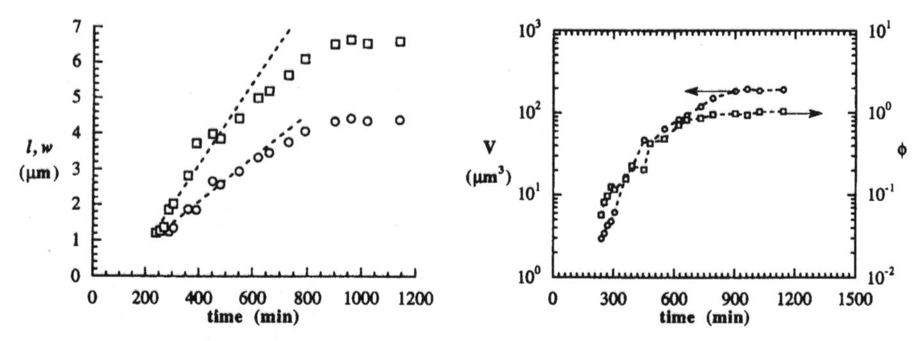

Figure 17.9. (Left) Evolution of the grain size parameters l (squares) and w (circles) after quenching of a poly(styrene-b-isoprene) diblock copolymer. The lines indicate the slopes used to extract the propagation velocities of the front. (Right) Evolution of the grain volume, V, and grain volume fraction, ϕ, during the same quench (Balsara 1998a).

in Figure 17.9, where the evolution of the grain size parameters l and w are shown for a quenched SI copolymer. In the same figure the evolution of the grain volume, V, and of the volume fraction, ϕ, occupied by grains is compared and display a similar dependence. This suggests that the evolution of the volume fraction of grains is entirely due to the growth of existing grains and not through the formation of new grains. This, again, suggests a heterogeneous nucleation and growth process in agreement with earlier results.

Following these studies a time-dependent Landau-Ginzburg model was proposed to relate the growth rate of the grains to molecular parameters. The growth rate is predicted to be proportional to the quench depth and chain dimensions and inversely proportional to the molecular relaxation time as:

$$v \approx \frac{\langle r^2 \rangle^{1/2}}{\tau} \left[\chi N - (\chi N)_{ODT} \right] g(f) \tag{17.13}$$

where r is end-to-end distance, τ is the longest relaxation time in the undercooled melt, and g(f) is a composition-dependent parameter. Eq. 17.13 suggests that the parameter providing the driving force for the front propagation is the difference in χN in the final and ODT states.

An interesting case has been reported for the ordering of cylinders arranged in a hexagonal lattice from the disordered melt (Balsara 1998b). The ordered phase has a liquid crystalline symmetry with liquid-like disorder along the cylinder axis and crystalline order in the hexagonal plane. When the kinetics of structure formation were studied by SAXS and depolarized light-scattering, the following was found; for small quench depths, the volume fractions obtained from the two experiments had a similar time-dependence, implying that the microstructure formation along the two directions are strongly correlated during the whole process, which, again, is suggesting a nucleation and growth mechanism. However, for deeper quenches, the

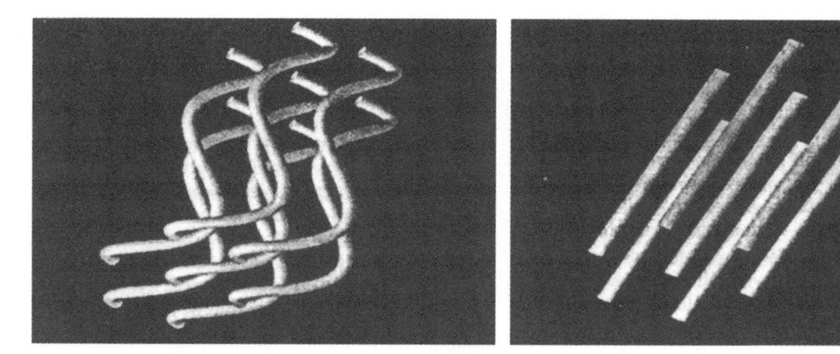

Figure 17.10. (Right) Cylindrical microstructure obtained at low undercoolings as opposed to the wormlike structure (left) obtained for deeper quenches (Balsara 1998b).

rate of growth of the hexagonal phase as detected by SAXS occurred more rapidly than the rate of growth of the ordered grains as detected by light-scattering. To explain this situation, the microstructure shown in Figure 17.10 was proposed. The proposed microstructure consists of an array of hexagonally packed, wormlike cylinders with a low persistence length. The different mechanism obtained for the deeper quenches may suggest that ordering proceeds by spinodal decomposition instead of nucleation and growth.

5. EFFECT OF BLOCK COPOLYMER ARCHITECTURE

The synthesis of nonlinear block copolymers allows the investigation of the copolymer architecture not only on the phase state but also on the ordering process. As an example we compare below the ordering kinetics in two miktoarm star copolymers (SI_2, SIB, S: polystyrene, I: polyisoprene, B: polybutadiene) and a four-arm star block copolymer $(SI)_4$ and two linear SI. Notice that these complex block copolymers order with the same mechanism (nucleation and growth). Figure 17.11 shows Avrami plots for the volume fraction of the ordered phase for the above cases. Notice that, in most cases, a good linearity is obtained, suggesting a common nucleation and growth mechanism. However, in comparing the ordering kinetics as a function of supercooling for the different architectures, there is a strong influence of the chain topology on the ordering times. The most intriguing feature of the kinetic times shown in Figure 17.12 is the freezing of the kinetics in the star diblock copolymer. Under the same undercooling, the star diblock needs about three additional decades of time-compared with the equivalent linear diblock-to develop the ordered structure. The ordering kinetics in block copolymers are related to the mobility of polymer chains and, therefore, to the presence of entanglements, which can result in the quenching of the star dynamics through the prefactor in eq. 17.11.

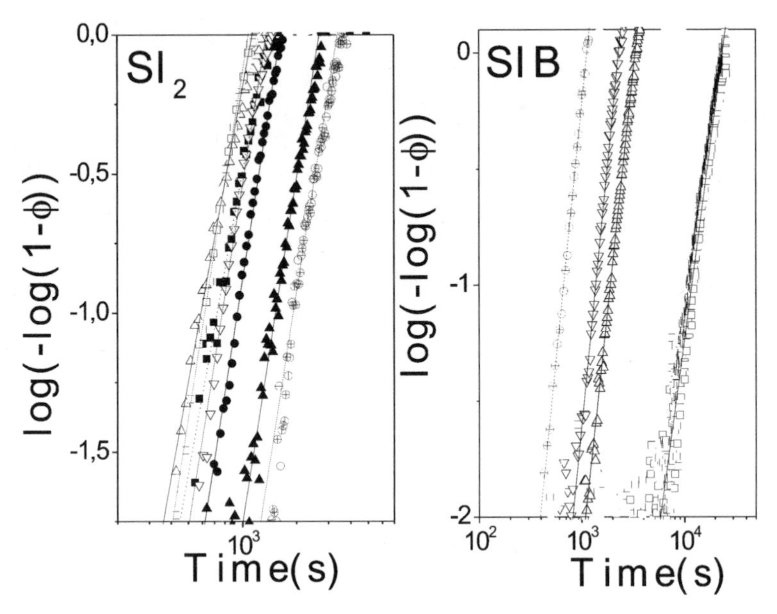

Figure 17.11. Time-evolution of the volume fraction of the ordered phase in block copolymers with different architectures. The solid lines are fits to the Avrami equation. The samples are SI_2: miktoarm star block copolymer ($M_n^{star} = 23800$, $w_{PS} = 32\%$), SIB: miktoarm star block copolymer of PS, PI, and PB ($M_n^{star} = 23800$, $w_{PS} = 34\%$).

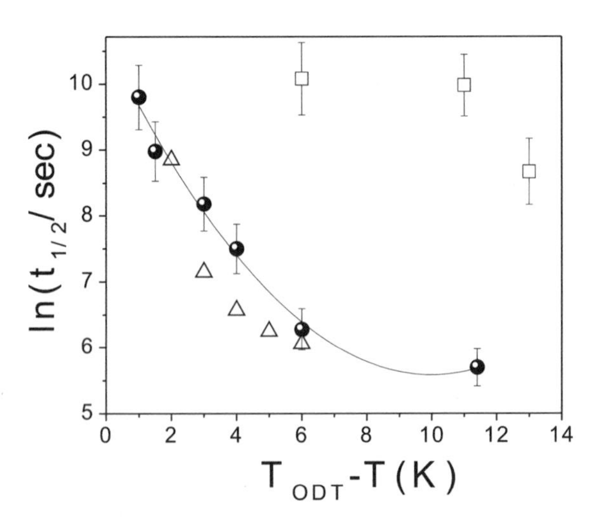

Figure 17.12. Characteristic ordering times $t_{1/2}$ plotted versus the temperature difference from T_{ODT}, for three block copolymers: circles: SI-85/65 symmetric diblock ($M_n = 12200$, $f_{PS} = 0.51$), triangles: SI asymmetric diblock ($M_n = 23800$, $f_{PS} = 0.22$), and squares: $(SI)_4$ star diblock copolymer ($M_n^{star} = 98000$, $f_{PS} = 0.25$) (Floudas 1996b).

6. TRANSITIONS BETWEEN DIFFERENT ORDERED STATES

Despite earlier work in surfactants and lipids, the kinetics of phase transformation between the different microphases, with consequences in processing and applications in nanotechnology, have started to be explored only recently. In exploring the polymorphism of some order-to-order transitions, there are many issues that deserve attention such as phase coexistence, nucleation sites, existence and nature of intermediate states, time scales, activation barriers, and the role of fluctuations.

On the theoretical side, Qi and Wang (Qi 1996 and 1998) employed a time-dependent Landau-Ginzburg approach and studied the Hex-to-Dis and Hex-to-bcc transformations. Laradji et al. (Laradji 1997a,b) employed a theory of anisotropic fluctuations and discussed the Lam-to-Hex transition. Goveas and Milner (Goveas 1997) studied the latter transformation by focusing on a subset of wave vectors of the two phases. Lastly, Matsen (Matsen 1998) applied a self-consistent field theory to the Hex-to-Gyroid transformation. In all of the above studies, epitaxy was either found or assumed. However, there are very few direct experiments that probe the kinetic pathways between amorphous ordered phases after sudden temperature changes. Below we review some of the experimental and theoretical efforts on this subject.

6.1. Hexagonal-to-Gyroid Transformation

A poly(ethyleneoxide-b-isoprene) diblock copolymer with $M_n = 9800$ and $f_{PEO} = 0.39$ was found (Floudas 1999) to undergo the following transitions by heating: $L_c \overset{325}{\rightarrow} Hex \overset{418}{\rightarrow} Gyroid \overset{486}{\rightarrow} Dis$, i.e., from the crystalline lamellar structure (L_c) to hexagonally packed cylinders (Hex) to the bicontinuous cubic phase with the $Ia\bar{3}d$ symmetry (Gyroid). The kinetics of the hexagonal-to-Gyroid transition were studied both by SAXS and rheology (Floudas 2000) by imposing sudden temperature changes from an initial temperature corresponding to the hexagonal structure (353 K) to different temperatures corresponding to the Gyroid phase.

Some representative SAXS spectra are shown in Figure 17.13, as well as the q*(T) for a transition from 353 K to 413 K. The strong q*(T) within the first 90 s originates largely by thermal expansion effects. After 150 s the spectra nearly locked in position, and a new reflection, corresponding to the (220) reflection of the Gyroid phase, appeared. The formation of the new phase was followed by subsequent growth at the expense of the Hex phase, and, after $2\text{x}10^3$ s, a stable Gyroid phase was formed as indicated by the (220) and (321) reflections (the latter differentiates the Gyroid from the fcc structure). The results revealed a short incubation time (~60 s) and a long-time phase coexistence, in favor of a nucleation and growth process (see the following text). The growth of the intensity at the $(4/3)^{1/2}q^*$ reflection corresponding to the (220) reflection of the Gyroid phase can be used as a probe of the time scale and activation barrier involved;

$$I\left(\sqrt{\frac{4}{3}}q^*, t\right) = I\left(\sqrt{\frac{4}{3}}q^*, 0\right) + \Delta I(1 - e^{-zt^n}) \tag{17.14}$$

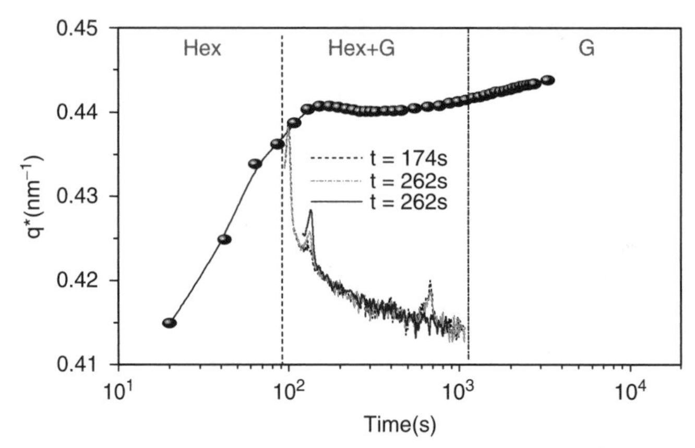

Figure 17.13. Time-evolution of the peak at q* (circles) following a temperature jump from 353 K (Hex) to 413 K (Gyroid). Some representative spectra are shown corresponding to the initial hexagonal structure, the final gyroid and the intermediate coexisting hexagonal and gyroid phases for $174 < t < 1800$ s (Floudas 2000).

where z and n are the usual Avrami parameters. The characteristic times (actually half-times) were obtained as $\tau = (\ln 2/z)^{1/n}$ as shown in the inset to Figure 17.14. In the same plot, the result of the same T-jump now performed with rheology is included. The evolution of $G'(t)$ shows an S shape, which, in view of the SAXS results, reflects the Hex-to-Gyroid phase transformation. The characteristic times in rheology can be obtained by using simple mechanical models of composite materials to extract the volume fraction of the newly formed phase $\phi_{Gyroid}(t)$ from the evolution of $G'(t)$ and $G''(t)$. In the limiting case of homogeneous distribution of strain (parallel model), the modulus of the two-phase system can be expressed as a linear combination of the moduli of the constituent phases as:

$$G* = \phi_{Hex} G^*_{Hex} + \phi_{Gyroid} G^*_{Gyroid} \tag{17.15}$$

where ϕ_{Hex} and ϕ_{Gyroid} ($=1-\phi_{Hex}$) are the volume fractions of the Hex and Gyroid phases, respectively. The volume fraction of the newly formed phase $\phi_{Gyroid}(t)$ is then fitted to the Avrami equation:

$$\phi_{Gyroid}(t) = 1 - e^{-zt^n} \tag{17.16}$$

The Avrami exponent from both experiments was 2 ± 0.2, implying an anisotropic growth of either rodlike objects from homogeneous nuclei or disklike objects from heterogeneous nuclei, the presence of interfaces being in favor of a heterogeneous process. The corresponding kinetic times from rheology are also plotted in the inset to Figure 17.14 and show a quantitative agreement with SAXS. The activation energy is about 50 kcal/mol, which corresponds largely to a collective

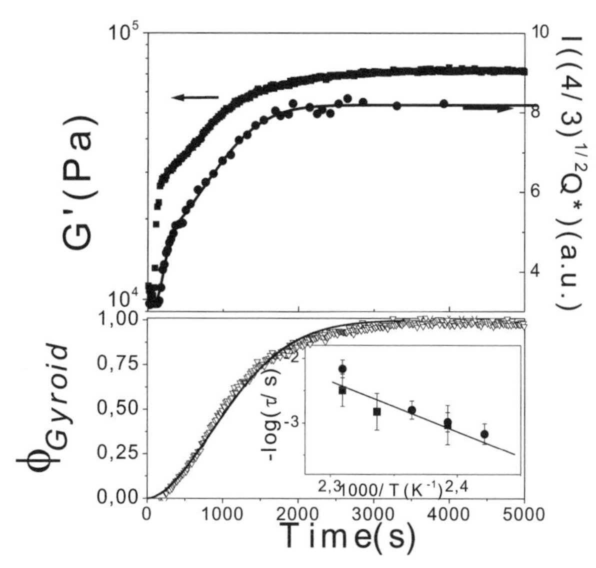

Figure 17.14. Top: Storage modulus during the isothermal (ω = 1rad/s, strain amplitude = 0.5%) kinetic experiment of the Hexagonal-to-Gyroid transition in the PEO-PI diblock (squares). The initial temperature was 353 K (Hex), and the final temperature 418 K. The intensity of the (220) SAXS reflection as a function of time for the same temperature jump is also shown (circles). Bottom: Evolution of the volume fraction of the Gyroid phase calculated at 418 K. The lines are fits to the Avrami equation. In the inset the characteristic times from SAXS (circles) and rheology (squares) are plotted in the usual Arrhenius representation.

process. It is worth mentioning that the homopolymer segmental and chain dynamics at these high temperatures (i.e., $T_g + 200$ K, both homopolymers have a glass transition temperature, T_g, of 208 K) have an apparent activation energy of only 7 kcal/mol. Despite the value of the activation energy being an order of magnitude higher than the energy associated with the chain relaxation, it is not excessively high and permits the transformation.

The experimental results for the Hex-to-Gyroid transition revealed a short incubation time with fluctuations within the hexagonal structure, nucleation, and phase coexistence with nearly epitaxial match of periodicity, highly anisotropic growth, and long time-scales associated with highly collective processes. The above experimental results can be compared with recent SCFT treating the Hex-to-Gyroid transformation (Matsen 1998a). The barrier for the Gyroid-to-Hex transition was found to vanish near the Hex/Sphere spinodal line while, in the forward direction (i.e., Hex-to-Gyroid), the system experiences two instabilities at the Hex-to-Lam and Hex-to-Gyroid spinodals. Notably, the spinodals (Figure 17.15) were well separated from the Hex/Gyroid boundary, and the separation was found to increase with increasing segregation, which implied a nucleation and growth mechanism. Furthermore, the small kinetic barrier suggested an epitaxial transformation, i.e., nearly complete match in orientation and domain size. A possible scenario for the

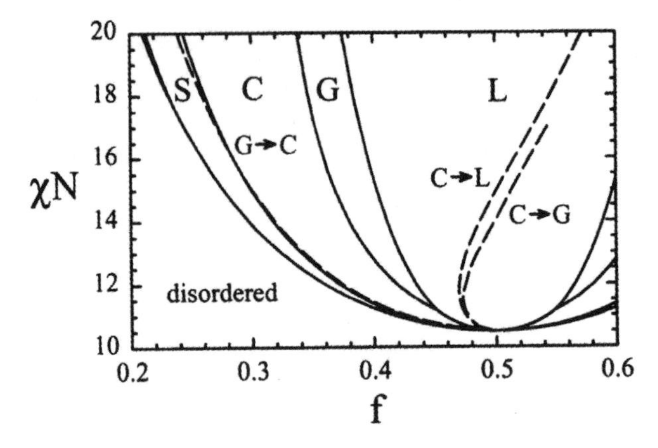

Figure 17.15. Phase diagram for diblock copolymer melts calculated with SCFT. The dashed lines are spinodal limits for the Gyroid-to-Hexagonal (G-C), Hexagonal-to-Lamellar (C-L), and Hexagonal-to-Gyroid (C-G) transitions (Matsen 1998a).

Hex-to-Gyroid transition shown in Figure 17.16 involves the formation of a nucleus of a single fivefold junction from three neighbor cylinders, which finally produce the required threefold junctions. The above scenario results in a highly anisotropic growth with the Gyroid phase growing predominantly along the cylinder axis and is consistent with the experimental results both for the anisotropy and for the highly collective process (nucleation and growth) involved. Furthermore, the transformation is nearly epitaxial, with only a 1% mismatch in length scale, which is well

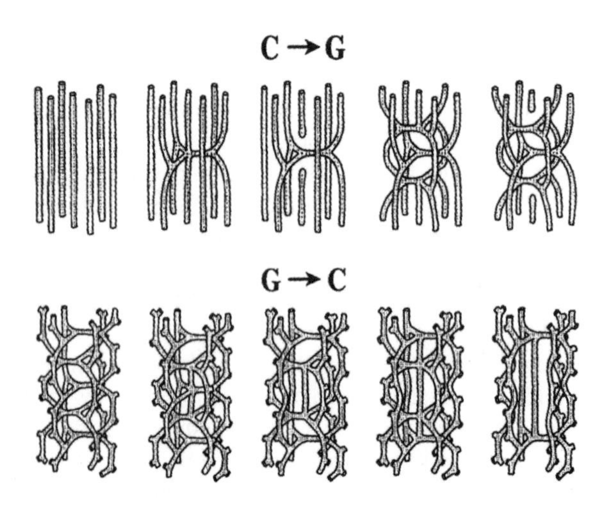

Figure 17.16. Schematic illustration of the nucleation and growth mechanism involved in the Hexagonal-to-Gyroid (C-G) and the reverse transformation (Matsen 1998a).

within the theoretical prediction and is responsible for the small energy barrier involved. The reverse transformation, i.e., Gyroid-to-Hex, was also studied experimentally and described as a near epitaxial nucleation and growth mechanism with very long (more than 2×10^3 s) coexistence between the Gyroid and Hex phases, partly due to mobility reasons. The theoretical scenario for this transformation involves the nucleation of a fivefold junction and the rapture of the original threefold junctions resulting in a slow unzip of the Gyroid phase to cylinders.

6.2. Lamellar-to-hexagonal Transformation

A time-dependent Landau-Ginzburg (TDLG) approach was used (Qi 1996 and 1998) to follow the kinetics of several order-to-order and order-to-disorder transformations after sudden temperature jumps in weakly segregated diblock copolymers. Direct numerical simulation of the TDGL equations indicated that, depending on the extent of the temperature jumps, some intermediate states between the initial and final equilibrium states were formed. An example of the microstructure evolution following a temperature jump from the lamellar to the hexagonal phase is shown in Figure 17.17.

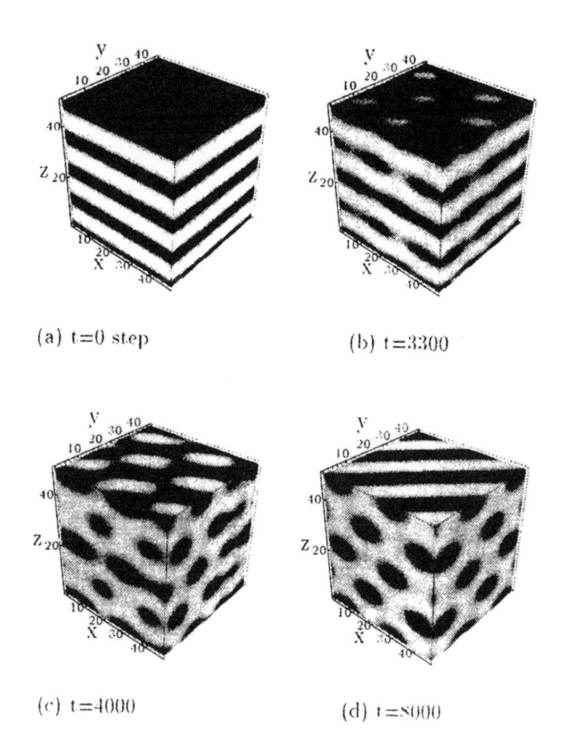

Figure 17.17. Evolution of microstructures after a temperature jump from the Lamellar to the Hexagonal phase at $f = 0.45$ and $(\chi N)_i = 12.0$ to $(\chi N)_f = 11.03$ at different stages of the simulation (Qi 1998).

The images reveal the formation of a metastable perforated lamellar (PL) structure, which has been seen in several experiments. The PL phase is considered as a kinetically trapped structure and, as such, is not an equilibrium structure. Similarly, a theory based on anisotropic fluctuations (Laradji 1997) found the PL phase to be unstable around the lamellar-to-hexagonal transformation. The stability of the double Gyroid structure was also discussed with the latter approach and was compared with the results from the earlier SCFT calculations (Matsen 1998b).

6.3. Cylinder to bcc Transformation

The existence of a thermally reversible transition between a hexagonal array of rods and a bcc array of spheres for asymmetric block copolymers is predicted by mean-field theory and has been shown experimentally (Koppi 1994, Sakurai 1996, Kimishima 2000). In the experiment an epitaxial relationship between the $\langle 001 \rangle$ axis of the cylinders and the $\langle 111 \rangle$ axis of spheres was demonstrated. In the first theoretical approach, anisotropic fluctuations were employed (Laradji 1997) to locate the spinodal lines for each ordered phase and the most unstable fluctuation modes. For the cylinder-to-bcc transformation, a real space representation of these fluctuations is given in Figure 17.18. The fluctuation modes have initially small amplitude and lead to the bulging and pinching of the cylinders. Stronger fluctuations result in the cylinder breakup to form spheres that are centered on a bcc lattice with a periodicity identical to that of the equilibrium spherical phase. Thus, an epitaxial relation between the cylindrical and spherical phases is predicted.

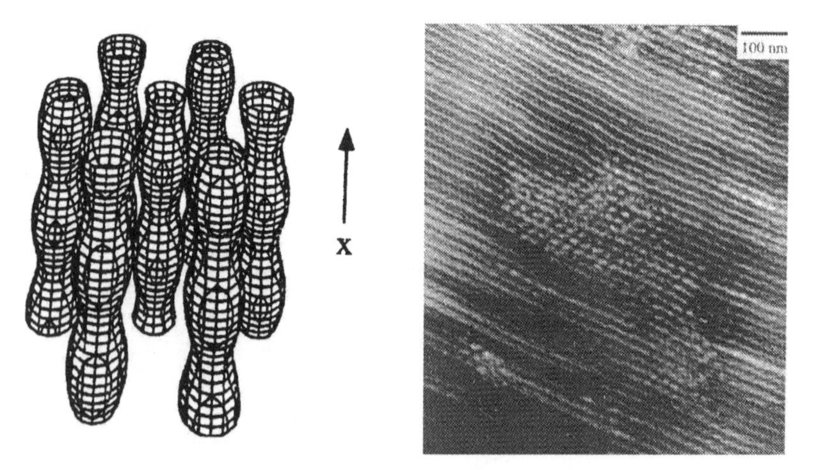

Figure 17.18. (Left) Three-dimensional contour plots of the Hexagonal phase showing the effect of fluctuations. (Right) TEM micrograph displaying a hexagonal arrangement of droplets with the underlying cylindrical orientation preserved. The image is taken from a shear-oriented poly(styrene-b-isoprene-b-styrene) triblock copolymer following annealing at high temperatures. TEM shows a grain of distorted cylinders with an internal structure of elongated droplets with a preserved underlying cylindrical orientation (Ryu 1999).

The second theoretical approach employed the TDLG theory (Qi 1996 and 1998) and predicts that the morphology evolves through periodic (peristaltic) undulations in the cylinder diameters with bcc symmetry. A direct evidence for the existence of fluctuations with cubic symmetry within the stable cylindrical phase has been provided by the TEM image shown in Figure 17.18 (Ryu 1998 and 1999).

REFERENCES

Adams J.L., Quiram D.J., Graessley W.W., Register R.A., Marchand G.R. (1996) Macromolecules 29, 2929.

Avrami M.J. (1939) J. Chem. Phys. 7, 1103.

Avrami M.J. (1940) J. Chem. Phys. 8, 212.

Avrami M.J. (1941) J. Chem. Phys. 9, 177.

Balsara N.P., Garetz B.A., Dai H.J. (1992) Macromolecules 25, 6072.

Balsara N.P., Garetz B.A., Chang M.Y., Dai H.J., Newstein M.C., Goveas J.L., Krishnamootri R., Rai S. (1998a) Macromolecules 31, 5309.

Balsara N.P., Garetz B.A., Newstein M.C., Bauer B.J., Prosa T.J. (1998b) Macromolecules 31, 7668.

Binder K. (1995) Physica A213, 118.

Dai H.J., Balsara N.P., Garetz B.A., Newstein M.C. (1996) Phys. Rev. Lett. 77, 3677.

Floudas G., Vogt S., Pakula T., Fischer E.W. (1993) Macromolecules 26, 7210.

Floudas G., Pakula T., Fischer E.W., Hadjichristidis N., Pispas. (1994a) Acta Polymer. 45, 176.

Floudas G., Hadjichristidis N., Iatrou H., Pakula T., Fischer E.W. (1994b) Macromolecules 27, 7735.

Floudas G., Fytas G., Hadjichristidis N., Pitsikalis M. (1995) Macromolecules 28, 2359.

Floudas G., Vlassopoulos D., Pitsikalis M., Hadjichristidis N., Stamm M. (1996) J. Chem. Phys. 104, 2083.

Floudas G., Pispas S., Hadjichristidis N., Pakula T., Erukhimovich I. (1996b) Macromolecules 29, 4142.

Floudas G., Pakula T., Velis G., Sioula S., Hadjichristidis N. (1998) J. Chem. Phys. 108, 6498.

Floudas G., Ulrich R., Wiesner U. (1999) J. Chem. Phys. 110, 652.

Floudas G., Ulrich R., Wiesner U., Chu B. (2000) Europhys. Lett. 50, 182.

Floudas G., Vazaiou B., Schipper F., Ulrich R., Wiesner U., Iatrou H., Hadjichristidis N. (2001) Macromolecules 34, 2947.

Fredrickson G.H., Binder K. (1989) J. Chem. Phys. 91, 7265.

Goveas J.L., Milner S.T. (1997) Macromolecules 30, 2605.

Hashimoto T., Sakamoto N. (1995) Macromolecules 28, 4779.

Hashimoto T., Sakamoto N., Koga T. (1996) Phys. Rev E 54, 5832.

Hohenberg P.C., Swift J.B. (1995) Phys. Rev. E 52, 1828.

Kasten H., Stühn B. (1995) Macromolecules 28, 4777.

Koppi K.A., Tirrell M., Bates F.S., Almdal K., Mortensen K. (1994) J. Rheol. 38, 999.
Kimishima K., Koga T., Hashimoto T. (2000) Macromolecules 33, 968.
Laradji M., Shi A-C., Desai R. C., Noolandi J. (1997a) Phys. Rev. Lett. 78, 2577.
Laradji M., Shi A-C., Noolandi J., Desai R. C. (1997b) Macromolecules 30, 3242.
Lodge T.P., Fredrickson G.H. (1992) Macromolecules 25, 5643.
Matsen M.W. (1998a) Phys. Rev. Lett. 80, 4470.
Matsen M.W. (1998b) Phys. Rev. Lett. 80, 201.
Qi S., Wang Z-G. (1996) Phys. Rev. Lett. 76, 1679.
Qi S., Wang Z-G. (1998) Polymer 39, 4639.
Ryu C.Y., Vigild M.E., Lodge T.P. (1998) Phys. Rev. Lett. 81, 5354.
Ryu C.Y., Lodge T.P. (1999) Macromolecules 32, 7190.
Sakamoto N., Hashimoto T. (1998) Macromolecules 31, 3815.
Sakurai S., Hashimoto T., Fetters L.J. (1996) Macromolecules 29, 740.
Stühn B., Vilesov A., Zachmann H.G. (1994) Macromolecules 2.

CHAPTER 18

BLOCK COPOLYMERS WITH STRONGLY INTERACTING GROUPS

Ion-containing polymers are materials with unique physical properties that have attracted extensive experimental and theoretical work. These materials are hydrocarbon polymers containing ionic groups, which are discussed as ionomers (usually less than 10%) (Eisenberg 1970) (MacKnight 1981) (Nyrkova 1993) or polyelectrolytes (more than 10%) (Earnest 1981). The first studies on ionomers were made with random ionomers where the ionic groups were randomly distributed along the polymer chain. Ionic groups are more polar than the hydrocarbon chains, and this creates an electrostatic driving force for association, which normally exceeds the elasticity of polymer chains and results in the formation of ionic aggregates. There is consensus that such ionic aggregates are responsible for the unique physical properties of ionomers. Evidence for ionic aggregation exists in: (i) the small-angle X-ray and neutron scattering patterns (Earnest 1981) with the new "ionomer-peak" and (ii) from rheology (Hird 1990) and dielectric spectroscopy (Hodge 1978) with the extension of the rubbery plateau and the existence of a new loss-peak at higher temperatures. Different structural models have been proposed (Eisenberg 1990) (Yarusso 1983) (MacKnight 1974) aiming to describe these results providing with useful insight, but a detailed structural description in terms of size, shape, internal structure, and spatial organization is still missing. This ambiguity originates mainly from the limited structure in the SAXS patterns, which show only a single and broad scattering peak.

A better understanding of the morphology and structure in ion-containing polymers can be brought about by studying model compounds where the position of ionic groups can be controlled. Telechelic ionomers (Williams 1986) (Register 1990), where the ionic groups are located at the chain-ends, and block copolymer ionomers (Gauthier 1987) (Weiss 1990) (Lu 1993), where one of the blocks is ionized, are two classes of such model compounds that recently have attracted attention. The latter system exhibits an interesting three-phase microstructure with ionic aggregates dispersed within one of the block copolymer domains.

The association behavior of end-functionalized polyisoprenes with the highly polar zwitterion group has been studied by SAXS (Shen 1991) and rheology (Davidson 1988) (Fetters 1988), both in solution and in the melt. The SAXS study from the lower molecular weight PI melts revealed the formation of two-dimensional lattices of close-packed aggregates where the core was made up of the zwitterionic heads. For the higher molecular weights (MW > 14,000), the aggregates formed a body-centered cubic lattice with long-range order. The rheological investigation revealed an extended aggregate structure for the zwitterion phase. This was supported by the enhancement of the viscosity-being larger than for a star-like structure-and from the anomalous strain dependence at low frequencies.

Functionalized block copolymers with ionic groups combine the strong enthalpic interactions of the hydrocarbon blocks favoring the local demixing of the unlike blocks with the electrostatic interactions of the polar groups favoring demixing of the polar and hydrocarbon material. These tandem interactions give rise to multiple levels of organization within the same material. The structure and dynamic response of such complex systems will be influenced by the choice of monomers, the volume fraction of the blocks, the volume fraction and polarity of the ionic groups, and the complex architecture, i.e., the location of the polar groups along the diblock copolymer chains. A step forward in this direction was the combination of the diblock copolymer microstructures with the ionic features with the synthesis of diblock and triblock copolymers with controlled positions of functional groups. The functional groups were either the dimethylamino and zwitterionic groups (Pispas 1994a,b) or quaternized ammonium and sulfonate groups (Schädler 1997 and 1998).

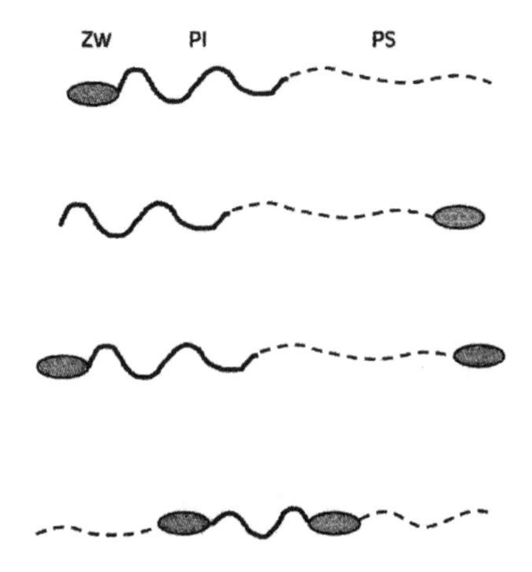

Figure 18.1. Schematic representation of the substituted SI copolymers with the zwitterion functional group. Four cases are shown: Zw-IS, Zw-SI, Zw-IS-Zw, and S-Zw-I-Zw-S. As we will see below, the position of the polar group can influence the stability of the microdomain structure, i.e., the Zw-SI copolymer is more ordered than the Zw-IS copolymer.

Herein we review the pertinent experimental findings from these systems. Furthermore, the presence and the exact location of the short, but strongly interacting, group in a diblock copolymer is expected to affect the phase state of the parent copolymer, and this case has been treated theoretically. Figure 18.1 gives a schematic representation of some of the cases that have been investigated.

1. CYLINDER-FORMING FUNCTIONALIZED SI DIBLOCK COPOLYMERS

The effect of the presence of the Zw group on the structure of the SI diblock copolymers can best be shown by examining the molecular weight dependence of the underlying structure (Floudas 1995). The SAXS patterns for the Zw-IS samples with MW in the range $6.2 \times 10^3 - 2.4 \times 10^4$ g/mol are shown in Figure 18.2. All spectra display three features: (i) an excess intensity at low q's, (ii) a microdomain peak, and (iii) an ionic peak, but the relative intensity among the three depends strongly on the MW. Decreasing the MW results in an increase of the relative intensity of the ionic peak because the volume fraction of ionic domains increases. For the lower MW sample (Zw-IS#6), the SAXS pattern is dominated by the aggregation peak because only a small peak due to the "correlation hole" effect exists in this disordered diblock.

This kind of twin microphase separation and aggregation in the ω-functionalized SI copolymers is also manifested in their rheological response. An example is

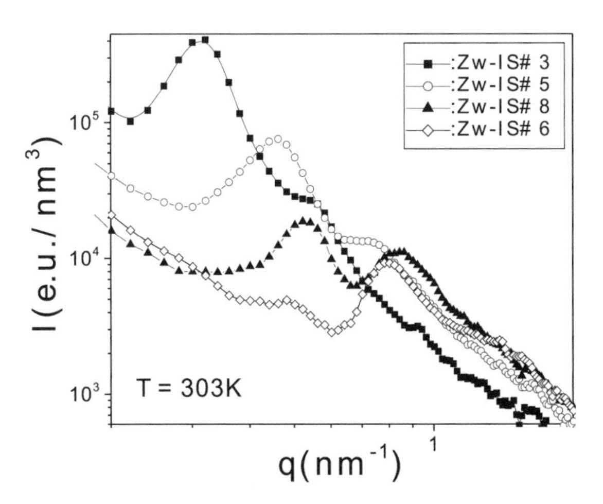

Figure 18.2. MW-dependence of the SAXS profiles for the zwitterionic functionalized IS copolymers (Zw-IS) compared at the same temperature (303 K). The weight fraction of PS is 28%, except for the sample: Zw-IS#6 ($w_{PS} = 16\%$). The different samples have the following molecular weights: Zw-IS#3: 24400, Zw-IS#5: 14100, Zw-IS#8: 11400, and Zw-IS#6: 6200 g/mol. Notice the interplay between the intensities of the microstructure and aggregation peaks, which can be tuned by changing the MW (Floudas 1995).

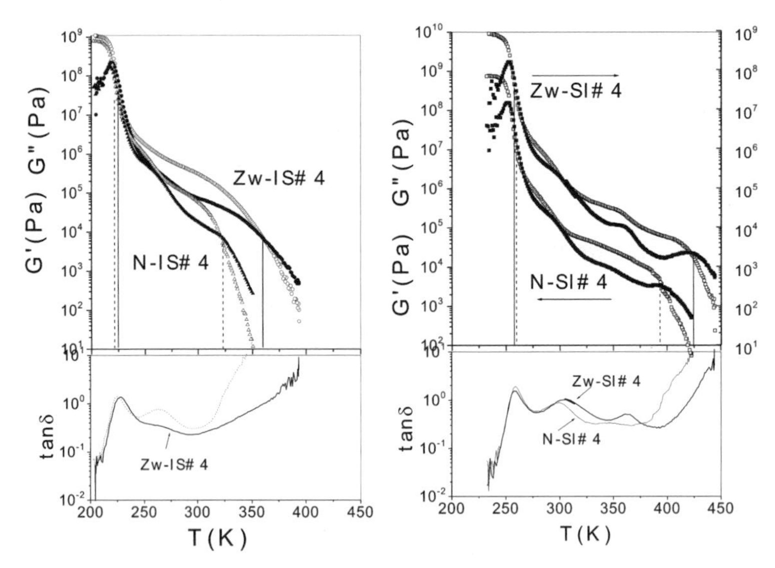

Figure 18.3. (Left): Temperature dependence of the storage (open symbols) and loss (filled symbols) moduli (at $\omega = 1$rad/s) for the ω-IS#4 copolymers (MW = 18800). (Right): The ω-SI#4 substituted with the dimethylamino (N) and zwitterion group (Zw). The loss tangent is also plotted (Floudas 1995).

shown in Figure 18.3, where the isochronal temperature scans of the same SI diblock, but substituted at different ends (i.e., Zw-IS vs. Zw-SI), is compared with the corresponding amino-substituted cases N-IS and N-SI. In general, a microphase separated diblock copolymer is expected to show four relaxations. Starting from low temperatures: (i) PI-segmental relaxation, (ii) PI-block relaxation, (iii) PS-segmental relaxation, and (iv) PS-block relaxation and/or relaxation of the aggregate as a whole. Additionally, a sudden drop of the moduli is expected for low frequencies at the order-to-disorder transition. Figure 18.3, for ω-SI#4, shows four relaxations (better seen in the loss tangent) for both samples. The low-T peak in tangent delta is associated with the PI-segmental relaxation and is in good agreement with the T_g^{PI} obtained from DSC. The relatively high T_g as compared with other polyisoprenes is the result of the higher 3,4 vinyl content. The second peak exists in both N-SI#4 and Zw-SI#4 samples and is slightly shifted to higher T in the latter. The third peak is much more pronounced in Zw-SI#4 and clearly shifted to higher T, which may signify an increase in the glass transition of PS. For the functionalized Zw-SI copolymers, ionic aggregation takes place within the PS phase, which stabilizes the microdomain structure over an extremely broad temperature range. The fourth relaxation clearly shifts to higher T in Zw-SI#4, which reflects the hindrance of the PS block relaxation due the presence of aggregates. In addition, this relaxation could also correspond to the flow of the structure as a whole in a similar way to large star molecules. Ionomers also show a high-T loss peak, which has been associated with the ionic structure. The storage modulus shows an extention of the plateau, which is reminiscent of the plateau in cross-linked systems. The extension of the rubbery plateau in the case of ionomers has been discussed in terms of the ionic cross-linking produced by the aggregates and/or of the aggregates acting as reinforcing filler particles. The same arguments can also been used for the Zw-SI system. Notice that there is no indication for an order-to-disorder transition from the G'(T) dependence (in agreement with SAXS), which supports a stabilized cubic microdomain structure up to very high temperatures.

When the functional group is placed on the PI chain-end (N-IS and Zw-IS), the dynamic mechanical response of the system is altered. In Figure 18.3 the storage and loss moduli of N-IS#4 and Zw-IS#4 at $\omega = 1$ rad/s are also shown as a function of T. The low T peak in the loss tangent is again associated with the I-segmental relaxation and occurs at lower T because of the smaller 3,4 vinyl content of the PI block. Furthermore, this low-T relaxation in N-IS#4 is only slightly shifted by the zwitterionic end-group. A pertinent feature is the suppression of the second peak in Zw-IS#4. There are two effects that can account for this suppression. First, the PI block relaxation can be significantly suppressed by the fixing of PI chain-ends to the PS cylinders from one side and to the ionic aggregates from the other. Because the aggregates are formed in the mobile phase, only partial subchain relaxation can occur that causes the suppression shown in Figure 18.3. Second, this peak could associate with a new process originating from PI segments of reduced mobility in the neighborhood of polar aggregates. In this picture aggregates act as large cross-links, which reduce the mobility of PI and create a higher T_g, which shows-up in the dynamic mechanical spectra as a peak of reduced amplitude. The above two alternatives are likely to be interrelated, and to differentiate between the two would require dynamic measurements over a broad frequency range. For this purpose dielectric spectroscopy can be employed, which selectively probes the PI-chain (Floudas 1995). Like the situation in the Zw-SI system, the terminal zone in Zw-IS is shifted by 40 K to higher T when compared with N-IS, notwithstanding the similar molecular weight for both samples. The extension of the rubbery plateau in the former constitutes evidence for the existence of ionic aggregates, which act as cross-links. At high T ($T > 360$ K), the rheological response becomes terminal, corresponding to a new structure where the polar end-groups are arranged between the PS cylinders with a liquid-like order.

The rich morphological behavior of the ω-functionalized diblock copolymers of PS and PI is depicted schematically in Figure 18.4. In some cases the morphology consists of two separate levels of microphase separation, one between the two blocks forming the microdomain structure and another between ionic and nonionic material. Because of the small difference in the dielectric constant of the two blocks, aggregates are consistently formed in the phase where the polar group is linked. Therefore, when the zwitterion is linked to the PI chain-end, aggregates are formed at low temperatures within the PI phase. The aggregation shows-up in rheology and dielectric spectroscopy with an extended rubbery plateau and with a new dielectric process associated with restricted PI segmental relaxation. At higher temperatures, the rheological response suggests a structure where the polar

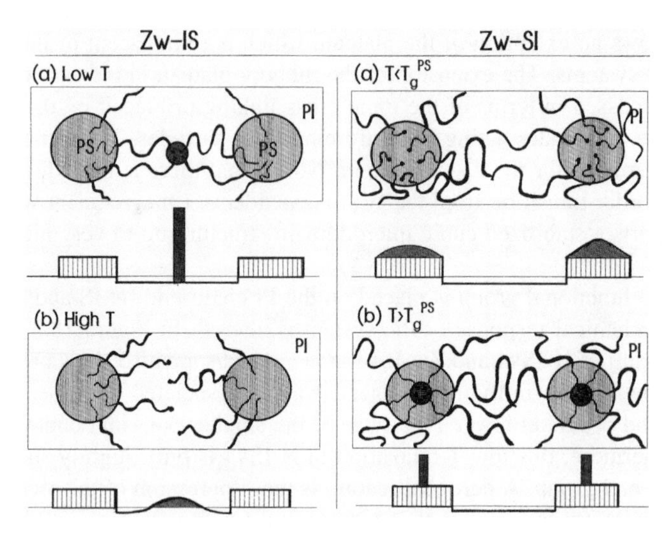

Figure 18.4. Schematic illustration of the microstructures in ω-functionalized SI block copolymers, showing the Zw-IS (left) and Zw-SI (right) cases at low (upper) and high (lower) temperatures. The corresponding electron density distributions are also shown (Floudas 1995).

end-groups are arranged between the PS cylinders with a liquid-like order. When the zwitterion is located on the PS, chain-end association takes place, at high temperatures, within the PS phase and leads to stabilization of the ordered structures and the formation of multiplets within the PS-phase. The latter case was also treated theoretically (Dormidontova 1997).

These results demonstrate that functionalization provides a new possibility of altering the phase behavior of diblock copolymers (Floudas 1995 and 1996). Depending on the position of the functional group with respect to the chain-ends, one can reduce or enhance the miscibility of the two blocks.

2. LAMELLAR-FORMING FUNCTIONALIZED DIBLOCK AND TRIBLOCK COPOLYMERS

The presence of the point-like functional group can have an influence on the experimentally obtained scattering patterns. Some insight on the scattering patterns from such complex materials can be gained by analyzing simplified block electron density profiles, which take into account the composition and geometry variations related to the actual cases and constructed their Fourier series representation (Floudas 2001). In the representative cases shown in Figure 18.5, a symmetric block composition ($f = 0.5$) and an ionic domain volume fraction of 0.04 and the following electron densities for PS, PI, and cyclopropanesultone: 340, 301, and 379 electrons/nm^3 were employed. Schematic representations of the assumed periodic density profiles for: (a) a symmetric copolymer lamellar and (b) a modified structure with aggregates within the high electron density phase (PS), corresponding to the

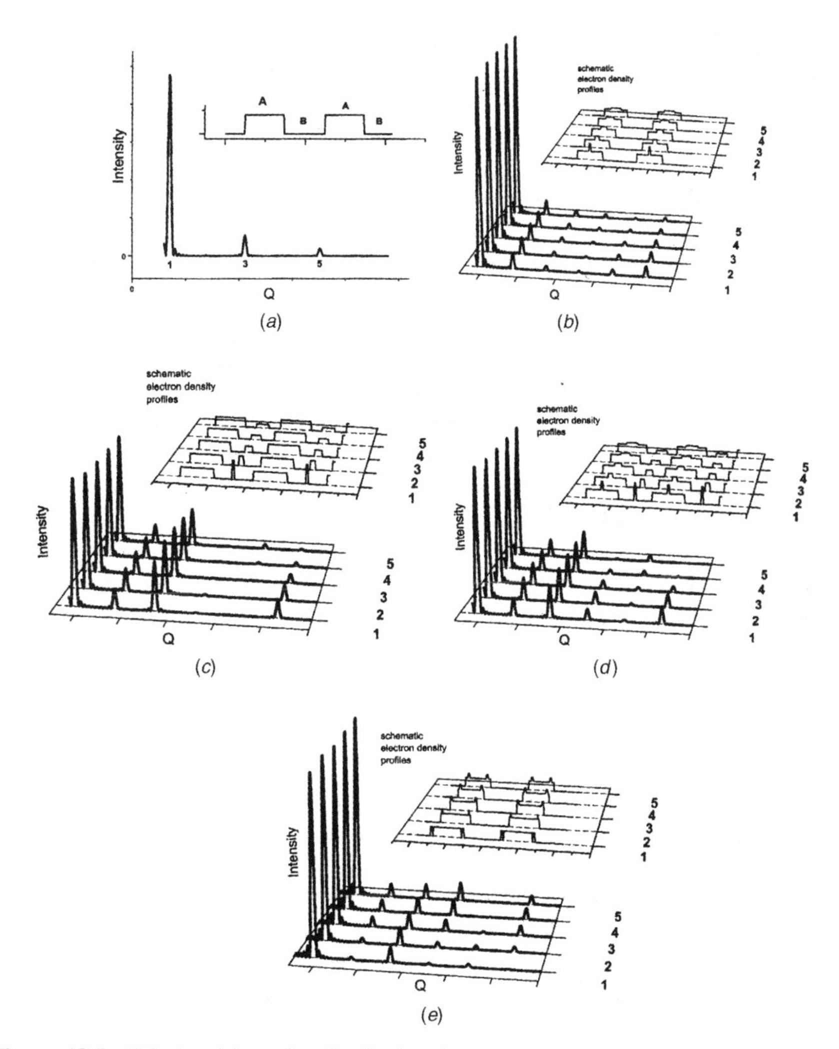

Figure 18.5. Calculated intensity distribution for (a) a symmetric lamellar-forming block copolymer (a) and for different block copolymer structures that are modified by additional electron density peaks: (b) in the middle of the high electron density phase (PS in this case), (c) in the middle of the low electron density phase (PI in this case), (d) in the middle of both phases (i.e., corresponding to the Zw-SI-Zw case), and (e) for various locations of the Zw group within the S-I interface (i.e., the S-Zw-I-Zw-S triblock case) (Floudas 2001).

Zw-SI case, are shown in Figures 18.5a and b, respectively. The calculated profiles correspond to perfect correlations along the domains, which explains the narrow peaks observed. In reality, however, correlations between the domains are limited, and instrumental effects tend to weaken and broaden the peak structure as well. The different electron density profiles and the corresponding scattering profiles in Figure 18.5b are made for various distributions of the aggregates within the high electron density phase (PS). As shown in the figure, the presence of the Zw groups alters significantly the scattering profile with peaks at relative positions with ratios 1:2:3 to the main peak. However, the main result from these scattering profiles is that the intensity of the higher order reflections is highly suppressed as compared with the main peak, which is in accord with the absence of strong reflections in the scattering profiles of real systems. Notice that the distribution of the functional groups within the PS phase has only a minor effect on the scattering profiles.

When the functional group is attached to the low electron density block (i.e., the PI block in this case), the situation is different. The SAXS spectra of Zw-IS revealed intense higher order reflections at relative positions with ratios 1:2:3, which could imply a well-ordered lamellar structure. However, this dramatic change of the scattering spectra is a simple consequence of the altered electron density profile related to the specific Zw distribution within the existing lamellar phase and does not require any improvement or alteration of the long-range order of the hydrocarbon blocks in the microdomain morphology. This effect is documented with the calculated profiles based on the assumed electron density profiles of Figure 18.5c. It is the *localization* of the functional groups-which possess the highest electron density-in the center of the PI domain (the low electron density phase) that gives rise to the specific intensity proportions of the higher order reflections. Again, the *distribution* of the polar groups within the PI phase makes nearly no difference to the relative intensity of the 2nd and 3rd order reflections; that is, even a weak association of the zwitterions within the PIphase would result in the same scattering patterns.

Similarly, intense higher order reflections are expected when the zwitterion is located in both chain-ends, i.e., the Zw-SI-Zw case (Figure 18.5d). The corresponding situation for the S-Zw-I-Zw-S case where the functional group is located within the interface of the SIS triblock copolymer is also shown in Figure 18.5e.

Overall, the SAXS spectra of the functionalized copolymers are very sensitive to the presence of the functional groups as they are able to modify sufficiently the electron density profiles. The effect is more pronounced when the functional group is located within the domains with the lowest electron density. On the other hand, the distribution of aggregates within the domains make nearly no difference to the scattering profiles.

3. ABC BLOCK COPOLYMERS WITH A SHORT BUT STRONGLY INTERACTING MIDDLE BLOCK

Recent theoretical efforts (Erukhimovich 1997) (Abetz 1996) have treated the case of ABC copolymers bearing a finite or a point-like central block B. It was found that the size and interactions of the central block play an important role in delineating the miscibility of the AC diblock. A long central block can induce a morphological transition from B cylinders at the AC lamellae interface to cylinders in a mixed AC matrix, i.e., the central B block can enhance the miscibility of the outer blocks. On the other hand, the case of a short but strongly interacting midblock is predicted to have a very different effect on the outer block miscibility.

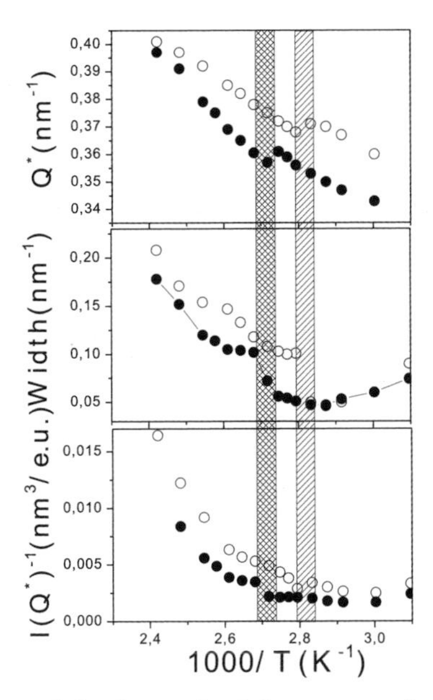

Figure 18.6. T-dependence of the first peak of the structure factor (peak position at Q*, width, and inverse peak intensity) for the SNI (open circles) and the S-Zw-I (filled symbols). The shaded areas indicate the order-to-disorder transition regimes (Pispas 1999).

In the case where the outer blocks have a small but finite incompatibility and the inner block is short but strongly interacting (i.e., $\chi_{AC}\#0$ and χ_{AB}, $\chi_{BC} \gg \chi_{AC}$), it is predicted that the critical temperature increases by 5/3 as compared with the AC diblock case.

These theoretical predictions have been tested experimentally (Pispas 1999) by employing junction point functionalized copolymers of polystyrene (PS) and polyisoperene (PI). The functional group was again the zwitterion (Zw), i.e., a short and strongly interacting ionic group.

In Figure 18.6 the results from the analysis of the peak features obtained by SAXS are compared for the same copolymers: SNI and S-Zw-I, with the only difference being that the amino group in the former is replaced by the Zw group in the latter. The inverse peak intensity, peak width, and peak position are discontinuous at the respective transition temperatures (356 K and 368 K, for the SNI and S-Zw-I, respectively). Notice that the T_{ODT} increases by about 12 K in the S-Zw-I copolymer.

The results presented in the figure demonstrate that when a short (point-like) block B is inserted in the AC junction, the A block, having a finite incompatibility with C, increases the incompatibility of AC. These results are in qualitative agreement with the theoretical predictions. The theory predicts that inserting an infinitely small central B block infinitely incompatible with both side blocks A,C (A and C having a small but finite incompatibility) results in a continuous microphase

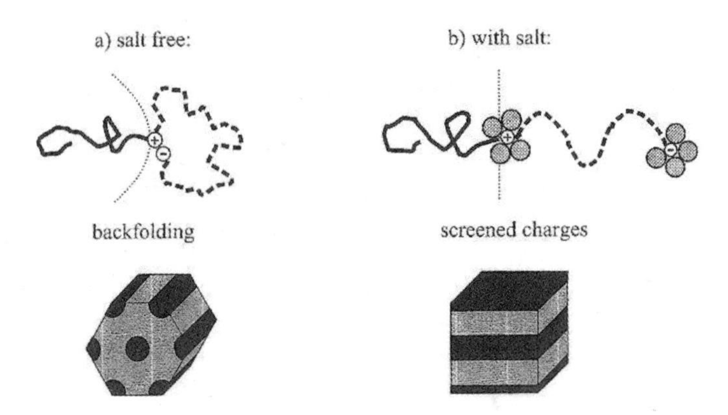

Figure 18.7. Schematic representation of the expected chain conformation of an α, ω-macrozwitterion (left) without salt, and (right) with added salt (Schöps 1999).

separation at a temperature $T_{ODT}^{ABC} = \frac{5}{3}T_{ODT}^{AC}$. However, the increase observed experimentally was only a few degrees, probably due to the finite incompatibility of B with the AC.

4. EFFECT OF SALT ON THE LAMELLAR SPACING AND MICRODOMAIN MORPHOLOGY

Quaternized ammonium and/or sulfonate groups have also been introduced at the chain-ends of lamellar SI diblock copolymers (Schädler 1997 and 1998) (Schöps 1999). These monofunctional species with their counterions can also be viewed as block copolymers with strong dipoles attached at one end. In such a study, an α, ω-macrozwitterion has been synthesized, which was obtained in a salt-free state by dialysis. It was shown that the periodicity of this block copolymer can be tuned by the concentration of added salt through the weakening of the Coulombic interactions.

Besides the variation in periodicity, it is also possible to induce a morphological transition to the copolymer by the addition of salt. This has been shown in a μ, ω-macrozwitterion, which was functionalized with a quaternized amino group at the junction point and with a sulfonate group at one chain-end. The structure in the salt-free sample, shown in Figure 18.7, is reminiscent of an I_2S miktoarm star copolymer. Salt was found to induce a transition from hexagonally packed cylinders to a lamellar structure through the suppression of the tendency for ionic aggregation. The microstructure control of block copolymers via ionic interactions (Russell 1988) (Horrion 1990) is an alternative approach to the known copolymerization routes and is likely to be explored further in the near future.

REFERENCES

Abetz V, Stadler R., Leibler L. (1996) Polymer Bull. 37, 135.

Davidson N.S., Fetters L.J., Funk W.G., Graessley W.W., Hadjichristidis N. (1988) Macromolecules 21, 112.

Dormidontova E.E., Khokhlov A.R. (1997) Macromolecules 30, 1980.

Earnest T.R., Higgins J.S., Handlin D.L., MacKnight W.J. (1981) Macromolecules 14, 192.

Eisenberg A. (1990) Macromolecules 3, 147.

Eisenberg A., Hird B., Moore R.B. (1990) Macromolecules 23, 4098.

Erukhimovich I., Abetz V., Stadler R. (1997) Macromolecules 30, 7435.

Fetters L.J., Graessley W.W., Hadjichristidis N., Kiss A.D., Pearson D.S., Younghouse L.B. (1988) Macromolecules 21, 1644.

Floudas G., Fytas G., Pispas S., Hadjichristidis N., Pakula T., Khokhlov A.R., (1995) Macromolecules 28, 5109.

Floudas G., Fytas G., Pispas S., Hadjichristidis N., Pakula T., Khokhlov A.R. (1996) Makromol. Chem. Macromol. Symp. 106, 137.

Floudas G., Pispas S., Hadjichristidis N., Pakula T. (2001) Macromol. Chem. Phys. 202, 1488.

Gauthier S., Eisenberg A. (1987) Macromolecules 20, 760.

Hird B., Eisenberg A. (1990) J. Polym. Sci.; Polym. Phys. 28, 1665.

Hodge I.M., Eisenberg A. (1978) Macromolecules 11, 283.

Horrion J., Jerome R., Teyssie Ph. (1990) J. Polym. Sci. Polym. Chem. 28, 153.

Lu X., Steckle W.P., Weiss R.A. (1993) Macromolecules 26, 5876.

MacKnight W.J., Earnest T.R. (1981) J. Polym. Sci., Macromol. Rev. 16, 41.

MacKnight W.J., Taggart W.P., Stein R.S. (1974) J. Polym. Sci., Polym. Symp. 45, 113.

Nyrkova I.A., Khokhlov A.R., Doi M. (1993) Macromolecules 26, 3601.

Pispas S., Hadjichristidis N. (1994a) Macromolecules 27, 1891.

Pispas S., Hadjichristidis N., Mays J.W. (1994b) Macromolecules 27, 6307.

Pispas S., Floudas G., Hadjichristidis N. (1999) Macromolecules 32, 9074.

Register R.A., Cooper S.L., Thiyagarajan P., Chakrapani S., Jerome R. (1990) Macromolecules 23, 2978.

Russell T.P., Jerome R., Charlier P., Foucart M. (1988) Macromolecules 21, 1709.

Shen Yi., Satinya C.R., Fetters L., Adam M., Witten T., Hadjichristidis N. (1991) Phys. Rev. A43, 1886.

Schöps M., Leist H., DuChesne A., Wiesner U. (1999) Macromolecules 32, 2806.

Schädler V., Wiesner U. (1997) Macromolecules 30, 6698.

Schädler V., Kniesse V., Thurn-Albrecht T., Wiesner U., Spiess H.W. (1998) Macromolecules 31, 4828.

Weiss R.A., Sen A., Pottick L.A., Willis C.L. (1990) Polym. Commun. 31, 221.

Williams C.E., Russell T.P., Jerome R., Horrion J. (1986) Macromolecules 19, 2877.

Yarusso D.J., Cooper S.L. (1983) Macromolecules 16, 1871.

CHAPTER 19

BLOCK COPOLYMER MORPHOLOGY

In this chapter we provide some of the recent novel morphologies obtained with complex block copolymer architectures such as rod-coil copolymers, ABC triblock terpolymers and miktoarm star terpolymers. We differentiate from the phase state chapter, discussed earlier with respect to the four known diblock copolymers systems, because the broad parameter space in some of these systems precludes, at least up to now, the construction of a complete phase diagram for each of the different architectures. Nevertheless, all these morphologies reflect a delicate free energy minimization that is common to all block copolymers and may provide with important applications as membranes for gas separation, waveguides, models in tissue engineering and biomaterials, solid lubricants, and catalysts, among others.

Before going to the more complex structures, we first recall that, in AB diblock copolymers, we found only four equilibrium structures, i.e., lamellae, hexagonally packed cylinders, spheres in a bcc lattice, and the double gyroid. It is instructive to compare these structures with the ones found in polyelectrolyte-surfactant complexes. The latter, together with their scattering peak position ratios, are summarized in Figure 19.1. In Figure 19.2 the same structures are shown together with their expected scattering curves. Apparently, the nanostructures encountered in diblock copolymers are a subset of the typical structures observed in polyelectrolyte-surfactant complexes.

1. ROD-COIL COPOLYMERS

When one of the blocks forming the diblock has an inherent capacity for self-organization (for example, liquid crystalline behavior) on top of the copolymer

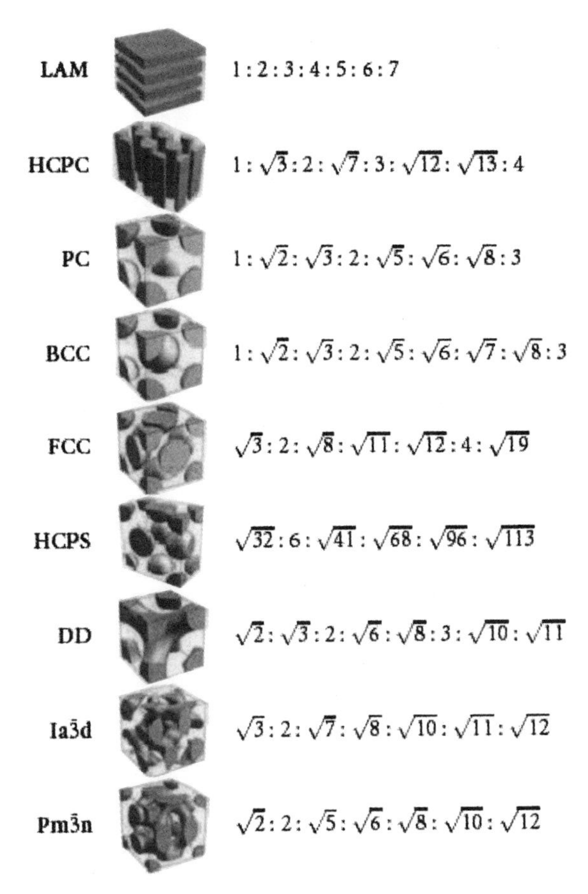

Figure 19.1. Typical nanostructures indicating the LAM: lamellar, HCPC: hexagonally packed cylinders, PC: primitive or simple cubic, BCC: body-centered cubic, FCC: face-centered cubic, HCPS: hexagonally closed packed spheres, DD: double diamond, $I\alpha\bar{3}d$ and $Pm\bar{3}n$ (adapted from Burger 2002).

assembly, then many unanticipated structures can be formed. In the case of an LC block, self-assembly is dictated by balancing the organizing forces such as conformational entropy and liquid crystallinity. The resulting material can show simultaneous organization in different length scales, ranging from few nanometers in the liquid crystal phase to 100 nm in the phase-separated microstructures of the block copolymer.

The phase state in such systems can be very complex. When the liquid crystal behavior dominates, then the intrinsic stiffness of the LC block gives rise to a variety of LC phases. An example is given in Figure 19.3, where some representative micrographs are shown from the poly(hexylisocyanate-b-styrene) (PHIC-b-S) rod-coil block copolymer system. Micrograph A shows a smectic C phase formed by alternating PHIC and PS layers arranged in a zigzag fashion. In this case the rod

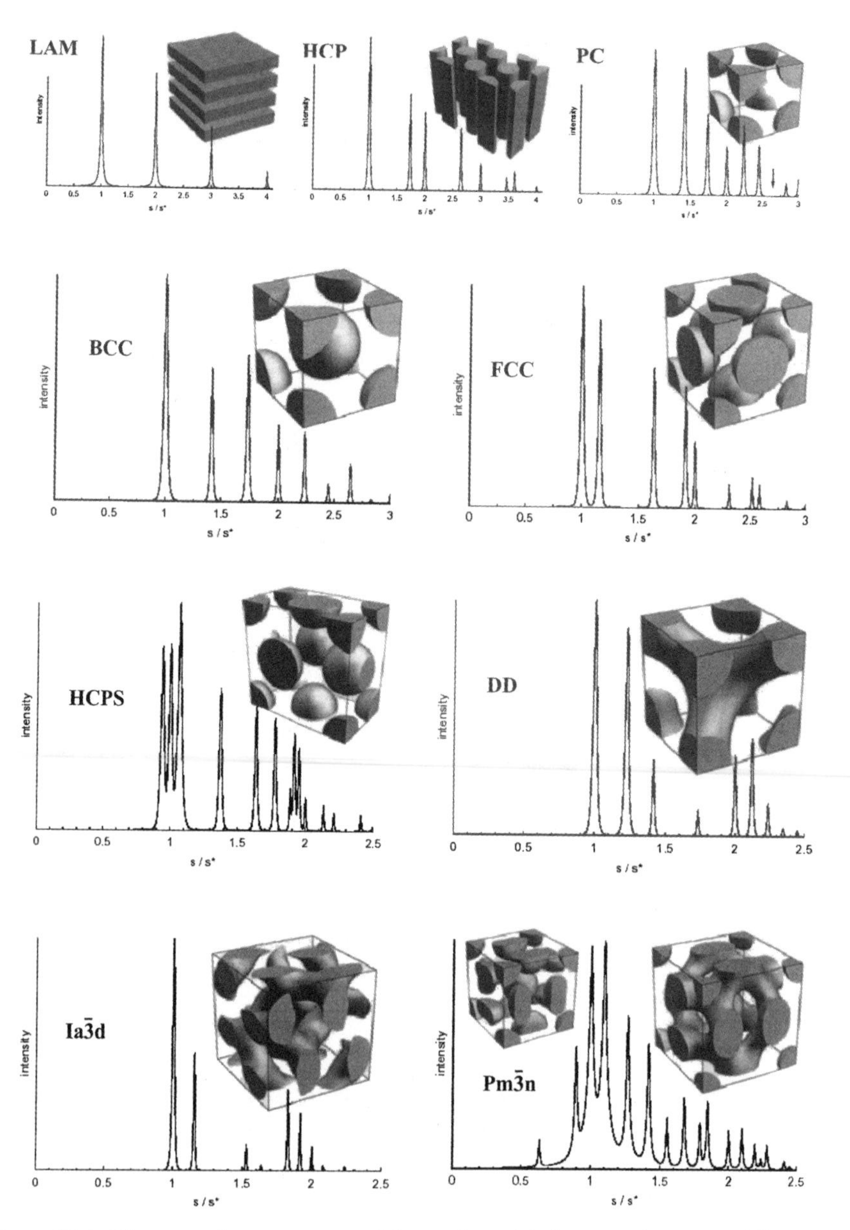

Figure 19.2. The same nine structure models together with the scattering curves. The x-axis is reduced as s/s*, where s* is the primary peak position (Burger 2002).

blocks are much longer than the PS block, and the rods tilt with respect to the layer normal and interdigitate forming the S_c phase. Micrograph B shows another novel morphology, obtained for very high volume fractions of the rod block called the "arrowhead" phase corresponding to a S_O phase.

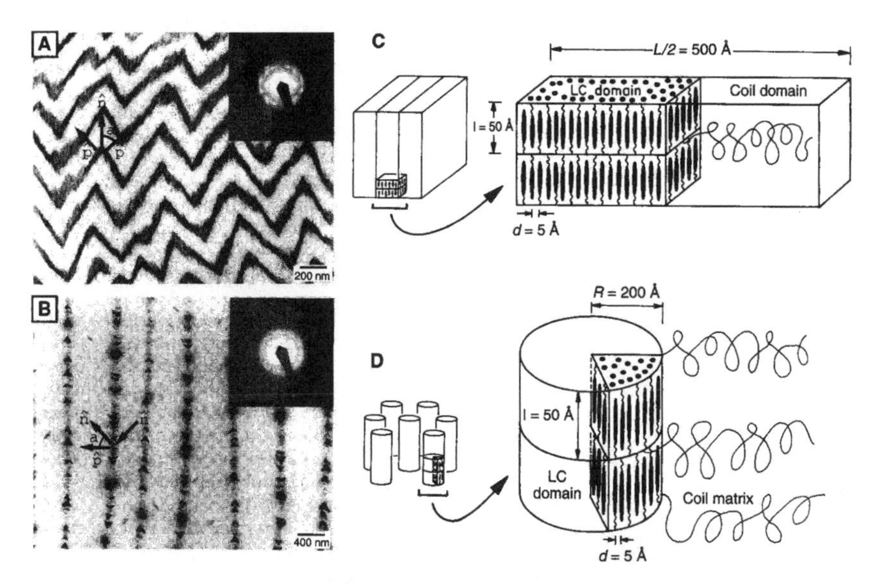

Figure.19.3. Different smectic phases in rod-coil diblock copolymers: (A) PHIC and PS layers arranged in a zigzag fashion. In the micrograph the PS domains appear dark because of the RuO_4 staining. The long PHIC rods are tilted with respect to the layer normal and interdigitate giving rise to the observed S_c structure. (B) This novel "arrowhead" morphology is formed when the volume fraction of the rods is very high (98%) giving rise to an S_O morphology where alternate domain boundaries comprise discrete PS regions pointing in opposite directions. (C) and (D): The preferential parallel alignment of the mesogens with respect to the interface can give rise to a lamellar or a cylindrical morphology (Chen 1996) (Mao 1997).

In this phase, the PS block forms arrowhead-shaped domains whose orientation flips by 180^0 between adjacent PS-rich layers.

On the other hand, when the driving force for phase separation between the coil and LC group is very large, then microphase separation occurs first and is followed by the development of the LC and crystalline ordering. An example is given in C and D (Figure 19.3), corresponding to the formation of a lamellar and a cylindrical structure of the LC blocks. In both cases, there is preference for a parallel alignment of the mesogens with respect to the interface.

Another possibility of self-assembly in rod-coil copolymers is through the formation of hollow spherical micelles when dissolved in a selective solvent for the flexible block (Jenekhe 1999). This self-assembly approach to form microporous solids is a nontemplate strategy. The resulting morphology consists of multilayers of hexagonally ordered arrays of spherical holes whose diameter, periodicity, and wall thickness depend on copolymer molecular weight and composition. The origin of this mechanism lies in the difficulty of efficient space-filling packing of rodlike blocks into a spherical or cylindrical core. In Figure 19.4 the rod-coil

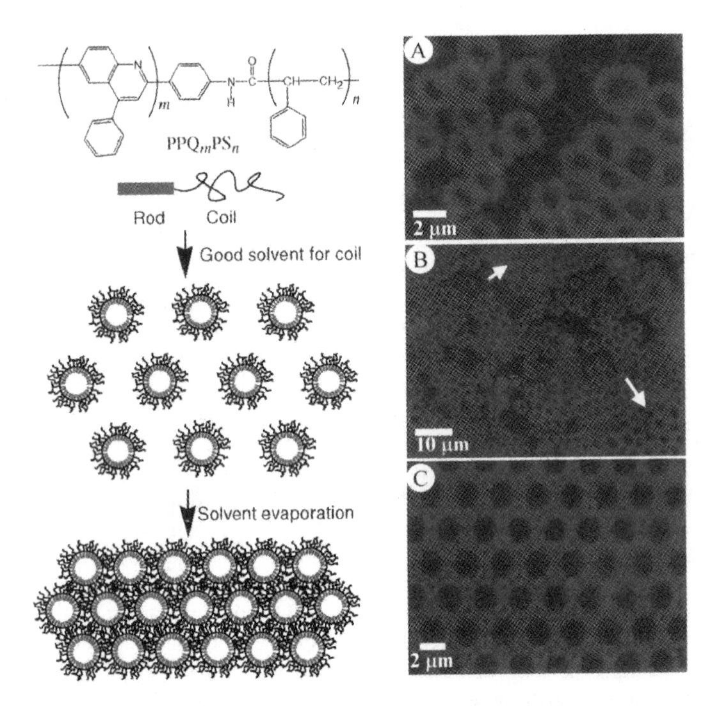

Figure 19.4. (Left) Molecular structure of rod-coil diblock copolymers and a highly schematic illustration of its hierarchical self-assembly into ordered microporous materials. (Right) Fluorescence photomicrographs of solution-cast micellar films revealing a two-dimensional hcp structure composed of air holes (Jenekhe 1999). See color plates.

copolymers overcome this steric problem, associated with the packing frustration of the rods into a sphere, by a new morphology: the hollow sphere.

A consequence of the stiffness asymmetry of the two blocks is that, by using different selective solvents, the system can organize into different colloidal particles: hollow hard spheres and hollow soft spheres. Furthermore, self-ordering these hollow soft spheres into two and three dimensions could be important in areas such as catalysis, size, and shape-selective separation media, etc.

As a final example of the unique structural possibilities given by rod-coil copolymers, we refer to miniaturized triblock copolymers with a rod-coil structure, which are shown to self-assemble into nanostructures that are highly regular in size and shape (Stupp 1997). Such nanostructures have interesting packing properties, which give rise to unanticipated morphologies. Some triblocks built from small-molecule precursors are indicated in Figure 19.5. These triblock molecules have a rod-coil architecture because the stiff rod-like segment is covalently connected to a flexible diblock segment that, in solution, adopts a coil conformation. The biphenyl ester segments (i.e., the rods) have an extremely high tendency to aggregate and order though the π-π overlaps. Subsequently, crystallization of the identical rod-like

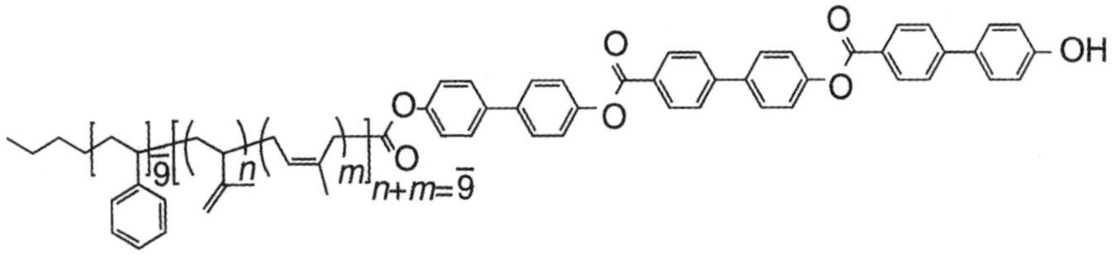

Figure 19.5. Miniature triblock copolymers composed from oligostyryl, oligoisoprene, and rod-like blocks.

segments excludes the oligostyryl and oligoisoprene blocks. The flexibility and small cross-sectional area of the oligoisoprene blocks serves as a structural buffer that can stretch and pack to accommodate the density of the crystallized rod-like segments.

WAXS data have shown that the rod-like segments pack into an orthorhombic unit cell with lattice parameters of 0.54 nm and 0.82 nm. Based on these unit cell parameters and the thickness of the nanostructures obtained from TEM, it is estimated that each nanostructure comprises about 100 molecules. The suggested model for the packing of these structures is indicated in Figure 19.6, which shows mushroom-like aggregates of about 100 triblock molecules. In the same figure, a schematic representation of a layered stacking of such mushroom-shaped nanostrcutures expected for a macroscopic film is also shown.

Evidently, the self-assembly of small-molecule precursors with a triblock copolymer sequence can give rise to supramolecular structures with well-defined shapes and sizes.

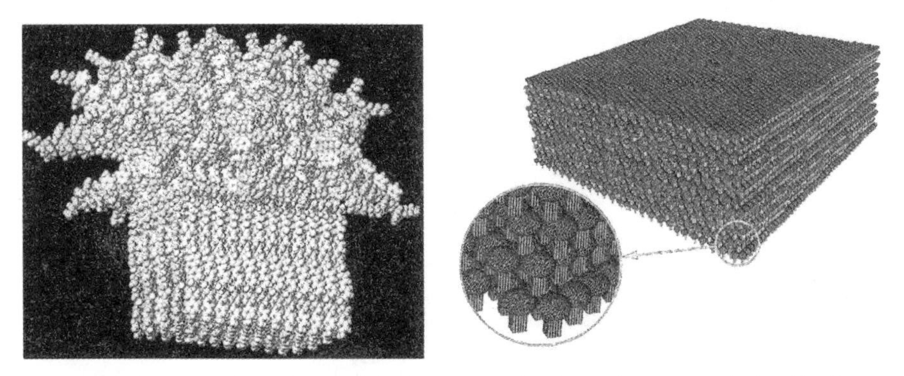

Figure 19.6. (Left) Molecular model of a supramolecular unit composed of 100 triblock copolymers with a rod-coil structure giving rise to mushroom-like nanostructures. (Right) Schematic representation of how these nanostructures of the triblock copolymers could organize to form a macroscopic film (Stupp 1997). See color plates.

2. ABC TRIBLOCK TERPOLYMERS

It was early recognized by R. Stadler and other groups that the parameter space greatly increases in going from an AB diblock to the ABC triblock case. As we have discussed in previous chapters, the product $\chi_{AB}N$ and the composition f_A are, at least to a first approximation, sufficient to locate a particular morphology in the AB diblock phase diagram. For an ABC triblock, however, there are three interaction parameters (χ_{AB}, χ_{BC}, χ_{AC}) and two composition variables (f_A, f_B) that are needed for positioning the particular microdomain morphology. Furthermore, the block sequence plays an important role and can strongly affect the phase behavior. As a result the four equilibrium morphologies known from AB diblocks are replaced by the more complex structures shown in Figure 19.7.

When the triblock is formed with nearly symmetric compositions, i.e., $f_A = f_B = f_C = 1/3$ and similar interaction parameters, i.e., $\chi_{AB} \approx \chi_{BC} \approx \chi_{AC}$, then the resulting microstructure is of a three phase lamellar (case a). When, however, $\chi_{AB} \ll \chi_{BC}$, then the system is driven by the interaction asymmetry to adopt the core-shell hexagonal morphology shown in b, so as to minimize the BC interfacial area relative to the AB interfacial area, resulting in a lower overall free energy.

Another set of novel morphologies results when $\chi_{AB} \approx \chi_{BC} \gg \chi_{AC}$ and for low f_B values. In this case the B layer becomes discontinuous, allowing for increased

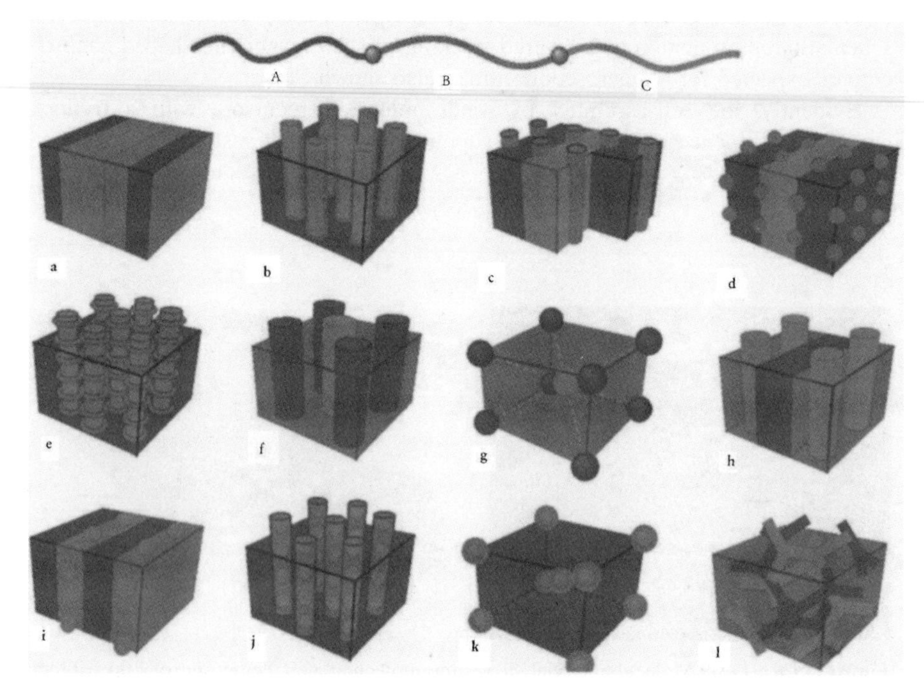

Figure 19.7. ABC linear triblock copolymer morphologies. Microdomains are colored following the code of the triblock molecule in the top (Stadler 1995, Zheng 1995, Bates 1999). See color plates.

contacts between the A and C blocks, giving rise to "cylinders at the wall" (case c), or "spheres at the wall" (case d), or even the decorated cylinders (case e), depending on the proportions of A and C. Increasing the content of the B-domains gives rise eventually to continuous B domains as in case a.

The more symmetric case, where $\chi_{AB} \approx \chi_{BC} < \chi_{AC}$ and $f_A \approx f_B$, can give rise to the structures shown in f, g, h, and l. In this case the system tries to minimize the unfavorable contacts between A and C and gives rise to tetragonal cylinders or to the tricontinuous Gyroid morphology of case l composed of two independent networks of the A and C domains. The above are some of the rich morphologies formed in the broad parameter space of ABC triblock terpolymers. These possibilities were first exploited by R. Stadler (Stadler 1995) and later by other groups both experimentally (Matsushita 1998) and theoretically (Zheng 1995) (Matsen 1998) and has been the subject of a recent review (Bates 1999).

3. MORE COMPLEXITY WITH ABCs

3.1. The "Knitting Pattern"

A triblock copolymer of poly(styrene-b-butadiene-b-methylmethacrylate) forming a double lamellar structure upon hydrogenation (SEBM) was found to transform to the complex pattern shown in Figure 19.8 known as the "knitting pattern." This striking complex morphology, discovered by R. Stadler and coworkers (Breiner 1998), was studied by TEM together with FFT techniques and SAXS, revealing the

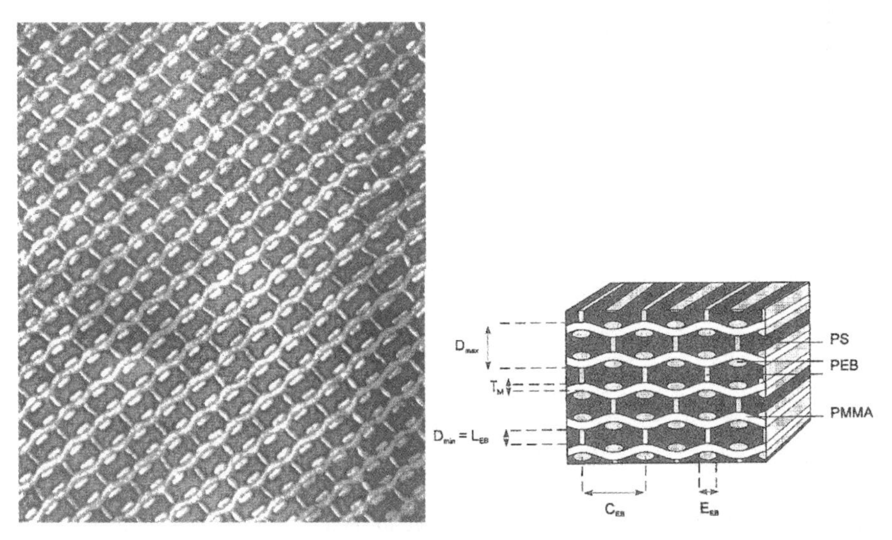

Figure 19.8. (Left) Bright field TEM of the SEM (35/27/38) triblock copolymer stained with RuO_4 showing the celebrated "knitting" pattern. (Right) Schematic description of the knitting pattern morphology (Breiner 1998). See color plates.

underlying c2mm space group symmetry. The dividing surfaces display highly nonconstant mean curvatures resulting from the nonuniform distribution of the A/B and B/C junctions.

3.2. The Core-shell Double Gyroid

This new morphology was found in a poly(isoprene-b-styrene-b-dimethylsiloxane) (ISD) triblock terpolymer with respective volume fractions of 0.40/0.41/0.19 called core-shell gyroid (Shefelbine 1999), as well as in blends of block copolymers (Goldacker 1999). Figure 19.9 depicts the unit cell in a highly schematic way, which consists of two D cores forming two lattices of three-fold connectors in a gyroid network. Each of the D domains is covered with an S shell, and the matrix is formed by the I blocks, resulting in five independent, three-dimensionally continuous domains. This triply periodic cubic structure is also called the pentacontinuous core-shell gyroid phase. In this system the sequence of interaction parameters is $\chi_{SD} \gg \chi_{ID} > \chi_{IS}$ and the structure formed is such that the D domain forms the core and the S domain completely separates the I and D domains. In this way the costly S-D interface is much smaller than the S-I interface.

Another example of this morphology is a linear poly(styrene-b-1,2 butadiene-b-2-vinylpyridine) (SBV) triblock copolymer as well as an SBV star terpolymer (Hückstädt 2000). Figure 19.10 gives an example of the TEM images obtained from the star terpolymer clearly indicating the core-shell structure. Such core-shell double gyroid morphologies can result, in general, in triblock copolymers with significantly different repulsive interaction between the middle and the two outer blocks.

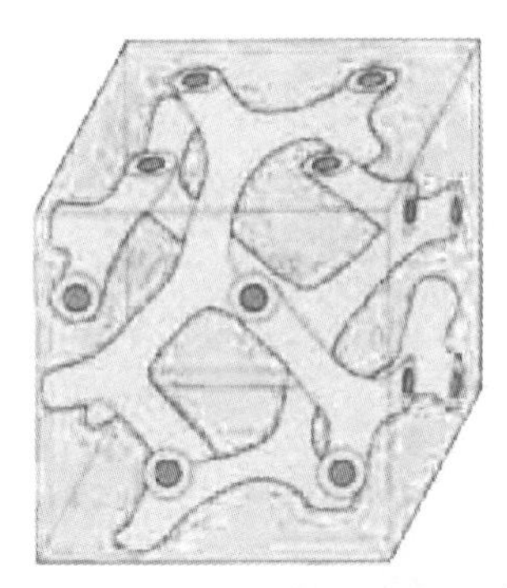

Figure 19.9. Schematic representation of the core-in-shell gyroid morphology found in the poly(isoprene-b-styrene-b-dimethylsiloxane) (ISD) triblock copolymer ($f_{PI} = 0.40$, $f_{PS} = 0.41$, $f_{PDMS} = 0.19$). Blue, red, and green regions correspond to I, S, and D domains, respectively (adapted by Shefelbine 1999). See color plates.

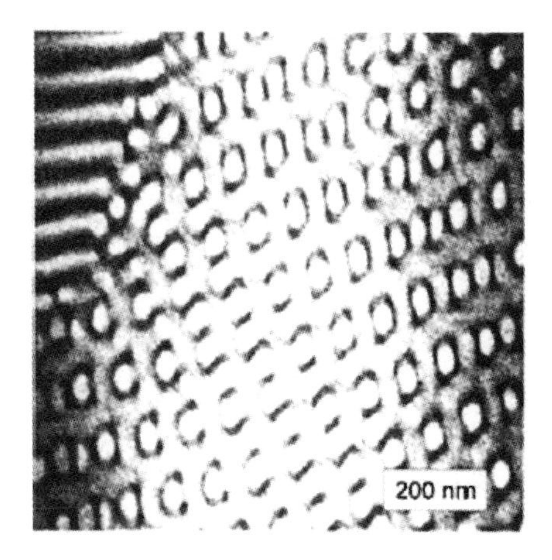

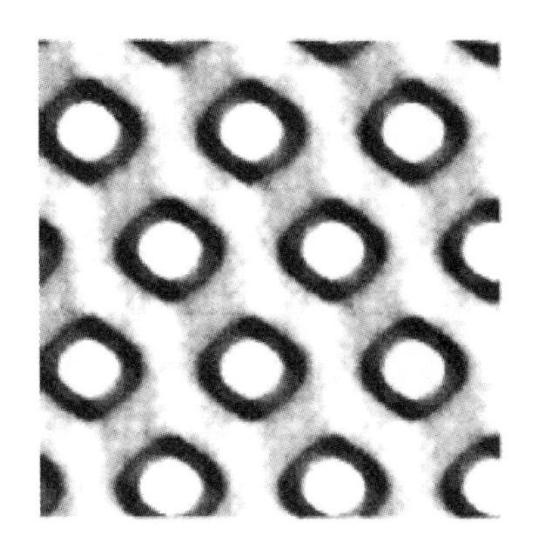

Figure 19.10. (Left) TEM micrograph of a star terpolymer composed of polystyrene, poly(1,2-butadiene), and poly(2-vinylpyridine) arms (SBV) (14/37/49) stained with OsO_4 and CH_3I together with the simulation of the [110] projection (Right) (Hückstädt 2000).

4. ABC MIKTOARM STAR TERPOLYMERS WITH AMORPHOUS BLOCKS

So far we have reviewed structures formed with linear ABC triblock terpolymers. The synthesis of ABC three-miktoarm star terpolymers enables a study of the same five-parameter space as with linear ABC triblocks where, however, the block sequence does not play a role. These star terpolymers differ from their linear analogues in that, in a star, there exists only one junction point instead of two.

The microdomain morphology of different three-miktoarm star terpolymers of PS, PI, and PMMA has been studied mainly by TEM (Sioula 1998a,b). For three of the samples, the molecular weight of the PMMA block was varied, while those of the PS and PI blocks were kept fixed. In a fourth sample (29.8/44.5/25.6), the PS and PMMA blocks were approximately of the same molecular weight, and the PI arm was the majority. These samples were investigated mainly by TEM using OsO_4 and RuO_4 selective staining. All four samples were found to have a three-microphase two-dimensional periodic microstructure of an inner PI column with a surrounding PS annulus in a matrix of PMMA. The TEM images and the inferred microdomain morphology is shown in Figure 19.11.

When the total volume fraction of PI and PS is 0.56 and 0.78, the PI-PS and PS-PMMA interfaces exhibited a nonconstant mean curvature diamond–prism shape with a c2mm plane group symmetry. This case is shown in Figure 19.12 together with a schematic representation of the morphology. In all of the above structures, the junction points of the stars were distributed over the PS/PI intermediate dividing surface, and the structures formed were such as to separate the more incompatible blocks (PI and PMMA) by forming a protective annulus of PS ($\chi_{IM} > \chi_{SI} \gg \chi_{SM}$).

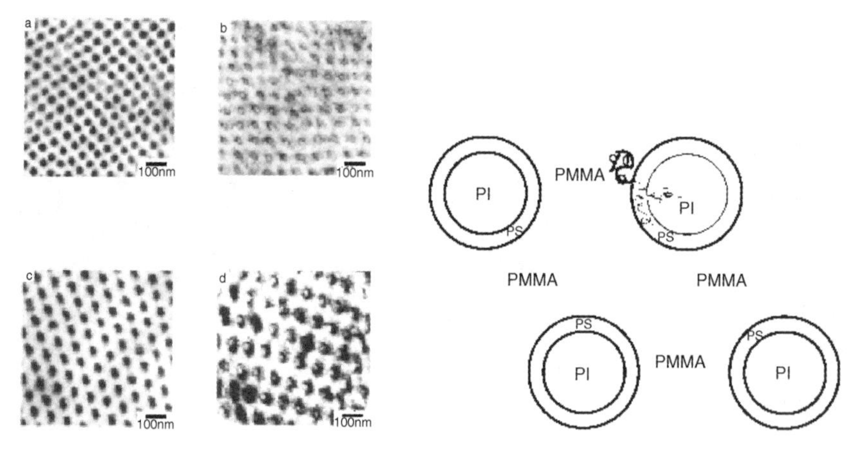

Figure 19.11. Bright field TEM images of (a) SIM star with composition 21.5/23/55.5 stained with OsO_4, (b) the same star now stained with RuO_4, (c) SIM with composition 19.6/20.9/59.5 stained with OsO_4, (d) the same star now stained with RuO_4. In the OsO_4 stained micrographs, the stained block is the PI, which forms hexagonally packed cylinders, whereas, in the RuO_4-stained micrographs, the stained block is the PS, and the PI phase appears gray. A schematic representation of the structure is shown, which is based on the TEM images (Sioula 1998a).

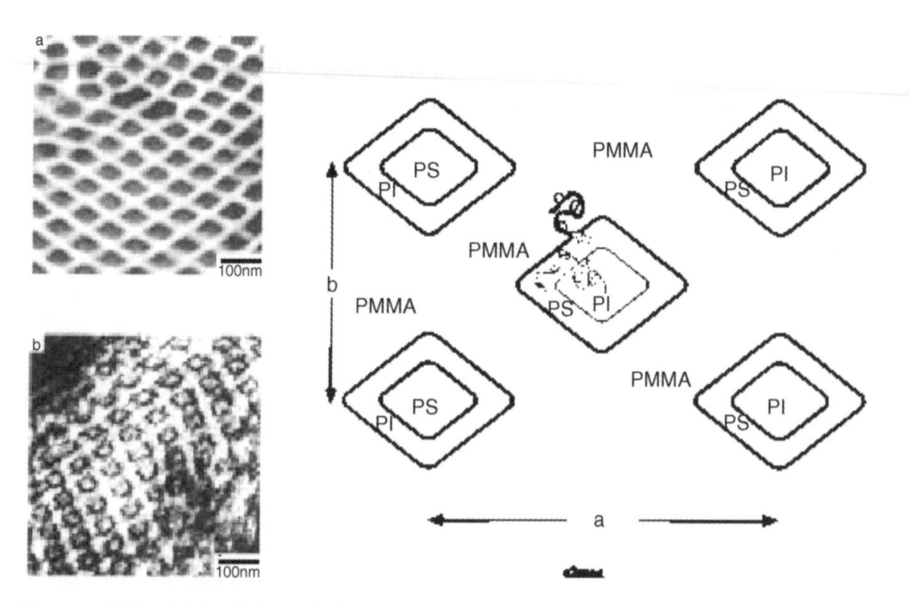

Figure 19.12. Bright field TEM images of a SIM star with composition (29.8/44.5/25.6) stained with OsO_4 (a) and with RuO_4 (b). In the OsO_4 case, the dark regions correspond to PI that forms rhomboid columns, whereas, in the RuO_4-stained micrograph, the dark regions correspond to the PS phase and the gray to the PI phase. A schematic representation of the unit cell is shown also in the figure (Sioula 1998a).

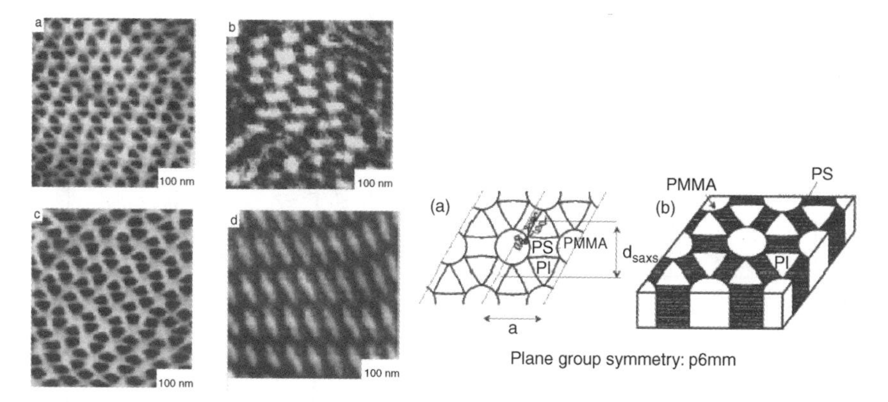

Figure 19.13. Bright field TEM images of: (a) a SIM star with composition (37.2/24.3/38.5) stained with OsO_4 and (b) the same miktoarm star stained with RuO_4, (c) a SIM star with composition (28.5/29.6/41.9) stained with OsO_4, and (d) with RuO_4. In the micrographs stained with OsO_4, the PI phase (dark regions) forms triangular prism-shape columns, whereas, in the micrographs stained by RuO_4, the dark regions are thought to correspond to the PI/PS phase boundary, the gray regions to the PS and PI phases, and the lightest regions to the PMMA phase. A schematic representation of the microdomain structure is given, which is based on the TEM study (Sioula 1998b).

Going into more symmetric SIM star terpolymers (SIM-28.5/29.6/41.9 and SIM-37.2/24.3/38.5) promoted the formation of new microdomain structures where the junction points are confined to lines where the three different types of microdomains intersect. This case is depicted in Figure 19.13, where bright field TEM images are shown from two such stars together with a schematic representation of the microstructure.

5. ABC STAR TERPOLYMERS WITH CRYSTALLIZABLE BLOCKS

Star terpolymers with two crystallizable blocks provide an example where the chain topology place additional constraints on the crystallization process and final morphology. An ABC star composed of two semicrystalline blocks exhibits order in different length scales: from the different unit cells of the crystallizable blocks, to the crystal sizes, to the microdomain structure, to the crystalline superstructure. Such ABC stars were recently synthesized based on polystyrene, polyethylene oxide, and poly(ε-caprolactone) (SEL), and their structure was investigated with SAXS/WAXS, optical microscopy, and AFM (Floudas 1998 and 2000). Figure 19.14 gives the wide-angle X-ray scattering profiles of two diblocks (PS-PEO and PEO-PCL) and of four SEL star terpolymers having the same PS and PEO arm lengths but different PCL arm length. Crystallization in the stars initiates from the homogeneous melt and drives the microphase separation. The different unit cells and specific area requirements of PEO and PCL crystals can be used as fingerprints of the block crystallization in the stars. The WAXS study revealed that both blocks

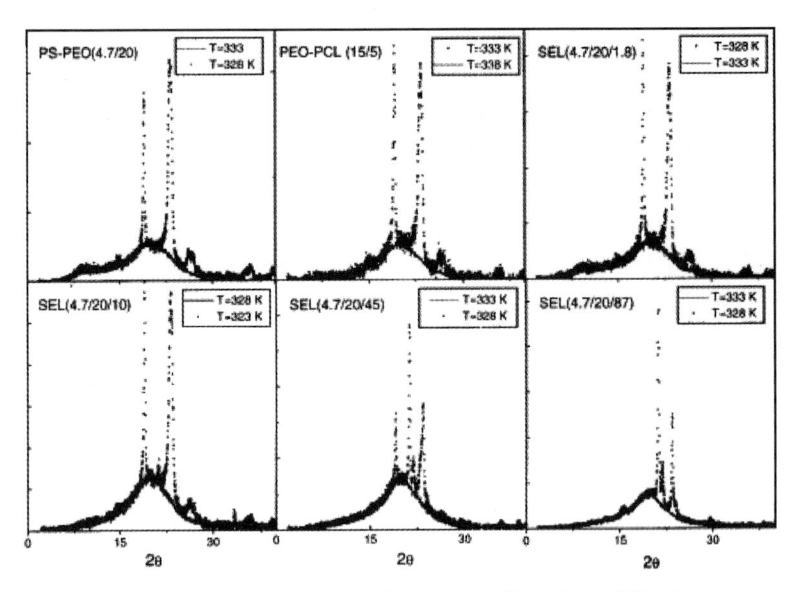

Figure 19.14. Wide-angle X-ray scattering intensity profiles of two diblock copolymers PS-PEO and PEO-PCL and four of the star terpolymers at temperatures corresponding to the melt and crystalline phases. The Bragg reflections for the SEL-4.7/20/87 (numbers give the number-average molecular weight of the three arms) sample reflect the orthorhombic unit cell of the PCL and those for PS-PEO, PEO-PCL, and SEL-4.7/20/1.8 to the monoclinic unit cell of PEO. An interesting case is for the SEL-4.7/20/10 and SEL-4.7/20/45, where the two sets of Bragg reflections correspond to the lattices of PEO and PCL (Floudas 1998).

crystallize in the stars provided that their length ration is less than 3. For the more asymmetric stars, the longer block was found to completely suppress the crystallization of the shorter one.

The optical microscopy images (Figure 19.15) revealed that, depending on the type of the crystallizable block (PEO vs. PCL), different superstructures were formed (spherulites vs. axialites). The nucleation sites of the two superstructures were completely independent, suggesting a heterogeneous distribution of star molecules with PEO or PCL crystals. This heterogeneous distribution implied that, upon crystallization, there is a mechanism of macrophase separation between unlike crystals. Furthermore, the AFM and optical microscopy studies indicated that the PS block is mixed with PCL and is responsible for the perforations within the PCL crystals. The AFM images have also shown some unanticipated complex morphologies in these systems. The kinetics of structure formation were subsequently investigated using synchrotron SAXS/WAXS (Floudas 2000).

6. ARCHITECTURE-INDUCED PHASE TRANSFORMATIONS

We have already discussed, in Chapter 14, that a phase transformation can be induced at a fixed volume fraction by changing the macromolecular architecture.

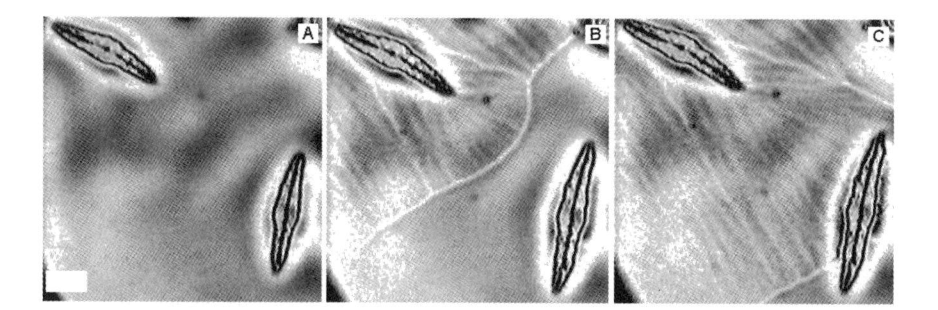

Figure 19.15. Optical micrographs showing the crystallization of PCL and PEO in the star SEL-4.7/20/10. The sample was crystallized at 313 K, and the images are taken after 11 min (A), 14 min (B), and 15 min (C). Two superstructures are formed differing in their shape and growth rates. The axialitic superstructures reflect the crystallization of the PCL block, whereas the spherulitic superstructure the crystallization of the PEO blocks in the stars (each image corresponds to a distance of about 150 μm) (Floudas 1998).

For example, while a symmetric AB diblock copolymer can only form a lamella structure, by increasing the number of arms corresponding to one of the blocks at a fixed composition, results in A_2B and A_3B miktoarm stars forming, respectively, cylinders and spheres.

Herein, we review experimental work on more complex stars, which revealed a tricontinuous cubic morphology in a symmetric starblock copolymer system at a symmetric composition. For this purpose different multiblock copolymers that contained several blocks of different molecular weight, while remaining symmetric in overall composition, have been synthesized (Tselikas 1996). The inverse starblock and linear tetrablock architecture is shown below (Figure 19.16).

Five of the copolymers investigated were found to form the usual lamellar morphology expected from the symmetric overall composition. However, a

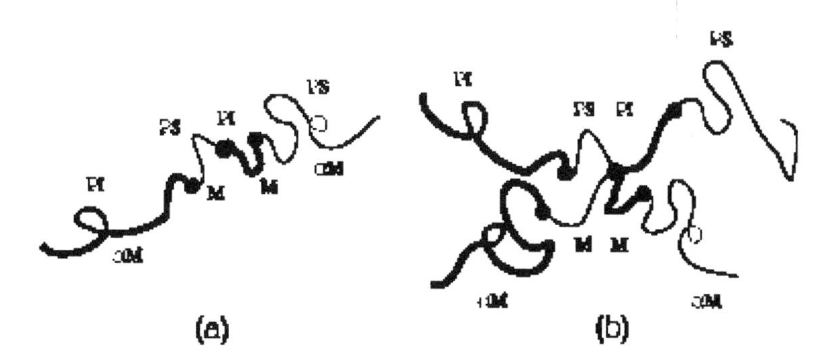

Figure 19.16. Schematic representation of the miktoarm inverse starblock copolymer and linear tetrablock copolymers. The parameter α gives the asymmetry and has values of 1, 2, and 4.

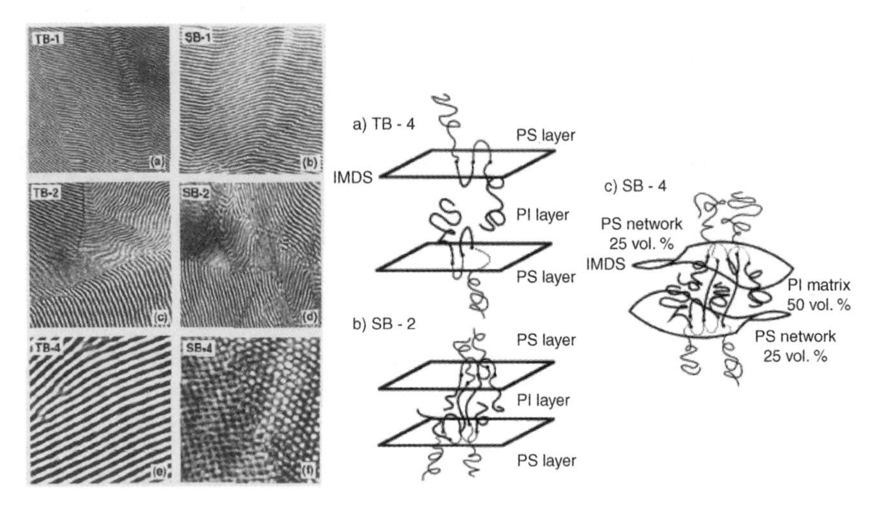

Figure 19.17. Bright-field TEM micrographs of the linear tetrablocks (indicated as TB) and the corresponding miktoarm starblock copolymers (indicated as SB). The dark and light regions correspond to the OsO_4-stained PI and the unstained PS components, respectively. In SB-4 there are regions exhibiting four-fold (left) and three-fold symmetry (right). A schematic of the proposed arrangements of the A/B junctions are also shown. Notice the triply periodic IMDS of constant mean curvature for the SB-4 case (Tselikas 1996).

miktoarm inverse star with the highest molecular weight asymmetry between the outer and inner blocks of the arms exhibited a nonlamellar morphology. Despite the SAXS investigation being inconclusive, the TEM images, shown in Figure 19.17, revealed both three- and four-fold rotational symmetry, which, together with the absence of birefringence, suggested a tricontinuous cubic symmetry.

The formation of the tricontinuous cubic morphology in this compositionally symmetric system is thought to relieve the overcrowding of the four peripheral PS-PI junctions by providing a curved intermaterial dividing surface, allowing for some bridging by the interior blocks of the miktoarm star.

REFERENCES

Bates F.S., Fredrickson G.H. (1999) Phys. Today 52, 32.

Breiner U., Krappe U., Thomas E.L., Stadler R. (1998) Macromolecules 33, 3757.

Burger C., Zhou S., Chu B. (2002), Handbook of Polyelectrolytes and Their Applications, Eds Tripathy, S.K., Kumar J., Nalwa H.S., Volume 3, Chapter 7, p125 (American Scientific Publishers).

Chen J.T., Thomas E.L., Ober C.K., Mao G. (1996) Science 273, 343.

Floudas G., Reiter G., Lambert O., Dumas P. (1998) Macromolecules 31, 7279.

Floudas G., Reiter G., Lambert O., Dumas P., Yeh F.-J., Chu B. (2000) Scattering from Polymers, ACS Series No. 732, 448.

Goldacker T., Abetz V. (1999) Macromolecules 32, 5165.

Hückstädt H., Goldacker T., Göpfert A., Abetz V. (2000) Macromolecules 33, 3757.

Jenekhe S.A., Chen X.L. (1999) Science 283, 372.

Mao G., Wang J., Clingman S., Ober C.K., Chen J.T., Thomas E.L. (1997) Macromolecules 30, 2556.

Matsen M.W. (1998) J. Chem. Phys. 108, 785.

Matsushita Y., Suzuki J., Seki M. (1998) Physica B 248, 238.

Shefelbine T.A., Vigild M.E., Matsen M.W., Hajduk D.A., Hillmyer M.A., Cussler E.L., Bates F.S. (1999) J. Am. Chem. Soc. 121, 8457.

Sioula S., Hadjichristidis N., Thomas E.L. (1998a) Macromolecules 31, 5272.

Sioula S., Hadjichristidis N., Thomas E.L. (1998b) Macromolecules 31, 8429.

Stadler R., Aushra C., Beckmann J., Krappe U., Voigt-Martin I., Leibler L. (1995) Macromolecules 28, 3080.

Stupp S.I., LeBonheur V., Walker K., Li L.S., Huggins K.E., Keser M., Amstutz A. (1997) Science 276, 384.

Tselikas Y., Hadjichristidis N., Lesanec R.L., Honeker C.C., Wohlgemuth M., Thomas E.L. (1996) Macromolecules 29, 3390.

Zheng W., Wang Z.-G. (1995) Macromolecules 28, 7215.

CHAPTER 20

BLOCK COPOLYMER DYNAMICS

Despite the large theoretical and experimental effort to understand and, ultimately, predict the phase state of block copolymers, the field of block copolymer dynamics started to be explored much later. Herein, we provide a current assessment of progress in this field. The issues covered include the collective relaxation, chain and self diffusion, local segmental, and global chain dynamics in relation to the phase state and topology of chains in block copolymers and in block copolymer/homopolymer blends.

1. DYNAMIC STRUCTURE FACTOR OF DISORDERED DIBLOCK COPOLYMERS

A full description of the dynamics requires space-temporal information of all possible pairs i and j on l, m polymers each consisting of N monomers as given in the dynamic structure factor (or intermediate scattering function):

$$S(q,t) = N^{-2}\left\langle \sum_{l,m,i,j} exp\left[iq\left(r_j^l(t) - r_i^m(0)\right)\right]\right\rangle \quad (20.1)$$

where q is the magnitude of the wave vector and $r_j^l(t)$ the position of segment j on chain l at time t; for $t = 0$, eq. 20.1 yields the static structure factor S(q).

In the mean field regime, the S(q,t) of disordered unentangled diblock copolymers is a two-step relaxation function (Erukhimovich 1986, Ackasu 1986):

$$S(q,t) = S_1(q)exp(-\Gamma_1(q)t) + S_2exp(-\Gamma_2(q)t) \quad (20.2)$$

where $\Gamma_1(q) = xg(1,x)/\tau_1 S_1(q)$ is the collective relaxation rate for the composition fluctuations $\phi_{q(t)}$ with $x \equiv q^2R^2$ (R being the radius of gyration of the copolymer AB chain and τ_1 being the AB chain longest relaxation time. The most probable ϕ_q have wave vector q* with $q^*R \sim 1$, and, hence, $S_1(q)$ peaks at a finite q*. The second contribution in the above equation was first experimentally found (Anastasiadis 1993) and then assigned (Jian 1994) to the chain self-diffusion detected due to the finite composition polydispersity $\kappa_o = \langle f^2 \rangle - \langle f \rangle^2$ of real AB polymers with average A-composition f:

$$S_2 = \kappa_o/(1 - 2\kappa_o \chi N), \Gamma_2 = q^2 D_s(1 - 2\kappa_o \chi N) \tag{20.3}$$

where χ is the monomer-monomer interaction parameter, and D_s is the AB self-diffusion coefficient.

For AB melts, the low q ($<q^*$) fraction of S(q,t) was first measured by photon correlation spectroscopy (PCS) operating at low q's but covering a very broad time range (Anastasiadis 1993). For a poly(dimethyl-b-ethylmethyl siloxene) (N = 1110, f = 0.49) melt, the experimental relaxation function ($\propto |S(q,t)|^2$ of Figure 20.1 over eight decades in the time domain displays the two contributions of eq. 20.2 with amplitude (upper inset) and rate (lower inset), both conforming to the theoretical predictions; at $x \ll 1$, $S_1(q) \propto a_1 \propto q^2$ and $\Gamma_1 \propto q^o$, whereas $S_2 \sim q^o$ and $\Gamma_2 \propto q^2$.

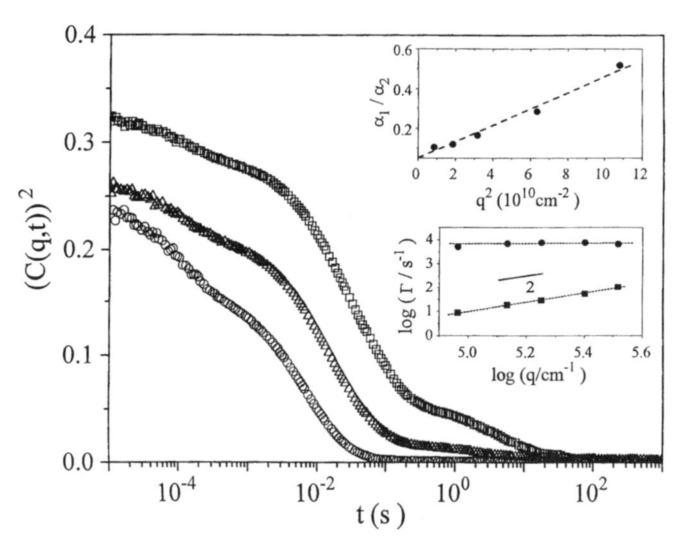

Figure 20.1. The square of the structural relaxation function for a disordered dimethyl-b-ethylmethyl siloxane diblock copolymer melt at 120°C at three scattering angles (□: 45°, △:60°, ○:90°) corresponding to x = 0.15, 0.2, and 0.29. The variation of the relaxation rates Γ_1(●) and Γ_2(■) and the intensity ratio $a_1/a_2 = S_1(q)/S_2$ of the fast and slow processes with q are shown in the insets. The solid line in the lower inset indicates a slope 2 for diffusive behavior (Anastasiadis 1993).

Access to the most important length scales, i.e., $qR \simeq O(1)$, would require either the high q's of the neutron and X-ray scattering or larger sizes and, hence, synthesis of AB with molecular mass $M_w \simeq O(10^6)$. While application of X-ray photon correlation spectroscopy is a task of the future, neutron echo lacks sufficient broad time range to study the dynamics of AB melts in the interesting $q \sim q^*$ range (Montes 1999). Therefore, nowdays, access to the $S(q \sim q^*, t)$ is feasible only for AB solutions in a common solvent using only PCS. Figure 20.2 shows the intermediate scattering function $C(q \simeq q^*,t)$ for a 2.2 wt% disordered semidilute solution of such a high M_w poly(styrene-b-isoprene) (S12M20) ($N = 25800$, $f = 0.26$) in the common solvent toluene; at 25°C this system enters the ordered regime with cylindrical morphology above 7.4%. The S(q) attains its peak at $q^* \simeq 0.033\ \text{nm}^{-1}$ within the light scattering q values (inset in the upper part of Fig. 20.2). The high quality of C(q,t) allows for an inverse Laplace transformation to yield the relaxation distribution function L(lnτ). As shown in Figure 20.2, L(lnτ)

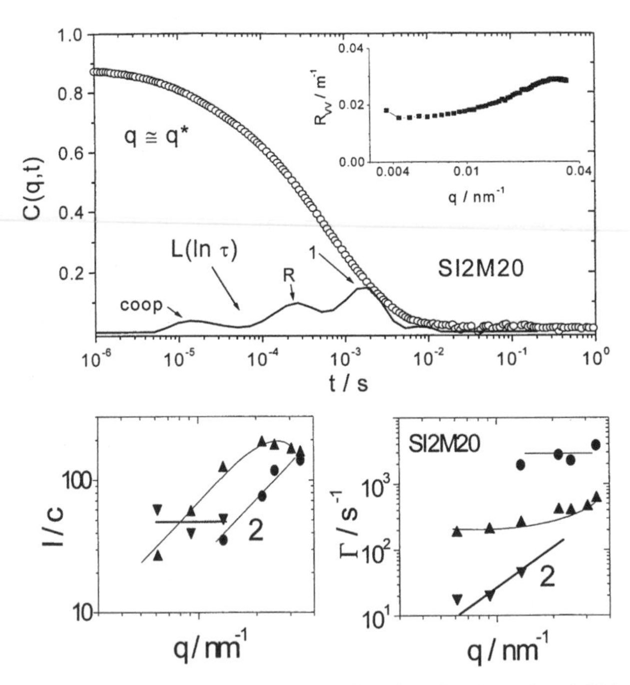

Figure 20.2. Upper part: Intermediate scattering function C(q,t) at $q^* \sim 0.034$ nm and 25°C for 2.2 wt% SI2M20 in the common solvent toluene along with the distribution L(lnτ) of relaxation times. The peaks of L(lnτ) are assigned to cooperative diffusion (Coop) and the two chain conformational motions (see text). The absolute light-scattering intensity distribution for this disordered solution is shown in the inset. Bottom: Reduced intensities and relaxation rates associated with the three modes of the dynamic structure factor ▲,● chain relaxation and ▼ self diffusion. Slope 2 indicates the q^2-dependence of the intensity I_R and the diffusive rate of the chain self-diffusion.

reveals the presence of different relaxation processes (three at $q \sim q^*$), the assignment of which leads to the underlying motional mechanisms.

The first peak of $L(\ln\tau)$ at short times is not AB-specific, as it relates to the cooperative diffusion responsible for the total AB concentration fluctuations driven by the solution osmotic pressure (de Gennes 1979). The main process (1) with intensity I_1 and rate Γ_1 shown in the lower part of Fig. 20.2 exhibits the characteristics of the chain relaxation (first exponential contribution in eq. 20.2). In the presence of topological constraints (N/N_e entanglements per chain), the conformational dynamics of a single chain undergoing one dimensional curvilinear diffusion (reptation) in a virtual tube and tube length fluctuations (Rouse motion) at shorter times $t/\tau_1 \ll 1$. The theoretical derivation (Sigel 1999) of S(q,t), using these two motional mechanisms, and the mean field theory (to include the interactions) can capture the characteristics of the additional chain relaxation (R in Fig. 20.2). These two chain conformational processes for entangled AB solutions are responsible for the composition fluctuations leading to the peak in S(q). The contribution to S(q) at lows q's arises from the slow diffusive process in S(q,t), with characteristic intensity and rate (triangles in the lower part of Fig. 20.2) conforming to the predictions of eq. 20.3.

The dynamic structure factor of sufficient monodisperse symmetric and asymmetric diblock copolymers far from the ordering transition in the disordered regime and $q \simeq O(q^*)$ can be represented by eq. 20.2, and the processes are characterized by the behavior of the lower part of Figure 20.2. It is the relative contribution and the q-dependence of the rates of the main processes that makes S(q,t) less q-dependent at $q \leq q^*$. However, the situation changes drastically when the composition polydispersity increases. The relaxation of the composition fluctuations with the most probable wavelength $1/q^*$ now proceeds via internal diffusion and not via chain overall motion. The change of S(q,t) from a pure relaxational to a pure diffusive character was demonstrated (Holmqvist 2002) in the case of partially hydrogenated SI diblocks and can theoretically be captured (Chrissopoulou 2001). The second process in eq. 20.2 dominates S(q,t) and is responsible for the maximum in S(q,t).

The proximity to the ordering transition has a pronounced effect on the internal diffusion at finite $q \sim q^*$ in contrast with the intermolecular polymer mixtures for which the macrophase separation manifests itself in the thermodynamic limit $(q = 0)$.

Figure 20.3 shows the counterintuitive slowing down of the internal diffusion $D = \Gamma_2(q)/q^2$ for the short-range composition fluctuations relative to its respective long-range value. The unfavorable enthalpic interactions between chemically dissimilar monomers greatly influence both static and dynamic behavior for length scales on the order of the characteristic dimension of the structure ($\sim 2\pi/q^*$). While the real phase morphology is well-established (Rosedale 1995), the dynamic map of this part of the phase diagram is in a premature state. Diblock copolymers near the ordering transition offer the possibility of studying the pretransitional dynamic behavior of systems exhibiting a weak first-order transition.

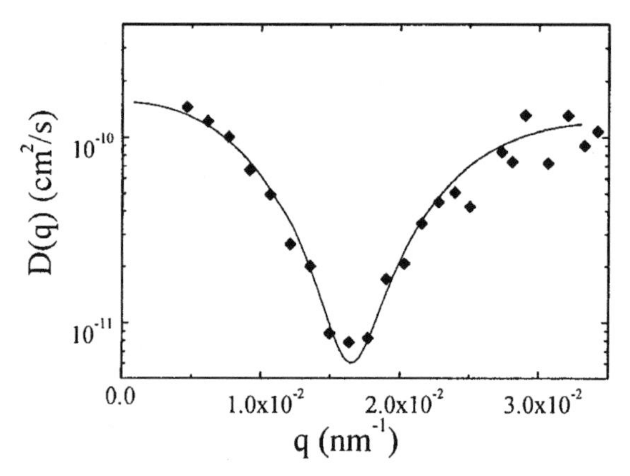

Figure 20.3. Wave vector dependence of the internal diffusion $D(q) = \Gamma_2(q)/q^2$ of a symmetric SI3M50 poly(styrene-b-isoprene) ($M_w = 3 \cdot 10^6$ g/mol) 2.5 wt% solution in toluene. $\Gamma_2(q)$ is the most probable decay rate of composition fluctuations (Boudenne 1996).

2. DYNAMIC STRUCTURE FACTOR OF ORDERED DIBLOCK COPOLYMERS

In addition to DLS, Forced Rayleigh scattering (FRS) (Fleischer 1993) (Ehlich 1993) (Zielinski 1995) and Pulsed-Field Gradient (PFG)-NMR (Hamersky 1998) (Fleischer 1999a,b) (Papadakis 2000) have been employed in the study of dynamics in ordered block copolymers.

The diffusion mechanism in a macroscopically ordered diblock copolymer should be anisotropic. The anisotropy should reflect the fact that the thermodynamic penalty for motion parallel and perpendicular to the lamellar planes is different (the expectation is that motion perpendicular to the lamellar planes should experience a higher thermodynamic penalty because a reptating AB diblock must necessarily pull its A block into the B-rich domain). This mechanism is referred to as activated reptation (Fredrickson 1990) to distinguish from the simple reptation mechanism in entangled homopolymers.

Figure 20.4 gives the self- and tracer-diffusion coefficients in macroscopically oriented poly(ethylenepropylene-b-ethylethylene) (PEP-PEE) diblock copolymers using forced Rayleigh scattering (FRS) (Lodge 1995).

Different symmetric copolymers have been studied with molecular weights: 3.1×10^4 (PEP-PEE-1), 5×10^4 (PEP-PEE-2), 8.1×10^4 (PEP-PEE-3), and 1.1×10^5 (PEP-PEE-4). D_0 denotes the "free" diffusivity, i.e., the mobility of a chain with $\chi = 0$. The self-diffusion coefficient parallel to the lamellar planes for the three ordered copolymers was found to scale as:

$$\frac{D_{par}}{D_0} = 6.06 \exp(-0.237\chi N) = \exp\{-0.237(\chi N - 7.6)\} \qquad (20.4)$$

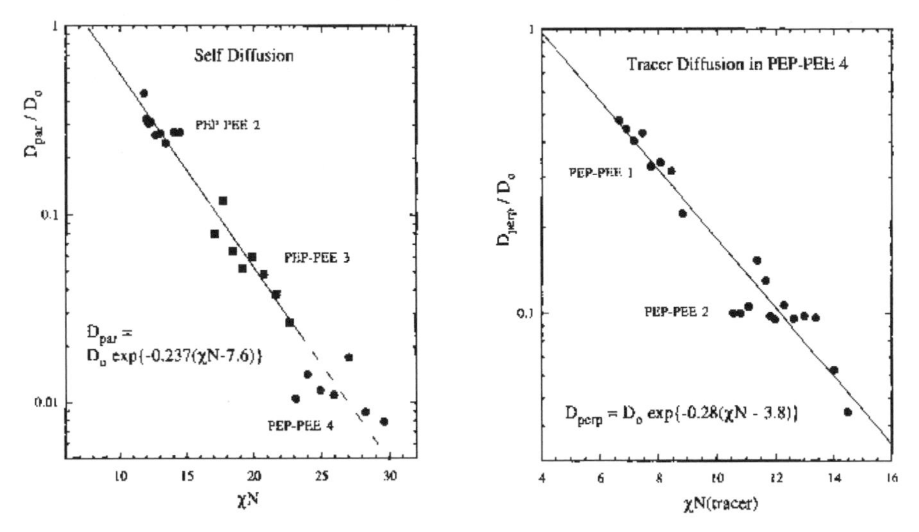

Figure 20.4. Self- (left) and tracer- (right) diffusion coefficients parallel and perpendicular to the lamellar planes for different poly(ethylenepropylene-b-ethylethylene) (PEP-PEE) with $f_{PEP} = 0.53$. Notice the similar reduction of the diffusion coefficients normalized to the corresponding free diffusivities (Lodge 1995).

which clearly shows that diffusion in the lamellar planes is retarded, and the barrier is proportional to the thermodynamic variable χN. Surprisingly, the effects of retardation would disappear at a very low value of $\chi N = 7.6$, corresponding well within the disordered regime. Furthermore, for $\chi N > 20$, the T-dependence of the same quantity was found to scale as

$$\frac{D_{par}}{D_0} = \exp\left(-0.17\gamma\frac{\mathrm{N}}{\mathrm{N}_e}\right) \tag{20.5}$$

independent of temperature, implying a crossover to a mechanism with an entropic activation barrier. A mechanism based on block-retraction analogous to the arm-retraction in branched polymers has been proposed to account for this effect.

The dependence of the normalized perpendicular diffusion coefficient (D_{perp}) is also plotted in the figure. The dependence can be parametrized as:

$$\frac{D_{perp}}{D_0} = 2.87\ \exp(-0.276\ \chi N) = \exp\{-0.276(\chi N - 3.8)\} \tag{20.6}$$

The expectation is that $D_{perp} \ll D_{par,}$ because, for the perpendicular diffusion, the PEP block would have to move through the PEE domain or vice versa, whereas activated reptation parallel to the layers would still be possible. The results, indeed, show that there is a different T-dependence in the two coefficients, but, surprisingly, the differences are small (Dalvi 1993a,b) (Dalvi 1994).

FRS and PFG-NMR have also been used to study block copolymer diffusion in the cylinder and Gyroid phases (Hamersky 1998). The sample was a poly(ethylene oxide-b-ethylethylene) (PEO-PEE) copolymer melt ($f_{PEO} = 0.42$, $M_w = 4100$ g/mol) undergoing the following transitions:

$$L_c \xrightarrow{45^\circ C} \text{Hex} \xrightarrow{80^\circ C} \text{Gyroid} \xrightarrow{175^\circ C} \text{Dis}$$

The large hysteresis loop permitted measurements in the cylinder and Gyroid phases at the same temperature (i.e., 60 °C). FRS results for both phases are shown in Figure 20.5.

Within the Gyroid phase, the signal decays linearly, indicative of a single exponential behavior. On the other hand, the diffusion in the cylinder phase shows a much slower nonsingle-exponential dependence. Two time scales are needed to describe this behavior (τ_{fast} and τ_{slow}), each decay constant exhibiting a linear dependence on d^2 (d is the grating spacing), confirming the diffusive origin of both relaxation processes. A shear-alighed cylinder sample was used to extract the diffusivity along (D_{par}) and across (D_{perp}) the cylinders. The mobility along the cylinders was two orders of magnitude larger, i.e., $D_{par}/D_{perp} \approx 80$, which is much higher than the anisotropy found in the lamellar phase. This result was consistent with the estimated enthalpic penalty ($D_{perp} \approx D_0 \exp(-f\chi N)$). Although the magnitude of the copolymer mobility was much lower than that of either constituent copolymer, it was found to be a smooth function of temperature through the order-to-disorder transition. Subsequently, the effect of composition fluctuations on the

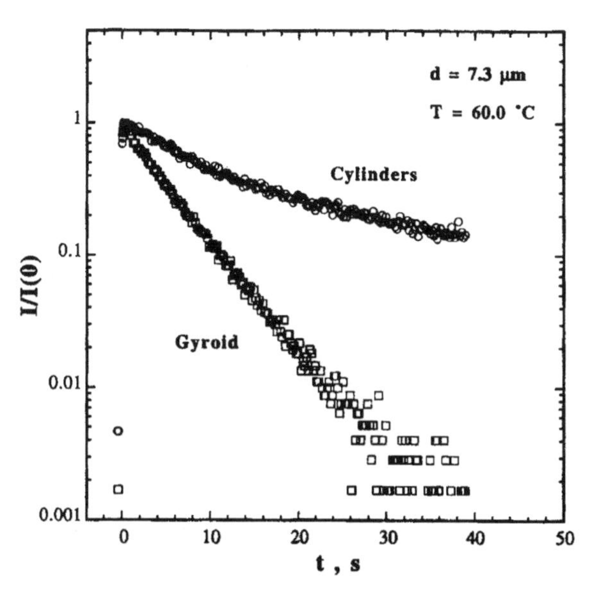

Figure 20.5. FRS decays obtained for a PEO-PEE, $f_{PEO} = 0.42$ copolymer at 60°C. Notice the single decay within the Gyroid phase as opposed to the "slower" nonlinear decay within the cylinder phase (Hamersky 1998).

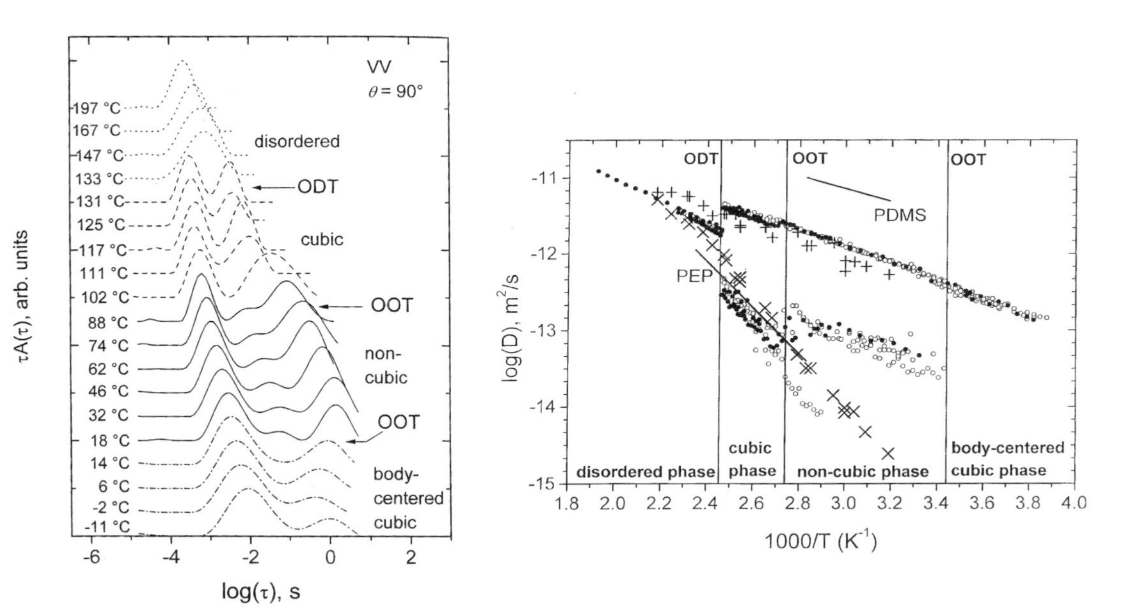

Figure 20.6. (Left): Distribution function obtained from DLS at different temperatures in the VV geometry at an angle of 90^0. (Right): Diffusion constants of the different modes obtained on heating (○) and cooling (●). The values correspond to the relaxation rates at an angle of 90^0:$D = \Gamma/q^2$; (+, X): fastest and most probable diffusion constants, D_{fast} and D_{max}, respectively, from PFG-NMR. The lines indicate the self-diffusion constants of pure PEP and PDMS for a molecular weight of 12500 g/mol (Papadakis 2000).

tracer diffusion was investigated in symmetric block copolymers (Kannan 1998). Fluctuations were found to suppress diffusion of both homopolymers and copolymer tracers.

The self-diffusion of the lamellar-forming nonentangled poly(ethylene propylene-b-dimethyl siloxane) (PEP-PDMS) diblock copolymers was also investigated by PFG-NMR (Fleischer 1999a). It was found that lowering the temperature from the disordered phase results in a sudden increase of the width of the distribution of self-diffusivities at the T_{ODT} due to the onset of anisotropic diffusion. In addition, the D(T) showed a slowing-down at the T_{ODT}, attributed to the effect of concentration fluctuations.

The bulk dynamics of an asymmetric PEP-PDMS diblock copolymer ($f_{PEP} = 0.22$, $M_w = 1.25 \times 10^4$ g/mol, $T_{ODT} = 134\,°C$) exhibiting different order-to-order transitions was studied with DLS and PFG-NMR (Fleischer 1999b) (Papadakis 2000). The pertinent experimental findings are summarized in Figure 20.6, below. In the disordered phase, the self-diffusion of chains dominates, but, in contrast with the symmetric case (Fleischer 1999a), the self-diffusivity, as well as its distribution width, changes only slightly at the T_{ODT}. At $T < T_{ODT}$, the sample forms a cubic phase with PEP micelles and exhibits two diffusion mechanisms: the mutual diffusion of micelles and of free chains experiencing the matrix friction (PDMS) and a slower process assigned either to the diffusion of micellar aggregates or to the mutual diffusion of the free chains and the ones bound to micelles, the latter process having the friction of the slow (PEP) component. The same modes persist at lower temperatures despite the intervening order-to-order transition except for a weak intermediate mode (see Fig. 20.6, left) with an activation energy of a PDMS homopolymer.

There is also evidence of layer undulation effects on the S(q,t) of lamellar diblock copolymers (Stepanek 2001).

3. DIELECTRIC RELAXATION IN DIBLOCK COPOLYMERS IN THE DISORDERED AND ORDERED PHASES

Dielectric spectroscopy (DS) is a very sensitive technique to both the local segmental modes as well as of the global chain dynamics in type A polymers (i.e., polymers having components of the dipole moment both perpendicular and parallel to the chain) (Stockmayer 1967) (Bauer 1965). Among the few type A polymers, polyisoprene (cis-PI) is the most (Adachi 1985) (Boese 1990) (Floudas 1999a) representative polymer studied by DS. The technique is very sensitive to the dipolar relaxations and has the required broad frequency range to capture the multiple relaxation processes present in block copolymers. Herein, we review the pertinent experimental findings on the local and global PI dynamics in disordered and ordered block copolymers. We will show that the spectrum of normal modes is very sensitive to the presence of ordered microdomains. Furthermore, DS can be employed as a probe of the interfacial width in microphase-separated copolymers of a particular architecture and is able to capture the importance of the relevant topological constraints on the dynamics of star block copolymers. Lastly, the same technique can be used in the investigation of the effect of confinement on polymer dynamics.

A typical example of dynamic processes affecting the dielectric loss spectra in a block copolymer is given in Figure 20.7 below. The system is a miktoarm star block terpolymer composed of polystyrene, polyisoprene, and polybutadiene arms forming PS cylinders in the matrix of I/B chains ($f_{PS} = 0.3$, $M_w = 24700$). The spectra reveal three processes: the "fast" segmental mode at high frequencies corresponding to the mixed I/B relaxation, the "slower" spectrum of normal modes reflecting the PI chain relaxation within the I/B matrix, and, at lower frequencies/higher temperatures, the conductivity contribution.

The segmental mode is very sensitive to the local environment, and, as such, it has been used with success to account for the proximity to the order-disorder transition in block copolymers (Quan 1991) (Alig 1992) (Kanetakis 1992) (Stühn 1992) (Floudas 1996). Pressure has also been used, and it was shown to affect the dynamics of thermodynamically miscible polymers forming a diblock copolymer by inducing dynamic miscibility (Floudas 1999b).

The effect of the order-to-disorder transition on longer length scales, such as of the chain relaxation, is quite spectacular. An example is given in Figure 20.8, where the dielectric loss curves for two symmetric diblock copolymers, one being in the disordered (top) and the other in the ordered state (bottom), are compared with the

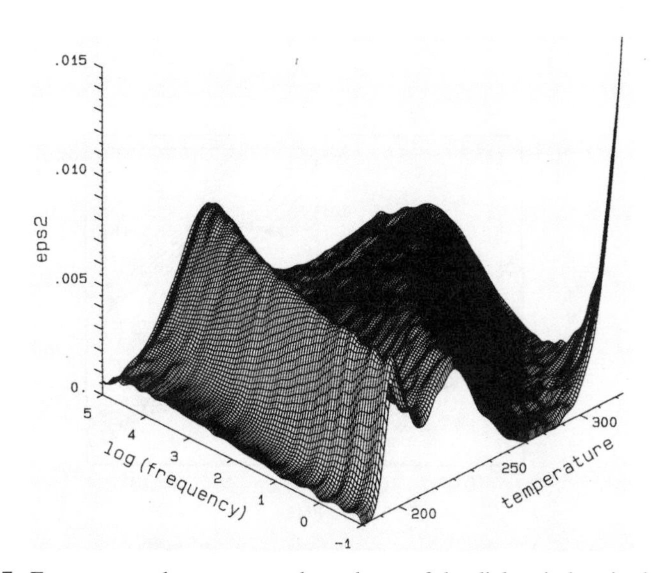

Figure 20.7. Frequency and temperature dependence of the dielectric loss in the star block copolymer SIB (composed of PS, PI, and PB arms) in the microphase-separated state ($T_{ODT} = 379$ K). Three processes are shown: the polyisoprene segmental relaxation in the I/B environment at low T, which is followed by a broad spectrum of normal modes and by the conductivity at higher temperatures (Floudas 1996).

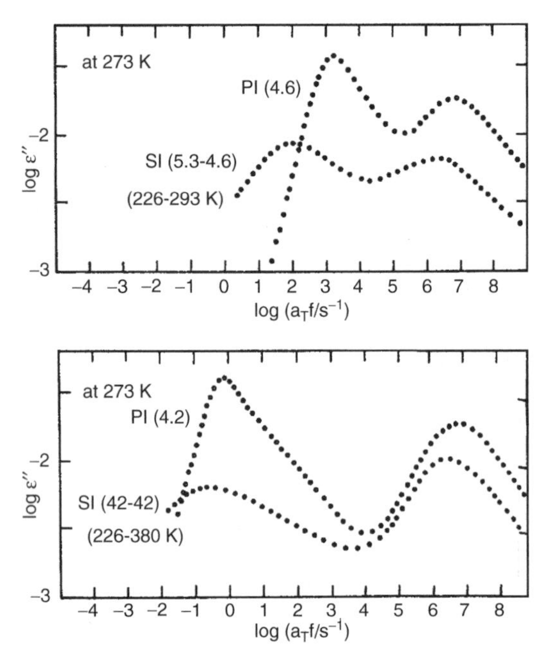

Figure 20.8. Dielectric loss curves for a symmetric SI diblock copolymer in the vicinity of the order-to-disorder transition (Top) as compared with the PI precursor. The corresponding loss curve is shown for a well microphase-separated SI diblock copolymer (Bottom) (Yao 1991).

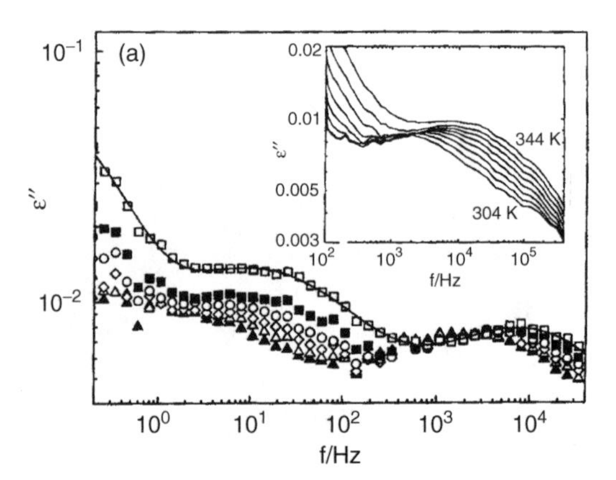

Figure 20.9. Dielectric loss curves of a nearly symmetric SI diblock copolymer ($f_{PS} = 0.52$, $M_n = 12\ 200$, $T_{ODT} = 359$ K) at different temperatures in the microphase-separated state: (open squares): 338 K, (filled squares): 333 K, (circles): 328 K, (rhombus): 323 K, (open triangles): 318 K, and (filled triangles): 313 K. Two modes are indicated. The "faster" corresponds to the spectrum of normal modes and the "slower" reflects the relaxation of the coherently ordered microstructure (Karatasos 1996).

corresponding curves of PI homopolymers of similar molecular weight with the PI block in the copolymers (Yao 1991). Evidently, the low-frequency spectrum of the global chain dynamics is retarded and broadened as compared with free PI chains. This is in agreement with the findings from rheology where, below the T_{ODT}, the moduli reflecting the microphase-separated state showed a nonterminal response. These differences in the DS spectra have been discussed in terms of the spatial and thermodynamic confinement.

The existence of the microdomain is also reflected in the DS spectra with the appearance of a new "slower" process reflecting the relaxation of the conformal interfaces formed in the ordered state (Karatasos 1996). The latter mechanism was discussed also, based on the results from computer simulations. As an example, we show in Figure 20.9 the existence of the slower process in a microphase-separated SI diblock copolymer. The amplitude of this mode was found to depend strongly on the preparation conditions and, thus, to the coherence of grains.

4. DYNAMIC INTERFACIAL WIDTH IN BLOCK COPOLYMERS

As we mentioned in Chapter 13, the interfacial thickness between two immiscible homopolymers depends on the interaction parameter, and, in the limit of long homopolymers, the work of Helfand and coworkers predicts (Helfand 1971): $\Delta_\infty = \frac{2\alpha}{\sqrt{6\chi}}$ where α is the statistical segment length (assuming $\alpha_A = \alpha_B = \alpha$) and χ is, as usual, the interaction parameter (see also eq. 13.1). The above equation correctly predicts the interfacial thickness in block copolymers only in the strong segregation limit ($\chi N \gg 10$). The effect of the finite molecular weight of the homopolymers is to increase Δ, which now depends on the block incompatibility (Broseta 1990)

$$\Delta \approx \Delta_\infty \left[1 + ln2\left(\frac{1}{\chi N_A} + \frac{1}{\chi N_B}\right)\right] \tag{20.7}$$

where N_A, N_B are the degrees of polymerization of the A and B homopolymers, respectively. Corrections due to the connectivity of blocks were found to be significant and resulted in a broader interface (Semenov 1993):

$$\Delta \approx \Delta_\infty \left[1 + \frac{1.34}{(\chi N)^{1/3}}\right] \tag{20.8}$$

Based on the above equations, a typical interfacial thickness for a phase-separated polystyrene-polyisoprene diblock copolymer with $N = 1000$, $\chi \approx 0.1$, and $\alpha \approx 0.68$ nm is about 2.2 nm.

Triblock copolymers composed of two glassy outer blocks and an elastomeric midblock provide the means of obtaining an estimate of the interfacial width in block copolymers using dielectric spectroscopy. Triblock copolymers, such as polystyrene-polyisoprene-polystyrene (SIS), can be considered as thermoplastic elastomers, a property that results from tethering both ends of the rubbery block to

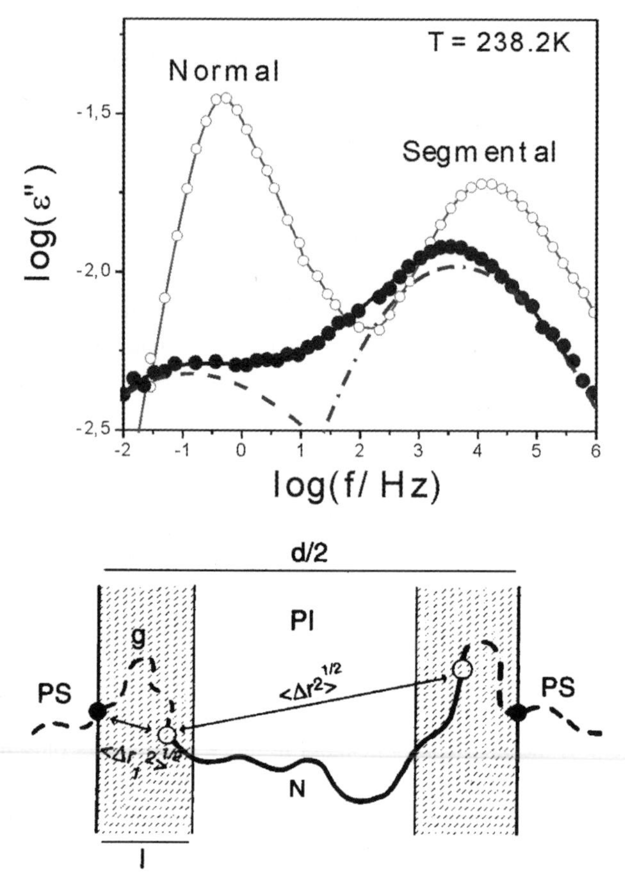

Figure 20.10. Frequency dependence of the dielectric loss of the SIS triblock copolymer ($f_{PS} = 0.6$, $M_n = 65000$) indicated by the filled circles in comparison with the normal and segmental modes of a PI homopolymer at 258 K. Besides the fast segmental mode, a slower mode exists in the triblock (top). In the bottom a highly schematic representation is given of the contribution from the junction point fluctuations in the same triblock, which gives rise to the slow process observed in the experiment (Alig 1997) (Floudas 1998). The intensity of this slow mode provides a direct estimate of the interfacial width.

the glassy PS domains. The rheological behavior of triblock copolymers has been examined for lamellar, cylindrical, and spherical structures. It is widely recognized that a factor that influences the viscoelastic properties of triblock copolymers is the ratio of loops versus bridges of the PI midblock. An answer to this problem was given recently by dielectric spectroscopy on SIS triblocks with a dipole inversion in the middle block (Watanabe 1995).

In a SIS triblock copolymer, the PI block chain-ends are attached to the glassy PS domains, and, thus, the chain dynamics are frozen. Despite this expectation, two recent DS studies in block copolymers with a basic triblock unit (Alig 1997)

(Floudas 1998) revealed a "slow" mechanism in addition to the expected "faster" mode due to the segmental relaxation. Figure 20.10 provides the dielectric loss spectrum of SIS and displays the "fast" segmental relaxation followed by a slower mode at lower frequencies.

For an infinitely thin interface without mobility of the junction points parallel to the interface (e.g., due to restrictions caused by entanglements), no dielectrically active relaxation of the end-to-end vector is expected because both ends are tethered. However, in block copolymers there is a finite interfacial thickness given by eqs. 20.7 and 20.8. This situation for a triblock copolymer is schematically shown in Fig. 20.10 (bottom). The junction point located at $\mathbf{r_1}$ (open circle) is allowed to fluctuate at the PS-PI interface (shadowed area). The portion of the PS chain, composed of an average number of g segments entering the interface, is effectively tethered on the one side by the glassy PS at the position $\mathbf{r_t}$ (filled circle). Assuming a Gaussian distribution for the subchain between $\mathbf{r_t}$ and $\mathbf{r_1}$ composed of g PS segments, we can calculate the mean-square distance between $\mathbf{r_1}$ and $\mathbf{r_t}$ as: $\langle \Delta r_1{}^2 \rangle = g a_{PS}{}^2$. Similarly, the mean-square end-to-end distance of the PI block is given by: $\langle \Delta r^2 \rangle = N a_{PI}{}^2$, where N is the number of PI segments. Thus, the value of $\langle \Delta r_1{}^2 \rangle^{1/2}$ can be considered as the characteristic length of fluctuations of the junction point within the interface. In general, there will be a distribution of PS segments entering the interface whose mean value is controlled by the interaction parameter ($g \sim \chi^{-1}$).

For each PI block of the SIS triblock copolymer, the fluctuations of both end points of the PI chain have to be considered, i.e., $\langle \Delta r^2 \rangle = 2\langle \Delta r_1{}^2 \rangle$. Then the characteristic length of the dynamically probed interface is related to the measured strength of this process by:

$$\langle \Delta r_1^2 \rangle = \frac{3k_B T M_{PI} \Delta\varepsilon}{4\pi C N_A \mu^2 2} \tag{20.9}$$

where C is the concentration of dipoles, N_A is Avogadro's number, μ is the dipole moment per contour length, and M is the PI molecular weight. It is evident that the ratio of the relaxation strength of this process to the normal mode of a free PI chain composed of N segments is

$$\frac{\Delta\varepsilon}{\Delta\varepsilon_N} \sim 2\frac{g}{N} \tag{20.10}$$

(under the usual mean-field assumption of equal segment lengths, i.e., $\alpha_{PS} \approx \alpha_{PI}$). Therefore, the expectation, borne out from the above equation, is that the intensity of this mode will be very much reduced as compared with a free PI chain.

We can estimate the theoretical interfacial thickness Δ and the number of PS segments, g, entering the interface from the interaction parameter χ. Because $g \sim \chi^{-1}$ and $\chi \sim 0.1$, the prediction is that 8-10 PS segments per junction point are entering the interface. The result for Δ_∞ with the corrections for the finite molecular weight and the connectivity of blocks is 2.4 nm. This estimate of the interfacial thickness compares favorably with the result from eq. 20.9, using the measured dielectric strength of the process under investigation, which, at $T = 300$ K, results in $\langle \Delta r_1{}^2 \rangle^{1/2} \approx$ 4-6 nm (the thickness of 2.4 nm refers to the half-width of the interface).

Similarly, the dynamics of the junction point will be dominated by the PS subchain entering the interface. The slowest relaxation time can be estimated from: $\tau_g \approx 4\zeta_{eff}\, g^2$, where ζ_{eff} is an effective friction coefficient created by the PI and PS segments at the interface, and the factor 4 is the usual factor for a tethered chain (Yao 1991). A calculation of the junction point dynamics relative to the normal mode time of a free PI chain composed of N segments is possible. This ratio is:

$$\frac{\tau_g}{\tau_N} \sim \left(\frac{\zeta_{eff}}{\zeta_{PI}}\right) \frac{g^2}{N^3/N_e} \tag{20.11}$$

Based on the estimation of g from the interaction parameter, the expectation is that $\tau_g/\tau_N \ll 1$. Thus, the relaxation times for this mode will generally be faster than those from the normal mode time of a free PI chain, which has also been observed experimentally.

The study of the dynamics in SIS triblock copolymers (Alig 1997) as well as in other nonlinear (Floudas 1998) microphase-separated block copolymers with a basic triblock unit revealed that DS can effectively be employed as a *dynamic probe of the interface*.

The dynamics associated with the chain–end relaxation is sensitive not only to the interface but also to the microstructures formed in ordered block copolymers. As we mentioned earlier, the confinement of blocks results in the broadening and retardation of the chain dynamics. Macromolecular architecture can be used to control the chain topology within the microdomains. Contrary to the local dynamics, the chain dynamics are very sensitive to the topology of chains within the microdomains. The investigation of the dynamics in star block copolymers of the $(SI)_4$ type with PI forming the core revealed a speed-up of the chain dynamics as a result of the localization of the star center within the center of the PI domain (Floudas 1997a).

5. DIELECTRIC RELAXATION IN BLOCK COPOLYMER/ HOMOPOLYMER BLENDS

There has been an increasing effort to understand, control, and, ultimately, predict the properties of liquids near surfaces, interfaces, and within porous media. These efforts are likely to have, apart from fundamental, technological importance in fields such as membrane separation, chromatography, and oil recovery. There is consensus that both the static and the dynamic properties of liquids are altered in the vicinity of an interface. For example, the melting temperature of liquids immersed in porous glasses is strongly suppressed (Jackson 1990), and the glass transition temperature is also affected as a result of the confinement. One disadvantage of the porous glasses used for the majority of the studies of liquid confinement is that they exhibit a broad pore-size distribution, and, in many cases, the pores have very irregular shapes. Therefore, these systems cannot be considered as model systems.

A way to overcome this difficulty is by new synthetic roots of inorganic materials with well-controlled sizes. Another way is by investigating polymer dynamics within the restricted geometries formed by microphase-separated block copolymers. It is the latter system that is of interest to us here. A certain advantage in using block copolymers as the confining medium is that the geometry (spheres, cylinders, lamellae) and size (5-500 nm) of the microdomains (hence, the "pores") can be controlled at the synthesis level. This observation brings forward a new possibility of studying the effect of restriction on the polymer dynamics. The only difference from other porous systems based on inorganic materials is that the confinement is provided here by the same chains that make up the domains.

The effect of confinement can be examined by immersing various homopolymers to the microdomains of diblock copolymers. The distribution of added homopolymer within the domains (Matsen 1995) (Floudas 1997b) depend on the block and homopolymer molecular weights and is determined by two opposing factors; the blocks expel the homopolymer towards the lamellar center, but entropic factors favor a uniform density and, thus, a more uniform distribution of the homopolymer. The thermodynamic field and the topological constraints are expected to alter the homopolymer dynamics in comparison with the bulk.

In a single such investigation (Floudas 1999c), the poly(styrene-b-isoprene) (SI) diblock copolymer system was employed and added homo-polyisoprenes of carefully chosen molecular weights and compositions to avoid macrophase separation. An example of the dynamic signature of the homopolymer addition to the SI microdomains is shown in Figure 20.11. The dielectric loss spectra are shown for the SI-22500/PI-1200 (80/20) blend at some temperatures in the range 238-258 K. There are clearly three peaks corresponding to the fast (single) segmental mode, to the PI homopolymer normal mode, and to the "slower" PI block relaxation. The data have been normalized to the segmental peak and reveal that, with decreasing temperature, the slow normal modes shift towards the segmental peak. This observation, that is, the different shift factors of the segmental and normal modes in type A-homopolymers, have been discussed in terms of the breakdown of the Rouse model in polymer melts (Plazek 1992, Nicolai 1998, Floudas 1999b).

The results for the segmental and homopolymer normal mode dynamics in the SI-22500/PI blends are summarized in Figure 20.12 at 320 K. Five data sets are shown, indicated with numbers. Data set #1 corresponds to the segmental mode of PI homopolymers in the bulk, data set #2 to the same mode in the SI-22500/PI blends. The dynamics of the segmental modes in the two cases have identical dependence on M_{PI}. With decreasing M_{PI} there is a speed-up of the segmental dynamics, which originates from the depression of the T_g^{PI} due to chain-end effects also seen in calorimetry. Data set #3 gives the longest normal mode of PI homopolymers in the bulk. Data set #4 corresponds to the homopolymer longer normal mode within the PI domain of the diblock copolymer.

The molecular weight dependence of this mode can be discussed in terms of three regimes. For $M < M_e$, the slow-down of the PI relaxation in the blends can be

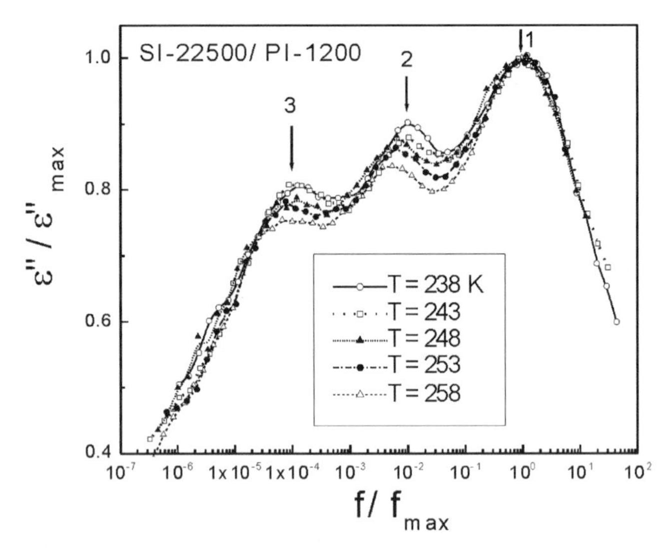

Figure 20.11. Representative dielectric loss spectra of the SI-22500/PI-1200 (80/20) blend shown for different temperatures as indicated. Three processes are shown: a "fast" segmental mode, an intermediate mode corresponding to the normal mode of PI-1200 in the blend, and a "slow" mode due to the PI block relaxation. The axes have been normalized to the fast segmental peak revealing that the two slow modes start to approach the segmental mode with decreasing temperature (see text) (Floudas 1999c).

accounted completely by the slow-down of the segmental mode, i.e., local friction effects. For $M \approx M_e$. the single normal mode in the blends originating from both the block and the homopolymer relaxation exhibit a weak dependence on the molecular weight. Finally, for $M > M_e$, the homopolymer chain relaxation in the blends becomes *faster* than the same mode in the bulk homopolymer. In the same figure, data set #5 gives the longest normal mode relaxation of the minority component in binary blends of PI-10800/PI-55000 (80/20) and PI-10800/PI-90000 (80/20) blends.

In the next section we will account for the different dynamic responses of the PI homopolymer and block in the SI/I blends, at least in a qualitative way. First, the case $M < M_e$. For such short chains, the homopolymer acts practically as a solvent and, in the length scale of the homopolymer, the slowing-down of the normal mode comes entirely from the local friction with a Rouse time $\tau_R = \tau_o N^2$, where $\tau_o = \zeta_o \alpha^2 / 3\pi^2 k_B T$ is the monomer time, and ζ_o is the monomeric friction coefficient. On the other hand, on the length scale of the block relaxation, the PI block speeds up. The PI block with a molecular weight of 10100 is (lightly) entangled. However, due to the localization of the junction point at the interface, the blocks can not reptate. Instead, the relevant mechanism for the block relaxation is arm retraction similar to a star polymer. The latter mechanism has intensity

$$I \approx N(\chi N)^{1/3} \tag{20.12}$$

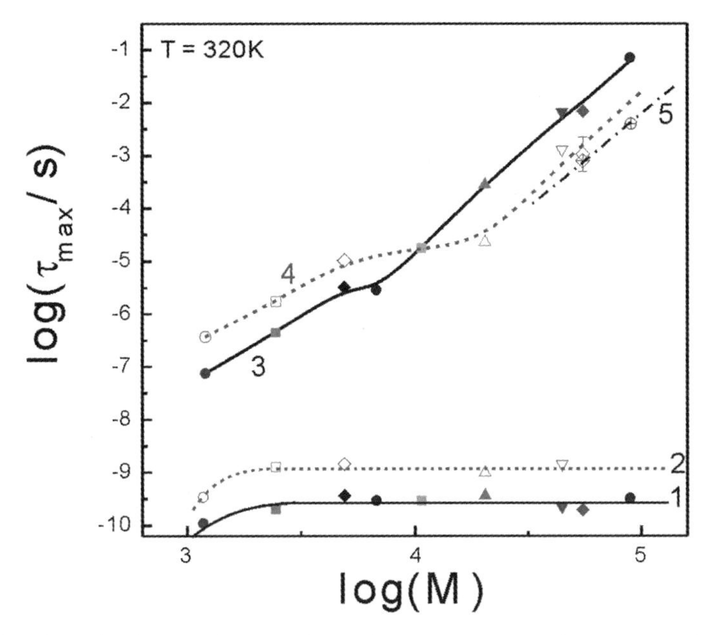

Figure 20.12. Molecular weight dependence of the segmental and longest normal modes in the SI-22500/PI (80/20) blends at 320 K. Five data sets are shown: set #1: segmental modes of PI homopolymers in the bulk; set #2 (short dash line): segmental modes of the same PIs in blends with SI-22500; set #3 (the solid line is a guide to the eye): longest normal mode of PI homopolymers in the bulk; set #4 (dashed line): chain relaxation of the same PIs in blends with SI-22500; set #5 (dash-dotted line): longest normal mode for the PI-55000 and PI-90000 in their blends with PI-10800. The different dynamic regimes and data sets are discussed in the text (Floudas 1999c).

and dynamics.

$$\tau \approx \left(\frac{\zeta_o \alpha^2}{3\pi^2 k_B T}\right) \frac{N^3}{N_e} \exp\left[\frac{N}{2N_e}\right] \tag{20.13}$$

Blending the SI diblock with short PI homopolymers results in a dilution of entanglements ($N_{e,\text{eff}} > N_e$) at the block length scale, and this can account for the observed speed-up.

When $M > M_e$, on the homopolymer length scale, Fig. 20.12 indicates a speed-up of the homopolymer normal mode. In this regime the relevant homopolymer dynamics are described by the reptation theory as:

$$\tau_{rep} = \left(\frac{\zeta_o \alpha^2}{3\pi^2 k_B T}\right) \frac{N^3}{N_e} \tag{20.14}$$

In the SI/I blend, N_e in eq. 20.14 should be replaced by an effective $N_{e,eff}$, which is now higher than the pure homopolymer case because blending of the long homopolymer with the shorter PI blocks of the SI copolymer results in a dilution of the entanglements, which qualitatively describes the experimental findings in Fig. 20.12. In the intermediate range ($M \approx M_e$), the two effects for the extreme cases produce nearly a flat line (i.e., molecular weight independent times). This range can be viewed as a transition regime between the low molecular weights (where local friction effects dominate) and high molecular weights (reptation with entanglement dilution).

Another effect that can influence the homopolymer chain dynamics within the SI microdomains is the possible localization (of the homopolymer) near the center of the PI domain. The expectation is that, the longer the PI, the lower the entropic effect and the more localized is the homopolymer near the lamellar center. Such an effect would tend to increase the relaxation times towards the bulk homopolymer values. To check this possibility, blends of a PI with molecular weight of 10800 (i.e., comparable to the PI in the SI-22500 diblock) with two PIs with molecular weights of 5.5×10^4 and 9.0×10^4 have been prepared. The relaxation times corresponding to the PI-90000 and PI-55000 longest normal modes in the blends are plotted in Fig. 20.12 (with the dash-dotted line) and give the fastest possible relaxation of the PI homopolymers; i.e., dynamics with the effect of dilution of entanglements but without the effect of localization. The difference between the two sets of data is small, but it signifies that the speed-up of the homopolymer dynamics in the blends (for $M > M_e$) is a composite effect originating mainly from the dilution of entanglements (speed-up) and to a smaller extent by the localization of PI near the center of the domain (slow-down). Clearly, the effect of homopolymer localization on the dynamics is only expected for very high molecular weights. Macrophase separation, however, precludes such an investigation. For short molecular weights, the entropic effects that favor a broader distribution dominate. These theoretical expectations are in good agreement with the homopolymer dynamics shown in Figure 20.12.

In conclusion, the thermodynamic field of microphase-separated block copolymers was found to impose structural changes, which reflect nicely on the homopolymer and block dynamics. The structure investigation revealed that, for long homopolymers, there is a tendency for localization within the center of the domain. This structural picture alters the topological constraints and has many consequences on the homopolymer and block dynamics.

REFERENCES

Ackasu A.Z., Benmouna M, Benoit H. (1986) Polymer 27, 1935.

Adachi K., Kotaka T. (1985) Macromolecules 18, 466.

Alig I., Floudas G., Avgeropoulos A., Hadjichristidis N. (1997) Macromolecules 30, 5004.

Alig I., Kremer F., Fytas G., Roovers J. (1992) Macromolecules 25, 5277.

Anastasiadis S.H., Fytas G., Vogt S., Fischer E.W. (1993) Phys. Rev. Lett. 70, 2415.

Bauer M.E., Stockmayer W.H. (1965) J. Chem. Phys. 43, 4319.

Boese D., Kremer F. (1990) Macromolecules 23, 829.

Boudenne N., Anastasiadis S.H., Fytas G., Xenidou M., Hadjichristidis N., Semenov A.N. (1996) Phys. Rev. Lett. 77, 506.

Broseta D., Fredrickson G.H., Helfand E., Leibler L. (1990) Macromolecules 23, 132.

Chrissopoulou K., Pryamitsyn V.A., Anastasiadis S.H., Fytas G., Semenov A.N., Xenidou M., Hadjichristidis N. (2001) Macromolecules 34, 2156.

Dalvi M.C., Lodge T.P. (1994) Macromolecules 27, 3487.

Dalvi M.C., Eastman C.E., Lodge T.P. (1993) Phys. Rev. Lett. 71, 2591.

Dalvi M.C., Lodge T.P. (1993) Macromolecules 26, 859.

de Gennes P.-G. (1979) Scaling Concepts in Polymer Physics Cornell Univ. Press, Ithaca, New York.

Ehlich D., Takenaka M., Hashimoto T. (1993) Macromolecules 26, 492.

Erukhimovich I.Ya., Semenov A.N. (1986) Sov. Phys. JETP 28, 149.

Fleischer G., Fujara F., Stühn B. (1993) Macromolecules 26, 2340.

Fleischer G., Rittig F., Kärger J., Papadakis C.M., Mortensen K, Almdal K., Stepanek P. (1999a) J. Chem. Phys. 111, 2789.

Fleischer G., Rittig F., Stepanek P., Almdal K, Papadakis C.M. (1999b) Macromolecules 32, 1956.

Floudas G., Hadjichristidis N., Iatrou H., Pakula T. (1996) Macromolecules 29, 3139.

Floudas G., Paraskeva S., Hadjichristidis N., Fytas G., Chu B., Semenov A.N. (1997a) J. Chem. Phys. 104, 5502.

Floudas G., Hadjichristidis N., Stamm M., Likhtman A.E., Semenov A.N. (1997b) J. Chem. Phys. 106, 3318.

Floudas G., Alig I., Avgeropoulos A., Hadjichristidis N. (1998) J. Non-Cryst. Solids 235–237, 485.

Floudas G., Reisinger T. (1999a) J. Chem. Phys. 111, 5201.

Floudas G., Fytas G., Reisinger T., Wegner G. (1999b) J. Chem. Phys. 111, 9129.

Floudas G., Meramveliotaki K., Hadjichristidis N. (1999c) Macromolecules 32, 7496.

Fredrickson G.K., Milner S.T. (1990) Mater. Res. Soc. Symp. Proc. 177, 169.

Hamersky M.W., Hillmyer M.A., Tirrell M., Bates F.S., Lodge T.P., von Meerwall E.D. (1998) Macromolecules 31, 5363.

Helfand E., Tagami Y. (1971) J. Chem. Phys. 56, 3592.

Holmqvist P., Pispas S., Sigel R., Hadjichristidis N., Fytas G. (2002) Macromolecules. In Press.

Jackson C.L., McKenna G.B. (1990) J. Chem. Phys. 93, 9002.

Jian T., Anastasiadis S.H., Semenov A.N., Fytas G., Adachi K., Kotaka K. (1994) Macromolecules 27, 4762.

Kanetakis J., Fytas G., Kremer F., Pakula T. (1992) Macromoleucles 25, 3484.

Kannan R.M., Su J., Lodge T.P. (1998) J. Chem. Phys. 108, 4634.

Karatasos K., Anastasiadis S.H., Floudas G., Fytas G., Pispas S., Hadjichristidis N., Pakula T. (1996) Macromolecules 29, 1326.

Lodge T.P., Dalvi M.C. (1995) Phys. Rev. Lett. 75, 657.

Matsen M.W. (1995) Macromolecules 28, 5765.

Montes H., Molkenbusch M., Willner L., Ratgeber S., Richter D. (1999) J. Chem. Phys. 110, 188.

Nicolai T., Floudas G. (1998) Macromolecules 31, 2578.

Papadakis C., Almdal K., Mortensen K., Rittig F., Fleischer G., Stepanek P. (2000) Eur. Phys. J. E 1, 275.

Plazek D.J., Schlosser E., Schönhals A., Ngai K.L. (1992) J. Chem. Phys. 25, 4915.

Quan X., Johnson G.E., Andreson E.W., Lee H.S. (1991) Macromolecules 22, 2451.

Rosedale J., Bates F.S., Almdal K., Mortensen K., Wignall G.D. (1995) Macromolecules 28, 1425.

Semenov A.N. (1993) Macromolecules 26, 6617.

Sigel R., Pispas S., Hadjichristidis N., Vlassopoulos D., Fytas G. (1999) Macromolecules 32, 8447.

Stepanek P., Nallet F., Almdal K. (2001) Macromolecules 34, 1090.

Stockmayer W.H. (1967) Pure Appl. Chem. 15, 539.

Stühn B., Stickel F. (1992) Macromolecules 25, 5306.

Watanabe H. (1995) Macromolecules 28, 5006.

Yao M.-L., Watanabe H., Adachi K., Kotaka T. (1991) Macromolecules 24, 2955.

Zielinski J.M., Heuberger G., Sillescu H., Wiesner U., Heuer A., Zhang Y., Spiess H.W. (1995) Macromolecules 28, 8287.

PART V

APPLICATIONS

CHAPTER 21

BLOCK COPOLYMER APPLICATIONS

Since the discovery of the living character of anionic polymerization in the mid-1950s (Szwarc 1956), about 50,000 references have been published on block copolymer synthesis, properties, and applications through the end of 2000 (SciFinder 2000). A total of 42% of these papers have appeared in the last few years (1995 through 2000). Despite this great level of academic and industrial interest, only very few applications of block copolymers have been commercialized thus far. This is clearly demonstrated by the very low (~0.6%) level of block copolymer consumption in the total worldwide polymers consumption, which for 2000 was estimated to be about 150 million metric tons (Ehrenstein 2001). In Figure 21.1 the increase in the consumption (Chem. Systems 1999, SRI International 1991 and 1993), as well as in publications (SciFinder 2000), per year are given.

Over the last few years, tremendous research efforts have been dedicated to the study of potential applications of block copolymers in advanced technologies, such as information storage, drug delivery, photonic crystals, etc. These studies have shown that block copolymers are very strong candidates for applications in these areas.

Conventional and potential high-technology applications of block copolymers are based on their ability to self-assemble, in bulk (see Chapter 19) or in selective solvents (see Chapter 11), into ordered nanostructures, with dimensions comparable to chain dimensions. By changing the molecular weight, chemical structure, molecular architecture, and composition of block copolymers, the size scale, the type of ordering, and the characteristics of these nanostructures can be manipulated.

The purpose of this chapter is to present the principles on which the applications of block copolymers are based. Therefore, in the following pages, only a few

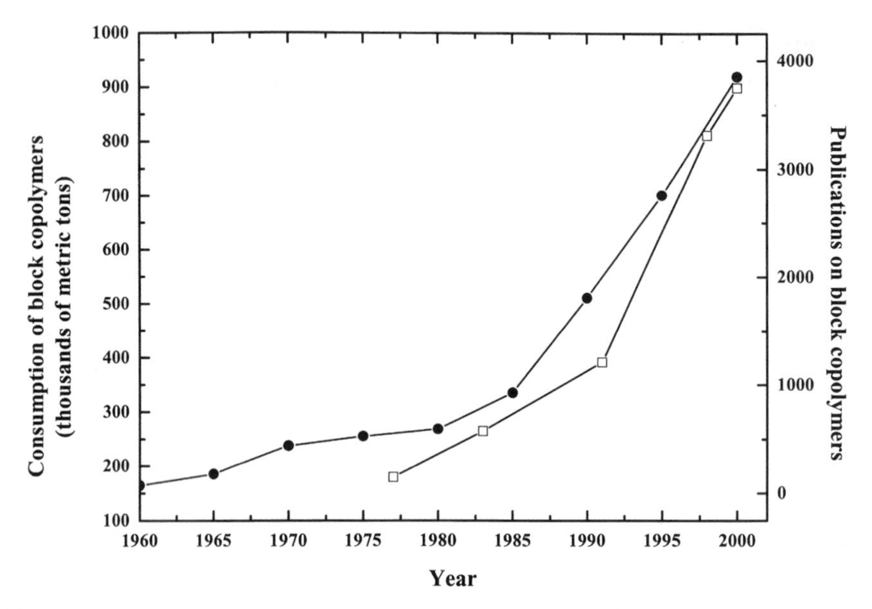

Figure 21.1. Block copolymers consumption (□) and number of references (●) per year.

examples of commercialized and potential applications will be discussed. For details, the reader should go to specialized literature on this subject.

1. COMMERCIALIZED APPLICATIONS

The most important and popular application of block copolymers is their use as thermoplastic elastomers (TPEs). These materials are so versatile that they can be used for wine bottle stoppers, jelly candles, outer coverings for optical fiber cables, adhesives, bitumen modifiers, or in artificial organ technology. Many books and reviews have been published on TPEs. Only a few of them are listed here (Holden 2001, Browmick 2001, Craver 2000, Mark 1994, Morton 1987, Legge 1987, Walker 1986, Folkes 1985, Goodman 1985, Meier 1983, Ceresa 1973, Allport 1973, Aggarwal 1970).

The first commercially available TPEs were produced by B. F. Goodrich Co. in the late fifties and were based on polyurethanes (TPU). The TPUs are linear multiblock copolymers (segmented block copolymers), made by condensation polymerization in two steps. In the first step, an α,ω-diisocyanate prepolymer is prepared by reaction of an excess of a diisocyanate monomer (i.e., 4,4′-diisocyanate-diphenylmethane) with a flexible long chain diol (i.e., α,ω-dihydroxy-polyethyleneoxide). In the second step, the α,ω-diisocyanate prepolymer is reacted with a short chain diol (i.e., 1,6-hexanediol) and a new quantity of diisocyanate monomer.

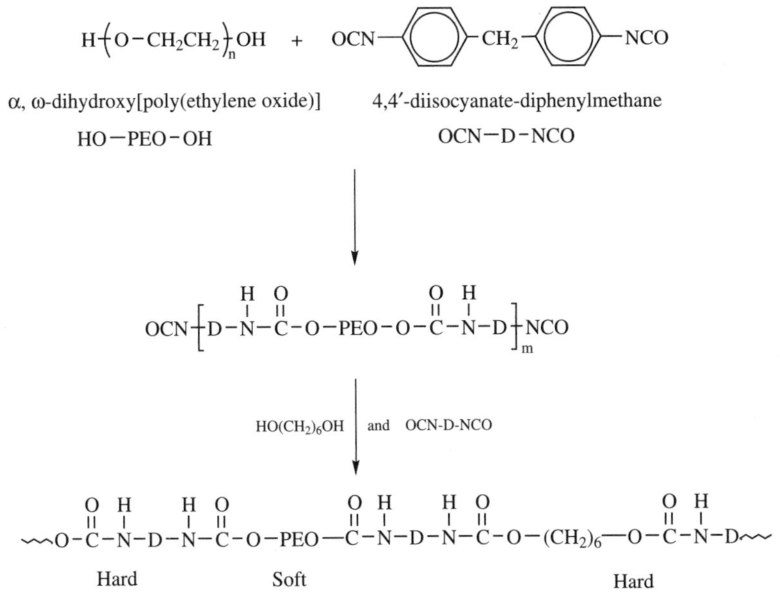

$$H\!-\!\!\left(O-CH_2CH_2\right)_n\!OH \quad + \quad OCN-C_6H_4-CH_2-C_6H_4-NCO$$

α, ω-dihydroxy[poly(ethylene oxide)] 4,4′-diisocyanate-diphenylmethane

HO−PEO−OH OCN−D−NCO

$$\downarrow$$

$$OCN\!\left[D-\overset{H}{N}-\overset{O}{\overset{\|}{C}}-O-PEO-O-\overset{O}{\overset{\|}{C}}-\overset{H}{N}-D\right]_m\!NCO$$

$$\downarrow \; HO(CH_2)_6OH \text{ and } OCN\text{-}D\text{-}NCO$$

$$\sim\!O-\overset{O}{\overset{\|}{C}}-\overset{H}{N}-D-\overset{H}{N}-\overset{O}{\overset{\|}{C}}-O-PEO-\overset{O}{\overset{\|}{C}}-\overset{H}{N}-D-\overset{H}{N}-\overset{O}{\overset{\|}{C}}-O-(CH_2)_6-O-\overset{O}{\overset{\|}{C}}-\overset{H}{N}-D\!\sim$$

Hard Soft Hard

Scheme 21.1. General reactions scheme for the synthesis of TPU.

The final product is an alternating block copolymer with two types of blocks (segments): those formed in the first step, consisting of long chain flexible polymer (soft segment), and those formed in the second step, based on the short diol and the urethane groups (hard segment). The general reactions are given in Scheme 21.1. PTUs are not well-defined block copolymers exhibiting high molecular weight and compositional polydispersity. Nevertheless, TPUs, due to the hydrogen bonding between the urethane groups, have excellent strength, wear, and oil resistance and are used in automotive bumpers, snowmobile treads, etc.

In 1965 Shell introduced the styrenic TPEs, under the trade name Kraton, which are polystyrene-b-polyisoprene(or polybutadiene)-b-polystyrene linear triblock copolymers, made by anionic polymerization. In contrast to the polyurethane multiblock copolymers, the styrenic PTEs are well-defined materials with low molecular weight and compositional heterogeneity. In Kraton TPEs the molecules are linear and the blocks pure homopolymers. Other styrenic PTEs, introduced later by other companies, have tapered blocks (i.e., Styroflex of BASF), or star block structure (i.e., Solprene, Phillips) (Scheme 21.2). The typical styrene content of TPEs is between 25 and 40 wt%. Their uses include footwear, bitumen modification, thermoplastic blending, adhesives, and cable insulation and gaskets.

Later on, TPEs based on polyesters, TPES (du Pont), and polyamides, TPA (Huls and Ato Chimie), have been introduced into the market. These are linear polyester or polyamide multiblock copolymers made by step polymerization under similar synthetic procedures as polyurethanes. The only difference is that the segments are linked together by ester (-CO-O-) or amide (-CO-NH-) linkages instead of urethane

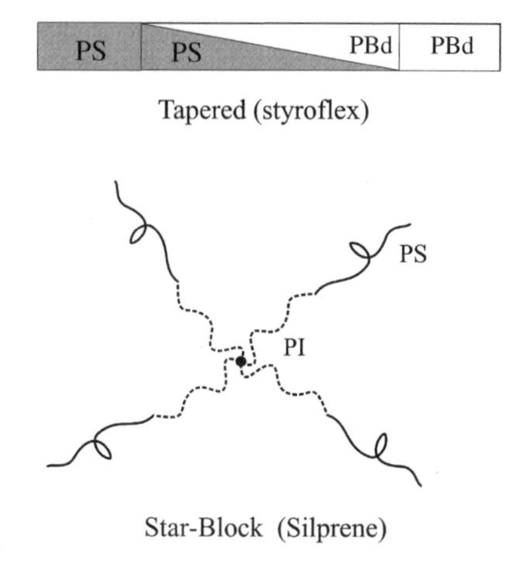

Scheme 21.2. Tapered and star block copolymers.

(-O-CO-NH-). Their applications include hose tubing, sport goods, automotive components, etc.

All TPEs exhibit properties characteristic of chemically crosslinked elastomers at room temperature but, at elevated temperatures, behave as thermoplastics (thermo from the Greek word θερμος meaning hot). Consequently, they can soften and flow and, thus, can be processed with high speed, efficiency, and economy on conventional thermoplastic equipment. Furthermore, unlike vulcanized rubber, they can be reused. Because of these advantages, TPEs were considered to be one of the breakthroughs in rubber technology.

"Thermoplasticity" and "elasticity" are two contradictory terms. On the molecular level, thermoplasticity is a consequence of noncrosslinked chains (i.e., linear or branched), whereas elasticity is consistent with crosslinked chains. Linear or branched chains can be melted if enough energy is imparted to break the intermolecular physical bonding that keeps them in the solid state. In contrast, flexible crosslinked macromolecules possess elasticity (primitive memory), but they cannot melt because the chains are connected into networks by chemical covalent bonds (Fig. 21.2, A).

These two contradictory behaviors can coexist in thermoplastic elastomers, which are two-phase systems. One of the phases is a hard polymer that does not flow at room temperature (Tg or Tm > room temperature) and plays the role of effectively crosslinking the flexible chains (Tg or Tm < room temperature), which constitute the other phase. In addition, the hard domains act as reinforcing "fillers". The driving force for phase separation is usually enthalpic. Crystallinity (TPES, TPA), hydrogen bonding (TPU, TPA), and van der Waals interactions (styrenic) all have been shown to cause microphase separation in these systems. The principle of thermoplastic elasticity is illustrated schematically in Figure 21.2, B for styrenic TPEs and in Figure 21.3 for multiblock TPEs.

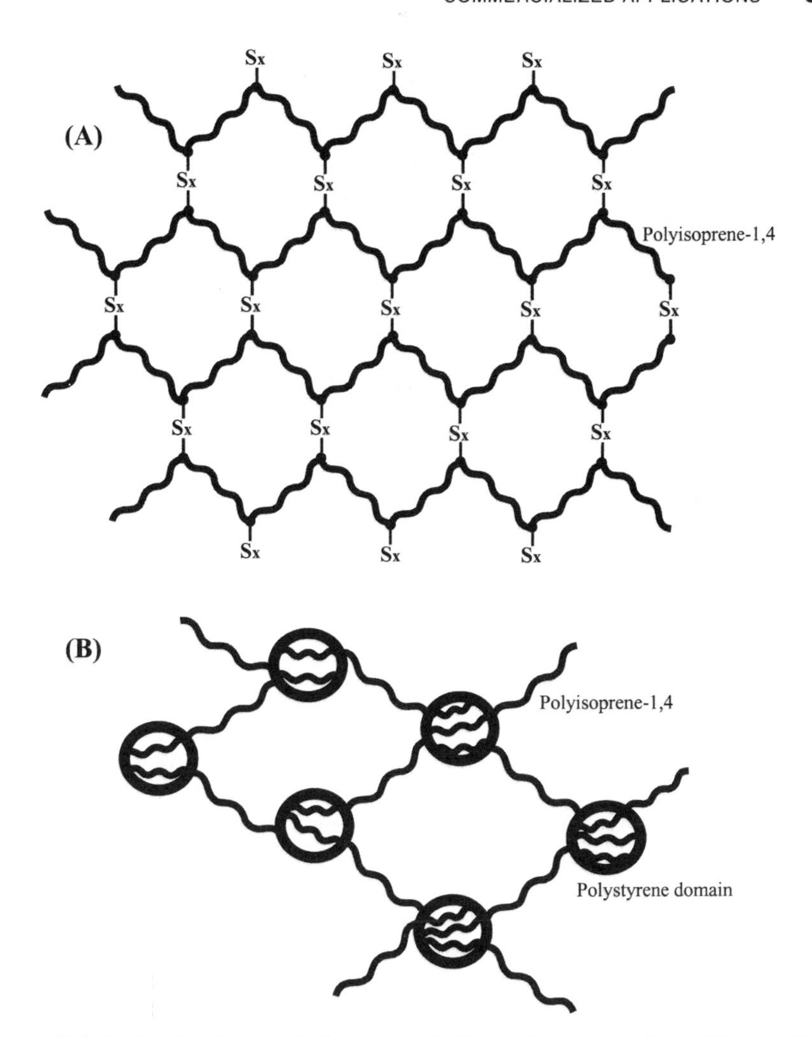

Figure 21.2. In (A) the elasticity is due to chemically nonthermoreversible sulfur crosslinks. In (B) the elasticity is due to physically thermoreversible crosslinks.

The difference between vulcanized rubber and styrenic TPEs can be seen in Figure 21.2. In TPEs the crosslinks are physical rather than chemical, which is the case of vulcanized rubber. Thus, the hard domains lose their strength when the material is heated (thermoreversible) or dissolved in a solvent (solvoreversible). When the TPE is cooled down or the solvent is evaporated, the domains became hard again, and the material regains its original properties. Consequently, if solvent resistance or high temperature service is needed, i.e., in radiator hose, drive belts, or automotive tires, TPEs are not as good as the corresponding crosslinked rubbers. In addition, TPEs exhibits high mechanical hysterisis, another factor that prevents TPEs from being used in automotive tires. However, Goodyear announced recently the use of special polyurethane-based thermoplastic elastomers for automotive tires.

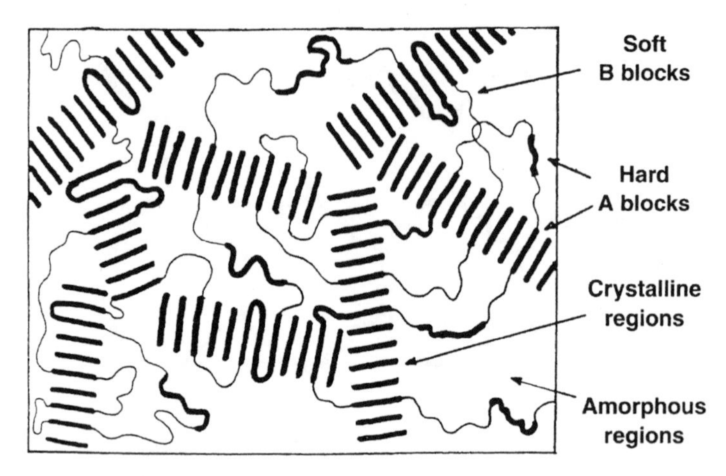

Figure 21.3. Morphology of multiblock polymers with crystalline hard segments.

Triblock copolymers with polydienes end blocks and diblock copolymers are not TPEs because the flexible chains are not immobilized at both ends by the glassy hard domains of PS and, consequently, cannot be elastic. This can be seen schematically in Figure 21.4.

The double bond of the diene-containing TPEs is quite reactive towards oxygen, ozone, and UV radiation and, thus, limits the stability of the products. To improve this property, the polybutadiene block may be transformed, by hydrogenation, to poly(ethylene-co-butylene), which is much more stable because it is a fully saturated polymer. The trade name of PS-b-PEB-b-PS (SEBS) is also Kraton. The Kraton family, SIS-, SDS-, and SEBS-type triblock copolymers, represents approximately 50% of all thermoplastic elastomers. Polyurethane, polyester, and polyamide-based TPEs are also relatively stable to oxidative degradation, although they can have problems with hydrolytic stability, due to the ester, amide, and urethane bonds. The polycondensation thermoplastic elastomers have approximately 20% of the market.

Another category of thermoplastic elastomers is the polyolefin TPEs, first commercialized by Uniroyal. Polyolefin TPEs are blends of polypropylene with either poly(ethylene-co-propylene) or EPDM (terpolymer of ethylene, propylene, and a diene monomer), which are often crosslinked or grafted during blending (dynamic vulcanization). These TPEs are characterized by excellent thermal stability and account for approximately 30% of the thermoplastic elastomer market. The good chemical resistance of the polyolefin TPE to hot water and detergents is very important for appliance applications (hose, seals, gaskets for washers, dryers, and refrigerators). The excellent combination of properties, wide service temperature range, low temperature flexibility, high temperature stiffness, and resiliency of these TPEs has permitted them to obtain rapid acceptance by the automotive industry for exterior body applications. Nevertheless, the polyolefin TPEs are not

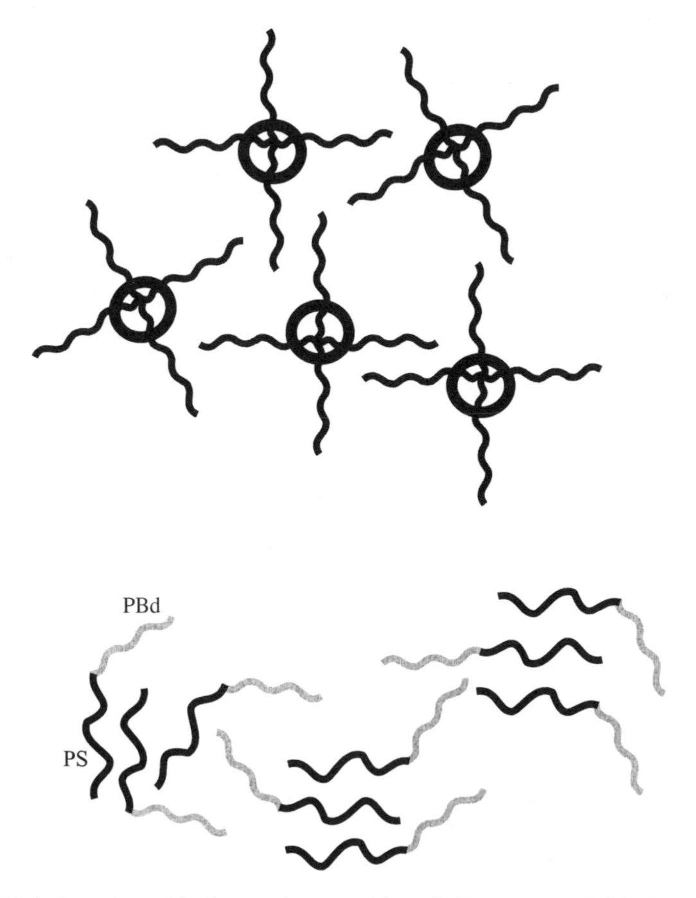

Figure 21.4. Styrenic triblock copolymer with polydiene external block or diblock copolymers (styrenic or PU) are not elastic.

considered as block copolymers and are not included in block copolymer consumption (Figure 21.1).

Styrenic thermoplastic elastomers are the most widely studied because they are well-defined materials and have low molecular weight polydispersity. The mechanical properties of TPEs depend on their morphology (Dair 1999), as can be seen in Figure 21.5, which shows the stress-strain curve, up to 100% strain of polygranular isotropic poly(styrene-b-isoprene-b-styrene), SIS, triblock copolymers with different morphologies.

As can be seen in Figure 21.5, the sample with the DG (double gyroid) morphology (Fig. 21.6) exhibits a stable neck, whereas the samples with spherical, cylindrical, and lamellar morphology do not exhibit necking. In addition, the yield stress of the DG is much higher than that of even the highest PS content sample.

Varying the architecture of the thermoplastc elastomer from linear to star branched can also impact properties. Figure 21.7 shows the stress-strain behavior

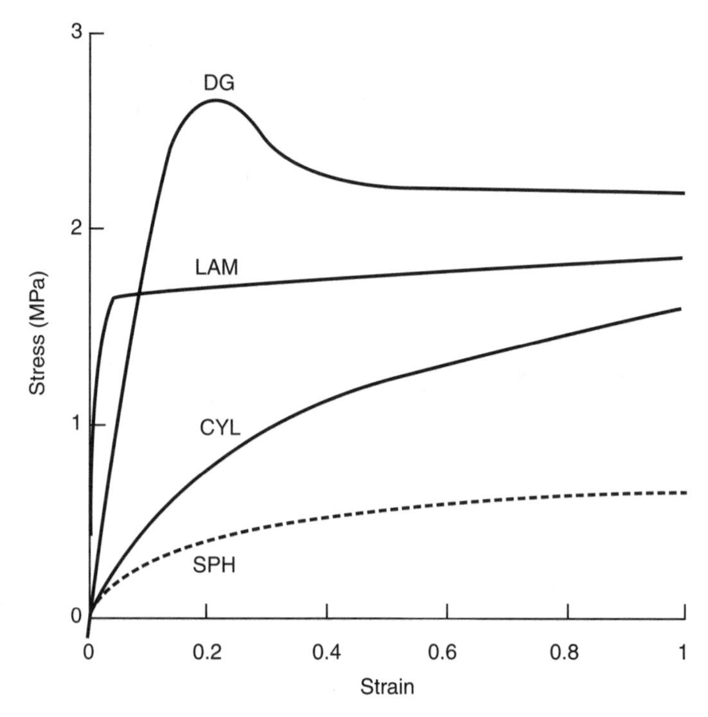

Figure 21.5. Stress-strain curves of isotropic, polygranular double gyroid (DG) overlayed with the classical microdomain morphologies of spheres (SPH), cylinders (CYL), and lamellae (LAM). Stress-strain behavior is focusing on the 0% to 100% strain region.

of a star block copolymer with 12 arms and DG morphology and a two-arm star block copolymer (linear triblock) with cylindrical morphology. In this case, again, the DG sample has improved mechanical properties over the cylindrical morphology. Table 21.1 gives the molecular characteristics corresponding to these materials.

Both Figures 21.5 and 21.7 demonstrate that the superiority of the mechanical properties of the DG samples over those of its classical counterparts is not an effect of volume fraction of the glassy component, the architecture of the molecule, or the molecular weight but is essentially due to the unique three-dimensional microdomain morphology of the double gyroid phase.

Recent results on tetrafunctional multigraft copolymers (Iatrou 1998, Beyer 2000, Weidish 2001), comprised of a polyisoprene backbone and two polystyrene branches at each regularly spaced branch point, with 22% PS content, show an exceptional combination of high strength and huge strain at break (Figure 21.8).

While the rubbery PI backbone provides a huge elasticity, the PS branches reinforce the coupling (physical crosslinking) between the rubbery and the glassy PS domains (Fig. 21.9).

Figure 21.8 shows, that with increasing the number of branch points, strain at break and tensile strength increases. Investigation of morphology by TEM and

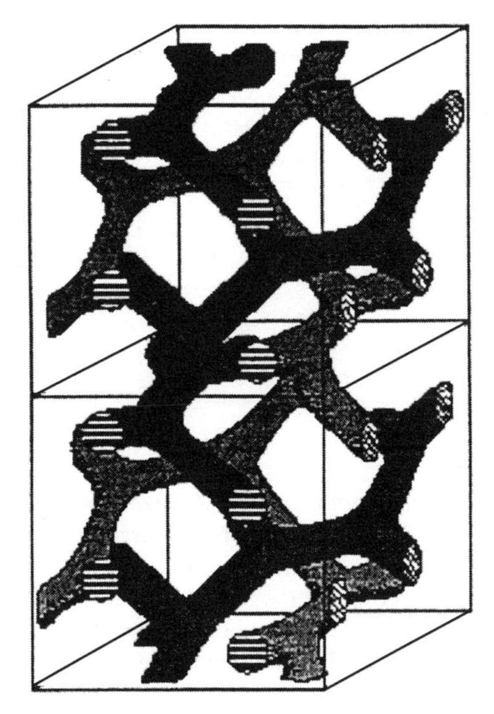

Figure 21.6. Schematic of the double gyroid morphology, shown with two unit cells. The two interpenetrating networks correspond to the PS domains in a matrix of PI.

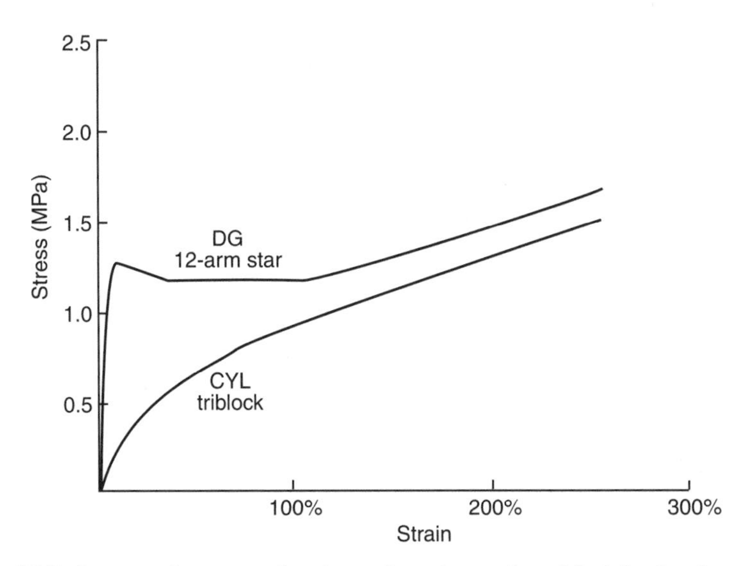

Figure 21.7. Stress-strain curves of an isotropic, polygranular triblock having the cylinder (CYL) phase and a starblock copolymer having the DG phase. Note that the DG phase exhibits yielding and necking phenomenon, but the cylinder phase does not.

TABLE 21.1. Molecular Characteristics of Triblock and Star Block Copolymers Corresponding to Figure 21.5 and Figure 21.7

Morphology	PS Content (vol %)	Total Molecular Weight (kg/mol)	Block Molecular Masses (kg/mol) PS/PI/PS	Polydispersity Index (M_w/M_n)
Spheres "SPH"	18	128	11.5/105/11.5	1.03
Cylinders "CYL"	30	97	14.5/68/14.5	1.03
Double Gyroid "DG"	36	74	13.6/46.4/13.6	1.04
Lamellae "LAM"	45	80	18/44/18	1.03
			Average functionality of stars	
DG (12-arm star)	30	396	11.70	
Cylinder (2-arm star)	30	66	2	

SAXS showed that these multigraft copolymers have a wormlike morphology with low long-range order (small grain size), which decreases with an increase of the number of branch points. In contrast, the Kraton sample studied reveals a hexagonal morphology with long-range order (large grain size). In the case of multigraft copolymers, a large number of branch points are necessary, rather than a well-ordered morphology. Existence of smaller grain size can also be correlated with an improved mechanical profile. Thus, multifunctional graft copolymers with tetra-

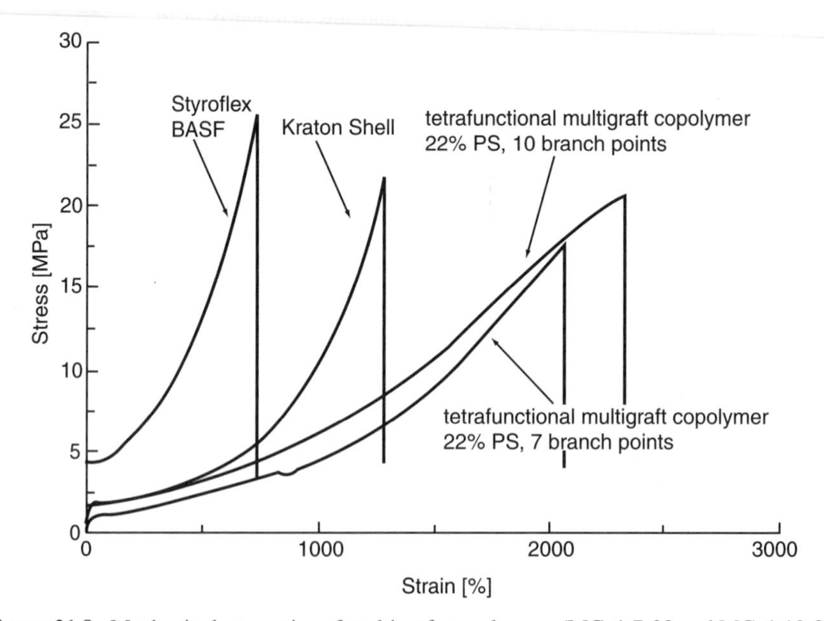

Figure 21.8. Mechanical properties of multigraft copolymers (MG-4-7-22 and MG-4-10-22) compared with commercial TPEs Kraton (20% PS) and Styroflex (58% PS), revealing the exceptional large strain at break of tetrafunctional multigraft copolymers.

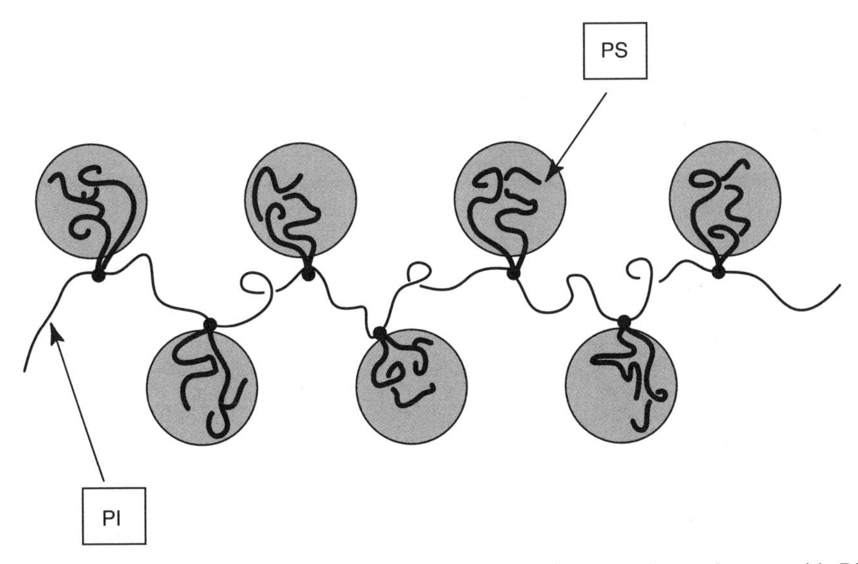

Figure 21.9. Schematic representation of tetrafunctional multigraft copolymers with PS domains in a PI matrix.

functional branch points are strong candidates for developing TPEs with exceptional mechanical properties.

In addition to linear block and symmetric star block structures, other structures, mainly star-shaped, have been commercialized. A typical example is the Styrolux of BASF (Knoll 1998), which is a mixture of linear and asymmetric star triblock copolymers (Fig. 21.10) with 70% PS. Styrolux is a stiff and tough resin, which retains its transparency when blended with general purpose PS. The highly unsymmetrical star block structure and the resulting complex morphology are contributing to the favorable mechanical properties.

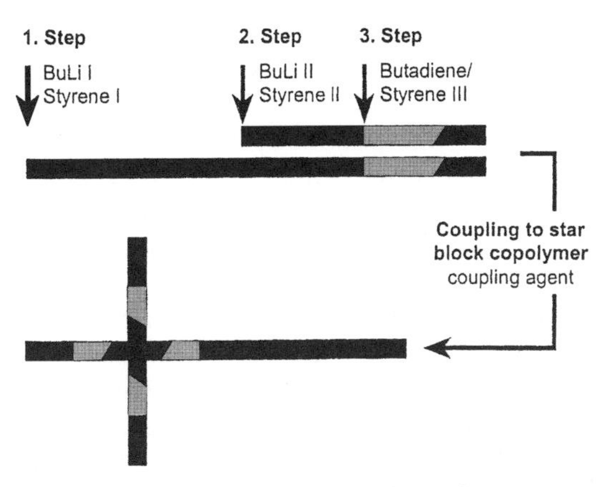

Figure 21.10. Synthesis of Styrolux.

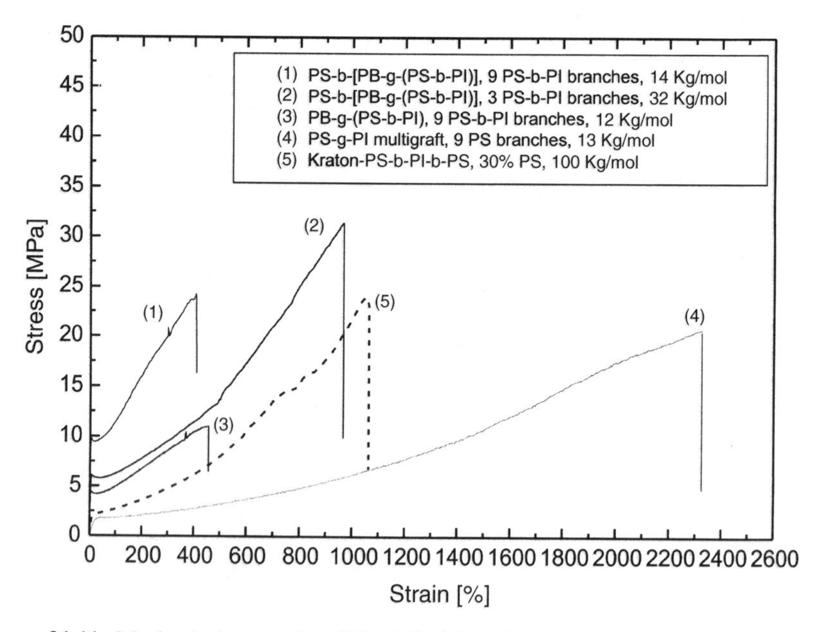

Figure 21.11. Mechanical properties of block double grafts, multigraft copolymers compared with commercial TPE's Kraton.

Block copolymers with more complex structures, for example, block graft copolymers (Velis 2000), exhibit better thermoplastic elastomer behavior than the multigraft block copolymers (Weidisch 2001), as can be seen in Figure 21.11. From the above results, it seems that macromolecular architecture will play a very important role in the future for developing novel TPEs.

Block copolymers have found commercial applications as adhesives (both hot-melt and solution) and sealants, as blending agents for use with different homopolymers for producing materials with desired characteristics, as surface modifiers of fillers and fibers in order to improve dispensability in the matrix, as surfactants for phase stabilization, as viscosity improvers of lubricating oils, as membranes for desalination, in biomedical applications, etc.

Block copolymers of ethylene oxide (EO) and propylene oxide (PO) with linear PEO-b-PPO-b-PEO (Pluronic) or star structure (PEO-b-PPO)4 (Tetronic), commercialized by BASF, are used widely in cosmetics, in the paper industry, in the petroleum industry, etc. These materials, due to the biocompatibility of PEO, are also of great interest in biomedical applications.

An interesting application of Kraton block copolymers is as gelling agents for organic liquids. An example is the jelly candles, which consist of a solid solution of $\sim$1% Kraton in a light fraction of petroleum. The hydrocarbons of the petroleum solubilize the middle block (PBd, PI, or ethylene/butene random copolymers) but not the external PS blocks and, thus, produce a gel (polymeric network). As in the case of thermoplastic elastomers, the crosslinks are the PS domains, which are not

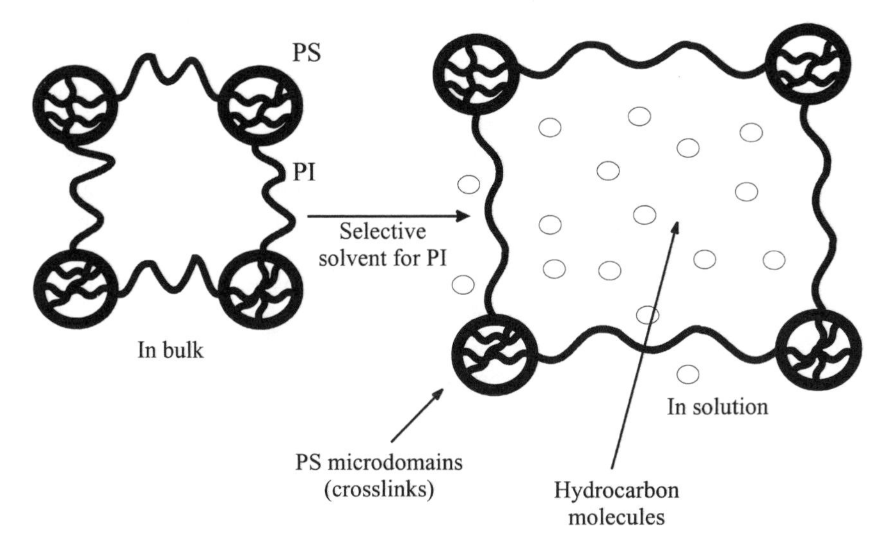

Figure 21.12. The addition of the selective solvent results in solvation of the PI phase, but the PS domains remain unaffected and act as crosslinks creating a swollen gel.

soluble in the liquid. The principle is illustrated in Figure 21.12. The same philosophy is used for electrical isolation of large cables. The wire is placed in the center of a plastic tube, and then the tube is filled with liquid hydrocarbon. Upon addition of Kraton, the liquid is solidified.

2. POTENTIAL APPLICATIONS

2.1. Block Copolymers in Selective Solvents

Block copolymers, when dissolved in liquids that are solvents for one block but nonsolvents for the other (i.e., selective solvents), self-associate usually in spherical micelles of nearly uniform size. The insoluble blocks form the core and the soluble blocks the shell of the micelles (Fig. 11.3). This phenomenon can be used for encapsulation and selective delivery or removal of organic/inorganic compounds. Examples are given below.

2.1.1. Drug Release in Target Cells. The selective delivery of drugs to malignant cells is very important in medical and pharmaceutical sciences because many drugs are toxic and produce side effects, or encounter solubility problems in the body, if they are released in non-target systems (Yokoyama 1996, Kabanov 1996).

Most drugs are hydrophobic compounds, and they are usually introduced into the patient through the blood stream. Blood is a connective tissue with a liquid matrix called plasma. It consists mostly of water and a wide variety of dissolved substances (enzymes, hormones, ions, respiratory gases, etc.). A block copolymer for use in drug delivery as a microcontainer device must consist of a water-soluble block (hydrophilic), in order to impart blood solubility of the microcontainer, and a water-insoluble block (hydrophobic) compatible with the drug to be carried. The water-soluble block, which will constitute the outer shell of the micellar carrier, should also be biologically inert to avoid foreign body interactions with the organism's antibodies. A synthetic block serving this purpose is poly(ethylene oxide), as noted above.

Block copolymers of ethylene oxide with propylene oxide (hydrophobic block), like PEO-b-PPO-b-PEO triblocks, or with β-benzyl-L-aspartate (Basp), like PEO-b-PBAsp, are good candidates for drug targeting using water-insoluble anticancer drugs, as for example Doxorubicin or Adriamycin (Kataoka 1994 and 2001). In aqueous solution this triblock copolymer forms micelles, with the hydrophobic PPO or PBAsp chain in the core, surrounded by the hydrophilic corona of PEO. The cmc of these copolymers is very low (usually in μg/mL order or below), and they can solubilize substantial amounts of drugs. The size of the micelles can be controlled by changing the molecular characteristics of the copolymer (molecular weight, composition). A convenient size range, which can be easily obtained, is 10 nm to 100 nm. This size range is much larger than the critical threshold of renal filtration or reticuloendothelial (RES) uptake yet smaller than that susceptible to nonspecific capture by the monocyte systems. Consequently, the micelles can maintain long-term circulation in the blood stream by escaping renal excretion, RES uptake, and avoiding the nonspecific capture, until they find the target tumor. The drug, which is physically trapped in the microcontainer, can be easily released in an active form, as a result of a micelle rearrangement (or destruction) that takes place during the micelle interaction with the components of the target tumor. To selectively target these micellar microcontainers to the specific cells in the organism, a "vector" molecule (e.g., galactose) should be covalently attached to the outer hydrophilic block (Ulbrich 1997). Accumulation of the drug at the tumor sites, possibly through extravasation, might be due to enhanced vascular permeability and retention effects in the tumor, as well as to the flexible nature of the micelle palisade (Fig. 21.13).

If the dimensions of the unimers, constituting the micelles, are designed to be lower than the critical value for renal filtration, the micelles, after delivering the drug, can decompose into unimers resulting in excretion from the renal route.

Finally, the size range allows polymeric micelles to be easily sterilized, before use, by filtration through common sterilization filters with submicron pores.

2.1.2. Removal/Recovery of Organic/Inorganic Compounds from Contaminated Waters. By using the micelle-forming ability of block copolymers, the removal/recovery of toxic organic compounds (as, i.e., halogenated and polyaromatic hydrocarbons) from contaminated water (coastal, surface, and ground) can be achieved. The organic compound is entrapped into the hydrophobic core of the micelles, which are stabilized in water by the external hydrophilic shell, followed by its removal and recovery.

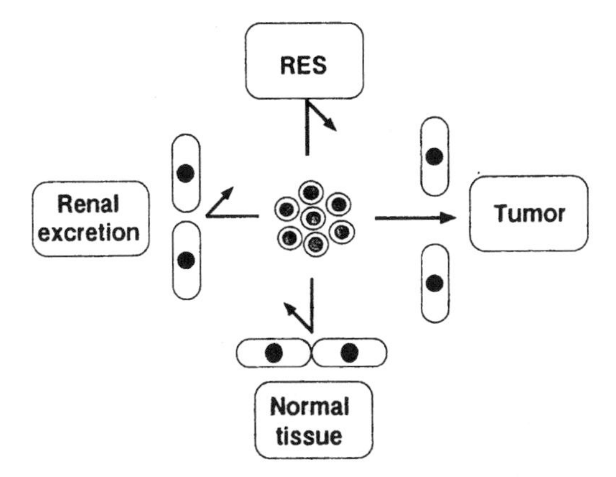

Figure 21.13. Accumulation of micelle-forming microcapsules in a tumor utilizing enhanced permeability of tumor vasculature.

Poly(2-cinnamoylethyl methacrylate)-b-poly(acrylic acid) (PCEMA)-b-(PAA) block copolymers are potential candidates for this purpose (Wang 1998, Henselwood 1998). In DMF/water solutions (~20% DMF by volume), the (PCEMA)-b-(PAA) diblock copolymers form spherical micelles with PAA in the shell and PCEMA in the core. Exposure of the micelles to UV light leads to the formation of a crosslinked PCEMA core and to the production of nanospheres. These nanospheres are stable in water and absorb large amounts of organic compounds such as toluene, perylene, and other hydrocarbons. By addition of a bivalent cation such as Ca^{++}, the nanospheres may be precipitated out without loosing the trapped organic compound. By addition of a complexing reagent, such as EDTA, or a precipitant for Ca^{++}, like CO_3^{--}, the nanospheres are redispersed and are ready to be used again.

Other amphiphilic block copolymers, i.e., of EO and PO (Hurter 1992), can be used to remove or recover toxic compounds, i.e. phenanthrene and naphthalene, from industrial and domestic effluents. It was found that the linear triblock copolymers PEO-b-PPO-b-PEO (Pluronic) are more efficient removers of polycyclic aromatic hydrocarbons than the corresponding star block copolymers (Tetronic). It seems that the configurational constrains on star structure renders these compounds less effective for solubilizing the aromatic hydrocarbons.

It is obvious that this method can be used for qualitative and quantitative analysis of organic compounds of domestic and industrial effluents.

Double-hydrophilic block copolymers can be used for removing calcium/magnesium carbonate from industrial washing baths (Sedlak 1998). Polyethylene-oxide-b-poly[(N-carboxymethyl)ethyleneimine], PEO-b-PEI, and polyethylene-oxide-b-polyaspartic, PEO-b-PasA, acids are two examples. The COOH-functionalized block interacts with the calcium or magnesium carbonate, whereas the PEO block keeps the microcrystals in solution by steric stabilization. The efficiency of the double-hydrophilic block copolymers is up to 20-fold that of commercial builders as poly[(acrylic acid)-co-(maleic anhydride)] or polyaspartic acid.

Double-hydrophilic block copolymers can also be used for volumetric sweeping of oil reservoirs. Poly(acrylic acid-b-acrylamide) and poly(2-acrylamido-2-methyl-propanesulfonic acid)-b-acrylamide) provide high viscosity in brine, making them desirable for mobility control in enhanced oil recovery applications (Wu 1985).

2.2. Block Copolymers in Bulk

Due to the miniaturization of electronic, optoelectronic, and magnetic devices, nanometer-scale patterning of materials is an important objective of current science and technology. Block copolymers, which have the ability to self-assemble into periodic ordered microstructures, are recognized to be promising candidates for patterning nanostructures. By changing the molecular weight, chemical nature, molecular architecture, and composition of the copolymer, one can control the size and type of ordering. A few examples of nanopatterning by using block copolymers are presented in the following text.

2.2.1. Nanopatterning. Nanopatterning is very important for lithography. Nanosizes greater than $\sim$150 nm can be routinely produced by photolithography techniques. The minimum size that can be achieved by photolithography is determined by the wavelength of light used in the exposure. Electron beam lithography is commonly used to access feature sizes between 150 nm and 30 nm. However, sizes less than 30 nm are not easily obtained by standard lithography. One way to overcome this problem is by using block copolymers.

Dense, periodic arrays of holes and dots have been fabricated on a silicon nitride-coated silicon wafer using block copolymers of styrene and butadiene, polystyrene-b-polybutadiene (Park 1997, Harrison 2000). The molecular weight of the PS block was 36,000 and of the polybutadiene block 11,000. In bulk this block copolymer microphase separates into a cylindrical morphology and produces hexagonally packed polybutadiene cylinders (20 nm across and 40 nm apart) in a matrix of polystyrene. A thin film of the block copolymer is coated onto the silicon substrate with the cylinders lying parallel to the substrate. The PBd cylinders are degraded and removed with ozone to produce a PS mask for pattern transfer by fluorine-based reactive ion etching (RIE). This PS mask of spherical voids was used to fabricate a lattice of holes. This technique accesses a length scale (3×10^{12} holes of approximately 20 nm wide, spaced 40 nm apart, and uniformly patterned on a three-inch wafer) difficult to produce by conventional lithography and opens new routes for the micropatterning. By staining the PBd with osmium tetroxide, the PBd domains become more resistant to etching than PS. This results in the fabrication of dots instead of holes, having the same nanodimensions with the block copolymers microdomain dimensions. The fabrication of hexagonal arrays of holes and dots are shown schematically in Figure 21.14.

By using the same principle, various dense nanometer patterns can be produced using block copolymers. For example, parallel lines can be produced either by a film of lamellae, which are oriented normal to the substrate or of cylinders that lie parallel to the surface.

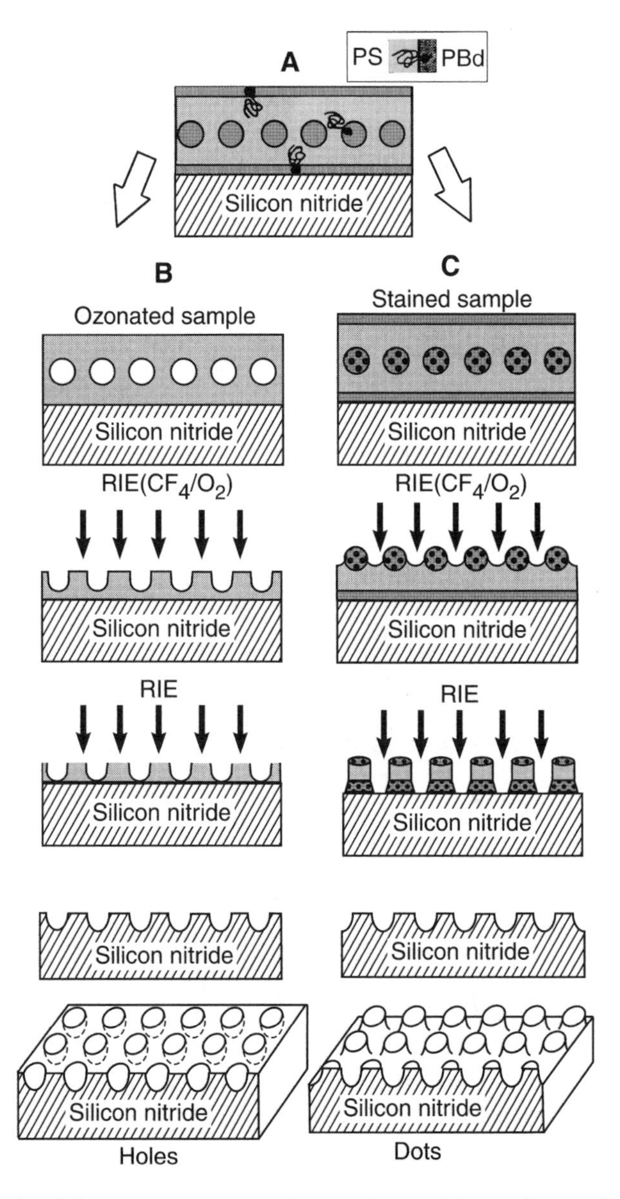

Figure 21.14. (A) Schematic cross-sectional view of a nanolithography template consisting of a uniform monolayer of PBd spherical microdomains on silicon nitride. PBd wets the air and substrate interfaces. (B) Schematic of the processing flow when an ozonated copolymer film is used, which produces holes on silicon nitride. (C) Schematic of the processing flow when an osmium-stained copolymer film is used, which produces dots in silicon nitride.

Diblock copolymers of styrene and methylmethacrylate were used to produce templates for dense nanowire arrays (Thurn-Albrecht 2000). The volume fraction of styrene (0.71) and the total molecular weight (39,600) of the diblock copolymer were chosen in order to produce 14-nm-diameter PMMA cylinders hexagonally packed in a PS matrix with a lattice constant of 24 nm. Films (~1 μm thick) were spin-cast from toluene solutions onto a conducting substrate (silicon, gold-coated silicon, or aluminized Kapton). Annealing the film for 14 hours at 165°C, above the glass transition temperature of both components, under an applied electric field, causes the cylindrical microdomains to orient perpendicular to the surface. The film was cooled to room temperature before the field was removed. Deep UV exposure (25 J/cm^2 dosage) degrades the PMMA domains and simultaneously crosslinks the PS matrix. The degraded PMMA is removed by rinsing with acetic acid. The resulting nanoporous PS film is optically transparent and contains 14-nm-diameter pores. Co and Cu nanowire arrays, with densities in excess of 1.9 10^{11} wires per

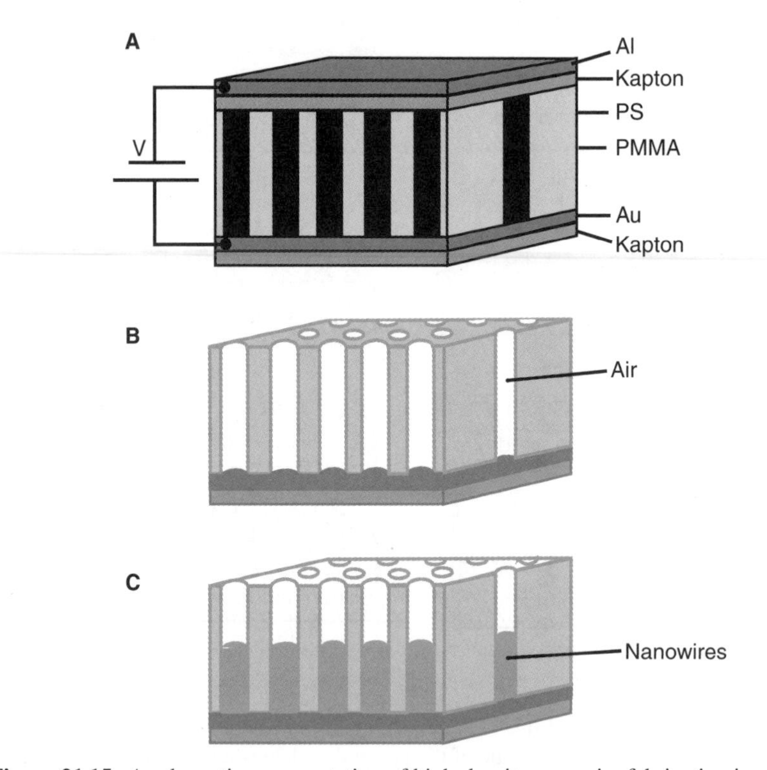

Figure 21.15. A schematic representation of high-density nanowire fabrication in a polymer matrix. (A) An asymetric diblock copolymer annealed above the glass transition temperature of the copolymer between two electrodes under an applied electric field, forming a hexagonal array of cylinders oriented normal to the film surface. (B) After removal of the minor component, a nanoporous film is formed. (C) By electrodeposition, nanowires can be grown in the porous template, forming an array of nanowires in a polymer matrix.

square centimeter, were prepared through direct current electrodeposition. The principle is illustarted in Figure 21.15.

The pores can be also filled with silicon tetrachloride, and, by hydrolysis with traces of water (nanoreactor), an array of silicon oxide posts in an organic matrix is produced (Kim 2001). In this case the organic support matrix is removed using RIE. A schematic diagram of the different steps used to fabricate SiO_2 nanoposts is shown in Figure 21.16. Such roughened surfaces have tremendous promise for sensory and on-chip separations application.

By using block copolymers, nanoporous and nanorelief ceramic films can be prepared with important applications as selective separation membranes, next generation catalysts, and photonic materials.

Two well-defined triblock copolymers of the A_1BA_2 type, where A is PI and B is poly(pentamethyldisilylstyrene) (P(PMDSS)), were prepared. One material had a combination of block lengths of 24/100/26 (kg/mol) and forms a double gyroid morphology of PI network (volume fraction of PI: 33%) in a matrix of P(PMDSS)

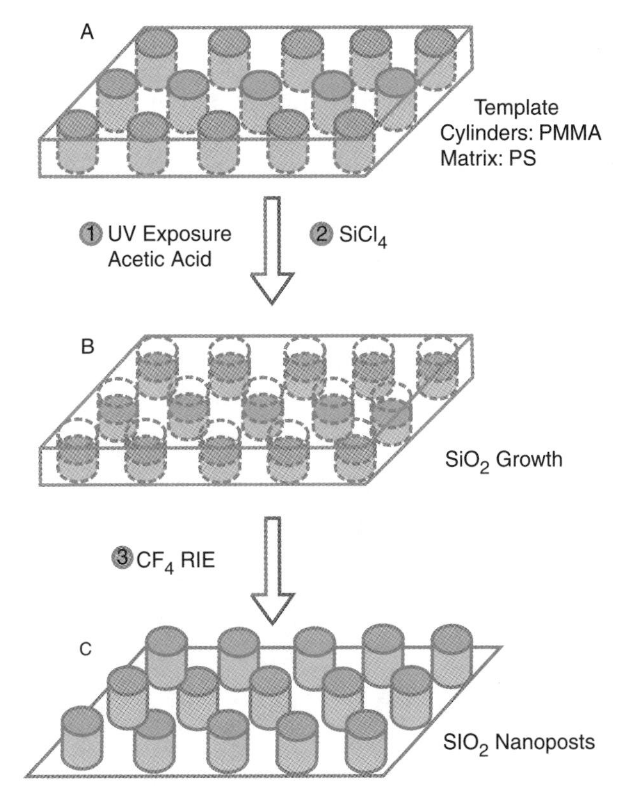

Figure 21.16. Schematic diagram of the steps required to generate SiO_2 nanoposts. (A) Block copolymer films having cylindrical microdomains oriented normal to the surface. (B) Growth of SiO_2 within the nanopores generated by selective elimination of PMMA cylinders. (C) Array of SiO_2 nanoposts after removing PS matrix with CF_4 RIE.

[referred as P(PMDSS)-DG] and is the precursor for the nanoporous structure. The other material has a combination of 44/168/112 (kg/mol) and forms the inverse double gyroid morphology of P(PMDSS) networks [volume fraction of P(PMDSS):51%] in a matrix of PI (designated PI-DG) and can be converted to a nanorelief structure. The spin-cast films of these materials were annealed for 2 days at 120°C, and exposed to a flowing 2% ozone atmosphere and 254 nm UV light simultaneously for 1 hour, and then soaked in deionized water overnight. Upon exposure PI was removed, and the P(PMDSS) was converted to silicon oxycarbide, as proved by ellipsometry, X-ray photoelectron spectroscopy, and Rutherford backscattering studies. The ceramic formed exhibits high-temperature chemical and dimensional stability.

These nanostructures could be used for applications where high-temperature stability, solvent resistance, or both are required. For example, the P(PMDSS)-DG could be used as high-temperature membranes with tailored monodisperse interconnected pores. The added advantage of these high-temperature membranes is that the redundancy of the interconnected pathways, characteristic of the DG structure, substantially decreases the likelihood of the membrane being clogged by the filtrate. By varying the molecular weight, a range of pore sizes and specific areas can be obtained, presenting opportunities for catalysis applications.

The ceramic network structure derived from PI-DG copolymer has potential use in iterconnects because of its low dielectric constant, high-temperature stability, and the inherent etch selectivity of this material to photoresist. These periodic and

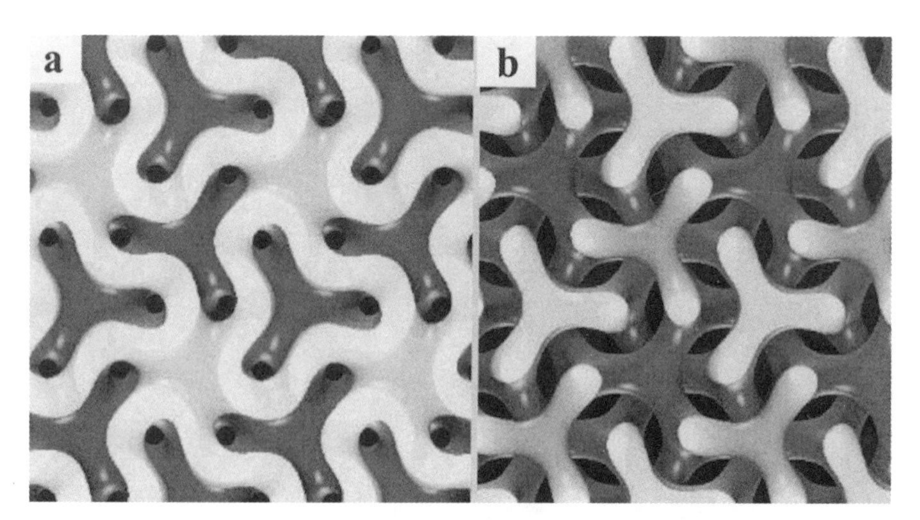

Figure 21.17. A comparison when removing the networks and when removing the matrix from a double gyroid (DG) morphology is exhibited. Image (a) corresponds to view direction along threefold axis, while image (b) corresponds to view direction along the twofold axis. The result of removing the networks is called porous (a), and the result of removing the matrix is called relief (b).

interconnected high dielectric/low dielectric ceramic/air structures also have potential as photonic band gap materials (Figure 21.17).

2.2.2. Organic-Inorganic Hybrid Mesostructures. By using an amphiphilic block copolymer, i.e., PI-b-PEO, as a structure-directing agent, and conventional sol-gel chemistry, organic-inorganic hybrid materials with nanoscale structures can be prepared (Templin 1997). Such materials could find applications in catalysis and separation technology. The procedure involves the selective swelling of the hydrophilic PEO phase by a mixture of two metal alkoxides, i.e., (3-glycidyloxypropyl)trimethoxysilane (GLYMO), and aluminum sec-butoxide. Hydrolysis and subsequent condensation of the alkoxides lead to the formation of organically modified aluminosilicate phase. The aluminum, on one hand, acts as a hardener of the organic-inorganic matrix and on the other hand, as a Lewis acid, catalyses the opening of the epoxy ring of the GLYMO. By changing the molecular characteristics of the block copolymer and the amount of the two alkoxides, various aluminosilicate-type mesostructures across the phase diagram of block copolymers can be produced. Two examples are given in Figure 21.18.

By thermal treatment of organic-inorganic hybrid material, single ceramic nano-objects of different shapes and sizes can be produced (Figure 21.19). This may open access to nanoengineering of ceramic materials through the sequence of synthesis-dissolution-manipulation-hardening (Ulrich 1999).

It is clear that block copolymers will play an important role in future high-technology applications.

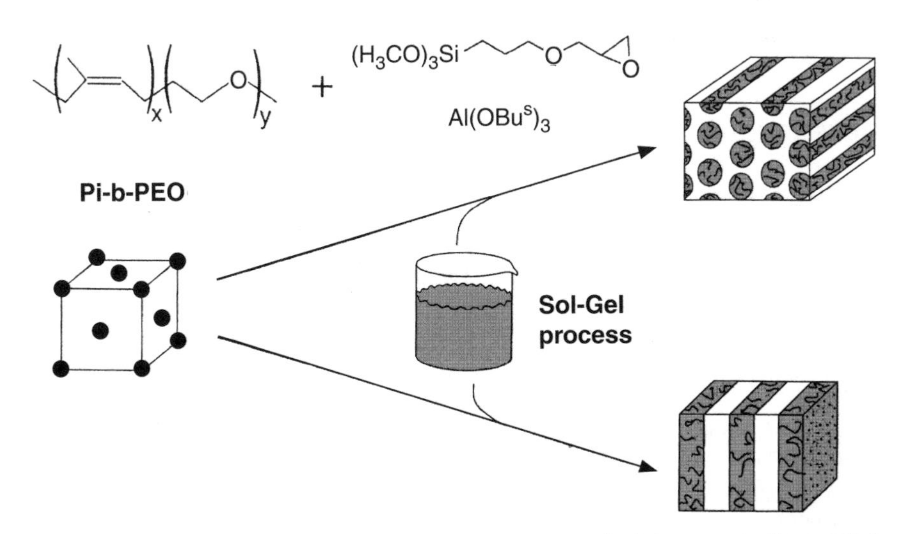

Figure 21.18. Schematic drawing of Wiesner approach for synthesizing organically modified silica mesostructures. (Left): The morphology of the precursor block copolymer. (Right): The resulting morphologies after addition of various amounts of the metal alkoxides.

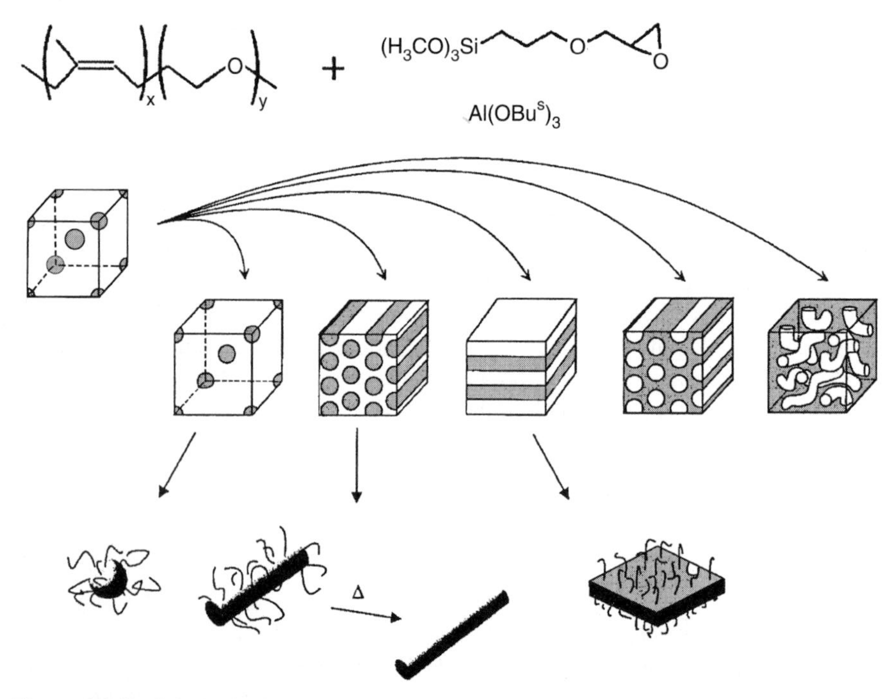

Figure 21.19. Schematic drawing of Wiesner approach for synthesis of single nano-objects with controlled shape, size, and composition. In the upper part, all themorphologies obtained from PI-b-PEO and different amounts of metal alkoxides are shown. As displayed in the lower part of the figure, the single "hairy" hybrid nano-objects of different shape are isolated by dissolution. As indicated for the case of the cylinders, the organic part can be removed by heat treatment, leading to a shrinkage of the aluminosilicate phase.

REFERENCES

Aggarwal S. L. (1970) Block Polymers, Plenum Press,

Allport D. C., Janes W. H. (1973) Block Copolymers, John Wiley and Sons, New York.

Beyer F. L., Gido S. P., Buschl C., Iatrou H., Uhrig D., Mays J. W., Chang M. Y., Garetz B. A., Balsara N. P., Tan N., Hadjichristidis N. (2000) Macromolecules 33, 2039.

Bhowmick A. K., Stephens H. L. (2001) Handbook of Elastomers, Marcel Dekker, Inc.,

Ceresa R. J. (1973) Block and Graft Copolymerization, John Wiley and Sons, New York, Vol. 1.

Chan V. Z.-H., Hoffman J., Lee V. Y., Iatrou H., Avgeropoulos A., Hadjichristidis N., Miller R. D., Thomas E. L. (1999) Science 286, 1716.

Chem. Systems. An IBM Co. (November 1999) Developments in Thermoplastic Elastomers 98/99S12.

Craver C. D., Carraher C. E. Jr. (2000) Applied Polymer Science. 21st Century, Elsevier Science Ltd,

Dair B. J., Honeker C. C., Alward D. B., Avgeropoulos A., Hadjichristidis N., Fetters L. J., Capel M., Thomas E. L. (1999) Macromolecules 32, 8145.

Ehrenstein G. W. (2001) Polymeric Materials. Structure-Properties-Applications, Hanser Gardner Publications, Inc.,

Folkes M. J. (1985) Processing Structure and Properties of Block Copolymers, Elsevier Applied Science Publishers Ltd.,

G. Holden (2001) Understanding Thermoplastic Elastomers, Hanser Gardner Publications, Inc,

Goodman I. (1985) Developments in Block Copolymers-2, Elsevier Applied Science Publishers Ltd.,

Guarini K., Black C. T., Tuominen M. T., Russell T. P. (2000) Science 290, 2126.

Harrison C., Adamson D. H., Cheng Z., Sebastian J. M., Sethuraman S., Huse D. A.,

Harrison C., Park M., Chaikin P. M., Register R. A., Adamson D. H. (1998) J. Vac. Sci. Technol. B 16, 544.

Henselwood F., Wang G., Liu G. (1998) J. Appl. Polym. Sci. 70, 397.

Hurter P. N., Hatton T. A. (1992) Langmuir 8, 1291.

Iatrou H., Mays J. W., Hadjichristidis N. (1998) Macromolecules 31, 6697.

Kabanov A., Alakhov V. Yu. (1996) Polymeric Material Encyclopedia, CRC Press, p. 757.

Kataoka K. (1994) J. Macromol. Sci.-Pure Appl. Chem. A31, 1759.

Kataoka K., Harada A., Nagasaki Y. (2001) Adv. Drug Deliv. Rev. 47, 113.

Kim H.-C., Jia X., Stafford C. M., Ha Kim D., McCarthy T. J., Tuominen M., Hawker C. J., Russell T. P. (2001) Adv. Mater. 13, 795.

Knoll K., Niessner N. (1998) Macromol. Symp. 132, 231.

Legge N. R., Holden G., Schroeder H. E. (1987) Thermoplastic Elastomers. A Comprehensive Review, Hanser Publishers,

Mark J. E., Erman B., Eirich F. R. (1994) Science and Technology of Rubber, Academic Press, Inc.

Meier D. J. (1983) Block Copolymers. Science and Technology, Harwood Academy Publishers,

Morton M. (1987) Rubber Technology, Van Nostrand Reinhold Co. Inc.,

Park M., Harrison C., Chaikin P. M., Register R. A., Adamson D. H. (1997) Science 276, 1401.

Register R. A., Chaikin P. M. (2000) Science 290, 1558.

SciFinder Central Databases (31 December 2000).

Sedlak M., Antonietti M., Colfen H. (1997) Macrom. Chem. Phys. 199, 247.

SRI International (June 1993) Progress Economic Program No. 104A.

SRI International (October 1991) Progress Economic Program No. 207.

Szwarc M. (1956) Nature 176, 1168.

Templin M., Franck A., Du Chesne A., Leist H., Zhang Y., Ulrich R., Schadler V., Wiesner U. (1997) Science 278, 1795.

Thurn-Albrecht T., Schotter J., Kastle G. A., Emley N., Shibauchi T., Krusin-Elbaum L., Guarini K., Black C.T., Tuominen M.T., Russell T.P. (2000) Science 290, 2126.

Ulbrich K., Pechar M., Strohalm J., Subr V. (1997) Macromol. Symp. 118, 577.

Ulrich R., Du Chesne A., Templin M., Wiesner U. (1999) Adv. Mater. 11, 141.

Velis G., Hadjichristidis N. (2000) J. Polym, Sci., Polym. Chem. 38, 1136.

Walker B. M. (1986) Handbook of Thermoplastic Elastomers, R. E. Krieger Publishing Co.,

Wang G., Henselwood F., Liu G. (1998) Langmuir 14, 1554.

Weidisch R., Gido S. P., Uhrig D., Iatrou H., Mays J., Hadjichristidis N. (2001) Macromolecules 34, 6333.

Weidisch R., Velis G., Hadjichristidis N. Unpublished results.

Wu M. M., Ball L. E. (1985) USA Patent 4,540,498.

Yokoyama M. (1996) Polymeric Material Encyclopedia, CRC Press, p. 754.

INDEX